James Sykes Gamble, Geore King

**Materials for a Flora of the Malayan Peninsula**

Vol. 1

James Sykes Gamble, Geore King

**Materials for a Flora of the Malayan Peninsula**
*Vol. 1*

ISBN/EAN: 9783337294113

Printed in Europe, USA, Canada, Australia, Japan

Cover: Foto ©berggeist007 / pixelio.de

More available books at **www.hansebooks.com**

# MATERIALS

FOR A

# FLORA

# OF THE MALAYAN PENINSULA.

BY

## GEORGE KING, M. B., LL. D., F. R. S., C. I. E.,

SUPERINTENDENT OF THE ROYAL BOTANIC GARDEN, CALCUTTA.

## THALAMIFLORÆ.

(No. 1 to 5 of the series.)

CALCUTTA:

PRINTED AT THE BAPTIST MISSION PRESS.

1889-1893.

# PREFACE.

—◦✦◦—

The following papers originally appeared in the *Journal* of the Asiatic Society of Bengal. For the convenience of botanists interested in the *Flora* of the region of which they treat, I had some spare copies of each printed off as it appeared, with the pages of all continuously numbered in the lower corner. As the whole of the Thalamifloral orders (in the sense of the *Genera Plantarum* of Messrs. Bentham and Hooker) have now been finished, I think it may be an additional convenience to publish an index to the species described under these orders. It will be observed that each page bears two numbers. The one at the top of the page is that of the volume of the *Journal* in which the paper originally appeared, and it is not noted in this index : it may, however, be used by any writer quoting these papers. The second number—*the one at the bottom of the page*—is the one used in the index.

The dates of the original publication of the five papers which cover the *Thalamiflora* are as follows :—

| | | | |
|---|---|---|---|
| No. 1 | 3rd July, | 1889. |
| ,, 2 | 5th February, | 1890. |
| ,, 3 | 1st April, | 1891. |
| ,, 4 | 13th June, | 1892. |
| ,, 5 | 7th June, | 1893. |

I may mention that the first paper contains no plants of the Andaman or Nicobar Islands ; for it was not part of my original scheme to include, within the area treated of, any of the islands except those which, like Penang and Singapore, lie close to the coasts of the Peninsula. Subsequently, however, I decided, as a matter of convenience, to include the Andaman and Nicobar groups, although the *Flora* of the Andamans is in character more Burmese than Malayan. I propose, as leisure permits, to continue the publication of these papers in the *Journal* of the same Society ; and, as each great group is finished, to supply an index of the species included in it.

In a Monograph of the Indo-Malayan species of Anonaceæ, published in the Annals of the Botanic Garden, Calcutta since the paper included in this volume was written, I have re-established Maingay's manuscript genus *Griffithia* ; and to it I have referred, as proposed by him, the plant here named *Polyalthia magnoliæflora*, H. f. and Th.

It may also be useful to mention that, in the following pages, the length given for a leaf is that of the blade only, the measurement of the petiole being given separately ; and that the breadth given for a leaf is that of its broadest part.

Royal Botanic Garden, Calcutta,
    *September*, 1893.

# INDEX.

The figures given in this index are those of the *lower outer* corners of the pages of the text.

# MATERIALS

FOR A

# FLORA OF THE MALAYAN PENINSULA.

BY

GEORGE KING, K. C. I. E., M. B., LL. D., F. R. S.,

SUPERINTENDENT OF THE ROYAL BOTANIC GARDEN, CALCUTTA.

*[Reprinted from the Journal of the Asiatic Society of Bengal, Vol. LVIII. Part II, No. 4, 1889.]*

## Nos. 1 and 2.

CALCUTTA :

PRINTED AT THE BAPTIST MISSION PRESS.
1898.

*Materials for a Flora of the Malayan Peninsula.—By* GEORGE
KING, M. B., LL. D., F. R. S., F. L. S., *Superintendent of the Royal
Botanic Garden, Calcutta.*

[Received and read July 3rd, 1889.]

As the Calcutta Herbarium contains a rich collection of Malayan
plants, I propose to publish from time to time a systematic account of
as many of them as are indigenous to British provinces, or to provinces
under British influence. In addition to the states on the mainland of
the Malayan Peninsula, these provinces include the islands of Singapore and Penang, and the Nicobar and Andaman groups. The classi-
fication which I propose to follow is that of the late Mr. Bentham and
Sir Joseph Hooker. It is unlikely that, with the scanty leisure at my
command, I shall be able, under several years, to complete even the
meagre account of the Flora of which the first instalment is now submitted. The orders will be taken up nearly in the sequence followed
in the *Genera Plantarum* of Bentham and Hooker, and in the Flora
of British India of the latter distinguished botanist. The natural orders
now submitted are *Ranunculaceæ, Dilleniaceæ, Magnoliaceæ, Menispermaceæ, Nymphæaceæ, Capparideæ* and *Violareæ.* The order *Anonaceæ*
should have come between *Magnoliaceæ* and *Menispermaceæ;* but, on
account of its extent and difficulty, I have been obliged to postpone its
elaboration pending the receipt of further herbarium material. It will
however, it is hoped, soon be taken up.

## ORDER I. RANUNCULACEÆ.

Annual or perennial herbs or shrubs. *Leaves* alternate or opposite.
Stipules 0, or adnate to the petiole, rarely free. *Flowers* regular or
irregular, 1-2-sexual. *Sepals* 5 or more, rarely 2 to 4, usually deciduous,
often petaloid, imbricate or valvate. *Petals* 0 or 4 or more, hypogynous,
imbricate, often minute or deformed. *Stamens* hypogynous; anthers
usually adnate and dehiscing laterally. *Carpels* usually many, free,
1-celled; stigma simple; ovule one or more, on the ventral suture, anatropous, erect with a ventral, or pendulous with a dorsal raphe. *Fruit* of
numerous 1-seeded achenes, or many-seeded follicles, rarely a berry.
*Seed* small, albumen copious; embryo minute. Distrib. Abundant
in temperate and cold regions: genera 30; known species about 310.*

* The above diagnosis of this order (copied from Sir Joseph Hooker's Flora of
British India) covers the entire order, which is usually sub-divided into five sub-
orders or tribes. Representatives of only one of these tribes (*Clematideæ*) have
hitherto been discovered in the region under review. But, as exploration of the
central mountain ranges proceeds, plants belonging to one or two of the other tribes

3

Tribe I. *Clematideæ.* Climbing shrubs. *Leaves* opposite. *Sepals* valvate, petaloid. *Carpels* 1-ovuled ; ovule pendulous. Fruit of many achenes.

Petals 0    ...    ...    ...    ... 1. *Clematis.*
Petals many, linear    ...    ...    ... 2. *Naravelia.*

### 1. CLEMATIS, Linn.

Woody climbers. *Leaves* opposite, simple or compound, exstipulate, *Sepals* 4 to 8, valvate. *Petals* 0. *Stamens* many. *Carpels* many, with long tails. *Ovule* solitary, pendulous.—Distrib. Temperate climates ; species about 100.

1. C. SMILACIFOLIA, Wall. in Asiat. Research. xiii, 414. *Leaves* simple, (rarely pinnate) ovate, blunt, with broad sub-cordate bases, boldly 5-nerved, coriaceous, glabrous, entire or remotely serrate, 3 to 10 in. long by 1·5 to 5 in. broad ; petioles nearly as long. *Panicles* axillary, few-flowered, 6 to 12 in. long. *Flowers* 1 to 1·5 in diam. *Sepals* 4 to 5, coriaceous, oblong, reflexed, outside dull brown tomentose, inside purple. *Filaments* linear, glabrous, the inner shorter with longer anthers. *Achenes* flat, pubescent, with broad margins and long feathery tails. A tall glabrous woody climber. DC. Prod., I., 10 ; Bot. Mag., t. 4259 ; H. f. et Th. Fl. Ind., i, 6 ; Hook. fil. Fl. Br. Ind., i, 3 ; Miq. Fl. Ind., Bat. I, Pt. ii, p. 2. *C. sub-peltata,* Wall., Pl. As. Rar. I, t. 20. *C. Munroana,* Wight Ill., i, 5, t. 1. ? *C. glandulosa,* Bl., Bijdr. i, 1.

Penang, Curtis ; but probably occurring also in the Central Range of mountains in the Malayan Peninsula.

2. C. GOURIANA, Roxb. Fl. Ind. ii, 670. An extensive climber, the young parts pubescent, adult glabrous. *Leaves* shortly petiolate, pinnate, 2-pinnate or 2-ternate, the *leaflets* shortly petiolulate, membranous, ovate to ovate-lanceolate, 5-nerved, sometimes sub-cordate, entire or irregularly dentate-serrate, 2 to 3 in. long by ·75 to 1·75 in. broad. *Panicles* many-flowered, longer then the leaves ; *flowers* small (·3 to ·5 in. in diam.) green-ish-white. *Achenes* narrowly oblong, pubescent, emarginate, with long silky tails. DC. Prod. i, 3 ; W. A. Prod. 2 ; Wight Ic. 933-4 ; H. f. et Th. Fl. Ind. 8 ; Hook. Fl. fil. Brit. Ind. i. 4 ; Miq. Ind. Fl. Bat. Vol. I, Pt. 2, p. 4. *C. cana* and *dentosa,* Wall. Cat. *C. javana,* DC. Prod. i, 7.

Not uncommon at low elevations in the Indo-Malayan region.

### 2. NARAVELIA, DC.

Climbing shrubs. *Leaves* 3-foliolate, terminal leaflet generally transformed into a tendril. *Sepals* 4 to 5. *Petals* 6 to 12, narrow, longer

may be found. I therefore think it better to let the diagnosis stand, than to modify it so as to include only the tribe *Clematideæ.*

4

than the calyx. *Achenes* long-stipitate, with long-bearded style.—Distrib.
Two E. Asiatic species.

N. LAURIFOLIA, Wall. Cat. Young parts puberulous, adult glabrous.
*Leaflets* broadly ovate, shortly acuminate, entire, boldly 5-nerved, 4 to 6 in.
long by 2·5 to 3 in. broad. *Panicles* longer than the leaves, many-
flowered; *petals* long, linear, whitish green. *Achenes* cylindric, glabrous,
with stout sericeous tails. Hook. fil. et Th. Fl. Ind. i, 3.; Hook. fil. Fl.
B. Ind. i, 7; Miq. Fl. Ind. Bat. I, pt. ii, 2. *N. Finlaysoniana*, Wall.
Cat. 468 (with diseased fruit). *Clematis similacina*, Bl. Bijdr. i, 1.

Common throughout the whole Indo-Malayan region to the Philippines.

ORDER. II DILLENIACEÆ.

Trees, shrubs or herbs, sometimes climbing. *Leaves* alternate, simple, entire or toothed (pinnatipartite in *Acrotrema*), exstipulate with
sheathing petioles, or more rarely with lateral deciduous stipules.
*Flowers* yellow or white, often showy. *Sepals* 5, imbricate, persistent.
*Petals* 5 (rarely 3 or 4) deciduous. *Stamens* many, hypogynous, many-
seriate; anthers innate, with lateral slits or terminal pores. *Carpels*
1 or more, free or cohering in the axis; styles always distinct; ovules
amphitropous, solitary or few ascending, or many and attached to
the ventral suture. *Fruit* follicular, or indehiscent and sub-baccate.
*Seeds* solitary or many, arillate, testa crustaceous, raphe short, albumen
fleshy; embryo minute, next the hilum.—Distrib. Chiefly tropical;
species about 210.

Tribe I. *Delimeæ*. Filaments thickened upwards; anthers short,
cells remote oblique.

Carpel solitary ... ... ... ... 1. *Delima.*
Carpels 2-5 ... ... ... ... 2. *Tetracera.*

Tribe II. *Dilleniæ*. Filaments not thickened upwards; anthers
with parallel cells.

Carpels 3; stemless herbs; leaves all radical, large... 3. *Acrotrema.*
Carpels 5-20; seeds arillate ... ... ... 4. *Wormia.*
Carpels 5-20; seeds not arillate ... ... 5. *Dillenia.*

1. DELIMA, Linn.

Woody climbers. *Leaves* parallel-veined. *Flowers* many, in terminal panicles, hermaphrodite, white. *Sepals* 5. *Petals* 2 to 5. *Stamens*
many; filaments dilated upwards, cells much diverging. *Ovary* solitary, subglobose, narrowed into a subulate style; ovules 2 to 3, ascending.
*Follicles* ovoid, coriaceous, 1-seeded. *Seed* with a cupular toothed aril.

5

1. D. SARMENTOSA, Linn. *Leaves* 3 to 5 in., obovate, ovate or broadly lanceolate, obtuse or acute, quite entire, serrate or crenate, appressed pilose; both surfaces; scabrid; nerves 9 to 11 pairs, straight, ascending, prominent: length 2·5 to 3·5 in., breadth 1 to 2 in., petiole ·4 to ·5 in. *Flowers* ⅓ to ⅓ in. in diam.; in tomentose or pilose spreading panicles that are often leafy. *Sepals* reflexed. DC. Prod. i. 69; Wall. Cat. 6632; Bot. Mag t. 3058; Miq. Fl. Ind. Bat. I, pt. ii, 7; Hook. Fl. Fl. B. Ind. I, 31. *D. intermedia*, Bl. Bijdr. *Tetracera sarmentosa*, Willd.; Roxb. Fl. Ind. ii. 645. *Leontoglossum scabrum*, Hance in Walp. Ann. ili 812.

Var. 1. GLABRA; fruit glabrous.

Var. 2. HEBECARPA; fruit hairy. *D. hebecarpa*, DC. Prod. i, 70, Deless. Ic. Sel. t. 72; Wall. Cat. 6633. *D. intermedia*, Blume. *Davilla hirsuta*, Teysm. et Binn. *Delimopsis*, *hirta*, Miq.

2. D. LÆVIS, Maingay MSS. *Leaves* oblong-lanceolate to narrowly elliptic, acute, entire, the base cuneate or rounded; nerves 8 to 9 pairs, ascending, prominent; upper surface smooth, shining, the lower puberulous, neither of them scabrid, length 5 to 7·5 in., breadth 2 to 3·5 in.; petiole ·8 in., broad. *Flowers* ·5 in. in diam., in narrow tomentose leafless panicles longer than the leaves. *Sepals* reflexed.

Malacca, Maingay No. 10.   Collected only by the late Dr. Maingay.

## 2. TETRACERA, Linn.

Climbing shrubs or trees, smooth, scabrid, or pubescent. *Leaves* with parallel lateral veins. *Flowers* in terminal or lateral panicles, hermaphrodite or partially l-sexual. *Sepals* 4 to 6, spreading. *Petals* 4 to 6. *Stamens* many, filaments dilated upwards, anther-cells distant. *Carpels* 3 to 5; ovules many, 2-seriate. *Follicles* coriaceous, shining. *Seeds* 1 to 5, with a fimbriated or toothed aril.—Distrib. All tropical; species about 25.

1. T. ASSA. DC. Prod. i. 68. Young branches striate, pubescent or sub-strigose. *Leaves* 2 to 5 in. long, ovate-lanceolate, acuminate, obscurely sinuate or serrate, glabrous except the nerves beneath. *Panicles* axillary and terminal, shorter than the leaves, few-flowered. *Follicles* several-seeded. W. and A. Prod. 5; Hassk. Pl. Rar. Jav. 177; Hook. fil. and Th Fl. Ind. i, 63; Hook. fil. Fl. B. Ind. I. 31; Miq Fl. Ind. Bat. I, Pt. ii. 8.

Common throughout Indo-Malaya, at low elevations.

2. T. EURYANDRA, Vahl. Symb. iii, 71. Young branches tomentose. *Leaves* rigid, 3 to 4 in. long, oblong or obovate-oblong, entire or obscurely sinuate, above glabrous except the midrib, below minutely tomentose when young. *Panicles* terminal and axillary, shorter than the leaves,

few-flowered. *Follicles* several-seeded. DC. Prod. I, 68; Roxb. Fl. Ind. ii, 646; H. f. et Th. Fl. Ind. i, 63; Hook. fil. Fl. Br. Ind. I, 32; Miq. Fl. Ind. Bat. Vol. I, pt. ii, 8. *T. lucida*, Wall. Cat.

Straits Settlements, at low elevations. Distrib. Moluccas and New Caledonia.

3. T. MACROPHYLLA, Wall. Cat. 6628. Young branches pubescent. *Leaves* broadly elliptic to obovate-elliptic, 5 to 7 in. long, margin sub-sinuate, scabrid on both surfaces. *Panicle* terminal, longer than the leaves, many-flowered. *Sepals* rotund, not ribbed. *Follicles* 1-seeded. Hook. fil. et Th. Fl. Ind. I, 63; Hook. fil. Fl. Br. Ind. I, 32; Miq. Fl. Ind. Bat. Vol. 1. pt ii, 8.

Straits Settlements, in tropical forests. Distrib. Sumatra.

4. T. GRANDIS, King. n. sp. A large tree. Young branches and inflorescence shortly velvety-tomentose. *Leaves* large, coriaceous, broadly elliptic, rarely slightly obovate, the apex truncate and minutely apiculate; the edges obscurely crenate or undulate toward the apex, entire below, recurved when dry; the base rounded or slightly narrowed; upper surface scabrous, lower minutely tomentose; nerves stout, 14 pairs, straight, erecto-patent; length of blade 8 to 10 in., width 5·5; petiole about 5 in., stout. *Inflorescence* in little-branched, lateral or terminal, panicles a foot or more long. *Flowers* shortly pedunculate. *Sepals* 5, broadly ovate, sub-acute, concave, ribbed and tomentose externally, ·5 in. long. *Petals* about as long as the sepals but narrower, glabrous. *Anthers* truncate, narrowed to the long slender filaments. *Follicles* with a slender curved beak.

Perak. Scortechini, No. 90*b*.

Said by Father Scortechini to be a very large tree.

### 3. ACROTREMA, Jack.

*Perennial* stemless herbs with woody rhizomes. *Leaves* large, parallel-nerved, with sheathing deciduous stipules. *Scape* short, axillary, bracteolate. *Flowers* large, yellow. *Sepals* 5. *Petals* 5. *Stamens* numerous, in 3 bundles which alternate with the carpels; filaments filiform, anthers erect, with longitudinal porous dehiscence. *Carpels* 3, slightly cohering; styles subulate, recurved, ovules 2 or more. *Fruit* of 3 irregularly dehiscing follicles. *Seed* with a membranous aril; the testa crustaceous, pitted.—Distrib. Ten species, of which 8 are endemic in Ceylon, 1 Peninsular-Indian, and 1 Indo-Malayan.

A. COSTATUM, Jack in Mal. Misc. ex Hook. Misc. ii, 82. Whole plant covered with stiff rufous or golden hairs, especially when young. *Leaves* obovate, the margins dentate-ciliate; the base narrowed, sagittate; upper surface strigose, often blotched with white, petiole short. *Racemes*

7

shorter than the leaves, 8 to 10-flowered, setose ; bracteoles minute, lanceolate. *Flowers* an inch in diam. *Stamens* about 15. Hook. fil. and Th. Fl. Ind. i, 65 ; Hook. fil. Fl. Br. Ind. I, 32 ; Miq. Fl. Ind. Bat. Vol. I, Pt. ii, 10. *A. Wightianum*, W. and A. Prod. 6 ; Wight Ill. t. 9.

Straits Settlements ; in damp shady spots at elevations of 500 to 2500 feet. Common.

## 4.  WORMIA, Rottb.

Trees, sometimes lofty ; or shrubs. *Leaves* broad, sub-coriaceous ; lateral nerves many, strong, parallel ; petioles usually with deciduous stipular wings. *Flowers* large, in terminal racemes or panicles. *Sepals* 5. *Petals* 5. *Stamens* indefinite, in several series, nearly free ; anthers linear, erect, cells opening by 2 pores. *Carpels* 5 to 10, scarcely cohering in the axis ; ovules numerous. *Fruit* of indehiscent or follicular 3- or more-seeded carpels. *Seeds* with a fleshy aril. Distrib. Tropical Asia and Australia, and one in Madagascar ; species about 9.

Sect. I. *Capellia*, Blume (genus). Inner row of stamens much longer than the outer and arching over them.

1.  W. SUFFRUTICOSA, Griff. Notul, iv. 706 ; Ic. iv. t. 649, f. 1.   A small tree. Young parts floccose. *Leaves* with short, broadly winged petioles ; broadly elliptic or sub-obovate-elliptic, blunt or acute, dentate, glabrous except the 12-20 pairs of nerves which are sparsely pilose beneath ; length of blade 7 to 9 in. ; breadth 4 to 5·5 in. ; petiole ·5 to 1·5 in. *Racemes* about as long as the leaves, leaf-opposed ; pedicels ·5 to 1 in. long. *Flowers* 3 to 4 in. in diam., yellow. *Sepals* broadly ovate, nerved, glabrous. *Petals* obovate, crenulate. *Carpels* 5 to 7, 3- to 5-seeded. Hook. fil. Fl. Br. Ind. I, 35. *W. excelsa*, H. f. and Th. Fl. Ind. I, 67 (not of Jack). *W. subsessilis*, Miq. Fl. Ind. Bat., Suppl. i. 618 ; Ann. Mus. Lugd. Bat. i. 315, t. 9.

Malacca, Singapore ; extends to Sumatra, Banka, Borneo and probably to other islands of the Archipelago.

2.  W. OBLONGA, Wall.   A tree.   Young branchlets thin, sparsely tomentose, the old glabrous. *Leaves* rather distant, on moderate, channelled (not winged) petioles, oblong to elliptic, acute at base and apex, entire or distantly serrate or sub-serrate, nerves 9 to 12 pairs ; length of blade 6 to 8 in., breadth 3 to 3·5 in. ; petiole 1·5 in. *Cymes* terminal or leaf-opposed, few-flowered, shorter than the leaves, tomentose ; pedicels about 1 in., thickened above, tomentose. *Flowers* 3 to 5 in. in diam. *Sepals* ovate-rotund, tomentose externally. *Petals* obovate, entire, veined, yellow. *Carpels* 8 to 10. H. f. & Th. Fl. Ind. i, 67 ; Hook. fil. Fl. Br. Ind. i, 35 ; Miq. Fl. Ind. Bat. Vol. 1. pt. ii, p 11.

Straits Settlements, in more or less dense forest. Distrib. Sumatra.

8

Sect. II. *Euwormia.* Filaments all erect and nearly equal in length.

3. W. PULCHELLA, Jack Mal. Misc. ex Hook. Comp. Bot. Mag. I, 221. A shrub. Young parts glabrous. *Leaves* on short narrowly winged petioles, obovate or obovate-oblong, obtuse, entire, truncate or retuse, sometimes mucronate, thickly coriaceous, glabrous ; nerves 5 to 7 pairs ; length of blade 2·5 to 4 in., breadth 1·5 to 2·5 in.; petiole ·75 in. *Flowers* solitary or in small cymes, axillary or terminal, 2 in. in diam.; pedicels 2 in. long. *Sepals* broadly ovate, glabrous. *Petals* ovate. *Carpels* about 5. *Seeds* few, with pulpy arillus. H. f. and Th. Fl. Ind. I, 68; Hook. fil. Fl. Br. Ind. I, 36; Miq. Fl. Ind, Bat. Vol. I, Pt. ii, p. 11.

Perak and Malacca; at elevations under 1,000 feet. Distrib. Sumatra.

4. W. MELIOSMÆFOLIA, King, n. sp. A small tree, the young parts and leaf-petioles softly fulvous-tomentose. *Leaves* crowded near the apices of the branches, coriaceous, obovate-lanceolate to obovate-elliptic, acute or acuminate, entire or minutely and remotely serrate, base acute, glabrous above except the 14 to 18 pairs of spreading pubescent nerves, under surface minutely tomentose ; length of blade 5 to 8 in., breadth 3 to 4 in., petiole about 1 in. *Flowers* about 3 in. in diam., axillary, solitary, on slender tomentose peduncles 2 in. long, or in few-flowered linear-bracteolate cymes ; peduncles 1 in. long. *Sepals* ovate-oblong, velvety-tomentose externally, glabrous internally. *Petals* oblanceolate, pale yellow, veined, wavy. *Stigmas* about 12, linear, recurved. *Follicles* with several compressed arillate seeds. *Dillenia meliosmae-folia*, Hook. fil. & Th. Fl. Br. Ind. I, 36.

Malacca, Perak.

Originally described as a *Dillenia* by Sir Joseph Hooker who had not seen the fruit.

5. W. SCORTECHINII, King, n. sp. A tree 60 to 70 feet. Branchlets thick, scarred, puberulous or glabrous. *Leaves* coriaceous, obovate-oblong, shortly acuminate, undulate-crenate, narrowed to the petiole ; upper surface shining, glabrous except the midrib and nerves which are minutely pubescent as is the under surface ; nerves prominent on the under surface, 26-30 pairs, each ending on the margin of the leaf in a tuft of hairs ; length of blade 4 to 6·5 in., breadth 2 to 3 ; petiole ·5 to 2 in., winged and expanded at the base. *Cymes* supra-axillary, sub-terminal, dichotomous, pubescent ; bracteoles oblong, ·3 in. long ; pedicels ·25 in. long. *Flowers* when expanded about 1 in. broad. *Sepals* 5, puberulous externally, broadly ovate, coriaceous. *Petals* 0. *Anthers* equal in length to the filaments, slightly hairy. *Ovaries* 3 to 5, usually 4. *Ripe fruit* unknown.

9

*l*

Perak.    Father Scortechini.

Collected only by Scortechini who left a MS. description of it under the name *Wormia apetala*. But, as that name is pre-occupied by a species of Gaudichaud, I have rechristened it after its lamented discoverer. In his description Father Scortechini says that, although he opened many buds, he never could find any trace of petals. In stamens this agrees with the *Euwormia* section of *Wormia* : but whether it is really a *Wormia* and not a *Dillenia* cannot be settled until ripe fruit is found.

*Species of which flowers are unknown.*

6.    W. KUNSTLERI, King, n. sp. Young branches thick, rugose, pubescent. *Leaves* on channelled petioles, obovate-oblong, blunt, the base narrowed, slightly sinuate-crenate, glabrous except the pubescent midrib and 18 to 22 pairs of nerves ; length of blade 6 to 8 in., breadth 3·5 to 4 in. ; petiole nearly 2 in. *Cymes* leaf-opposed, few-flowered, tomentose ; peduncles about 1 in. *Sepals* broadly ovate, glabrous, thick. Young *seeds* arillate.

Perak.    King's Collector No. 5905. The only specimens have unripe fruit.

## 5.    DILLENIA, Linn.

Characters of *Wormia* except that the flowers are more often solitary and are sometimes white ; the carpels are rather more numerous, are never dehiscent, cohere in the axis, and are enveloped in the thickened accrescent calyx, while the seeds are exarillate.

1.    D. INDICA, Linn. A tree ; the young branches tomentose. *Leaves* crowded at the ends of the branches, coriaceous, lanceolate or ovate-lanceolate; sharply serrate, glabrous above, pubescent beneath especially on the 30 to 40 pairs of stout nerves ; length of blade 8 to 12 in., breadth 3 to 4 in., petiole 1 to 1·5 in., channelled, sheathing at the base. *Flowers* 6 to 9 in. in diam., solitary, axillary, on short tomentose pedicels 2 to 3 in. long. *Sepals* orbicular, concave, fleshy. *Petals* white, obovate-oblong, undulate. Inner *stamens* longer than the outer and arching over them. *Pistils* about 20 ; the stigmas lanceolate, recurved, radiating. *Carpels* 1-celled with many reniform hairy seeds. Ham. Linn. Trans. XV, 99 ; H. f. and Th. Fl. Ind. I. 69; Hook fil. Fl. Br. Ind. I, 36 ; Martelli in Malesia III, 154. *D. speciosa*, Thunbg. Linn. Trans. i, 200 ; DC. Prod. i, 76 ; Roxb. Fl. Ind. ii, 651 ; W. & A. Prod. 5 ; Wight Ic. 823 ; Miq. Fl. Ind. Bat. Vol. I, Pt. ii, 11. *D. elliptica*, Thunbg. Linn. Trans. I, 200.

Tropical Forests of the Indo-Malayan Region. Distrib. India, Ceylon.

2.    D. OVATA, Wall. A tree, the branchlets as thick as a quill, softly tomentose. *Leaves* coriaceous, ovate or obovate-rotund, apex sub-

10

acute, rounded or emarginate ; minutely denticulate or sub-entire, glabrous above except the 16 to 20 pairs of sub-horizontal pubescent nerves, fuscous-tomentose beneath ; blade 3·5 to 7 in. long by 2·75 to 3·5 in. broad, petiole about 1 in., tomentose. *Flowers* solitary, 6 to 8 in. in diam. ; peduncles stout, tomentose, 2·5 in. long. *Sepals* reflexed, ovate, concave, thick, minutely adpressed-tomentose in the middle externally, the margins and inner surface glabrous. *Petals* obovate, yellow, veined. *Stigmas* about 15, long, linear, recurved. H. f. and Th. Fl. Ind. I, 70 ; Hook. fil. Fl. Br. Ind. I, 36 ; Miq. Fl. Ind. Bat. Vol. I, Pt. ii, 12.

Penang and Perak. In tropical forests.

I have seen no ripe fruit of this, nor have I found any description of the ripe fruit. I am therefore unable to say whether the sepals and seeds are those of a *Dillenia* or a *Wormia*. This comes near *D. aurea*, but the panicles are much laxer and the petioles longer.

3. D. AUREA, Sm. Ex. Bot. 93, t. 92. A tree, the branchlets as thick as the little finger, glabrous, scabrid ; young parts rufous-sericeous. *Leaves* obovate, blunt or acute, crenate-dentate, glabrous above except the 20 pairs of pubescent, bold, sub-transverse nerves ; under surface softly puberulous ; length of blade 5 to 8 in. : breadth 3 to 4·5 in. ; petiole 1·25 to 2 in., sheathing at the base. *Flowers* solitary, from shortened branches, 4 to 5 in. in diam. ; peduncles stout, 1 in. or more. *Sepals* oblong, obtuse, concave, fleshy, adpressed-pubescent externally when young, when old glaucous. *Petals* obovate, yellow, veined. *Stigmas* about 10, linear, radiate. DC. Prodr. i. 76 ; Wall. Cat. 6624 ; H. f. & T. Fl. Ind. 70 ; Hook. fil. Fl. Br. Ind. I, 37 ; Miq. Fl. Ind. Bat. Vol. I, pt. ii, 12 ; Martelli in Malesia III, 155. *D. ornata*, Wall. Pl. As. Rar. i, 21, t. 23 ; Cat. 947. *D. speciosa*, Griff. Notul. iv, 703. *Colbertia obovata*, Bl. Bijdr. 6.

Throughout the Indo-Malayan region, at low elevations.

4. D. RETICULATA, King, n. sp. A large tree. *Leaves* very coriaceous, obovate-elliptic, obtuse ?, dentate ; the base truncate, cordate ; upper surface glabrous except the impressed midrib and nerves which are pubescent ; under surface pubescent, the midrib and nerves very strong ; reticulations minute, very bold, areolate ; nerves 32 to 36 pairs, sub-horizontal ; length of blade 10 to 18 in., breadth 7 to 12 in., petiole 2 to 3 in., winged, much expanded at the base. *Cymes* lateral and terminal, short, branching, 10- to 12-flowered, pubescent ; pedicels stout, bracteolate. *Sepals* thick, broadly elliptic to orbicular, 1 in. in diam., pubescent externally. *Petals* oblong, concave, 1·5 in. long. *Stamens* in many series ; the outer inflexed, introrse ; the inner erect and extrorse. *Pistils* about 9, puberulous ; stigmas linear, radiating, ovules ex-arillate.

11

Perak ; Father Scortechini.

*Ripe fruit* is unknown, and it is possible that the seeds may be arillate. It has, however, the facies of a *Dillenia* rather than of a *Wormia*. In leaf this resembles *D. eximia*, Miq., but that plant has broader leaves much more narrowed to the base. It also resembles *D. grandifolia*, Wall., but is distinguished by its closer nerves and remarkable areolar reticulations.

A species imperfectly known.

5. D. GRANDIFOLIA, Wall. *Leaves* oblong-lanceolate, acute, serrate, softly hairy above, tomentose beneath ; nerves about 40 to 50 pairs, transverse ; length of blade 24 in., breadth 9 in. ; petioles stout, 5 to 6 in. long, densely tomentose as is the midrib. H. f. & Th. Fl. Ind. I, 71 ; Hook. fil. Fl. Br. Ind. I, 38 ; Miq. Fl. Ind. Bat. Vol. I, pt. ii, 12.

Penang and Malacca.

Known only by a few imperfect specimens in Herb. Wallich. *D. eximia*, Miq., from Sumatra, is a species also known only by a few leaf specimens which I have examined. The two differ as above described and may belong to *Wormia*.

## ORDER III. MAGNOLIACEÆ.

Trees or shrubs, sometimes climbing, often aromatic, wood-tissue with glandular markings. *Leaves* alternate, quite entire or toothed, stipulate or not. *Flowers* axillary and terminal, often showy, white, yellow or red, sometimes unisexual. *Sepals* and petals very deciduous, hypogynous, arranged in whorls of 3. *Stamens* indefinite, hypogynous, filaments flattened or terete, free or monadelphous ; anthers basi-fixed, adnate cells bursting longitudinally. *Carpels* indefinite, free or partly cohering in one whorl, or in several on an elongate axis ; styles short or rarely long, stigmatose on the inner surface ; ovules 2 or more, on the ventral suture, anatropous or amphitropous. *Fruit* baccate, or follicular, or of woody dehiscent carpels, which are sometimes arranged in a cone. *Seeds* solitary or few, sometimes pendulous from a long funicle, testa single and crustaceous, or double, the outer fleshy ; albumen granular or fleshy and oily ; embryo minute, cotyledons spreading ; radicle short, blunt, next the hilum.—Distrib. Chiefly natives of the tropical and temperate Asiatic mountains and United States, a few are Australian ; species about 80.

Tribe I. *Magnoliæ.* *Flowers* bisexual. Erect shrubs or trees. *Stipules* conspicuous, convolute and embracing the leaf-buds, deciduous. *Carpels* on an elongated axis.

Carpels not separating from the carpophore,
dehiscing dorsally.

12

Carpophore sessile, carpels closely packed.
Fruit elongate, cylindric, ovules 2 ... 1. *Magnolia*.
„ ovoid, ovules 6 ... ... 2. *Manglietia*.
Carpophore stalked, carpels distant ... 3. *Michelia*.
Carpels when ripe separating from the sessile
carpophore and dehiscing ventrally ... 4. *Talauma*.
Tribe II. *Winteriæ*. Flowers usually bisexual. Shrubs
or small trees. Stipules 0. Carpels in one whorl 5. *Illicium*.
Tribe III. *Schizandreæ*. Flowers unisexual. Climb-
ing shrubs. Leaves exstipulate.
Carpels of fruit capitate ... ... ... 6. *Kadsura*.

1. MAGNOLIA, Linn.

Trees or shrubs. *Leaves* evergreen or deciduous ; buds enveloped
in the convolute stipules which are connate in pairs. *Flowers* large,
terminal. *Sepals* 3. *Petals* 6 to 12, 2- to 4-seriate. *Stamens* numerous,
many-seriate, filaments flat ; anthers adnate, introrse. *Gynophore*
sessile. *Carpels* many, imbricated on a long axis, 2-ovuled, persistent ;
stigmas decurrent on the ventral suture. *Fruit* an elongated axis, with
persistent adnate 1-2-seeded dorsally dehiscing follicles. *Seeds* pendulous from the carpels by a long cord ; outer walls of testa fleshy,
albumen oily.—Distrib. Temp. N. America ; temp. and trop. E. Asia and
Japan ; species about 16.

M. MAINGAYI, King, n. sp. A tree. Young branches and inflorescence densely clothed with pale shaggy hair. *Leaves* membranous,
oblanceolate-oblong or obovate, acumiuate, narrowed to the rounded
base ; upper surface glabrous, the lower glaucous, the midrib sparsely
villous ; nerves 16 to 20 pairs ; length of blade 6 to 8 in., breadth 2·5 to 3
in., petiole ·3 in. *Flowers* terminal, solitary, shortly pedunculate, buds
ovoid, the stipular hood villous like the pedicel. *Sepals* 3, oblong.
*Petals* about 6, of the same shape as the sepals but smaller, yellowish-
white, glabrous, 1 to 1·3 in. long. *Ovaries* villous.. *Ripe fruit* cylindric, 1·5 in. long by ·6 in. diam. ; the individual carpels ovoid, not beaked,
·35 in. long, villous.

Malacca, (Maingay No. 17). Perak, Penang ; in dense low forest up
to 2,500 feet.

2. MANGLIETIA, Blume.

Trees ; foliage and inflorescence of *Magnolia*. *Sepals* 3. *Petals* 6 or
more, 2- or more-seriate. *Stamens* very numerous, many-seriate ;
anthers linear, adnate, introrse. Gynophore sessile. *Ovaries* many,
cohering in an ovoid head ; stigma decurrent on the ventral suture ;

13

ovules 6 or more. *Fruit* ovoid ; carpels persistent, dehiscing dorsally. *Seeds* as in *Magnolia.*—Distrib. Mountains of tropical Asia ; species 5.

1. M. SEBASSA, Miq. Ann. Lugd. Bat. IV, 71. A shrub. Adult branches with pale shining bark ; young tawny-villous, as are the petioles and spathoid hood of calyx. *Leaves* coriaceous, oblong or oblanceolate, acute, much narrowed at the base, glabrous and shining on both surfaces, bullate ; nerves 14 to 20 pairs, prominent ; length of blade 11 to 14 in., breadth 4·5 to 7 in. ; petiole 1 in., much thickened at base. *Flowers* solitary, terminal, on villous peduncles 2 to 3 in. long. Spathoid hood densely adpressed fulvous-sericeous. *Sepals* and petals about 9, fleshy, pale yellow, from obovate to oblong, abruptly contracted into a claw at the base, nearly 2 in. long. *Stamens* with very stout filaments. *Pistils* about 20. *Ripe carpels* (*fide* Miquel) oblong, lenticellate. Miq. Fl. Ind. Bat. Suppl. 367.

Perak, Kunstler ; in dense jungle, at elevations of about 1500 feet : only once collected. Distrib. Sumatra.

The specimens collected by the late Mr. Kunstler are without fruit. They agree in other respects with the Sumatran specimens on which Miquel founded the species.

2. M. GLAUCA, Bl. Bat. Verh. IX, 149. A tall glabrous tree. *Leaves* coriaceous, oval or obovate-oblong, slightly acuminate, the edges slightly recurved when dry, glaucescent beneath ; nerves 12 to 14 pairs, not prominent ; length of blade 5 to 7 in., breadth 2·5 to 3 in. ; petiole ·75 to 1·5 in. *Flowers* terminal, solitary, on annulate peduncles 1 to 1·5 in. long, yellowish, about 1·5 in. long. Stipular hood broadly ovate, glabrous. *Sepals* and petals 9, greenish-yellow, oblong, sub-acute, those internal smaller. *Filaments* short. *Pistils* numerous. *Ripe fruit* ovoid, the size of a hen's egg, glabrous. *Seeds* with a red fleshy arillus. Bijdr. 8 ; Fl. Jav., *Magnol.* 22, t. 6 ; Miq. Fl. Ind. Bat. Vol. I, pt. 2, p. 15.

Perak at 3,000 feet. Distrib. Java.

3. M. SCORTECHINII, King, n. sp. A tree. The young branches stipules and under surfaces of the leaves minutely rufous-pubescent. *Leaves* thinly coriaceous, lanceolate, narrowed to base and apex, the latter subacute ; upper surface shining, reticulations minute, distinct on both surfaces ; nerves about 12 pairs ; length of blade 3·5 to 4 in., breadth 1·25 in., petiole ·3 to ·4 in. *Flowers* solitary, axillary, on annulated pedicels shorter than the petioles ; stipular hood rufous, silky. *Sepals* and petals about 12, similar, ·5 in. long, linear-lanceolate. *Stamens* linear. *Fruit* ovoid (young ·5 in. long) on a short gynophore, rufous-pubescent. *Carpels* 6 to 8.

Perak. Father Scortechini.

The only known specimens of this have unripe fruit. The ovaries 14 have only 1 ovule; but apparently it is a *Manglietia* and I put it into this genus provisionally. The small rufous leaves, shining on the upper surfaces, and small axillary flowers mark it as distinct from any other Indo-Malayan *Magnoliad.*

## 3. MICHELIA, Linn.

Trees. *Leaves* as in *Magnolia.* *Flowers* axillary, solitary or terminal. *Sepals* and petals similar, 9 to 15 or more, 3- or more-seriate. *Stamens* as in *Magnolia.* Gynophore stalked. *Carpels* in a loose spike, stigma decurrent; ovules 2 or more. *Fruit* a lax or dense elongate spike of coriaceous dorsally dehiscing carpels. *Seeds* of *Magnolia.*—Distrib. About 15 species; temp. and trop. Mountains of India.

1. M. CHAMPACA, Linn. A tall tree, the branchlets pubescent. *Leaves* membranous, ovate-lanceolate, acuminate, rather abruptly narrowed to the base, shining above, pale and glabrous or puberulous beneath; main nerves thin, 12 to 16 pairs; length of blade 4·5 to 9 in.; breadth 2·25 to 3·5 in.; petiole 1 to 1·5 in. *Flowers* yellow, solitary, axillary or terminal, shortly pedunculate; buds narrowly ovoid, the stipular hood silky. *Sepals* oblong. *Petals* about 15, narrowly oblong, 1 in. long. *Ripe fruit* 3 to 6 in. long; individual carpels ovoid, lenticellate, woody. DC. Prodr. i, 79; Roxb. Fl. Ind. ii, 656; W. & A. Prod. i, 6; Wight Ill. i, 13; Blume Fl. Jav., *Magnol.* t. 1; Bijdr. 7; H. f. & T. Fl. Ind. 79; Hook. fil. Br. Ind. I, 42; Miq. Fl. Ind. Bat. Vol. I, pt. ii, 16. *M. rufinervis,* DC. l. c. 79; Bl. Bijdr. 8. *M. Doldsopa,* Ham. ex DC. l. c.; Don Prodr. 226; Wall. Tent. Fl. Nep. t. 3. *M. aurantiaca,* Wall. Cat.; Plant. As. Rar. t. 147. *M. Rheedii,* Wight Ill. i. 14, t. 5, f. 6. *M. pubinervia,* Bl. Fl. Jav., *Magnol.* p. 14, t. 4.

In temperate forests in the Straits Settlements, but not common. Distrib. India.

2. M. MONTANA, Bl. in Verh. Bat. Gen. IX, p. 153. A glabrous tree. *Leaves* thinly coriaceous, shining, obovate or obovate-rotund, shortly and abruptly apiculate, rather suddenly narrowed to the base; nerves 10 to 12 pairs, thin, spreading; length of blade 6-7·5 in., breadth about 4 in.; petiole, slender, ·75 in. *Flowers* white, 1·5 in. in diam., solitary, terminal or axillary, on annulate peduncles about ·5 in. long. Buds cylindric. *Sepals* and petals about 8, oblanceolate or lanceolate. *Pistils* 3 to 4. *Carpels* usually single, sub-globular, 1·5 in. long, the walls lenticellate, woody, ·5 in. or more thick. Bl. Bijdr. 7; Fl. Jav., *Magnol.* p. 15, t. 5; Miq. Fl. Ind. Bat. Vol. I, Pt. ii, 17.

Perak, at low elevations. Java, on the mountains. Distrib. Eastern Himalaya.

Specimens from the E. Himalaya have less obovate leaves, and rather

15

larger flowers on longer peduncles ; but in other respects they agree with the Java plant. This species is readily distinguished by its enormously large, solitary, woody carpels.

### 4.    TALAUMA, Juss.

Trees or shrubs.  *Leaves* and inflorescence of *Magnolia.  Sepals* 3. *Petals* 6 or more, in 2 or more whorls.  *Stamens* very numerous, many-seriate ; anthers linear, introrse.  *Gynophore* sessile.  *Ovaries* indefinite, 2-ovuled, spiked or capitate ; stigmas decurrent.  *Carpels* woody, separating from the woody axis at the ventral suture, and dehiscing so as to leave the seeds suspended from the axis by an elastic cord.  *Seeds* of *Magnolia.*  Distrib.  Tropics of Eastern Asia, and South America ; Japan.  Species about 18.

1.    T. LANIGERA, Hook. fil. & Th. Fl. Br. Ind. I, 40.    A small tree. Young leaves, petioles and branches, the peduncles, outer surface of the stipular involucre enveloping the calyx, and the ovaries densely fulvous-tomentose.    *Leaves* sub-coriaceous, oblong or oblanceolate, abruptly and shortly acuminate, narrowed at the base, when adult shining and glabrous except on the lower half of the midrib below ; length of blade 9 to 12 in., breadth 2·5 to 4 in. ; petiole 1 to 1·5 in., thickened below. *Sepals* and petals white, ovate, tomentose at the very base.    *Ripe fruit* 3 to 4 in. long by 2 in. broad ; the carpels glabrescent when quite ripe, beaked, 1·5 in. long.

In open forest on low hills ; Perak and Malacca.

From Miquel's description of his *T. villosa*, (Fl. Ind. Bat. Suppl. 366), that species and this must be very closely allied ; and, if they are identical, Miquel's name, dating 1860, must take precedence of H. f. & T.'s, which was published in 1875.

2.    T. ANDAMANICA, King, n. sp.    A glabrous shrub or small tree. *Leaves* sub-coriaceous, oblanceolate-oblong, rarely lanceolate, the apex (usually rather abruptly) acute, below gradually narrowed to the petiole ; both surfaces shining ; main nerves 10 to 14 pairs, thin but prominent below ; length of blade 7 to 10 in., breadth 2·5 to 3·5 in. ; petiole ·5 to 1 in., expanded at the base.    *Flowers* solitary, sub-globose, 1·5 in. long, on thick terminal annulated pubescent peduncles 1 in. long ; stipular hood of calyx glabrous.    *Sepals* 3.    *Petals* 6, ovate or obovate.    *Ripe fruit* globose, pyriform, 1·5 in. long ; the individual carpels rhomboid, 6 in. long and nearly as broad, shortly beaked ; seeds ·4 in. long.

Andaman Islands, on Mount Harriet.

In leaf this closely resembles *T. Rabaniana*, H f. and Th., but has smaller flowers and fruit.  The individual carpels of this are not more than a third the size of those of *T. Rabaniana* which are more than 1 in. long, narrowly oblong and not rhomboidal.

16

3. T. MUTABILIS, Bl. Fl. Jav., *Magnol.* p. 35, t. 10, 11, 12, fig. B. A glabrous shrub. *Leaves* oblong-lanceolate, acute at either end, slightly unequal at the base, sub-coriaceous, shining on both surfaces; nerves about 12 pairs, spreading; length of blade 5 to 8 in., breadth 2·25 to 2·75 in.; petiole ·75-8 in., thickened at the base, minutely muriculate when dry as in the midrib. *Flowers* solitary, terminal, on annulate peduncles about 1·5 in. long; stipular hood fuscous-villous. *Sepals* 3, broadly ovate. *Petals* 6, in 2 whorls, broader than the sepals, sometimes obovate, concave, connivent, passing from rosy green to reddish brown. *Ripe fruit* ovoid, 1·5 to 2 in. long, pubescent at first, ultimately glabrous; individual carpels, rhomboid, lenticellate, with blunt recurved beaks ·75 to 1·25 in. long. Korth. in Ned. Kruik. Arch. II, 98; H. f. and Th. Fl. Ind. I, 74; Hook. fil. Fl. Br. Ind. I, 40; Miq. Fl. Ind. Bat. Vol. I, pt. ii, 14. *Manglietia Candollei*, Wall. Cat. (not of Bl.).

In the Straits Settlements, in shady damp spots near water. Distrib. The Malayan Archipelago.

A very variable shrub of which Blume distinguishes 3 varieties. All the specimens I have seen from the Straits have glabrous leaves; but Blume and others describe the leaves as often pubescent or even pilose below.

4. T. KUNSTLERI, King, n. sp. A tree, 25 to 30 feet high; glabrous, except the peduncle and unripe carpels. *Leaves* oblong-lanceolate, acuminate at base and apex, thinly coriaceous, both surfaces shining, nerves 10 to 14 pairs, length of blade 6 to 9 in., breadth 1·75 to 2·5 in.; petiole ·5-1·25 in., slender, the base much thickened. *Flowers* terminal, solitary, ovoid, scarcely expanding, ·85 in. long, on erect pubescent annulate peduncles 1 in. long. *Sepals* 3 and petals 6, scarcely exceeding the stamens, broadly elliptic, fleshy, glabrous, waxy white. *Anthers* sessile, more than ·5 in. long. *Pistils* 6 to 8, linear, pubescent. *Ripe fruit* ovoid, pointed, 1·25 in. long and ·75 in. in diam.; individual carpels ·75 to 1 in. long, with short stout sub-terminal beaks.

Perak, in dense forest at elevations of from 3,500 to 4,000 feet.

I here subjoin a description of a new species from Sumatra.

5. TALAUMA FORBESII, King, n. sp. A small tree or shrub; glabrous except the peduncles which are adpressed-villose. *Leaves* oblong-lanceolate, acuminate both at base and apex, green and shining on both surfaces, thinly coriaceous, nerves 12 to 15 pairs, length of blade 4·5 to 6 in., breadth 1 to 1·5 in. *Flowers* terminal, solitary, erect, ·75 in. long, on stout peduncles. *Stipular hood* of calyx densely covered with adpressed, fulvous silky hair; buds pointed. *Sepals* and petals about the same length, white, nearly glabrous. *Ripe fruit* 1·25 in. long; the individual carpels ·6 in. long, ovate, rugose, with short terminal beak. *Seeds* ·4 in. broad, by ·3 in. long, the base compressed.

17

Sumatra, on Kaiser's Peak, &c., at elevations of 5,000 to 6,500 feet. Forbes, Nos. 1853, 2066 and 2204.

This resembles *T. pumila,* but its leaves are not glaucous beneath and they have more nerves; moreover the flowers are smaller and not drooping, the petals ovate and not obovate, and the carpels are only about half the length of those of that species.

## 5.  ILLICIUM, Linn.

Evergreen aromatic shrubs or small trees. *Leaves* quite entire, pellucid-dotted. *Flowers* bi-sexual or unisexual, solitary or fascicled, yellow or purplish. *Sepals* 3 to 6. *Petals* 9 or more, 3- or many-seriate. *Stamens* indefinite, filaments thick; anthers adnate, introrse. *Ovaries* indefinite, 1-seriate, 1-ovuled; style subulate, recurved. *Fruit* of spreading compressed hard follicles. *Seeds* compressed, testa hard, shining albumen fleshy. Distrib. North America, China, Indo-Malaya; species about 6.

1.  I. CAMBODIANUM, Hance in Journ. Bot. 1876, p. 240. A small glabrous tree. *Leaves* opposite or in whorls of 3 or 4, coriaceous, oblanceolate or obovate-lanceolate, rarely lanceolate, acuminate, entire; length of blade 3 to 4·5 in., breadth 1 to 2 in., petiole less than ·5 in. *Flowers* red to white, 4 in. in diam., on long, slender, axillary pedicels, solitary or in groups of 3 or 4. *Sepals* 3 or 4, rotund. *Petals* about 9, diminishing in size inwards, ovate-oblong, blunt. *Stamens* 9 to 13 in a single row, the filaments about as long as the anthers. *Follicles* 8 to 12, beaked, radiate. Pierre Flore Forestiere de la Cochin Chine, t. 4.

Perak, in dense forests to elevations of from 3,600 to 7,000 feet.

There is some variability as to the shape and size of the sepals; sometimes they are triangular and much smaller than the petals, in other specimens they resemble the petals both in size and shape. The stamens also vary in number, but they never form more than a single row. The texture of the leaves in some plants is thin and membranous, in others almost coriaceous. It is possible there may be two species included in this.

2.  I. EVENIUM, King, n. sp· A small glabrous tree. *Leaves* very coriaceous, opposite or in whorls of 3, oblanceolate or obovate-oblong, the apex with an abrupt blunt short acumen, the base elongate-cuneate, gradually narrowed to the short thick petiole; nerves undistinguishable (when dry); length of blade 3·5 to 5 in., breadth 1·25 to 2 in.; petiole ·3 in. or less. *Flowers* globular, 2 in. in diam., pedicellate, solitary or in 2- to 3-flowered racemes; pedicels with a few minute bracteoles near the apex, about 1 in. long. *Sepals* and petals 8 or 9, rotund, fleshy, similar, or the former a little smaller. *Stamens* 30 to 50, in several rows,

18

filaments shorter than the anthers; connective of outer stamens narrow, of the inner broad. *Female flowers* unknown.

Perak, Scortechini.

This species was collected only once by the late Father Scortechini. All the flowers which I have seen are male; and, as no unisexual species of this genus has hitherto been described, I have modified the generic definition accordingly. This differs from *T. Cambodianum* in having very coriaceous leaves with shorter petioles, smaller flowers, and much more numerous stamens.

## 6. KADSURA, Kaempfer.

Climbing glabrous shrubs. *Leaves* exstipulate. *Flowers* unisexual, white, yellow or reddish, axillary, or in the axils of scales near the base of short lateral leafy branches. *Sepals* and *petals* 9 to 12, imbricate in about 3 series. ♂ *Stamens* 5 to 15 or more, in a spiral series; filaments very short, free or subconnate, often fleshy; anthers free or subimmersed in a fleshy head of confluent filaments, cells small, remote. ♀ *Ovaries* many, densely imbricated; stigma sessile; ovules 2 to 4. *Fruit* a globose head of indehiscent, fleshy, 1-seeded carpels. *Seeds* 1 or 2, suspended, albumen fleshy, testa crustaceous; embryo minute.—Distrib. Temp. and subtrop. Asia; species about 7.

1. K. SCANDENS, Bl. Fl Jav., *Schizandreæ,* p. 9, t. 1. A woody glabrous climber 30 to 40 feet long; the bark dark-coloured, irregularly striate. *Leaves* coriaceous, broadly ovate to ovate-oblong, shortly acuminate, entire, the base rounded very slightly decurrent towards the petiole; main nerves about 5 pairs, sub-erect; length of blade 4 to 6 in., breadth 2 to 4 in.; petiole about 1 in., expanded at the base. *Flowers* axillary, with a few from the trunk below the leaves, solitary, on bracteolate peduncles, those of both sexes similar. *Calyx* of 3 triangular sepals much shorter than the petals and united at the base. *Petals* ovate-oblong, blunt, fleshy, erecto-patent, sub-concave. *Male flowers* with numerous short, cuneate, equal stamens densely packed on an ovoid, fleshy receptacle; the connective fleshy, the anthers sub-truncate, 2-celled. *Female flowers* without stamens, the ovaries numerous, compressed; the stigmas sessile, elongate, fleshy. *Ripe fruit* sub-globose, 2 to 2·5 in. in diam.; the individual carpels sessile, globose or subcompressed, fleshy, mucronate or beaked, less than ·5 in. in diam. Miq. Fl. Ind. Bat. Vol. I, Pt. ii, 19; *K. cauliflora,* Bl. 1. c. p. 11, t. 2; *Sarcocarpon scandens,* Bl. Bijdr. 21.

Perak, Penang; at low elevations. Distrib. Java, Sumatra, and probably in the other islands of the Archipelago.

2. K CAULIFLORA, Bl. Fl Jav., *Schizand.* 11, t. 2. A stout woody

19

climber, 30 to 40 feet long; bark of young shoots dark, smooth; that of the main stem corky, furrowed, lenticellate. *Leaves* ovate-rotund or broadly ovate, slightly and rather abruptly acuminate, the base broad and rounded; nerves 5 to 7 pairs, ascending; length of blade 6 to 7 in., breadth 3·4 to 4·5 in , petiole 1 to 1·5 in. *Flowers* usually from the stem below the leaves, solitary or fascicled, on bracteolate peduncles; rarely axillary. *Sepals* and *petals* as in the last, but larger and more ovate. *Carpels* on pedicels ·5 in. long. Miq. Fl. Fl. Ind. Bat. Vol. I, pt. 2, p. 19.

Perak, Scortechini and King's Collector.

This is very near *K. scandens*, but has corky bark, larger leaves, larger and less orbicular petals, and the flowers are mostly on the old wood and rarely axillary.

3.   K. ROXBURGHIANA, Arn. in Jard. Mag. Zool. and Bot. II, 546.  A glabrous woody and stout climber, with rough bark. *Leaves* membranous, ovate, obovate or oblong, acute or shortly acuminate, entire, the base narrowed; main nerves 7 to 8 pairs, not prominent; length 4 to 6 in., breadth 1·75 to 2·5 in., petiole ·5 in. *Flowers* ·5 in. in diam., axillary, solitary, on bracteolate pedicels ·5 in. or more long. *Sepals* and *petals* rotund, concave, fleshy. *Filaments* connate into a column, the upper the smaller. *Ripe fruit* globose, 1 to 2 in. in diam. *Carpels* cuneate with rounded tops. Hook. fil. & Thoms. Fl. Ind. I, 83; Hook. fil. Fl. Br. Ind. I, 45. *Kadsura japonica*, Wall. Tent. Fl. Nep. t. 12. *Uvaria heteroclita*, Roxb. Fl. Ind. ii, 663.

Andamans. Distrib. The base of the Eastern Himalaya and Khasia Hills.

4.   K. LANCEOLATA, King, n. sp.  A slender woody climber, 20 to 30 feet long. *Leaves* thinly coriaceous, lanceolate or ovate-lanceolate, slightly unequal-sided, acuminate, the base cuneate or rounded, edges slightly recurved when dry; nerves obscure, about 12 pairs; length of blade 3 to 4 in., breadth 1·5 in., petiole ·35 in. *Flowers* solitary, axillary, globular, 3·5 in. in diam., on minutely bracteolate peduncles shorter than the petioles. *Sepals* about 3, triangular, much smaller than the petals. *Petals* about 9, rotund or broadly ovate, cream-coloured, fleshy, slightly concave. *Stamens* as in *K. scandens*. *Ripe fruit* globular, ·75 in. in diam. or less. *Pistils* numerous, the stigmas minute, subterminal. *Carpels* ovoid, the apices truncate, the minute lateral stigmas persistent.

Perak, at elevations of from 500 to 1,000 feet.

A smaller plant than *K. scandens*, with smaller truncate carpels.


ORDER IV.   ANONACEÆ.   To be taken up subsequently.

20

ORDER V. MENISPERMACEÆ.

Climbing or twining, rarely sarmentose, shrubs. *Leaves* alternate, entire or lobed, usually palmi-nerved; stipules 0. *Flowers* small or minute, solitary fascicled cymose or racemed, diœcious, sometimes 3-bracteolate. *Sepals* 6 (rarely 1 to 4, or 9 to 12), usually free, imbricate in 2 to 4 series, outer often minute. *Petals* 6 (rarely 0 or 1 to 5), free or connate. ♂ *Flowers*: *Stamens* hypogynous, usually one opposite each petal, filaments free or connate; anthers free or connate, 2-celled. Rudimentary carpels small or 0. ♀ *Flowers*: *staminodes* 6 or 0. *Ovaries* 3 (rarely 1, or 6 to 12); style terminal, simple or divided; ovules solitary (2 in *Fibraurea*), usually amphitropous. *Ripe carpels* drupaceous, with the style-scar subterminal, or by excentric growth sub-basal. *Seed* usually hooked or reniform, often curved round an intrusion of the endocarp (condyle Miers), albumen even or ruminate or 0; cotyledons flat or semi-terete, foliaceous or fleshy, appressed or spreading.—A large tropical order: genera 32; species about 100.

Tribe I. *Tinosporeæ.* *Flowers* 3-merous. *Ovaries* usually 3. *Drupes* with a subterminal rarely ventral or sub-basal style-scar. *Seed* oblong or subglobose; albumen copious or scanty; cotyledons foliaceous, usually spreading laterally

Drupes with a terminal or subterminal style-scar.
    Sepals 6; petals 6; filaments free ...  ... 1. *Tinospora.*
    Sepals 9; petals 6; filaments free ...  ... 2. *Tinomiscium*
    Sepals 6; petals 0; filaments free ...  ... 3. *Fibraurea.*
Drupes with a sub-basal style-scar.
    Sepals 6; filaments all connate  ...  ... 4. *Anamirta.*
    Sepals 9; outer filaments free  ...  ... 5. *Coscinium.*

Tribe II. *Cocculeæ.* *Flowers* 3-merous. *Ovaries* usually 3. *Drupe* with a sub-basal rarely subterminal style-scar. *Seed* horse-shoe shaped, albumen copious; embryo slender, cotyledons linear or slightly dilated.

Sepals 3 to 10, all imbricate; petals 4 to 6, stamens
    6 to 10, ovaries 3 to 6, style canaliculate, sub-
    3-lobed    ...    ...    ...    ... 6. *Hypserpa.*
Sepals 9, the inner 3 valvate; petals 3 or 6,
    ovaries 3, style compressed    ...    ... 7. *Limacia.*
Petals 6; ovaries 3 to 6; styles subulate    ... 8. *Cocculus.*
Petals 6; ovaries 3; style forked    ...    ... 9. *Pericampylus.*

Tribe III. *Cissampelideæ.* *Flowers* 3 to 5-merous. *Ovaries* usually solitary. *Drupe* with a sub-basal style-scar; endocarp dorsally muricate or echinate. *Seed* horse-shoe-shaped, albumen scanty; embryo linear, cotyledons appresed.

Sepals 6 to 10, free; petals of ♂ and ♀ 3-6, free... 10. *Stephania.*
Sepals 4, free; petals of ♂ 4 connate, of ♀ 1 ... 11. *Cissampelos.*

21

Sepals 4 to 8, connate; petals of ♂ 4 to 8 connate,
of ♀ 1, anthers sessile on a central column    ... 12. *Cyclea.*
Tribe IV. *Pachygoneæ. Flowers* usually 3-merous. *Ovaries* usually 3. *Drupes* with a sub-basal or ventral style-scar. *Seed* curved,
hooked or inflexed, albumen 0; cotyledons thick fleshy.
Sepals 8; petals 2; stamens 4 or 8 ...        ... 13. *Antitaxis.*

### 1. TINOSPORA, Miers.

Climbing shrubs. *Flowers* in axillary or terminal racemes or panicles. *Sepals* 6, 2-seriate, inner larger, membranous. *Petals* 6, smaller.
*Male flower; stamens* 6, filaments free, the tips thickened; anther cells
obliquely adnate, bursting obliquely. *Female flower; staminodes* 6, clavate.
*Ovaries* 3; stigmas forked. *Drupes* 1-3, dorsally convex, ventrally flat;
style-scar subterminal; endocarp rugose, dorsally keeled, ventrally
concave. *Seed* grooved ventrally or curved round the intruded sub-
2-lobed endocarp; albumen ventrally ruminate; cotyledons foliaceous,
ovate, spreading.—Distrib.   Species about 8, tropical Asiatic and
African.

1.  T. CRISPA, Miers Contrib. III, 34.  Young shoots glabrous, the
older bark warted. *Leaves* membranous, glabrous, ovate-cordate or
oblong-acuminate, entire or repand, sometimes sub-sagittate; length of
blade 2 to 6 in., breadth 1 to 4 in., petiole 1 to 3 in. *Racemes* from the old
wood, solitary or fascicled. *Flowers* 2 to 3, in the axils of ovate fleshy
bracts, ·15 in. long, campanulate, green. *Stamens* adnate to the base of
the petals, anthers quadrate. *Drupe* elliptic-oblong, pale yellow, about
1·5 in. long or less. Hook. fil. Fl. B. Ind. I, 96; H. f. & T. Fl. Ind.
183; Miq. Fl. Ind. Bat. I, pt. i, 78; Kurz For. Flor. Burmah, I, 52.
*Menispermum crispum,* L. *M. verrucosum,* Roxb. Fl. Ind. iii, 808. *M.
tuberculatum,* Lamk. *Cocculus crispus,* DC. Prodr. i. 97; Hassk. Pl.
Jav. Rar. 166. *Cocculus coriaceus,* Bl. Bijd. 25. *C. verrucosus,* Wall.
Cat. 4966 A. B.

In all the provinces, but apparently not very common.  Distrib.
Malayan Archipelago: tropical British India.

2.  T. ULIGINOSA, Miers Contrib. iii, 35.  All parts glabrous, the
branches terete, the loose brown bark bearing many 4-lobed warts. *Leaves*
subcoriaceous, remote ovate or ovate-oblong, acuminate sub-repand-
sinuate or entire, the base cordate, 5-nerved; venation prominent; length
of blade 3 to 4 in., breadth 1·5 to 2 in.; petiole slender, swollen at the base,
1 to 2 in. long. *Racemes* slender axillary, longer than the leaves;
pedicles 1-flowered. *Drupe* as in *T. crispa,* but with thinner endocarp.
Hook. fil. Fl. B. Ind. i. 97 ; Hook fil. & Thoms. Fl. Ind. 105. *Cocculus
petiolaris,* Wall. Cat.

22     .

Malacca, Maingay. Distrib. Java and Borneo.

A species of which I have seen no good specimen. The foregoing description is chiefly copied from Miers.

## 4. TINOMISCIUM, Miers.

A scandent shrub, juice milky. *Flowers* racemed. *Sepals* 9, with 3 bracts. *Petals* 6, oblong, margins incurved. *Male flower; stamens* 6, filaments flattened; anthers oblong, adnate, bursting vertically. *Rudimentary carpels* 3. *Female flower* unknown. *Drupes* much compressed, ovoid-oblong, style-scar terminal; endocarp much compressed, dorsally convex, ventrally flat or slightly concave, not intruded. *Seed* almost flat, oblong; cotyledons quite flat, nearly as broad as the thin layer of albumen, very thin, closely appressed; radicle short, cylindric.—Distrib. 3, all E. Asiatic.

T. PETIOLARE, Miers Contrib. iii, 45, t. 94. Young shoots and rachises of inflorescence brown-tomentose; bark of older branches nearly glabrous, pale brown, deeply striate, very sparsely verrucose. *Leaves* membranous, glabrous, ovate-oblong, obtuse or shortly and suddenly acuminate, entire, the base rounded or truncate, 5-nerved, the nerves all sparsely pubescent and 2 of them small; length of blade 4 to 8 in., breadth 2 to 4·5 in.; petiole 2 to 5 in., slender. *Racemes* fasciculate on stem tubercles, 4 to 8 in. or even 12 in., long. *Flowers* ·35 in. in diam.; sepals puberulous. *Drupe* elongate-ovoid, compressed, 1·25 in. long, and ·75 in. broad; endocarp rugose, woody. Miq. Fl. Ind. Bat. i, pt. i, 87; Hook. f. Fl. B. Ind. i. 97.

Common in the Straits Settlements. Distrib. Sumatra.

## ANAMIRTA, Miers.

Climbing shrubs. *Flowers* in panicles. *Sepals* 6, with 2 adpressed bracts. *Petals* 0. *Male flower; anthers* sessile, on a stout column, 2-celled, bursting transversely. *Female flower; staminodes* 9, clavate, 1-seriate. *Ovaries* 3, on a short gynophore; stigma sub-capitate, reflexed. *Drupes* on a 3-fid gynophore, obliquely ovoid, dorsally gibbous, style-scar sub-basal; endocarp woody. *Seed* globose, embracing the sub-globose hollow intruded endocarp; albumen dense, of horny granules; embryo curved; cotyledons narrow, oblong, thin, spreading.

1. A. LOUREIRI, Pierre Flore Forest. Cochin Chine, t. 110. Glabrous; bark of the younger branches brown, that of the older pale and slightly striate. *Leaves* sub-coriaceous, shining, ovate-rotund to broadly elliptic, abruptly and shortly acuminate, entire, the base sometimes minutely cordate, 4-nerved, and with 4 small pits between the nerves at

23

their junction with the petioles; length of blade of 4 to 5 in., breadth 3·5 to 4 in.; petiole 3 to 3·5 in. slender, swollen and bent at the base. *Racemes* extra-axillary, slender, branched, 3 to 6 in. long. *Male flowers* sessile, 1 in. in diam., anthers 9. *Female flower* unknown ; ripe inflorescence stout, woody, a foot or more long; pedicel of ripe fruit much thickened, clavate, rugose ; *ripe drupe* transversely ovoid, sub-compressed, 1·25 long by nearly 1·5 broad ; pericarp thick, fibrous, pulpy ; endocarp woody, rugose.

Malacca, Maingay No. 116, 115 (in part). Perak, common. Distrib. Cochin-China.

An enormous climber, often with a stem 4 in. in diameter. The leaves and male flowers are excellently figured by M. Pierre, who however does not appear to have seen the ripe fruit.

FIBRAUREA, Loureiro.

Glabrous climbing shrubs. *Leaves* ovate to oblong, 3-nerved. *Flowers* in panicles. *Sepals* 6, with 3 minute bracts, inner larger. *Petals* 0. *Male flower; stamens* 6, filaments clavate ; anthers terminal, adnate; cells spreading, bursting vertically. *Female flower; staminodes* 6. *Ovaries* 3, ovoid, 2-ovuled ; stigma sessile, punctiform. *Drupes* 1-seeded, oblong, terete, style-scar subterminal ; endocarp oblong, dorsally convex, ventrally flattened and channelled, hardly intruded. *Seed* oblong, terete, reniform on transverse section ; albumen copious, horny ; cotyledons foliaceous, longitudinally curved, oblong, very thin ; radicle short, cylindric.

1. F. CHLOROLEUCA, Miers Contrib. iii, 42. Glabrous, the branches striate ; bark of the young branches brown, of the old pale. *Leaves* coriaceous, ovate-oblong, shortly acuminate ; the base rounded, 5-nerved (2 of the nerves joining the central one half an inch above the base) ; venation inconspicuous ; length of blade 5 to 9 in., breadth 2·5 to 4 in.; petiole 2 to 4 in., rather slender but swollen and curved towards the base. *Panicles* extra-axillary, or from the stem below the leaves, slender, lax, their branches horizontal, 3 to 8 in. long. *Male flowers* ·2 in. in diam., shortly pedicellate ; filaments clavate, flattened. *Female flower* unknown ; the ripe female inflorescence slender, stouter than the leaf-petioles, about 1 foot long. *Ripe drupes* on rather slender terete pedicels which are capitate at the apex and ·5 in. long, ovoid, smooth, pulpy, rather more than 1 in. long, ·75 in. in diam., endocarp smooth.

Malacca, Perak ; not uncommon. Distrib. Sumatra, Borneo.

This appears to me to be a different plant from Loureiro's *F. tinctoria*, the type of which is in the British Museum.

24

### 7. COSCINIUM, Colebrooke.

Climbing shrubs. *Flowers* in dense globose heads. *Sepals* 6, with a bract, orbicular. *Petals* 3, large, spreading, elliptic. *Male flower ; stamens* 6, filaments cylindric, 3 inner connate to the middle ; anthers adnate, outer 1- the inner 2-celled, bursting vertically. *Fem. flower ; staminodes* 6. *Ovaries* 3-6, subglobose ; styles subulate, reflexed. *Drupes* globose ; endocarp bony. *Seed* globose, embracing a globose intrusion of the endocarp ; albumen fleshy, ruminate in the ventral face ; embryo straight; cotyledons orbicular, spreading, thin, sinuate, laciniate, or fenestrate.—Distrib. Species 2 ; tropical Asiatic.

1. C. FENESTRATUM, Colebrooke in Trans. Linn. Soc. xiii, 65. Young shoots faintly striate, shortly tomentose, often ferruginous. *Leaves* coriaceous, very slightly peltate, rotund-ovate, acute or shortly acuminate, the base truncate and sometimes sub-sinuate, shining above, yellow-tomentose beneath except the 7 stout glabrous nerves ; reticulations prominent ; length of blade 5 to 7 in., breadth 4 to 6 in. ; petiole 2 to 3 in., swollen and bent at base. *Flowers* in small pedunculate heads, in extra-axillary racemes shorter than the leaves. *Petals* orbicular and, like the sepals, persistent. *Ripe drupes* on stout pedicels with capitate apices, globose, tomentose, ·75 in. in diam. ; cotyledons laciniate. Miers in Hook. Bot. Mag. t. 6458; Contrib. iii. 22, t. 88; H. f. & T. Fl. Ind. 178; Hook. fil. Fl. Brit. Ind. i. 99. *C. Maingayi,* Pierre Fl. Coch. Chinc. *C. Wallichianum* and *Wightianum,* Miers in Tayl. Ann. Ser. 2, vii. 37 ; Contrib. iii. 23. *Menisp. fenestratum,* Gærtn. ; DC. Prod. i. 103 ; Roxb. Fl. Ind. iii. 809. *Cocculus Blumeanus,* Wall. Cat. 4971, partly. *Pereira medica,* Lindl. Fl. Med. 307.

Straits Settlements, at low elevations, not so common as the next. Distrib. Ceylon, and perhaps some of the Malayan Islands.

The Ceylon specimens have larger leaves and a more condensed inflorescence than the Malayan ; but the flowers are alike. Pierre's species *C. Maingayi* is founded on Maingay's Malacca specimens (Kew. Distrib. 117 ) but I cannot see that they differ specifically from his No. 118, or from Wallich's.

2. C. BLUMEANUM, Miers Contrib. iii, 23. Young shoots sub-striate, tawny-tomentose. *Leaves* coriaceous, peltate, oblong, elliptic, rarely ovate-rotund, obtuse or acute, the base rounded or truncate, sometimes sub-sinuate, shining above, white-tomentose beneath, the 7 nerves bold and prominent on lower surface as are the reticulations, length of blade 8 to 12 in., breadth 4 to 7 in. ; petiole 4 to 6 in. swollen at base and apex. *Male inflorescence* 5 in. long, racemose, densely ferruginous-tomentose ; the flower heads ·35 in. in diam. *Female inflorescence* from the stem, 8 in. long, its branches horizontal ; drupes globular, tomen-

25

Materials for a Flora of the Malayan Peninsula

tose. Hook, fil. & Thoms Fl. Ind. 179; Hook; fil. Fl. B. Ind. i. 99 ; Miq. Fl. Ind. Bat. i. Pt. 1, 77 ; *Cocculus Blumeanus*, Wall. Cat. 5971, partly. Climbing like the last on high trees. I have not seen specimens of the ripe drupe.

### LIMACIA, Loureiro.

Climbing shrubs or small trees ; flowers in racemes or panicles. *Sepals* 9, tomentose, in three series ; the two outer smaller, ovate ; the inner large, rotund, concave, valvate. *Petals* 3 or 6, small, (as large as the outer 3 sepals), obovate, clawed, glabrous, embracing the stamens *Stamens* 3 or 6, free, sub-equal ; filaments short, erect, thickened upwards ; anthers connivent, cordate, 2-lobed. *Fem. flower ; sepals* and *petals* as in male; *staminodes* 6, clavate. *Ovaries* 3, hirsute. *Style* short. *Drupes* 3, usually 1 by abortion, obovoid or reniform, fleshy ; style-scar sub-basal ; endocarp 3-celled, the 2 lateral cells empty. *Seed* elongate, embracing the intruded endocarp ; embryo slender, the cotyledons linear, plano-convex. Distrib. Tropical Asia.

1.   L. TRIANDRA, Miers Contrib. iii, 112. Branches closely striate, puberulous when young, glabrous when adult. *Leaves* membranous, ovate-lanceolate, acuminate, 3-nerved above the rounded base, length of blade 3 to 5 in., breadth 1·5 to 2 in., petiole about ·5 in. *Racemes* supra-axillary, shorter than the leaves, usually solitary, puberulous ; bracts and flowers minute. *Stamens* 3, free, the filaments cuneate ; anthers large, the cells divergent. Hook. fil. Fl. B. Ind. i, 100 ; Miq. Fl. Ind. Bat. i. pt. 1, 80 ; Kurz For. Flor. Burm. i, 55 ; H. f. & T. Fl. Ind. 188. *L. Amherstiana* and *Wallichiana*, Miers l.c. 112, 113. *Menisp. triandrum*, Roxb. Fl. Ind. iii. 816. *Cocculus triandrus*, Colebrooke in Trans. Linn. Soc. xii. 64 ; Wall. Cat. 4962, 4959 C. 4958 L.

Penang. Distrib. Burmah.

2.   L. OBLONGA, Miers Contrib. iii. 109. Scandent or shrubby ; branches minutely striate, rusty-puberulous when young. *Leaves* membranous, glabrous, oblong-lanceolate or broadly elliptic, acute or mucronate ; the base rounded, 3-nerved ; reticulations few, bold, pubescent like the nerves and petiole ; length of blade 4 to 7 in., breadth 2·25 to 3·5 in., petiole ·75 to 1·25 in. *Male panicles* slender, extra-axillary, usually in pairs, much longer than the leaves, branched, rusty-pubescent. *Stamens* 6, filaments thickened upwards but not cuneate. *Female inflorescence* much shorter. *Drupe* transversely reniform, compressed, glabrous, pulpy, rugose when dry, ·75 in. long, stylar scar on a sub-basal projecting horn. Hook. fil & Th. Fl. Ind. 189; Hook. fil. Fl. B. Ind. i. 100 ; Miq. Fl. Ind. Bat. i. Pt. i, 80. *Cocculus oblongus*, Wall. Cat.

Common in all the Settlements.

26

Miers remarks that Wallich describes the species as a bush. Mr. Kunstler, who collected it at various places in Perak, describes one set of his specimens (No. 6184) as bushes of 8 to 10 feet; the other he describes as climbers. Between the male flowers of these two I can detect no difference. The species is at once recoguised by the length of the panicles of male flowers.

3. L. VELUTINA, Miers Contrib. iii. 110. Whole plant, but especially the young branches, olivaceous-tomentose. *Leaves* thinly coriaceous, ovate-oblong, often slightly obovate, obtuse, acute or shortly and finely acuminate, the base acute or rounded; when adult glabrous and shining above except the midrib; beneath olivaceous-pubescent; 3-nerved; length of blade 3·5 to 5·5 in., breadth 1·5 to 2·25 in.; petiole ·5 to ·75 in., stout, terete. *Cymes* shorter than the petiole, umbellate, in axillary fascicles of 2 to 6. *Male flowers* small. *Stamens* 6, filaments much thickened upwards; anthers large, 2-celled. *Drupes* 1 or 2, transversely reniform, very little compressed, sparsely tomentose; otherwise as in the last. Hook. fil. Fl. Br. Ind. I. 100; Miq. Fl. Ind. Bat. i, Pt. i, 80; Kurz For. Flor. Burmah I, 55; H. f. & T. Fl. Ind. 189. *L. inornata*, Miers l. c. iii. t. 109. *Cocculus velutinus*, Wall. Cat. 4970.

Var. GLABRESCENS, leaves nearly glabrous. *L. distincta*, Miers Contrib. iii, 111, t. 109.

In Forests in the Straits Settlements. Distrib. Sumatra.

A slender climber 15 to 20 feet long; readily recognised by the yellowish olivaceous tomentum and short cymose inflorescence.

4. L. KUNSTLERI, King, nov. spec. Branches, petioles, and nerves of leaves softly pubescent. *Leaves* sub-coriaceous, glabrous, pale beneath, shortly petiolate, lanceolate, mucronate, the base cuneate, 3-nerved; reticulations wide, distinct; length of blade 1·5 to 2·5 in., breadth ·5 to 1 in., petiole 2 in. *Cymes* pedunculate, axillary, solitary, little longer than the petioles; or in terminal racemes. *Male flowers* small. *Stamens* 6, the filaments short, clavate; anthers large, cordate, 2-celled, connivent.

North Coast of Singapore near the Sea; King's Collector, No. 70.

This very distinct species has been gathered only once. Only the male flowers are known, but they are unmistakeably those of a *Limacia*. The species is distinguished by its shortly petiolate small leaves, and numerous short cymes which (toward the end of the branches) are arranged in racemes.

HYPSERPA, Miers.

Climbing shrubs. *Flowers* in short axillary cymes. Parts of flower varying in number. *Sepals* in three rows, the outer 3 bractiform and hairy ; the inner 6 to 8 much larger ; all imbricate, glabrous. *Petals* 4 to 6, smaller than the inner sepals, oblong, incurved at the apex, rather fleshy. *Stamens* 6 to 10, free, in two series ; filaments compressed, thickened upwards ; anthers ovate, 2-celled. *Female flower* ( *fide* Miers) *Sepals* 8, oblong, all imbricate. *Petals* 5 or 6, oblong, concave. *Staminodes* 6, clavate. *Ovaries* 6, rarely 3. Style very short. *Drupes* 2-3, transversely ovate, fleshy ; endocarp bony, sub-globose, slightly compressed, with radiating grooves, and with a single lunate cell ; embryo terete, slender ; radicle as long as the cotyledons.

1. H. TRIFLORA, Miers Contrib. iii, 102. Branches striate, the youngest pubescent. *Leaves* small, sub-coriaceous, shining, glabrous, oblong-lanceolate, tapering gradually to the rather blunt minutely mucronate apex ; the base rounded, 3-nerved ; reticulations fine but rather obscure, as are the nerves ; length of blade 1·5 to 3·5 in., breadth ·5 to 1·25 in., petiole ·3 to ·5 in. *Cymes* about 3-flowered, little longer than the petioles, usually solitary, axillary and supra-axillary, the female shorter.

Malacca ; Griffith, Maingay (Kew Distrib. 123). Perak ; Scortechini, King's Collector. Distrib. Sumatra.

The Perak specimens agree with the types of Miquel's *Limacia microphylla* from Sumatra in Herb. Calcutta; and they do not agree in externals with the Indian *L. cuspidata.* Ripe fruit of this is unknown. But the flowers are so different from those of *Limacia* that I think it ought not to be included in that genus, and I further venture to think that Miers's genus *Hypserpa* has a sufficiently sure basis on the structure of the flowers alone.

## 10. COCCULUS, DC.

Climbing or sarmentose shrubs, rarely suberect. *Petiole* not dilated at the base. *Flowers* in panicles. *Sepals* 6, 2-seriate, outer smaller. *Petals* 6, smaller, usually auricled. *Male flower ; stamens* embraced by the petals ; anthers sub-globose, cells bursting transversely. *Female flower ; staminodes* 6 or 0. *Ovaries* 3 to 6 ; styles usually cylindric. *Drupes* laterally compressed ; endocarp horse-shoe-shaped, dorsally keeled and tubercled, sides excavate. *Seed* curved, albumen fleshy ; embryo annular ; cotyledons linear, flat, appressed.—Distrib. All warm climates.

1. C. KUNSTLERI, King, n. sp. Glabrous ; the branches striate, pale. *Leaves* membranous, with long petioles, peltate, rotund, acute ; nerves 9, radiating from the petiolar insertion, thin but prominent on the pale under surface ; length of blade 3·25 to 4 in., breadth about ·25 in. or less ; petiole slender, terete, about 3 in. long. *Panicles* in fascicles of 2 to 4, from flat warty tubercles on the stem, narrow, the lateral branches only about ·5 in. long, few-flowered. *Sepals* imbricate, glabrous. *Petals* 6,

28

each embracing a stamen. *Filaments* free, clavate; anthers broad, 4-celled. *Ripe drupes* 1 to 1·25 in. long, and about ·6 in. broad, narrowly sub-obovoid, compressed; pericarp of a thin pulp; endocarp horny, narrowly horse-shoe-shaped, the edge boldly ridged, the sides with deep radiating grooves and the central part with a deep vertical hollow: embryo sausage-shaped, bent along the circumferential chamber of the endocarp.

Perak, Ulu Bubong; King's collector, Nos. 4417 and 10282.

This has the flower of *Cocculus*, but the fruit of *Stephania*. It comes nearest to the Indian *O. macrocarpus*, which has a similar though smaller fruit, and, like that species, would belong to Miers' genus *Diploclisia*. It must be near *D. pictinervis* of that author.

### 11. Pericampylus, Miers.

A climbing shrub. *Leaves* subpeltate; petioles slender, articulate. *Flowers* in axillary cymes. *Sepals* 6, with 3 bracts, outer smaller, inner spathulate. *Petals* 6, cuneate. *Male flower*: stamens 6, filaments cylindric; anthers adnate, bursting transversely. *Female flower*: Staminodes 6, clavate. *Ovaries* 3; styles 2-partite, segments subulate. *Drupes* subglobose; endocarp horse-shoe-shaped, dorsally crested and echinate, sides excavated. *Seed* curved; cotyledons elongate, flat, scarcely broader than the radicle.

1. P. incanus, Miers Contrib. iii. 118, t. 3. Young branches minutely tomentose, not striate. *Leaves* membranous, orbicular-reniform obtuse or acute, sometimes slightly retuse, mucronulate, the base truncate or sub-cordate; upper surface pubescent or glabrescent, lower tomentose; nerves usually 5; length of blade 2 to 4 in., breadth about ·5 in. or more; petiole 1 to 2 in. *Cymes* pedunculate, axillary, in fascicles of about 4, 2-3-chotomous. *Flowers* minute, crowded. *Petals* 6, obovate, larger than the sepals. *Ripe drupe* the size of a pea. Hook. fil. Fl. Br. Ind. i. 102; Hf. & Th. Fl. Indica, 194; Miq. Fl. Ind. Bat. i. Pt. 1, 83. *P. aduncus, assamicus,* and *membranaceus,* Miers *l. c.* 119-122. *Cocculus incanus,* Coleb. in Trans. Linn. Soc. xiii. 57. *Cissampelos mauritiana,* Wall. Cat. 4980 (not of DC.). *Menisp. villosum,* Roxb. Fl. Ind. iii, 812 (not of Lamk.).

A common climber. Distrib. British India, Java, Sumatra, and probably in other parts of the Malayan Archipelago.

### 12. Stephania, Loureiro.

Climbing shrubs. *Leaves* usually peltate. *Flowers* in axillary, cymose umbels. *Male flower*; sepals 6 to 10, free, ovate or obovate. *Petals* 3 to 5, obovate, fleshy. *Anthers* 6, connate, encircling the top of the staminal column, bursting transversely. *Female flower : sepals* 3 to 5. *Petals* of the male. *Staminodes* 0. *Ovary* 1 ; style 3- to 6-partite. *Drupe* glabrous ; endocarp compressed, horse-shoe-shaped, dorsally tubercled, sides hollowed and perforated. *Seed* almost annular; cotyledons long, slender, ½-terete, appressed.—Distrib. Tropics of the Old World.

1. S. HERNANDIFOLIA, Walp. Rep. i, 96. Young branches striate, glabrous. *Leaves* membranous, broadly ovate-rotund, acute or acuminate, rarely obtuse, peltate ; the base truncate, emarginate or sub-cordate ; glabrous or sparsely pubescent; nerves about 10 radiating from the petiolar insertion, dark-coloured on the pale or glaucous under surface, reticulations open ; length of blade 2·5 to 5 in., breadth 2·25 to 3 in. ; petiole 1·75 to 2 in. *Umbels* on long slender peduncles, many-flowered. *Petals* 3 to 4. *Drupes* red, pisiform, compressed. Hook. fil. Fl. Br. Ind. i, 103; Hf. & T. Fl. Ind. 196 ; Miq. Fl. Ind. Bat. i. Pt. 1, 83 ; Miers Contrib. iii, 222. *S. intertexta, latifolia,* and *hypoglauca,* Miers *l. c.* 224, 226, 227. *Cissampelos hernandifolia,* Willd.; DC. Prodr. i. 100 ; Roxb. Fl. Ind. iii. 842 ; Wall. Cat. 4977 D, E, F, G, H, K. *C. discolor,* DC. *l.c.* i. 101 ; Bl. Bijdr. 26. *C. hexandra,* Roxb. *l.c.* iii. 842. *Clypea hernandifolia,* W. &. A. Prodr. i, 14 ; Wight Ic. t. 939. *Steph. discolor,* Hassk. Pl. Jav. rar. 168.

Common in shady places. Distrib. The Malayan Archipelago, British India, Australia, Africa.

13.  CISSAMPELOS, Linn.

Suberect or climbing shrubs. *Leaves* often peltate. *Male flowers* cymose. *Sepals* 4, 5 or 6, erose. *Petals* 4, connate, forming a 4-lobed cup. *Anthers* 4, connate, encircling the top of the staminal column, bursting transversely. *Female flowers* ; racemed, crowded in the axils of leafy bracts. *Sepals* 2 (or sepal and petal 1 each), 2-nerved, adnate to the bracts. *Staminodes* 0. *Ovary* 1 ; style short, 3-fid or 3-toothed. *Drupe* ovoid, style-scar sub-basal ; endocarp horse-shoe-shaped, compressed, dorsally tubercled, sides excavated. *Seed* curved ; embryo slender ; cotyledons narrow, ½-terete, appressed.—Distrib. All hot climates.

1.  C. PAREIRA, Linn. Young branches pubescent. *Leaves* usually peltate, membraneous, orbicular-reniform or cordate, obtuse and mucronate, rarely acute, base truncate to cordate, above glabrescent, below pubescent to tomentose ; length 1·8 to 3 in., breadth rather greater, petiole 1 to 3 in. *Male cymes* 2 or 3, axillary, slender. *Female racemes* with large reniform or orbicular bracts. *Ripe drupes* scarlet, sub-globose, hirsute, ·2 in. in diam. Hook. fil. Fl. Br. Ind. i. 104 ; H. f. & Th. Fl. Indica 198 ; Miq. Fl. Ind. Bat. i. Pt. 1, 85 ; DC. Prod. i. 100 ; Miers Contrib. iii. 139 ; *C. caapeba,* Linn. ; Roxb. Fl. Ind. iii. 842. *O. convo-*

30

*lrulacca,* Willd., Wall. Cat. 4979; W. &. A. Prod. i. 14; Roxb. *l. c.*
*C. orbiculata, discolor* and *hirsuta,* Ham. DC. *l. c.* 101. *C. diversa, grallatoria, eriantha, elata* and *delicatula,* Miers *l. c.* 187–189. *C. sub-peltata,*
Thw. Enum. 13 & 399; Miers *l. c.* 195, *Menispermum orbiculatum,*
Linn.

A common climber in all parts of the Settlements. Distrib. Everywhere in the Tropics.

## 14. CYCLEA, Arnott.

Climbing shrubs. *Leaves* usually peltate. *Flowers* in axillary panicles; *Male flower; sepals* 4–8, connate into an inflated 4–5-lobed calyx. *Petals* 4 to 8, more or less connate into a 4- or 8-lobed corolla. *Anthers* 4 to 6, connate, crowning the staminal column, bursting transversely. *Female flower; sepal* 1, oblong. *Petal* 1, orbicular. *Ovary* 1; style short, 3- to 5-lobed, lobes radiating. *Drupe* ovoid, style-scar subbasal; *endocarp* horse-shoe-shaped, dorsally tubercled, sides convex, 2-locellate (as in *Limacia*). *Seed* curved; cotyledons slender, $\frac{1}{2}$-terete, appressed. —Distrib. Tropical Asia.

1. C. PELTATA, H. f. & Th. Fl. Indica, 201. Branchlets striate, reflexed, pubescent or glabrous. *Leaves* coriaceous, peltate, deltoid or orbicular-ovate, acute or acuminate, often mucronate, the base truncate to cordate; above glabrous or glabrescent, beneath pubescent to tomentose, the 9 nerves rather prominent, length of blade 4·5 to 5·5 in., breadth 3·5 to 4·5 in.; petiole 2 to 2·5 in., reflexed, pubescent or tomentose, striate. *Panicles* usually longer than the leaves, the males often much branched and spreading and a foot long, the females smaller. *Calyx* campanulate, 4-lobed, glabrous or pilose externally. *Corolla* much smaller. *Drupe* pisiform, pilose; endocarp much tuberculate. Hook. fil. Fl. Br. Ind. i. 104; Miq. Fl. Ind. Bat. i. Pt. 1, 86; Miers Contrib. iii. 236. *C. barbata, Arnottii, versicolor, laxiflora* and *pendulina,* Miers *l. c. Menisp. peltatum,* Lamk. *Cocculus peltatus,* DC. Prod. i. 96. *Clypea Burmanni,* W. & A., in part. *Cyclea Burmanni,* Arnot in Wight Ill. i. 22. *Rhaptomeris Burmanni,* Miers in Tayl. Ann. Ser. 2, vii. 41.

Not common in the Straits Settlements. Distrib. Java, British India, Ceylon.

2. C. ELEGANS, King, nov. spec. Young branches spirally striate, puberulous, as are the petioles and panicles; otherwise glabrous. *Leaves* slightly peltate, membranous, shining on both surfaces, the reticulations minute but distinct, ovate to ovate-oblong, shortly acuminate, the base rounded or cordate, 7-nerved (4 of the nerves minute); length of blade 3 to 4 in., breadth 1·5 to 2 in., petiole about 1 in. *Male* and *female*

*panicles* sub-equal, slender, solitary, axillary, shorter than the leaves. *Male flowers* crowded, minute; anthers about 4, broad. *Drupes* pisiform, slightly compressed, pulp thin, endocarp boldly tubercled.

Perak; at elevations of from 1500 to 2000 feet; King's collector, Scortechini. A slender creeper 15 to 25 feet long; not common.

## 15. ANTITAXIS, Miers.

Climbers or shrubs with peuni-nerved leaves. *Pedicels* 1-flowered, numerous, in axillary fascicles, flowers diœcious. *Male flower; sepals* eight, in decussate pairs, the two outer pairs oblong, pubescent; the two inner pairs rotund, concave, glabrous, imbricate, all increasing in size inwards. *Petals* 2, smaller than fourth row of sepals, rotund, concave· *Stamens* 4 or 8, filaments clavate, anthers sub-globose. *Female flower* unknown. *Drupes* 1-3 (usually 1) sub-globose, or pyriform; endocarp brittle, thin, sub-reniform, 1-celled. *Seed* sub-globular, concave ventrally, albumen none; cotyledons oblong, semi-terete, thick, incurved; radicle minute.—Distrib. Eastern Archipelago.

1. A. LUCIDA, Miers Contrib. iii. 357. A glabrous climber, bark of young shoots dark and smooth, that of old shoots pale and warted. *Leaves* coriaceous, shining, oblong or sub-obovate-oblong, acute or acuminate, the base slightly narrowed; nerves about 6 pairs, obscure, as are the reticulations; length of blade 3 to 3·5 in., breadth 1·25 to 1·5 in., petiole ·5 in. *Female flowers* (male unknown) in fascicles. *Drupes* 1 to 3, (usually solitary) pyriform, glabrous, shining, about ·5 in. long, pericarp pulpy; endocarp thin, brittle.

On Ulu Bubong in Perak, King's collector. Distrib. Java.

A slender creeper from 40 to 60 feet long. *Male flowers* of this are unknown, and I put it into this genus on account of the structure of the fruit and from its general resemblance to *A. fasciculata*, Miers, which however differs in being non-scandent and in having tomentose drupes. Kurz's species *A. calocarpa* has 8 stamens (although he describes it as having only 4), and is also a climber with glabrous drupes. I have modified Miers' description of the genus as to the number of stamens, and in other particulars.

## ORDER VI. NYMPHÆACEÆ.

Aquatic perennial herbs. *Leaves* usually floating, often peltate, margins involute in vernation. *Scapes* 1-flowered, naked. *Floral-whorls* all free, hypogynous or adnate to a fleshy disc that surrounds or envelops· the carpels. *Sepals* 3 to 5. *Petals* 3 to 5, or many. *Stamens* many. *Carpels* 3 or more, in one whorl, free or connate, or irregularly sunk in pits of the disc; stigmas as many as carpels, peltate or decurrent; ovules few,

32

or many and scattered over the walls of the cells, anatropous or orthotropous. *Fruit* formed of the connate carpels, or of separate and indehiscent carpels, or of the enlarged turbinate flat-topped disk with the nut-like carpels sunk in its crown. *Seeds* naked or arilled ; albumen floury or 0 ; embryo enclosed in the enlarged amniotic sac.—Distrib. Temperate and tropical ; genera 8, species 30-40.

Suborder 1. *Nymphææ. Sepals* 4-6. *Petals and stamens* indefinite. *Carpels* confluent with one another or with the disk into one ovary. *Ovules* many. *Seeds* albuminous.

Sepals, petals and stamens ½-superior, inserted on the
disk which is confluent with the carpels ... 1. *Nymphæa.*
Sepals inferior ; petals superior ; carpels sunk in the
torus ... ... ... ... 2. *Barclaya.*

Suborder II. *Nelumbieæ. Sepals* 4-5. *Petals and stamens* indefinite. *Carpels* irregularly scattered, sunk in pits of the turbinate disk. *Ovules* 1-2. *Seeds* exalbuminous... ... ... 3. *Nelumbium.*

## 1. NYMPHÆA, L.

Large herbs ; rootstock creeping. *Flowers* expanded, large, floating on long radical scapes. *Sepals* 4, adnate to the base of the disk. *Petals* in many series, inner successively transformed into stamens, all adnate to the disk. *Filaments* petaloid ; anthers small, linear, introrse. *Ovaries* many, 1-seriate, sunk in the fleshy disk and forming with it a many-celled syncarp crowned by connate, radiating, stigmas : ovules many, anatropous. *Fruit* a spongy berry ripening under water. *Seeds* small, buried in pulp.—Distrib. Species 20, most temperate and tropical regions.

1. N. STELLATA, Willd. *Leaves* elliptic, deeply cordate, entire or with obtuse shallow sinuate teeth, often blotched with purple below, 6 to 8 in. long. *Flowers* 1·5 to 9 in. in diam., blue (white, or pink in varieties), petals 10 to 30, linear-lanceolate. *Stamens* 10 to 50, anthers with apical appendages. *Stigmatic rays* 10 to 25. *Fruit* 1·5 to 2 in. in diam. *Seeds* sub-striate. Hook. fil. Fl. Br. Ind. i. 114 : Hook. fil. & Th. Fl. Ind. i. 243 : Wight Ic. ⁴⁷⁵/₅ : Miq. Fl. Ind. Bat. i. Pt. ii. 90.

Common throughout the warmer parts of the Indo-Malayan region. Distrib. Australia.

Var. 1. VERSICOLOR, Hf. & Th. *l. c. Flower* and *leaves* intermediate between this and the next ; the former white, rose or blue. *N. versicolor*, Roxb. Hort. Beng. 41 ; Fl. Ind. ii. 577 ; Bot. Mag. t. 1189 ; Wall. Cat. 7257. *N. punctata*, Edgew. in Trans. Linn. Soc. xx. 29. *N. Edgeworthii* and *N. Hookeriana*, Lehm. der Gatt. Nymph. 7 and 21.

Var. 2. PARVIFLORA, Hf. & Th. *l. c. Leaves* and *flowers* much smaller than in the last, the latter usually blue and sometimes not more than 1 to 2 in. in diam.  *N. stellata,* Willd., W. & A. Prod. i. 17.

## 2.  BARCLAYA, Wallich.

Aquatic herbs with short villous root-stocks and floating leaves. *Peduncles* elongate, sometimes extra-axillary.  *Flowers* pink or claret-coloured.  *Sepals* 5, inserted at the base of the ovary.  *Petals* numerous, 3-seriate, united below into a tube which is confluent with the carpels. *Stamens* in many series, inserted within the corolla tube; filaments slender, short, reflexed ; the anthers pendulous, the outer imperfect.  *Ovaries* about 6 to 8, confluent, the apex conical ; styles tri-angular, connivent into a 10-rayed cone stigmatiferous within ; ovules numerous, ortho-tropous, parietal.  *Berry* globose, pulpy, crowned with the corolla-tube and annular torus.  *Seeds* elliptic, albumen floury, embryo small.

1.  B. MOTTLEYI, Hook. fil. in Trans. Linn. Soc. xxiii. . 157, t. 21. *Leaves* rotund, the apex sometimes bluntly apiculate, the base deeply cordate, glabrous on the upper surface, otherwise tomentose as are the peduncles ; length 3 in., breadth 3·5 in., petiole 3 to 7 in.  *Sepals* tomen-tose externally with long glabrous sub-apical tails.  *Petals* linear, pink or red.  *Seeds* echinate.

In ponds at low elevations in the Forest.  Malacca, Griffith ; Maingay. Perak ; Scortechini, Wray, King's Collector.  Distrib. Borneo, Motley, Lobb.

Var.  KUNSTLERI, King  *Leaves* ovate-rotund, cordate, the under-surfaces and petioles pubescent or glabrous, as are the peduncles ; *petals* claret-coloured; *seeds* rugose, occasionally echinate.

In similar situations with the last.  Perak ; King's collector, Scor-techini, Wray.  The leaves of this are thinner in texture than those of the typical form.

2.  B. LONGIFOLIA, Wall. in Trans. Linn. Soc. xv. 442, t. 18.  *Leaves* oblong, obtuse, the base cordate, glabrous or glabrescent, length 6 to 8 in., breadth 1 to 1·5 in., petiole 4 to 8 in.  *Sepals* glabrous or glabrescent, with short apical tails.  *Petals* oblong, reddish within, green externally. *Seeds* echinate.  Hook. fil. Fl. B. Ind. i. 115 ; Hook. Ic. Pl. t. 809, 810 ; Griff. Notul. 218, t. 57 ; H. f. & T. Fl. Ind. 246.

Andamans.  Distrib. Burmah.

## 3.  NELUMBIUM, Juss.

An erect large water-herb with milky juice ; rootstock stout, creep-ing.  *Leaves* raised high above the water, peltate.  *Flowers* rose-red white or yellow.  *Sepals* 4-5, inserted on the top of the scape, caducous. *Petals* and stamens many, hypogynous, many-seriate, caducous.  *Anthers*

34

with a clubbed appendage. *Ovaries* many, 1-celled, sunk in the flat top of an obconic fleshy torus, attachment lateral; style very short, exserted; stigma terminal, dilated; ovules 1-2, pendulous. *Carpels* ovoid, loose in the cavities of the enlarged spongy torus; pericarp bony, smooth. *Seed* filling the carpel, testa spongy, albumen 0; cotyledons fleshy, thick, enclosing the large folded plumule.—Distrib. Species 2, one Asiatic and Australian; the other W. Indian.

1. N. SPECIOSUM, Willd. *Leaves* 2 to 3 feet in diam., concave, glaucous. Peduncles and petioles 3 to 6 feet long, smooth, or with small scattered prickles. *Flowers* 4 to 10 in. diam; petals elliptic, rose, rarely white. *Fruiting torus* flat-topped, 2 to 4 in. diam. *Ripe carpels* ovoid, about •5 in. long. Wight & Arn. Prodr. i. 16; Roxb. Fl. Ind. II. 647; Wight Ill. i. t. 9; H. f. & T. Fl. Ind. 247; Miq. Fl. Ind. Bat. i. Pt. 2, p. 91. *N. asiaticum*, Rich. in Ann. Mus. xxii. 249, t. 9. *Nelumbo Indica*, Poir. Encycl. iv. 453. *Nelumbo*, Smith Exot. Bot. i. 59, t. 31, 32. *C. mysticus*, Salisb. Ann. Bot. ii. 75. *Nymphœa Nelumbo*, Linn.

In stagnant water throughout the Indo-Malayan region. Distrib. Persia, China, Japan and tropical Australia.

## ORDER VII. CAPPARIDEÆ.

Herbs, shrubs or trees, erect or climbing. Leaves simple or palmately 3- to 9-foliolate; stipules 2 or 0, sometimes spinescent. *Inflorescence* indefinite; flowers solitary, racemed, corymbose or umbelled, regular or irregular, usually 2-sexual. *Sepals* 4, free or connate, valvate or imbricate, rarely open in bud. *Petals* 4 (rarely 2 or 0), hypogynous or seated on the disc, imbricate or open in bud. *Stamens* 4 or more, hypogynous or perigynous, or at the base of or on a long or short gynophore. *Disc* 0, or tumid, or lining the calyx-tube. *Ovary* sessile or stalked, 1-celled; style short or 0; stigma depressed or capitate; ovules indefinite, on 2 to 4 parietal placentas, amphi- or campylo-tropous. *Fruit* capsular or berried. *Seeds* angled or reniform, exalbuminous; embryo incurved.—Distrib. Genera 23, species 300, chiefly tropical.

## 1. CLEOME, Linn.

Herbs. *Leaves* simple or digitately 3- to 9-foliolate. *Flowers* solitary or racemed, yellow, rose or purple. *Sepals* 4, spreading. *Petals* 4, regular or ascending. *Stamens* 6 to 20, sessile on the disc. *Ovary* sessile or with a short gynophore; style short or 0; ovules many, on 2 parietal placentas. *Capsule* oblong or linear, valves 2, separating from the seed-bearing placentas. *Seeds* reniform. Distrib. Species about 80, chiefly tropical.

35

1. C. HULLETTII, King, n. sp    A much-branched, sub-decumbent small shrub; the stem striate, puberulous, and with a few short prickles in distant pairs. *Leaves* dimorphous; those of the lower part of the stem petiolate, trifoliolate, the leaflets obovate; those of the upper part simple, sessile, ovate; all pubescent and from ·5 to ·75 in. long. *Flowers* solitary, axillary, about ·5 in. in diam. *Stamens* 6. *Pedicels* slender, much longer than the leaves. *Capsules* terete, striate, glabrous, about 1·5 in long; seeds large, muricate.

Singapore, in dry place by road-sides.

This is allied to the Peninsular Indian species *C. aspera*, Koen, and *C. Burmanni*, W. & A., but differs from both in its dimorphous leaves.

2. C. VISCOSA, Linn.    An erect, glandular-pubescent, viscid herb. *Leaves*; the lower with long petioles, the upper sometimes sub-sessile, 3-to 5-foliolate; leaflets obovate or ovate. *Flowers* in terminal corymbs, on long pedicels. *Petals* yellow, reflexed, about ·5 in. long. *Stamens* 12 to 20. *Capsule* glandular-pubescent, striate, narrowed to the apex, 2 to 3·5 in. long. *Seeds* small, reniform, transversely ridged. Hook. fil. Fl. Br. Ind. i. 170; Miq. Fl. Ind. Bat. i. Pt. 2, 97; Bl. Bijdr. 52; DC. Prodr. i. 242; Wall. Cat. 6968. *Polanisia icosandra*, W. & A. Prodr. 22; Wight Ic. t. 2.—Rheede Hort. Mal. ix. t. 33.

A common weed at low elevations in the tropics.

## 2. GYNANDROPSIS, DC.

An annual, glandular-pubescent or glabrate herb. *Leaves* 5-foliolate, long-petioled. *Flowers* racemed. *Sepals* 4, spreading. *Petals* 4, spreading, long-clawed, open in bud. *Stamens* 6, filaments adnate below to the slender gynophore, spreading above. *Ovary* stalked, ovules many. *Capsule* elongate, stalked; valves 2, separating from the seed-bearing placentas. *Seeds* reniform, black, scabrous.

1. G. PENTAPHYLLA, DC. Prod. i. 238.    An erect, glabrous or pubescent, spreading herb. *Leaves* on long petioles, quinate, the leaflets sessile, obovate or cuneate, acute or obtuse, entire or serrulate, 1 to 1·5 in. long. *Flowers* whitish or purple, in terminal racemes, ·35 to ·8 in. in diam., bracts 3-foliolate. *Capsules* cylindric, pointed, striate, nearly glabrous, 2 to 4 in. long. Hook. fil. Fl. Br. Ind. i. 171; Miq. Fl. Ind. Bat. i. Pt. 1, 96; W. & A. Prod. 21. *G. affinis*, Bl. Bijdr. 51. *Cleome pentaphylla*, Linn. Roxb. Fl. Ind. ii. 126.

Abundant in waste ground all over the tropics.

## 3. CAPPARIS, Linn.

Trees or shrubs, erect, decumbent or climbing, unarmed, or with stipular thorns. *Leaves* simple, rarely 0. *Flowers* white or coloured, often showy. *Sepals* 4, free, imbricate in 2 series, or 2 outer subvalvate. *Petals* 4, sessile, imbricate. *Stamens* indefinite, inserted on the torus at the base of the long gynophore. *Ovary* stalked, 1-to 4-celled; stigma sessile; ovules many, on 2 to 6 parietal placentas. *Fruit* fleshy, rarely bursting by valves. *Seeds* many, imbedded in pulp, testa crustaceous or coriaceous; cotyledons convolute.—Distrib. Species 125, natives of all warm climates, except N. America.

*Synopsis of Species.*

Fruit globose or sub-globose :—
    Flowers solitary, axillary ... ... 1. *C. Larutensis.*
      ,,    umbellate ... ... ... 2. *C. sepiaria.*
      ,,    racemose ... ... ... 3. *C. Scortechinii.*
    Flowers supra-axillary, in rows of 2 or 3 :—
      Leaves 3 to 4 in. long... ... 4. *C. micracantha.*
        ,,   5 to 7 in. long ... ... 5. *C. pubiflora.*
Fruit much elongate, cucumber-shaped :—
    Fruit 4 to 7 in. long ... ... 6. *C. Finlaysoniana.*
      ,,  3 in. long ... ... ... 7. *C. cucurbitina.*

1.  C. LARUTENSIS, King, n. sp.  Scandent, the young branches and petioles densely covered with minute, rusty, sub-deciduous tomentum. *Stem* sub-striate; prickles in pairs, short, hooked. *Leaves* glabrous, coriaceous, oblong-lanceolate, obtuse or retuse, the midrib prominent, the 5 or 6 pairs of nerves obscure; length 1 to 1·25 in., breadth 4 in., petiole ·2 in. *Flower-pedicels* solitary, glabrous, more than half as long as the leaves. *Flowers* ·75 in. in diam., white to pink. *Sepals* fleshy, glabrous, the outer pair ovate; the inner rotund, concave. *Petals* oblong, glabrous. *Stamens* about 12. Gynophore slender, nearly 2 in. long. *Fruit* globose, with an apical beak.

Perak, at Laroot, on trees.  King's Collector, No. 5103.

A woody climber, 30 to 40 feet long. The petals change from white to pink. Ripe fruit is unknown. This must be near *C. erythrodasys*, Miq.

2.  C. SEPIARIA, Linn.  A scrambling shrub. Branches divaricate, with rather distant pairs of short recurved thorns, sub-striate, the younger puberulous. *Leaves* membranous, shortly petiolate, ovate to oblong, pubescent or glabrescent, nerves 4 or 5 pairs. *Flowers* ·35 to ·5 in. in diam., in terminal umbels, the pedicels slender, ·5 in long; buds globose. *Sepals* · oblong. *Petals* narrow, white. *Ovary* apiculate, gynophore ·25 to ·5 in. long. *Fruit* pisiform, black. Hook. fil. Fl. Br. Ind. i. 177; Miq. Fl. Ind. Bat. i. pt. 2, 101; DC. Prod. i. 247; Roxb. Fl. Ind. ii. 568; W. & A. Prod. 26; Camb. in Jacq. Voy. Bot. t. 22; Dalz. & Gibs. Bomb. Flora, 10; Kurz For. Fl. Burm. i. 66.

In hot dry places at low elevations, but not common in the Straits Settlements. Distrib. India, Ceylon, Philippines, Timor, Australia.

3. C. SCORTECHINII, King, n. sp. Scandent; young parts, and the inflorescence at all stages, densely and minutely ferruginous-tomentose ; older branches with 1 or 2 striæ, glabrescent ; thorns stipular, in pairs, much shorter than the petioles, hooked. *Leaves* coriaceous, broadly lanceolate, acuminate, much narrowed at the base, smooth and shining above and the nerves obsolete ; under surface ferruginous-tomentose when young, but becoming glabrescent, the midrib and 6 pairs of nerves bold and prominent ; length 5 to 8 in., breadth 2 to 3 in., petiole ·5 in. *Flowers* 1·5 in. in diam., on short lateral, leafly, bracteate, woody racemes. *Bracts* petiolate, ovate-acuminate, ·75 in. long, deciduous. *Sepals* rotund, concave, tomentose externally. *Petals* pink, broadly elliptic, sub-obovate, blunt, notched, the base cuneate, glabrous. *Stamens* more than 20. *Fruit* globose, 3·5 to 4 in. in diam. ; the gynophore transversely wrinkled, 2·5 in. long and ·6 in. in diam.

Batang Padang district in Perak. King's Collector No. 8083, Scortechini, 191.

The fruit of this has been only once collected. The species is a fine creeper 15 to 20 feet long. It closely resembles *C. trinervia*, Hf. Th. in many respects ; but the leaves are not triple-but pinnate-nerved.

4. C. MICRACANTHA, DC. Prod. i. 247. Shrubby, not scandent, glabrous. Branches minutely striate ; the spines in pairs, stipular, one-third the length of the petioles, divergent, polished, not hooked. *Leaves* thinly coriaceous, shining, minutely reticulate, broadly lanceolate to oval, acute, mucronate ; the base narrowed or rounded ; midrib stout, nerves 7 to 8 pairs ; length 3 to 4 in., breadth 1·5 in., petiole ·3 in. *Flowers* on short pedicels, 2 or 3 in a line, supra-axillary, 1 to 1·5 in. in diam. *Sepals* and *petals* oblong ; the former puberulous, the latter white. *Stamens* 15 to 20, shorter than the gynophore. *Fruit* sub-globose, smooth. Hook. fil. Fl. Br. Ind. i. 179 ; Miq. Fl. Ind. Bat, i. pt. 2, 99 ; Blume Bijdr. 52. *C. conspicua* and *C. Finlaysoniana*, Wall. Cat. 6991 and 6992 A (not B).

Generally diffused at low elevations throughout the Malayan region. Distrib. Burmah. Sir Joseph Hooker gives the size of the ripe fruit (which I have not seen) as 2 to 3 in. in diam.

5. C. PUBIFLORA, DC. Prod. i. 246 ; var. *Perakensis*, Scortechini, MSS. A straggling shrub 15 feet long, branched from the ground. Branchlets compressed, striate, glabrous ; thorns in pairs, minute. *Leaves* membranous, elliptic-oblong, shortly acuminate, the base narrowed or rounded, glabrous on both surfaces, nerves (9 to 10 pairs) and reticulations rather prominent ; length 5 to 7 in., breadth 2 to 2·5 in. ; petiole ·25 in., stout,

28

channelled. *Flowers* 1 in. in diam. on long slender pedicels, 2 or 3 in a line, supra-axillary. *Sepals* oblong, concave, the two inner recurved, the two outer pubescent externally. *Petals* white, oblong, the two posterior united at the base by a gland so as to form a short spur. *Stamens* numerous. *Gynophore* about 1 in. long, pubescent as is the ovary.

Perak. Scortechini.

Collected only once, and without fruit. In Father Scortechini's field notes, he remarks that the petals are pubescent above and round the margin of the gland, and have a purple blotch.

A straggling but non-scandent shrub, almost unarmed, the thorns being very small. This variety differs from the type as described by De Candolle, and by Decaisne from Timor (Nouv. Ann. du Muséum, ii, 436) in having the venation of its leaves more straight and erect; otherwise it agrees.

6.  C. FINLAYSONIANA, Wall., Hook. fil. Fl. Br. Ind. I, 179. Scandent, glabrous; the spines stipular, nearly straight, in pairs, very short, with broad bases and blackish rather blunt tips. *Leaves* coriaceous, shortly petiolate, broadly lanceolate to elliptic, shortly acuminate, slightly narrowed to the base; the under surface pale (yellow when dry) the midrib and 6 pairs of nerves very bold; reticulations minute, distinct on the upper surface; length 6 to 7·5 in., breadth 2·25 to 3 in., petiole under ·5 in. *Flowers (fide* Hook. fil.) solitary or in pairs, supra-axillary, larger than in *C. micrantha.* *Sepals* lanceolate, acute, glabrous. *Ripe fruit* solitary, on a long stout stalk of which 1·5 in. is pedicel and the remaining 1·5 in. carpophore, cylindric, tapering to the apex, 4 to 7 in. long, and 1 to 1·5 in. in diam., yellowish-red, glabrous. *Seeds* ovoid, smooth, ·4 in. long.

Singapore, Wallich: Ulu Bubong in Perak.

Sir Joseph Hooker, who describes this species in the Flora of Brit. India from Wallich's Singapore specimens (which are accompanied by no field notes), is in doubt whether this is erect or scandent. Kunstler's field notes on the Perak specimens show it to be a creeper 20 to 30 feet long. It does not appear to be a common plant.

*Species of which the flowers are unknown.*

7.  C. CUCURDITINA, King, n. sp. Scandent; branchlets finely striate, nearly glabrous; the thorns stipular, in pairs, hooked, very sharp, much shorter than the petiole. *Leaves* glabrous, shining, more or less broadly lanceolate or oblong-lanceolate, shortly acuminate, the base narrowed or rounded; main nerves 8 or 9 pairs, anastomosing in bold intramarginal arches, the secondary nerves bold as is the midrib,

39

396      *Materials for a Flora of the Malayan Peninsula.*

the reticulations distinct; length 5·5 to 7·5 in., breadth 1·75 to 2·25 in. ; petiole under ·5 in., slender. *Flowers* unknown. *Fruit* cylindric, tapering to each end, the apex shortly beaked; when ripe orange-coloured, 3 in. long, and 1·5 in. in diam.; gynophore nearly ·75 in., pedicel 1·25 in., slender. *Seeds* ovoid, smooth, ·4 in. long.

Ulu Bubong, Perak ; King's Collector, Nos. 10027 and 10795.

A creeper, 20 to 30 feet long ; allied to *C. Finlaysoniana*, Wall. by its curious cucumber-like fruit, but with different leaves. Flowers have not as yet been collected.

8.   CAPPARIS KUNSTLERI, King, n. sp.   Scandent, the branches glabrescent ; thorns stipular, in pairs, hardly ·1 in long.   *Leaves* membranous, oblong-lanceolate or oblanceolate, acute, the base narrowed, both surfaces glabrous, midrib and 7 pairs of sub-horizontal curving nerves prominent below ; length of blade 4·5 to 5·5 in., breadth 1·5 to 2·25 in., petiole ·5 in. *Fruit* axillary, solitary, globose, 1·5 to 2 in. in diam., deep yellow ; gynophore about 2 in. long, stout ; pedicel rather shorter.

Gunong Bubu, in Perak, at an elevation of 800 feet, King's Collector, No. 8337.

A creeper, 40 to 60 feet long.   Only fruiting specimens have been collected.

### 4.   ROYDSIA, Roxb.

Large unarmed woody climbers, branches spotted white. *Leaves* simple. *Flowers* yellow, racemed or panicled. *Calyx* 6-partite, segments 2-seriate, tips a little imbricate. *Petals* 0. *Stamens* indefinite, inserted above the base of the short cylindric gynophore. *Ovary* ovoid, 3-celled from the prolongation of the placentas; styles 3, subulate, or single and undivided ; stigmas small, terminal ; ovules many, 2-seriate in the angles of the cells. *Fruit* fleshy, with a woody 3-valved, 1-celled endocarp, 1-seeded. *Seed* erect, cotyledons fleshy, unequal, longtitudinally folded, the larger embracing the smaller.—Distrib. Species 3, tropical Asiatic.

1.   R. PARVIFLORA, Griff. Notul. iv. 578; Ic. Pl. Asiat. t. 607, f. 1. A semi-scandent shrub, 4 to 8 feet high, glabrous, except the puberulous inflorescence and tomentose sepals. *Leaves* membranous, oblanceolate to obovate-elliptic or sub-rotund, with an abrupt short blunt acumen ; both surfaces shining, the midrib bold ; primary nerves 5 or 6 pairs, prominent on the under surface as are the intermediate nerves and reticulations; length 3·5 to 4 in., breadth 1·5 to 2·25 in.; petiole ·5 to ·6 in., slightly thickened in the upper half. *Flowers* in long naked racemes arranged in a terminal leafless panicle much longer than the leaves, shortly pedicellate, ·2 in. in diam. *Sepals* 6, valvate, linear-oblong, sub-acute, densely tomentose on both surfaces, reflexed. *Stamens*
40

20; filaments equal, not compressed; anthers innate. *Pistil* as long as the stamens; gynophore shorter than the glabrous, 2-celled, ovoid ovary. *Stigmas* 3, globular, minute. *Fruit* unknown. Hook. fil. Fl. Br. Ind. i. 409.

Perak; King's Collector, No. 1611. Distrib. Burmah.

2. R. SCORTECHINII, King, n. sp Scandent, glabrous except the minutely pubescent inflorescence and sepals. *Leaves* membranous, elliptic or obovate-elliptic, shortly and abruptly acuminate, narrowed to the base; upper surface shining; the lower pale, the midrib and 6 pairs of arching main nerves prominent; length of blade 5·5 to 6·5 in., breadth 3 in.; petiole 1·25 in., thickened in the upper half and bent in the middle. *Flowers* in terminal or axillary panicles or racemes, shortly pedicellate, ·25 in. in diam. *Sepals* 6, slightly imbricate in bud, united at their bases, linear-oblong, sub-acute, minutely tomentose on both surfaces, reflexed. *Stamens* 30; the filaments unequal, slender, compressed, united by their bases; anthers innate. *Pistil* as long as the stamens, the gynophore shorter than the ovary, puberulous. *Ovary* glabrous. *Style* twice as long as the ovary, cylindric; stigmas 3. sessile, ovate, small. *Ripe fruit* ovoid, smooth, yellow, 1·5 in. long and 1 in. in diam.; endocarp membranous.

Perak; Scortechini, King's Collector, Nos. 8464 and 4225; in open rocky places from 500 to 1200 feet.

In his field note on No. 8464, the collector describes this as "a splendid creeper 80 to 100 feet long"; in that on No. 4225, he says, "a tree 40 to 50 feet high." Fr. Scortechini's specimens have no notes. From the flexuose appearance of the dried twigs, I believe this is a creeper, and not a tree. This and the last belong to the section of *Roydsia* characterised by having an undivided style, for which Sir J. D. Hooker, (F. B. I. i. 409), proposes the sectional name of *Alytostylis*, but with an expression of doubt as to whether it should not be separated off as a genus. To this group belongs also the Philippine species *R. floribunda*, Planch. An undescribed species from Burmah in the Calcutta Herbarium (Gallatly No. 499) also falls into this section.

5. CRATÆVA, Linn.

*Trees. Leaves* 3-foliolate. *Flowers* large, yellow or purplish, polygamous. *Sepals* 4, cohering below with the convex lobed disk. *Petals* 4, long-clawed, open in bud. *Stamens* indefinite, adnate to the base of the gynophore. *Ovary* on a slender stalk, 1-celled; stigma sessile, depressed; ovules many, on 2 parietal placentas. *Berry* fleshy. *Seeds* imbedded in pulp.—Distrib. Species about 6, tropical and cosmopolitan.

1. C. MACROCARPA, Kurz in Journ. Bot. 1874, p. 195, t. 148, figs. 8 to

41

10. A small glabrous tree with pale smooth bark. *Leaflets* sessile or nearly so, sub-coriaceous, obliquely elongate-oblanceolate, the middle one narrower than the outer, rather bluntly acuminate, the base much narrowed; upper surface shining, lower pale, dull, with the midrib and nerves prominent; length 4 to 5 in., breadth 1·75 in. to 2·25. *Flowers* hermaphrodite, in terminal corymbose racemes, 2 to 3 in. in diam. *Petals* obovate, obtuse, unguiculate. *Stamens* 10 to 15, longer than the petals; anthers small, lanceolate, obtuse. *Ovary* ovoid, glabrous, the gynophore as long as the filaments; stigma sessile, discoid; placentas 2, parietal, multi-ovulate. *Ripe fruit* on a stout lenticellate carpophore 3 inches long, ovoid, smooth, purple spotted with grey, 2·5 in or more long by 2 in. in diam. *Seeds* embedded in pulp, compressed, ovoid-reniform, smooth on the sides, shortly tuberculate along the edge, nearly ·5 in. long by ·35 in. broad.

Malacca; Maingay (No. 125 Kew Distrib.), Scortechini No. 1771, King's Collector, No. 10461.

A species distinguished by its sessile leaflets and flat ovate-reniform tubercle-edged seeds. Kurz's figure of the seeds is bad; as he confessedly worked with imperfect material in describing this plant. I have seen no authentic specimens of *C. magna*, DC, or of *C. membranifolia*, (Miq. Fl. Ind. Bat. Suppl. 387) but, judging from the descriptions, they probably refer to this plant; in which case the older name (*magna*) would stand.

2. C. HYGROPHILA, Kurz J. A. S. B. Part II, 1872 p. 292; Journ. Bot. 1874, 196 tab. 148, figs. 6, 7. A small glabrous tree, with pale striate lenticellate bark; the youngest branches dark-coloured. *Leaflets* membranous, very shortly petiolulate, obliquely lanceolate, acuminate at both base and apex, the lower surface glaucescent; nerves about 6 pairs, sub-horizontal; length 3 to 4·5 in., breadth 1 to 1·25 in., petiole ·1 in. *Fruit* axillary, solitary or in fascicles of 2 or 3, cylindric, the apex with a blunt beak when ripe, brownish, spotted with grey, 4 to 5 in. long and 1·5 in. or more in diam.; the carpophore and pedicel each about 2 in. long, dark-coloured and faintly lenticellate. *Seeds* embedded in pulp, ·5 in. in diam., compressed but not flat, reniform, shortly muricate over the greater part of the surface.

Trang; King's Collector No. 1412. Distrib. Burmah.

Kurz founded this species on specimens from Burmah (in young fruit) which are now in the Calcutta Herbarium. The Malayan specimens (in mature fruit) agree with these. Kurz's drawing of the seed is misleading. For it was made from a young seed which had neither acquired its full size, nor its characteristic tubercles. Flowers of this species are as yet unknown; but the ripe fruit shows that they cannot

42

be in racemes as in the other species. The seeds approach in appearance those of *C. lophosperma*, Kurz, but are more tuberculate.

Besides the foregoing, there are in the Calcutta Herbarium specimens from Perak (King's Collector No. 818) of a *Crataeva* with leaves and flowers like *C. Narvala*, Ham. It is, however, described as having a thorny stem,—a character, so far as I am aware, not known in this genus. This is probably a new species, but, in the absence of fruit, I do not venture to describe it. The seeds appear to me to afford in this variable genus safer characters than any other part.

## Order VIII. VIOLACEÆ.

Herbs or shrubs. *Leaves* alternate, entire or serrulate, stipulate. *Flowers* regular or irregular, 2-bracteolate. *Sepals* 5, persistent, equal or unequal, imbricate in bud. *Petals* 5, hypogynous, equal or unequal, imbricate or contorted in bud. *Stamens* 5, filaments short, broad; anthers free or connate, their cells often with apical processes; connective broad, produced beyond the cells. *Ovary* sessile, 1-celled; style simple; stigma capitate truncate or cupular, entire or lobed; ovules many, on 3 parietal placentas, anatropous. *Fruit* a 3-valved capsule. *Seeds* small, albumen fleshy; embryo straight, cotyledons flat.—Distrib. Genera 21, species 240; natives of temp. and trop. regions.

Tribe I. *Violeæ. Corolla* irregular; lower petal dissimilar. *Staminodes* 0. *Capsule* loculicidal.

Sepals produced at the base ... ... 1. *Viola.*

Tribe II. *Alsodeieæ. Corolla* regular. *Staminodes* 0. *Fruit* a loculicidal capsule.

2. *Alsodeia.*

## 1. VIOLA, Linn.

Herbs, rarely shrubby below. *Flowers* on 1- rarely 2-flowered peduncles, often dimorphic, some large-petalled which ripen few seeds, others small petalled or apetalous and very prolific. *Sepals* produced at the base. *Petals* erect or spreading; lower largest, spurred or saccate at the base. *Anthers* connate, connectives of two lower often spurred at the base. *Style* clavate or truncate, tip straight or oblique; stigma obtuse, lobed or cupular. *Capsule* 3-valved. *Seeds* ovoid or globose.—Distrib. Species about 100, all temp. regions.

V. SERPENS, Wall. in Roxb. Fl. Ind. Ed. Wall. ii. 449 (not of Cat.), and DC. Prodr. i. 296; hirsute or glabrous, stolons or stems usually long, leafy and flowering; *leaves* ovate-cordate, obtuse or acute, crenate-serrate; stipules toothed or fimbriate, spur saccate; *sepals* acute; capsules

43

globose, few-seeded; glabrous or pubescent. Stigma very oblique or quite lateral, often minute and perforated. Hook. fil. Fl. Br. Ind. i. 184; Miq. Fl. Ind. Bat. i. pt. ii. 113; Roylo Ill. 74, t. 18, f. 1; W. & A. Prodr. 32. *V. Wightiana*, var. pubescens, Thwaites Euum. 20. *V. pilosa*, Blume Bijd. 57; Miq. Fl. Ind. Bat. i. pt. ii. 113.

Perak; on Ulu Batang Padang; L. Wray, Junior. Distrib. mountain ranges of India, and of the Malayan Islands.

## 2.  ALSODEIA, Thouars.

Trees or shrubs. *Leaves* alternate (rarely opposite), distichous; secondary nerves often numerous and parallel. *Stipules* rigid. *Flowers* small, axillary or terminal, solitary, fascicled, cymose or racemose, regular; peduncles with many bracts. *Sepals* 5, subequal, rigid. *Petals* 5, subequal, sessile. *Stamens* 5, inserted inside or upon an annular disc; with long or short often broad dorsal membranous connectival appendages, the cells of the anthers sometimes with apical linear processes. *Ovary* ovoid; style straight, stigma terminal; ovules few or many. *Capsule* 3-valved, few-seeded. *Seeds* glabrous in the E. Ind. species.— Distrib. Species about 50, chiefly tropical American.

Sect. I. *Prosthesia*, Bl. (genus). Anthers with a subulate appendage from the apex of each cell, and a broad (usually dorsal) appendage from the connective.

1. A. WALLICHIANA, Hook. fil. and Th. Fl. Br. Ind. I, 187. A glabrous shrub. *Leaves* membranous, oblong-lanceolate to elliptic, shortly acuminate, entire or slightly serrulate, the base rounded or slightly narrowed; nerves 10 to 15 pairs, arching, prominent, their axils beardless; length of blade 9 to 12 in., breadth 2·5 to 5 in., petiole ·5 to 1·5 in.; yellowish when dry especially on the under surface; stipules linear-lanceolate, glabrous, ·75 in long. *Racemes* shorter than the petioles, with many deciduous linear bracteoles. *Flowers* 4 to 8, pedicellate. *Perfect male flowers*; sepals acute, erect, lanceolate, equal to or longer than the petals. *Petals* oblong. *Filaments* short, attached to a 5-lobed fleshy disc. *Anthers* ovate with a small apical process on each lobe in front, and a single large orbicular hooded membranous appendage rising from the dorsum. *Pistils* rudimentary, or none. *Perfect female flowers*; sepals spreading, ovate-acute, shorter than the petals. *Petals* erect, oblong, obtuse, their apices recurved. *Filaments* longer than in the perfect male, the anthers without pollen. *Ovary* sessile, ovoid-conic, smooth; style cylindric. *Fruit* sub-globular, obtusely 3-angled, granular, ·35 in. long, dehiscing by 3 blunt valves. *Seeds* mottled.

Penang; Wallich. Perak; King's Collector, Scortechini.

The flowers in this species are practically unisexual and apparently

44

more frequently diœcious than monœcious. In flowers where perfect stamens occur the ovary is either absent or rudimentary; and in plants with a well developed ovary the stamens, although in most cases perfectly formed, contain no pollen. These sexual differences are accompanied by slight differences in the leaves, those of the male plants being oblong-lanceolate, narrowed to the base and serrulate, while those with female flowers have entire elliptic leaves with rounded or slightly narrowed bases. Specimens of the former, collected in Penang, were issued by Wallich as No. 4024 of his Catalogue under the name *Pentaloba macrophylla*; while specimens of the female were issued as Nos. 7501 and 7513 (un-named, but with the notes by R. Brown). These notes are as follows; on No. 7501, " *Urticeae habitu; arborescens ;*" and, on No. 7513, " *Indeterminata fruticosa, decumb.; foliis alternis integerrimis, coriaceis, impunctatis, glaberrimis, pedunculis axillaribus.*"

Wallich was wrong in referring this plant to the genus *Pentaloba* of Loureiro, for that author describes no appendages to its stamens. It belongs most certainly to the genus *Prosthesia* of Blume (Bijd. 866.)

2. A. KUNSTLERIANA, King, n. sp. A glabrous shrub or small tree ; the branchlets striate, sometimes lenticellate. *Leaves* subcoriaceous, oblong-lanceolate, acuminate or caudate-acuminate, sometimes minutely and obscurely serrulate, very much narrowed to the base ; upper surface smooth and shining; lower dull, rough from the numerous short transverse secondary nerves and 14 to 16 pairs of prominent ascending main nerves ; the midrib bold and subrugose ; the reticulations minute and distinct; length of blade 6 to 10 in.; breadth 2·25 to 3 in; petiole ·25 to ·5 in. *Stipules* lanceolate, ·25 in. long. *Female flowers* in axillary, often crowded, fascicles or very short racemes of 3 to 8, bracteolate, the pedicels longer than the leaf-petioles. *Sepals* ovate, obtuse, imbricate, strongly nerved, the edges ciliate, shorter than the petals. *Petals* erect, the tips not reflexed, ovate-acute, rigid. *Stamens* with short flat filaments, each inserted into the apex of a lobe of the deeply 5-lobed disc. *Anthers* (without pollen) broad, adpressed to the ovary, each with 2 linear anterior and one large dorsally-attached halbert-shaped membranous appendage, the latter conniving into a cone round the upper part of the ovary. *Ovary* sessile, ovoid-conic ; the style exserted, cylindric. *Capsules* ovoid, glabrous, shining, smooth, ·5 in. long, dehiscing into three narrow compressed pointed valves ; *seeds* one in each valve, ovoid, white, shining.

Singapore; Wallich, King's Collector. Perak; Scortechini, King's Collector; at low elevations. This species is more often practically monœcious than *A. Wallichiana*, to which it is closely allied. It differs, however, from that species in its much more acuminate rougher leaves, and also in its capsules and seeds.

3.   A. MAINGAYI, Hook. fil. Fl. Br. Ind. i. 188.  A small tree, glabrous except the inflorescence.  *Leaves* membranous, nearly sessile, elliptic, acute or sub-acuminate, serrulate, the base rounded; main nerves 10 to 12 pairs, prominent below as are the transverse secondary nerves; length 5 to 6 in., breadth 2·5 to 2·75 in., petiole ·15 in.; stipules lanceolate, ·25 in. long.  *Umbels* axillary, solitary, on peduncles ·5 in. long, 8- to 16-flowered; the bracteoles, small, ovate.  *Sepals* imbricate, ovate-rotund, or broadly ovate, obtuse, villous in the middle externally.  *Petals* longer than the sepals, ovate, concave, villous in the middle externally and with a villous line along the midrib internally.  *Stamens* with short, thick, densely tomentose filaments inserted on a thick, sub-glabrous, fleshy disc; connective tomentose behind.   Anthers elongate-ovate, with 2 ovate setose anterior, and 1 broad sub-terminal ovate, dorsal appendages.  *Ovary* sub-globular, style thick, both densely villous-tomentose.  *Capsule* ovoid, sparsely strigose, the valves acute.  *Seeds* with a white spongy caruncle.

Malacca, Griffith.

4.   A. MEMBRANCEA, King, n. sp.  A tree or shrub, the young branches shortly pubescent or tomentose.  *Leaves* thin when dry, obliquely obovate-elliptic, shortly and rather abruptly acuminate, serrate, the base narrowed, rather unequal-sided, both surfaces glabrous, except the midrib and 6 to 8 pairs of pubescent arching nerves, the reticulations wide; length 5 to 7 in., breadth 2·5 to 3 in.; petiole pubescent, ·5 in. long; stipules subulate, pubescent, ·2 in. long.  *Racemes* axillary, condensed, sessile, shorter than the petioles, few-flowered.  *Sepals* 5, unequal, the outer 2 rather smaller than the inner, all broadly ovate-obtuse, pubescent externally.  *Petals* 5, obovate-oblong, obtuse, the margins ciliolate with a few adpressed hairs on the back.  *Stamens* 5, glabrous, the filaments very short, rising from a fleshy 5-lobed disc.  *Anthers* broadly cordate or sub-reniform, with 2 small subulate processes on the apices of the cells and a large dorsal, cordate-acuminate, brown, membranous appendage as wide as the anther.  *Ovary* sessile, ovoid, villose; style cylindric, glabrous; stigma cup-shaped.  *Capsule* ovoid, glabrous, the valves in dehiscence blunt; seeds sub-globular with a beaked caruncle.

Perak at low elevations.  King's Collector, Scortechini.

A shrub or small tree.  This comes near *A. dasycaula*, Miq. in externals; but has fewer-nerved, more glabrous leaves.

5.   A. HOOKERIANA, King, n. sp.  A small glabrous tree, the branchlots lenticellate.  *Leaves* membranous, shining, shortly petiolate, elongate-oblanceolate or lanceolate, apex shortly and rather bluntly acuminate, entire or obscurely serrulate, gradually narrowed below the middle to the base; nerves 7 to 9 pairs, arching, slightly prominent; length 5

46

to 8 in., broadth 1·5 to 2 in., petiole ·25 in.; stipules ovate, only ·1 in. long. *Racemes* axillary, and on the older branches from the axils of fallen leaves, numerous, rather dense when young, afterwards sparse and open, from ·75 to 1·5 in. long, bracteoles linear, pedicels as long as or longer than the flowers. *Sepals* ovate, equal, pubescent or glabrous. *Petals* lanceolate with linear blunt apex, longer than the sepals, glabrous, or pubescent along the midrib externally. *Stamens* glabrous, the filaments as long as the rather deep disc. *Anthers* broadly ovate, the base cordate, almost sub-reniform, the cells each with an apical point and with a broad membranous ovate acute dorsal appendage wider than the anther. *Ovary* sessile, pubescent; the style cylindric, glabrous; the stigma cup-shaped, truncate. *Capsule* ovoid, compressed, obtusely angled, glabrous, reticulate, ·5 in. long; the valves unequal, obtuse, compressed, sub-falcate; seeds sub-globose, pale, minutely mottled at the apex, the base with a papillate pitted caruncle.

Perak; at low elevations, Scortechini, Wray, King's Collector.

A tree from 20 to 30 feet in height; readily distinguished by its open, comparatively long, racemes. This closely resembles Blume's *Prosthesia Javanica.*

6. A. WRAYI, King, n. sp. A sub-glabrous shrub, the branchlets with pale brown puberulous bark, rarely lenticellate. *Leaves* membranous, glabrous, shining, shortly petiolate, oblanceolate, shortly acuminate, distinctly serrulate, narrowed to the base, length 3·5 to 4·5 rarely 6 to 7 in., breadth 1 to 2 in., petiole ·1 to ·2 in.; stipules linear, only ·1 in. long. *Racemes* very short, crowded, axillary or extra-axillary, ·25 in. long; bracteoles broadly ovate, pubescent. *Sepals* lanceolate, tomentose externally. *Petals* oblong, obtuse, thickened and tomentose along the midrib. *Stamens* from the inside of a disc which is as deep as the filaments are long. *Anther* ovate-cordate, with a dorsal, ovate-acute, membranous appendage as broad as itself, and a terminal apical process on each cell. *Ovary* sessile, densely villous as is the base of the cylindric style; stigma sub-capitate. *Capsule* minutely fulvous-velvety when young, glabrous when old, ovoid with obtuse angles, 1·25 in. or more long and ·75 in. in diam.; the valves blunt, narrow. *Seeds* ovoid, brown, mottled, with a sub-apical papillate pitted caruncle.

Perak; at low elevations; Scortechini, Wray, King's Collector.

A shrub 8 to 10 feet high. In respect of leaves very like *A. Hookeriana,* but smaller. The very short racemes, more hairy flowers and larger velvety capsules distinguish it, however, from that species.

7. A. CINEREA, King, n. sp. A glabrous shrub or small tree, the branchlets whitish, sparsely lenticellate. *Leaves* membranous, elliptic-ovate or lanceolate, sometimes oblanceolate, acuminate, narrowed at the base,

47

serrulate, pale when dry; nerves 8 to 10 pairs, slightly prominent below; length 4 to 6 in., breadth 1·5 to 2·5 in., petiole ·2 to ·4 in.; stipules scarious, pale, ovate-acute, striate, puberulous, ·25 in. long. *Racemes* terminal, 4 to 6 in. long, bearing numerous 2- to 5-flowered cymules; bracts broadly ovate, scarious, concave, striate. *Sepals* ovate, ciliolate. *Petals* oblong, obtuse, the apex undulate, erose or toothed, edges ciliolate. *Stamens* from the edge of a deep fleshy disc; filaments very short, glabrous; anthers ovate, with a broadly ovate appendage from the middle of the back curving over the apex, the cells divergent at the apex and each with a subulate terminal appendage. *Ovary* sessile, glabrous, globose; Style cylindric, with a few white adpressed hairs; stigma obliquely truncate, cup-shaped. *Capsules* ovoid, bluntly angled, glabrous, reticulate, ·75 in. long; valves blunt; seeds globose with an ovoid beaked hilum, pale, smooth.

Perak, at low elevations.   King's Collector.

Var. *hirsutiflora*, King.   *Sepals* tomentose externally; filaments sparsely villous; disc small; the cymules larger and the bracteoles longer and narrower than in the typical form.

Perak; Changkat Jerin, L. Wray, junior.

The whole plant when dried has a characteristic grey colour, and from this circumstance I have given its specific name.

Sect. II. *Pentaloba.* Anthers with a broad, usually terminal, appendage from the connective; but none from the cells.

8.   A. LANCEOLATA, Wall.   (*Pentaloba*) Hook. fil. Fl. Br. Ind. i. 188. All parts, except the inflorescence, quite glabrous; bark of the young branches pale.   *Leaves* shortly petiolate, elongate-lanceolate, bluntly acuminate, the base much narrowed; nerves 9 to 12 pairs, sub-erect; slightly curved, prominent especially below, secondary venation transverse; length 5 to 8 in., breadth 1·25 to 1·75 in., petiole ·2 in.   *Racemes* about ·5 in. long, 4- to 6-flowered, minutely bracteolate.   *Flowers* on short pubescent pedicels.   *Sepals* ovate, obtuse, thick, pubescent, about half as long as the petals.   *Petals* lanceolate, acuminate, sparsely villous towards the middle.   *Filaments* as long as the anthers, slender, glabrous, rising from a small glabrous disc; anthers linear-lanceolate with a single lanceolate terminal appendage.   *Ovary* rudimentary in many flowers, sub-globose and, like the cylindric style, villous.   *Fruit* sub-globose, minutely pubescent, ·35 in. long, valves in dehiscence beaked.   Oudem. in Ann. Mus. Lugd. Bat iii. 68. *Vareca lanceolata*, Roxb. Fl. Ind. i. 648.   *Pent. lanceolata*, Arn. in Jard. Mag. Zoo. Bot. ii. 544.

Penang; Wallich, Stolickza, Curtis, King's Collector.

A shrub 6 to 8 feet high; apparently confined to Penang. This species is much more frequently truly hermaphrodite than some of

48

the others. The Sumatran species *A. dasypyxis*, Miq. comes very near this, but has longer racemes and more hairy fruit.

In *Alsodeia lanceolata*, Wall. there is a transition from *Pentaloba* to *Prosthesia*. Many of the specimens of *A. lanceolata* have the single terminal lanceolate appendage from the apex of the connective; others (Curtis's Penang specimen) have this appendage ovate and broader, while from the apex of each anther there is a rudimentary apical appendage, thus approaching *Prosthesia*.

9. A SCORTECHINII, King, n. sp. A small glabrous tree, the branchlets usually pale brown. *Leaves* membranous, shortly petiolate, obliquely elliptic-lanceolate or oblanceolate, shortly and bluntly acuminate, irregularly serrulate; main nerves about 12 or 13 pairs, rather bold; secondary nerves transverse, slightly prominent below; length 7 to 9 in., breadth 2 to 3 in., petiole ·2 in. *Racemes* sessile, about ·5 in. long, several together, axillary or extra-axillary, about 3- to 5-flowered, with ovate bracteoles. *Flowers* pedicellate. *Sepals* puberulous, broadly ovate, much shorter than the petals. *Petals* 5, oblanceolate, with long bluntly acuminate exserted apices. *Stamens* glabrous, shorter than the petals; the filaments twice as long as the anthers, slender, rising from a deep, 5-lobed, slightly-notched disc; the anthers short, ovate, with a single very small terminal appendage; ovary sessile, globose, glabrous, warted; style long, cylindric, glabrous or puberulous. *Ripe capsule* ovoid, pointed, rather more than ·5 in. long, glabrous, lenticellate; valves compressed, pointed. *Seeds* 2 in each valve, sub-rotund, whitish, carunculate.

Perak; King's Collector, Scortechini, Wray; at low elevations.

A large shrub or tree 20 to 25 feet high. Externally this much resembles *A. Maingayi*, but the flowers, and especially the anthers, differ much.

10. A. CONDENSA, King, n. sp. A glabrous tree; the older branchlets pale, lenticellate. *Leaves* membranous, shortly petiolate, inequilateral, elliptic to elliptic-oblong, sub-acute, serrulate, gradually narrowed below the middle to the acute unequal base; shining above, darker and dull beneath; midrib and 13 to 15 pairs of prominent main nerves pale beneath, and sub-erect secondary nerves transverse; length 8 to 14 in., breadth 3 to 4·5 in., petiole ·25 to ·35 in.; stipules subulate, ·35 to ·5 in. *Panicles* axillary, crowded, much branched, spreading, 1 to 2·5 in. long, (longer in fruit) puberulous or glabrescent; the bracteoles numerous, ovate, acute. *Sepals* unequal; the outer 2 or 3 larger, rotund; the inner 3 or 2 ovate, pubescent on the back. *Petals* ovate, a little longer than sepals, rhomboid, with pale edges, villous on the back externally and along the midrib internally. *Stamens* glabrous, the filaments rather short, from a fleshy disc; anthers cordate, with a single terminal white

49

ovate membranous appendage. *Ovary* sessile, globose, glabrous. Style cylindric; stigma cup-shaped. *Capsule* ovoid, pointed, glabrous, not lenticellate; valves compressed, pointed; seeds sub-globular, carunculate.

Perak, Scortechini, King's Collector; at low elevations.

A tree 30 to 40 feet high, approaching *A. Scortechini* in externals, but with different flowers.

11. A. FLORIBUNDA, King, n. sp. A shrub or tree, the young branches minutely fulvous-tomentose. *Leaves* membranous, shortly petiolate, oblong-lanceolate, ovate-lanceolate, sometimes elliptic, acute or acuminate, more or less obscurely serrulate; the base rounded, rarely acute; upper surface glabrous except the pubescent midrib and nerves; under surface minutely and softly tomentose, the midrib and 14 to 16 pairs of rather straight nerves and the transverse secondary nerves prominent; length 4·5 to 7 in., breadth 1·5 to 2·5 in., petiole ·1 to ·2 in.; stipules lanceolate, pubescent on the midrib, ·25 in. long. *Cymes* axillary, on peduncles 1 to 1·5 in. long, much branched, dichotomous, spreading, many-flowered; bracteoles oblong, obtuse, pubescent. *Sepals* unequal, the outer 3 ovate-rotund, the inner 2 ovate, all obtuse and pubescent. *Petals* ovate-oblong, obtuse, longer than the sepals, the apices usually reflexed. *Stamens* from a deep, pilose, 10-lobed disc; filaments expanded and pilose towards the apex, contracted and glabrous below; anthers elongate-ovate, with a single connectival ovate terminal appendage. *Ovary* sessile, villous, tomentose; style cylindric, puberulous; stigma truncate, cup-shaped. *Capsule* ovoid, obtusely angled, adpressed-pubescent, ·35 in. long; the valves blunt; seeds sub-globose, angled; caruncle long, narrow.

Perak, at low elevations; very common.

Distrib. Sumatra; Lampongs, Forbes, 1719; Padang, Beccari. P. S. 683.

Usually a tree, and sometimes attaining the height (*fide* Kunstler) of 70 feet. But also, according to the same collector, found as a shrub 6 to 8 feet high. This is allied to the Burmese species *A. mollis*, H. f. and Th., which however, besides having the anthers of a *Prosthesia*, has much smaller cymes, and broader bracteoles and sepals.

12. A. ECHINOCARPA, Korth. in Ned. Kruidk. Arch. II, 360. A small tree, the young branches fulvous or ferruginous-tomentose. *Leaves* membranous, shortly petiolate, obovate or ovate-elliptic, abruptly and shortly acuminate, boldly and unequally serrate, the base rounded or narrowed; upper surface glabrous except the pubescent midrib and nerves; the lower softly pubescent, the midrib, 11 to 15 pairs of nerves and transverse secondary nerves pale and prominent; length 6·5 to 9 in., breadth 2·25 to 3·25 in., petiole ·1 to ·2 in.; stipules ·25 in. long, ovate

acute, pubescent, the margins scarious. *Cymes* axillary and extra-axillary, sessile, condensed, 3- to 6-flowered; bracteoles lanceolate, keeled, pubescent. *Sepals* slightly unequal, ovate, obtuse, ribbed, tomentoso externally, nearly as long as the petals. *Petals* oblong, obtuse, pubescent externally, glabrous internally. *Stamens* from a short glabrous disc; filaments longer than the anther, glabrous. *Anthers* narrowly ovate, the base cordate, with a single small terminal white appendage. *Ovary* sessile, densely villous, style sparsely villous; stigma truncate, cup-shaped. *Capsule* when ripe from 1 to 2 in. across, densely covered with brownish, tomentose, branched, felted fibres; valves compressed, blunt, ·75 in. long; seeds sub-globose, compressed, smooth, the caruncle ovate. Hook. fil. Fl. Br. Ind. i. 188; Miq. Fl. Ind. Bat. i. pt. 2, 116; Oudem. Ann. Mus. Lugd. Bat. iii. 79; Miq. l. c. iv. 216; Pl. Jungh. i. 122.

Singapore, Malacca, Perak, Penang, at low elevations. Distrib. Sumatra, Bangka.

Usually a small tree 20 to 30 feet high. Sometimes shrubby.

13. A. CAPILLATA, King, n. sp. A small shrub, the young branches rufous-tomentose. *Leaves* membranous, shortly petiolate, lanceolate, acuminate both at apex and base, serrulate; upper surface glabrous except the pubescent nerves and midrib; lower rufous-pubescent; the nerves about 11 pairs, bold, as are the transverse veins; length 6 to 7 in., breadth 1·5 to 1·75 in , petiole ·25 in.; stipules lanceolate, pubescent externally. *Flowers* in small, sub-sessile, 3- to 5-flowered, axillary cymes. *Sepals* narrowly oblong, obtuse, tomentose externally. *Petals* linear-oblong, the apex sub-acute and reflexed, hairy along the midrib externally. *Stamens* alternating with the lobes of a deep, 5-lobed, glabrous disc; filaments slender, glabrous, longer than the anthers. *Anthers* small, ovate, each with an ovate acute small terminal appendage. *Ovary* sessile, globular, villous; style long, cylindric, sub-villous; stigma truncate, cup-shaped. *Capsule* ·5 in. long, rusty-pubescent externally and densely covered with unbranched, often hooked, soft, pubescent spines about ·5 in. long and not felted. *Seeds* ovoid, smooth, pale, with dark semi-circumferential band and an oblong carunculus.

Laroot in Perak, King's Collector No. 2462. A small bush 4 to 8 feet high. This comes near *A. echinocorpa* and *A. comosa*, but differs notably in its flowers and seeds.

Section III. Anther cells each with a terminal subulate appendage; no appendage from the connective.

14. A. COMOSA, King, n. sp. A shrub or small tree, the young branches densely ferruginous-tomentose. *Leaves* membranous, sub-sessile, oblong-oblanceolate, caudate-acuminate, serrulate, the base rounded; upper

51

surface glabrous; the lower pubescent, especially on the prominent midrib and 11 to 14 pairs of lateral nerves; length 5·5 to 7·5 in., breadth 1·5 to 2·5 in., petiole ·2 in.; stipules subulate, ·3 in. long. *Flowers* in dense axillary bracteolate glomeruli. *Sepals* ovate, obtuse, tomentose. *Petals* lanceolate, acuminate, pubescent externally, and (like the sepals) with an apical tuft of hairs, glabrous internally. *Stamens* from a short minutely toothed glabrous disc, the filaments shorter than the anthers; anthers narrowly ovate, each cell with a terminal apical seta, but without any appendage from the connective. *Ovary* elongate, sparsely villous; style pubescent; stigma truncate, cup-shaped. *Capsule* ·5 in. long, flocculent-tomentose, densely covered with unbranched, subulate, soft, pubescent spines about ·5 in. long, not felted. *Seeds* pale, ovoid, smooth, with a sub-terminal papillate caruncle.

Perak, Wray No. 3299 and 1254; King's Collector Nos. 406 and 554.

*Species imperfectly known.*

15.  A. PACHYCARPA, King, n. sp.  A small tree; the young branches pale, glabrous, sparsely lenticellate. *Leaves* membranous, oblong-lanceolate to elliptic-lanceolate, sub-acuminate, minutely and rather irregularly serrulate, the base slightly narrowed; both surfaces glabrous, the lower darker in colour; nerves 11 to 14 pairs, thin, but prominent below; length 5 to 7 in., breadth 2 to 3 in., petiole ·4 in. *Sepals* rotund, pubescent, with thin glabrous edges. *Capsules* on short axillary branches, usually solitary, about 1·25 in. long; the valves boat-shaped, compressed, separating when ripe into two layers, the outer dark-coloured and pubescent, the inner pale, smooth, cartilaginous, and bearing the angular smooth carunculate seeds.

Perak; King's Collector No. 10235; Scortechini (without number). A tree 20 to 25 feet high.  Fresh flowers being unknown, the section of the genus to which this belongs cannot be determined.  The capsules, however, show that it is a distinct species.

In addition to the foregoing, there are in the Calcutta Herbarium specimens of what appear to be five distinct species of this genus.  The materials are, however, insufficient for accurate determination.

ORDER IX. BIXINEÆ.

Trees or shrubs with alternate minutely stipulate or exstipulate leaves. *Flowers* regular, 1-2-sexual. *Sepals* 4 or 5 (rarely 2 to 6) imbricate, free, or connate and bursting irregularly, usually deciduous. *Petals* 4 or 5, or absent, imbricate or contorted, deciduous, often with basal scales. *Stamens* hypogynous or sub-perigynous, (united into a column in *Ryparosa*); anthers 2-celled with porous or longitudinal dehiscence. *Disc* thick, often glandular. *Ovary* free, usually 1-celled, the placentas parietal. *Styles* and *stigmas* free or united. *Fruit* dry with valvular dehiscence, the seeds along the middle of the valves; or fleshy and indehiscent. *Seeds* arillate, albumen fleshy, embryo axile, straight or curved; cotyledons foliaceous. Distrib. Chiefly tropical; genera 30; species about 170.

Tribe I. *Bixineæ.* Petals broad, contorted, without
    basal scales; anthers elongate, opening by termi-
    nal pores or short slits.
        Capsule with parietal placentas, 2-valved,
          softly muricate   ...   ...   ... 1. *Bixa.*
Tribe II. *Flacourtiæ.* Petals small and imbricate,
    or absent. Anthers short, opening by slits.
        Flowers hermaphrodite; petals 4 to 6.
          Stamens numerous   ...   ... 2. *Scolopia.*
           ,,    5 or 6   ...   ... 3. *Erythrospermum.*
        Flowers diœcious; petals 0.
          Ovary 2- to 8-celled   ...   ... 4. *Flacourtia.*
Tribe III. *Pangiæ.* Flowers diœcious, petals with
    an adnate basal scale or appendage; fruit large,
    indehiscent.
        Sepals free.
          Sepals 5, imbricate. Petals 5. Stamens
            5 to 8. Stigmas 3 to 6 ...   ... 5. *Hydnocarpus.*
          Sepals 4. Petals 8, in 2 rows; Stamens
            20 to 30. Stigma 1   ...   ... 6. *Taraktogonos.*

Sepals combined into a cup, its mouth entire
at first, but irregularly toothed on expansion.
Flowers large ; stamens numerous, free 7. *Pangium.*
Flowers small ; stamens united in a
column bearing 5 anthers          ... 8. *Ryparosa.*

1. BIXA, Linn.

A tree. *Leaves* simple ; stipules minute. *Flowers* in terminal
pan        2-sexual. *Sepals* 5, imbricate, deciduous. *Petals* 5, contorted
in bud. *Anthers* opening by 2 terminal pores. *Ovary* 1-celled ; style
slender, curved, stigma notched ; ovules many, on 2 parietal placentas.
*Capsule* loculicidally 2-valved, placentas on the valves. *Seeds* many,
funicle thick, testa pulpy ; albumen fleshy ; embryo large, cotyledons flat.
1. B. ORELLANA, Linn. A small tree. *Leaves* cordate, acuminate,
glabrous ; length 5 to 7 in., breadth 3 to 5 in , petiole 1·5 to 2·5.
*Flowers* in short terminal branched cymes, 2 in. in diam., purple or
white. *Capsule* compressed-ovoid, softly prickly, 1·5 in. long ; seeds covered with coloured pulp. Bl. Bijdr. 55 ; Roxb. Fl. Ind. II, 31 ; Miq.
Fl. Ind. Bat. I, Pt. 2, p. 107 ; Hook. fil. Fl. Br. Ind. I, 190.
Cultivated widely in the tropics on account of the dye (Arnatto)
yielded by the testa of its seeds.

2.   SCOLOPIA, Schreber.

Trees, spinous in India, spines often compound. *Leaves* alternate,
entire ; stipules minute or 0. *Flowers* small, racemed, axillary, 2-sexual.
*Sepals* 4-6, slightly imbricate in bud. *Petals* 4-6, subsimilar, imbricate
in bud. *Stamens* many with a row of glands outside them ; anthers
ovoid, opening by slits, connective produced into a terminal appendage.
*Ovary* 1-celled ; style erect, stigma entire or lobed ; ovules few, on 3 or 4
parietal placentas. *Berry* 2-4-seeded. *Seeds* with long funicles, testa
hard ; cotyledons foliaceous.—Distrib. Species about 15 ; Australian,
Asiatic, and African.

S. RHINANTHERA, Clos in Ann. Sc. Nat. Ser. IV, Vol. 8, p. 252. A
tree ; young branches puberulous. *Leaves* sub-coriaceous, ovate-lanceolate
to lanceolate, shortly acuminate, obscurely and minutely glandular-toothed, the base usually rounded, glabrous, shining ; nerves about 7 pairs,
faint ; length 3·5 to 5 in. ; breadth 1·75 in. to 2·5 ; petiole biglandular at
the apex, ·35 long. *Racemes* axillary and terminal, pubescent, bracteolate,
3-4 in., long. *Flowers* on tomentose bracteolate pedicels. *Sepals* 4, ovate-
lanceolate, tomentose externally. *Petals* 4, larger than the sepals, rotund,
54

tomentose on the edges and along midrib. *Stamens* indefinite, connective glabrous. *Ovary* cylindric. *Stigma* hemispheric, *Fruit* pisiform, 2-6-seeded. Hook fil. Fl. Br. Ind. I, 190 ; Miquel Fl. Ind. Bat. I, pt. 2, 107. *Phoberos rhinanthera*, Benn. Pl. Jav. Rar. 187, t. 39. *P. macrophylla*, W. & A. Prodr. 30. *Flacourtia inermis*, Wall. Cat. 6673 G, H, only.

Malacca, Griffith ; Penang, Curtis. Distrib Java, Borneo.

2. S. ROXBURGHII, Clos in Ann. Sc. Nat. Ser. IV, Vol. 8, 250. A glabrous shrub or small tree with spiny stem. *Leaves* sub-coriaceous, shining above, ovate, ovate-lanceolate to oblong-lanceolate, shortly acuminate, sub-entire or faintly and remotely crenate ; the base rounded or slightly narrowed, 3- to 5-nerved ; lateral nerves about 3 pairs, bold ; length 4·5 to 6·5 in., breadth 1·75 to 3·5 in. ; petiole biglandular at the apex, ·35 in. long. *Racemes* pubescent, axillary, about 1 in. long, 2-6-flowered, bracteolate. *Flowers* on tomentose pedicels. *Sepals and petals* 5 or 6 each, densely tomentose externally, broadly ovate. *Stamens* indefinite, the connective ciliate. *Ovary* ovate, *style* cylindric, *stigma* 3-lobed. *Fruit* baccate, the size of an olive. *Seeds* few. Hook. fil. Fl. Br. Ind. I, 190; Miq. Fl. Ind. Bat. I, pt. 2, 107. *Phoberos Roxburghii*, Benn. Pl. Jav. Rar. 192. *Ludia spinosa*, Roxb. Fl. Ind. ii. 507. *Flacourtia stigmarota*, Wall. Cat. 6678, (in part).

Penang, Curtis ; Perak, King's Collector. Distrib. Burmah, Sumatra.

3. S. CRENATA, Clos in Ann. S. Nat., Ser., IV, Vol. 8, 250. A tree, glabrous except the inflorescence. *Leaves* coriaceous, shining above, ovate to oblong-lanceolate, obtusely or sharply acuminate, obscurely glandular-crenate ; the base narrowed, rarely rounded, obscurely 3-5-nerved ; lateral nerves about 5 pairs, faint ; length 2 to 5 in., breadth 1 to 1·75 in., petiole ·25 to ·35 in. *Racemes* axillary or terminal, pubescent or tomentose, bracteolate, 1 to 3 in. long. *Flowers* pedicelled. *Sepals* and *petals* 4, rarely 5 or 6, the former tomentose and smaller than the petals. Connective of anthers glabrous. *Ovary* globular, smooth. *Style* cylindric. *Stigma* discoid. *Fruit* globose, about ·75 in. in diam. Hook fil. Fl. Br. Ind. I, 191; Miq. Fl. Ind. Bat. I pt. 2, p. 167. *S. pseudo-crenata*, *acuminata*, *chinensis*, *lanceolata*, and *crassipes*, Clos l. c. *S. sæva*, Hance in Ann. Sc. Nat. Ser. 4, xviii, 182. *Phoberos crenatus*, W. & A. Prodr. 29 ; Dalz. & Gibs. Bomb. Fl. 11. *P. lanceolatus* and *P. Wightianus*, W. and A. Prodr. 30. *P. acuminatus*, *Hookerianus*, and *Arnottianus*, Thwaites Enum. 17 and 400.

Penang, Curtis ; Perak, King's Collector. Distrib. Brit. India and Ceylon, China, Philippines.

In the young state this is thorny. It is a very variable species indeed, and too near *S. rhinanthera*.

### 3. ERYTHROSPERMUM, Lamarck.

Trees or shrubs. *Leaves* alternate, quite entire. *Flowers* racemed, fascicled or panicled, 2-sexual. *Sepals* 4-6, imbricate in bud. *Petals* 4-6, usually small. *Stamens* 4-6; anthers lanceolate-sagittate, connective dilated. *Ovary* 1-celled; style short, stigma entire or 3-4-fid; ovules many, on 3-4 parietal placentas. *Capsule* coriaceous, 3-4-valved; valves bearing the seeds on the middle. *Seeds* few, testa coriaceous or fleshy; embryo incurved.    Distrib. Species about 8, of which 6 are Mascarene, one is from Ceylon, and the following Malayan.

E. SCORTECHINII, King n. sp. A small glabrous tree, the branchlets lenticellate. *Leaves* thickly membranous, broadly oblanceolate, abruptly shortly and bluntly acuminate, faintly crenate-serrate, the base slightly narrowed; nerves 5 to 6 pairs, thin, anastomosing ·25 in. from the margin; length 4 to 6 in., breadth 2 to 2·5 in.; petiole ·5 in.; *Stipules* caducous. *Racemes* two to four in a lax terminal panicle, 3 to 4 in. long in flower, and twice as long in fruit. *Ovary* glabrous, 12—20- ovuled; style glabrous; stigma 3-lobed. *Capsules* on thin pedicels ·5 in. long, globular, smooth, ·35 in. in diam., crowned by the conical style with 3-cleft stigma, 3-valved, 1-seeded. *Seed* sub-globular with red pulp.

Perak; Scortechini.

This species was collected only once by Father Scortechini; and he found no flowers. He describes it as a tree 30 to 40 feet high. No species of the genus has hitherto been described from any Malayan province, Ceylon being the nearest country in which one is indigenous.

### 4. FLACOURTIA, Commers.

Trees or shrubs, often spinous. *Leaves* toothed or crenate. *Flowers* small, diœcious, rarely 2-sexual. *Sepals* 4-5, small, imbricate. *Petals* 0. *Stamens* many; anthers versatile. *Ovary* on a glandular disc; styles 2 or more, stigmas notched or 2-lobed; ovules usually in pairs on each placenta. *Fruit* indehiscent; endocarp hard, with as many cells as seeds. *Seeds* obovoid, testa coriaceous; cotyledons orbicular. Distrib. About 12 species, natives of the Old World, some being cultivated in various tropical countries.

FLACOURTIA RUKAM, Zoll. et Moritzi Verz. 33. A tree; the young branches puberulous and lenticellate. *Leaves* ovate or ovate-lanceolate, membranous, shortly acuminate, slightly and remotely crenate-serrate, the base narrowed, glabrous except the puberulous petiole and midrib; nerves 7 to 8 pairs; length 4 to 5·5 in., breadth 2 to 2·5 in., petiole ·3 in. *Racemes* three times as long as the petioles, axillary, pubescent, bracteolate,

4- to 8-flowered. *Flowers* diœceous, pedicelled. *Sepals* 4, reniform, tomentose internally. *Male flower* with a circle of glands outside the numerous stamens ; pistil none. *Female flower* with a sub-entire flattish fleshy disc at the base of the globular glabrous ovary ; *styles* 6 to 8, distinct to their bases, stout, spreading ; *stigmas* discoid with a mesial groove. *Fruit* sub-globular, ·5 to ·75 in. long, its pericarp succulent, when dry 6-8-ridged ; Hook. fil. Fl. Br. Ind. I, 192; Clos in Ann. Sc. Nat. Ser. iv. Vol. 8, p. 216 ; Miq. Fl. Ind. Bat. I, Pt. 2, 104. *F. cataphracta,* Bl. (not of Roxb.) Bijdr. 55, (probably).

Perak. Common at low elevations. Malacca, Griffith. Distrib. Burmah, Sumatra and the Malayan Archipelago generally ; Philippines.

This species is badly represented in collections and is not well understood, all published descriptions of it being very brief. Clos diagnoses it by its having 5 sepals ; but I do not find that this character holds at all. It approaches *F. inermis,* Roxb. very closely in foliage and fruit. According to Roxburgh, who originally described *F. inermis* from plants from the Moluccas cultivated at Calcutta, its flowers are hermaphrodite ; and in that respect they differ from those of the other species of the genus. The only authentic specimens of *F. inermis* which I have seen were cultivated in the Bot. Garden, Calcutta, and these are undoubtedly hermaphrodite. The styles are moreover very short and united, and the 5 stigmas form a radiating star on the apex of the ovary, each stigma being cuneate-emarginate. The stigmas of *F. Rukam* are quite different ; inasmuch as they are discoid and the styles are distinct to the very base. Forbes's Sumatra specimens No. 1206ª appear to belong to *inermis,* and they are the only uncultivated ones which I have seen. The fruit of *Rukam* as well as of *inermis* is eatable, although sour. I have not seen an authentic specimen of Blume's *F. cataphracta ;* but I can readily believe that it is *F. Rukam,* which is a common Malayan plant. The plants issued as Wall. Cat. 6673 belong (as regards many of the sheets) in my opinion to this, and not to *F. inermis,* Roxb.

2. FLACOURTIA CATAPHRACTA, Roxb. in Willd. Sp. Pl. iv. 830 ; Cor. Pl. iii. t. 222 ; Fl Ind. iii. 834. A small tree, often thorny when young. Branchlets glabrous, lenticellate. *Leaves* membranous, oblong or oblong-lanceolate, bluntly acuminate (the older sometimes blunt) obscurely crenate-serrate, narrowed to the base ; both surfaces glabrous, shining ; the 3-4 pairs of nerves thin, sub-erect ; the reticulations minute ; length 3 to 4 in., breadth 1·25 in., petiole ·3 in. *Flowers* in axillary racemes shorter than the leaves, small, (·15 in. diam.) ; ovary flask-shaped ; stigmas 4-6, capitate. *Fruit* the size of an olive, purple. Hook. fil. Fl. Br. Ind. I, 193 ; Clos in Ann. Sc. Nat. Ser. IV, Vol. 8, p. 216 (not of Roth, Blume, or Dalzell). *F. Jangomas,* Gmel. Syst. ; Miq. Fl. Ind.

118     *Materials for a Flora of the Malayan Peninsula.*

Bat. Vol. I, pt. ii, 105. *Stigmarosa Jangomas*, Lour. *Roumea Jangomas,*
Spreng. *Spina spinarum*, Rumph. Amb. Cap. 43, p. 38, xix, t. 1, 2.
In all the provinces. Distrib. British India, China. Often cul-
tivated.

5. HYDNOCARPUS, Gærtner.

Trees. *Leaves* alternate, serrate or entire; transverse venules numerous; stipules deciduous. *Flowers* solitary, or in irregular axillary
few-flowered racemes or fascicles, monœcious or diœcious. *Sepals* 5,
equal or unequal, imbricate in bud. *Petals* 5, with a scale opposite
each. FL. ♂; *Stamens* 5–8; anthers reniform, connective broad. *Ovary*
0 or rudimentary. FL. ♀; *Stamens* as in the ♂ but without pollen, or
reduced to staminodes. *Ovary* 1-celled; stigmas 3–6, sessile or subsessile, spreading, dilated, lobed; ovules many, on 3–6 parietal placentas.
*Berry* globose, many-seeded, rind hard. *Seeds* many, imbedded in pulp;
testa crustaceous, striate; albumen oily; cotyledons very broad, flat.
Distrib. Species about 12, tropical Asiatic.

1. HYDNOCARPUS CASTANEA, Hf. and Th. Fl. Br. Ind. I, 197. A glabrous tree 50 to 60 feet high. Branches and young shorts brown. *Leaves*
coriaceous, narrowly elliptic to oblong, gradually narrowed to the shortly
acuminate apex; the base unequal, rounded at one side, contracted at the
other; both surfaces shining and pale brown when dry; nerves 4–9
pairs, sub-erect, thin but prominent as are the reticulations; length 7
to 14 in., breadth 2·5 to 4·5 in.; petiole thickened at both ends and bent
at the apex, ·75 to 1 in. long. *Flowers* in axillary clusters of 2–6, male
and female alike and about equal in number, both on tawny-pubescent
pedicels 1·25 in long. *Sepals* obovate, imbricate, shorter than the petals,
the exposed parts tomentose. *Petals* ·6 in. long, linear-oblong, the scales
linear-obtuse, short. *Stamens* with thick subulate filaments; anthers
ovate-cordate; rudimentary ovary small, hispid. *Female flowers* like
the male, the stamens barren. *Ovary* ovoid, acuminate, tomentose;
stigmas sessile; ovules numerous. *Fruit* on a pedicel 1·25 to 1·54 in.
long, globular, 1 in. to 1·5 in. diam., minutely rugose, densely covered
with short fulvous tomentum; stigma persistent, hemispheric. *Seeds*
large, angular. Kurz F. Flora B. Burmah, I, 77.

Malacca; Perak; common. Distrib. Burmah.

2. HYDNOCARPUS NANA, King, n. sp. A shrub or small tree; the
branches and young shoots glabrous or (var. *pubescens*) pubescent. *Leaves*
subcoriaceous, from ovate-lanceolate to oblong-lanceolate, inequilateral,
subfalcate, shortly acuminate, remotely and minutely mucronate-serrulate,
narrowed and unequal at the base, shining and glabrous except the
midrib and nerves which, on both surfaces, are usually more or less
58

pubescent; nerves 5 to 8 pairs, spreading or sub-erect, thin but pro-minent beneath; length 2·5 to 5 in., breadth ·75 to 2·5 in., petiole ·25 to ·35 in.; stipules persistent, linear-lanceolate, pubescent, about as long as the petioles. *Male inflorescence* small, supra-axillary, 1- to 4-branched, uniparous, tomentose, bracteolate cymes not much longer than the petioles; flowers ·25 in. in diam. *Sepals* 5, rotund, the 3 external slight-ly imbricate, pubescent; the 2 inner much imbricate, glabrescent. *Petals* 5, smaller than the sepals, fleshy, with long white silky hairs externally, and each internally with a small oblong scale. *Filaments* short, thick, sericeous, the connective reniform; the anther cells small, remote from each other. Ovary 0. *Female flowers* solitary, supra-axillary, on glabrous pedicels ·5 in. long. *Sepals* and *petals* as in the male; stamens without pollen ; ovary ovoid, tomentose; *stigmas* 3, large, flat, bifid, reflexed. *Fruit* on a pedicel ·5 in. long, solitary, axillary, depressed-globular, minutely rugose, and velvety tawny-tomentose; about 1 in. in diam., or less; pericarp dry, thin. *Seeds* 3 or 4, plano-convex, 5 in. long.

Penang, Curtis, 854; Perak; King's Collector, Scortechini, Wray.

This varies considerably as to size of leaf and fruit and in the amount of pubescence. In some specimens of the male plant the leaves towards the apices of the branches are much reduced in size. The form which has larger more pubescent leaves may be separated as a variety, and farther acquaintance with it may prove that it is separable as a species.

Var. *pubescens.* Young parts, branchlets, and lower surfaces of adult leaves pubescent.

Perak; at Goping, King's Collector, No. 761.

3. HYDNOCARPUS CURTISII, King, n. sp. A glabrous shrub or small tree. Young branches slender, pale brown when dry. *Leaves* coriaceous, shining on both surfaces, oblong-lanceolate, rarely ovate, slightly inequi-lateral, gradually narrowed to the acuminate apex ; the base unequally narrowed, rarely rounded ; nerves 7 to 11 pairs, thin, spreading ; reticula-tions obscure on the upper surface; length 6 to 12 in., breadth 2·25 to 3 in. ; petiole less than ·5 in., thick. *Male flowers* in small, axillary, branch-ed, bracteolate, uniparous cymes not much longer than the petioles, ·75 in. in diam ; pedicels scurfy-tomentose, ·75 in. long. *Sepals* re-flexed, ovate, blunt, imbricate, pale, minutely pubescent, shorter than the petals. *Petals* 5, narrowly oblong, blunt, concave at the apex, ·65 in. long, glabrous; the gland nearly as long, linear. *Anthers* much longer than the filaments, cordate at the base. Ovary 0. *Female flowers* on shorter, grooved pedicels ; ovary elongate-ovoid, tawny-tomentose ; the stigmas 3, fleshy, bifid, spreading. *Fruit* on a stout pedicel nearly ·5 in. long, globose with long apical papilla, minutely rugose and velvety, vertically ridged ; the stigmas persistent ; nearly 1·5 in. long and 1 in. in diam. *Seeds* few, plano-convex, ·4 in. long.

Penang; Curtis, 800, 1534.   Perak; King's Collector, Scortechini.
No specimen that I have seen has female flowers showing anything
besides the ovary.   Complete female flowers are much wanted.

4.  HYDNOCARPUS SCORTECHINII, King, n. sp.   A tree, all parts except
the sepals glabous.   Branchlets pale brown when dry, angular.   *Leaves*
sub-sessile, coriaceous, shining on both surfaces, slightly inequilateral,
elliptic or elliptic-oblong, tapering to the acuminate apex, the edges slightly recurved when dry; the base rounded, slightly unequal ; nerves 7-8
pairs, thin, spreading; the reticulations minute and distinct on both
surfaces ; length 5 to 7 in., breadth 2·5 to 3·5 in., petiole about ·2 in.
*Cymes* small, monœcious, axillary or extra-axillary, on the young branches,
about three times as long as the petioles, densely bracteolate, 2-3-branch-
ed.   *Male flowers* on pedicels ·75 in. long.   *Sepals* elliptic, blunt, their
apices incurved, puberulous.   *Petals* smaller than the sepals but of the
same shape; the gland nearly as long, linear.   *Anthers* narrow, elongate ;
filaments short, conical.   *Ovary* none.   *Female flowers* like the males,
but on short pedicels and the stamens barren; ovary ovoid below, its
upper half cylindric, ridged, pale coloured, glabrous; stigmas large,
fleshy, reflexed, shortly bifid.   *Fruit* (young) ovoid, minutely rugose,
glabrous.

Dinding Islands ; Scortechini, Curtis.

This species bears a general resemblance to *H. Curtisii.*   But it differs
from that species in having broader leaves on shorter petioles, much
broader and shorter petals, and a glabrous ovary.   Ripe fruit of this is
unknown.

5.  HYDNOCARPUS CUCURBITINA, King, n. sp.   A tree 60 to 80 feet high ;
very young branches and leaves with minute ferruginous mealy tomentum ; otherwise glabrous.   *Leaves* thinly coriaceous, slightly inequilateral
and contracted at the base on one side, elliptic-oblong, tapering to either
end, the apex with a short rather blunt acumen, the edge very slightly
recurved when dry ; both surfaces, but especially the lower, shining and
with the transverse veins and minute reticulations very distinct ; main
nerves 5 to 6 pairs, sub-erect, thin ; length 3·5 to 5 in., breadth 1·5 to
2·25 in., petiole ·25 in.   *Cymes* diœcous, (the female flowers few)
axillary, three times as long as the petioles, bracteolato, 3- to 6-branched.
*Male flowers* on pedicels ·35 in. long, about ·3 in. in diam.   *Sepals* broadly ovate, blunt, pubescent-tomentose externally.   *Petals* ovate-rotund,
glabrous, thin, each with a fleshy scale with white ciliate edges and
nearly as large as itself.   *Anthers* ovate-cordate, glabrous ; the filaments
short, conical ; *Ovary* rudimentary, sericeous.   *Female flowers* like the
males, but on slightly shorter pedicels and with smaller barren stamens.
*Ovary* cylindric, densely sericeous-tomentose ; *stigmas* elongate, fleshy,

60

bifid at the apex, not reflexed when young. *Fruit* narrowly obovoid, cylindric, mamillate at the apex and contracted at the base; minutely rugose, smooth, dark brown when ripe and from 3 to 5 in. long; carpophore and pedicel about ·5 in. each, or more. *Seeds* one or two, obovoid, smooth, about 1 in. long.

Perak, up to elevations of 1,000 feet. Common.

Distinguished from every hitherto described species of this genus by its elongate cucumber-shaped fruit. The scales of the petals are also much larger and broader than is usual in Hydnocarpus.

.6. HYDNOCARPUS WRAYI, King, n. sp. A small sub-glabrous tree. Young branches with pale brown, minutely lenticellate, puberulous bark. *Leaves* sub-coriaceous, elliptic, shortly and abruptly acuminate, the edge slightly recurved when dry; the base rounded, sometimes narrowed and unequal; the reticulations on both surfaces very prominent; upper surface glabrous, shining, minutely pustulate when dry; the lower of a pale brown when dry, glabrous except the pubescent midrib and 8-9 pairs of bold sub-erect nerves; length 8 to 10 in., breadth 3·5 to 5 in.; petiole less than 5 in., stout. *Male flowers* nearly 5 in. in diam., in very minute, axillary, pedicelled, few-flowered cymes. *Sepals* 5, slightly imbricate, rotund, pubescent, larger than the petals. *Petals* 5, of the same shape as the sepals but smaller, each with a fleshy roughly cuneate scale the apex of which is irregularly toothed and ciliate. *Stamens* 15, the filaments glabrous, much thickened at the base; anthers broadly ovatecordate. *Female flowers* unknown. *Fruit* narrowly ovoid, tapering at both ends, often 3 in. long. and 1·75 in. in diam., minutely fulvousvelvety; the apical mamilla ·75 in. long, with its apex depressed and crowned by the 3 fleshy bifid stigmas; one-celled, several-seeded. *Pedicel* short, stout. *Seeds* embedded in a little pulp, elongate, plano-convex, ·75 in long.

Perak. King's Collector, No. 3800; Wray, No. 2608.

This species has more stamens than are usual in the genus *Hydnocarpus*. In this respect it appears to form a connecting link with *Taraktogenos*; but in shape the anthers do not agree with those of that genus.

### 6. TARAKTOGENOS, Hassk.

Trees with entire alternate leaves and minute fugaceous stipules. *Flowers* in more or less dense, short, axillary, few-flowered cymes; a few hermaphrodite, but the majority staminiferous only. *Staminiferous flower; sepals* 4, in decussate pairs, much imbricate, rotund, concave; *petals* 8, in two rows, smaller than the sepals, imbricate, each with a gland at its base; glands less than half as large as the petals, fleshy,

61

cuneate, plano-convex, ridged, the apex often irregularly toothed and with 2 or 3 cylindric pits    *Stamens* 20 to 32, the anthers deeply cordate. *Female flowers* like the males, but the sepals often only 3, the petals 6, and the stamens 16 or 17; *ovary* elongate-ovoid, sulcate, divided above into 4 oblong, divergent, reflexed lobes, each bearing a stigmatic surface internally; 1-celled with 4 multi-ovulate parietal placentas. *Fruit* large, globular or ovoid, with hard fibrous or woody rind, and several large seeds embedded in a scanty pulp. *Seeds* with thick hard testa, copious, albumen, and straight central embryo; the cotyledons large, cordate, foliaceous, 3-nerved.    Species probably about 8; all Malayan.

    *Note.*—This genus was founded by Hasskarl (Retzia, i. 127) on the plant named *Hydnocarpus heterophyllus* by Blume (Rumphia, iv, 22, t. 178 B., fig. 1, and Mus. Bot. i, 16).    Until now that plant has been the only known species.    But the following have been discovered by Messrs. Kunstler and Wray in Perak.    And, from the similarity in externals to *Hydnocarpus*, and from the imperfect nature of the Herbarium materials of the latter, it appears to me extremely probable that several things now referred to *Hydnocarpus* really belong to *Taraktogenos*.    In the Calcutta Herbarium, there are imperfect materials of, at least, 8 undescribed species which belong either to one or other of these two genera.

    1.  TARAKTOGENOS SCORTECHINII, King, n. sp.    A large glabrous tree; young branches with dark-coloured bark. *Leaves* coriaceous, shining, inequilateral, oblong-lanceolate, oblong or elliptic, with a short abrupt rather blunt acumen and slightly waved edges; the base slightly narrowed and unequal, 3-nerved; the upper surface smooth, the lower rough from the prominent reticulations and 4 to 5 pairs of ascending nerves; length 3·5 to 7 in., breadth 1·5 to 2·75 in.; petiole ·5 to ·75 in. . *Cymes* trichotomous, 1 in. in diam., on pedicels as long as the petioles, solitary, axillary, few-branched, uniparous. *Male flowers* ·5 to ·6 in. in diam.; pedicels ·25 to ·35 in. *Petals* densely sericeous externally; the basal scales less than half their length. *Stamens* 20 to 24, filaments hirsute, anthers sagittate. *Female flowers* and fruit unknown.

    Perak; Scortechini, No. 833; Wray, 1169.

    Var. *gracilipes*, King; petioles longer (·75 to 1 in.) and more slender; leaves smaller, 2·5 to 4 in. long, by 1·25 to 1·5 in. broad.

    Perak; Bujong-Malacca; Scortechini, No. 1894.

    2.  TARAKTOGENOS KUNSTLERI, King, n. sp.    A sub-glabrous tree 40 to 60 feet high.    Young branches fulvous-puberulous. *Leaves* coriaceous, unequal-sided, oblong-lanceolate to elliptic, shortly acuminate; the base narrowed and unequal, 3-nerved; both surfaces shining, the lower rough from

the prominent nerves and reticulations; lateral nerves 3 to 5 pairs on the narrower and 4 to 7 pairs on the wider side, sub-erect, prominent; length 4·5 to 6 in., breadth 1·5 to 3 in.; petiole ·3 to ·5 in., puberulous. *Cymes* dense, many-flowered. *Male flowers* as in the last, the scales half as long as the petals, their apices erose, glabrous. *Stamens* 32; the filaments short, subulate, sericeous; anthers elongate, deeply cordate. *Female flowers* like the males, but sepals 3, petals 6, and stamens 17 only. *Ovary* ovoid, glabrous, deeply sulcate, with 4 radiating reflexed oblong stigmas, 1-celled, with 4 multi-ovulate parietal placentas. Fruit solitary, globular, smooth, 2·5 in. in diam.; the pericarp thick, the outer layer fibrous, the inner woody. *Seeds* embedded in scanty pulp, plano-convex, ·75 in. or more in length.

Perak; in dense forest at low elevations; King's Collector, Nos. 6042 and 8183; Wray, 3389.

3. TARAKTOGENOS TOMENTOSA, King, n. sp. A tree 60 to 80 feet high. Young branches fulvous-tomentose. *Leaves* coriaceous, often inequilateral, ovate-oblong, abruptly and very shortly acuminate, the base rounded and slightly unequal; the reticulations prominent on both surfaces, upper surface smooth, shining; the lower fulvous-tomentose; lateral nerves 6 to 7 pairs, bold, sub-erect; length 5 to 7 in., breadth 2·5 to 3 in.; petiole ·25 to ·5 in., tomentose. *Cymes* woody, dense, short. *Fruit* ovoid, smooth; when ripe 3 in. long; the pericarp nearly ·5 in. thick, the outer layer fibrous, the inner thin and woody.

Perak; at an elevation of 500 feet; King's Collector, No. 7795.

Flowers of this are unknown. It is readily distinguished from the former two species by its tomentose leaves, but in other respects it much resembles them.

I subjoin a description of the Burmese species referred to *Hydnocarpus heterophyllus* by Kurz.

TARATOGENOS KURZII, King. A tree 40 to 50 feet high. Youngest branches, leaves and inflorescence tawney-pubescent, otherwise glabrous; older branches grey, minutely lenticellate. *Leaves* sub-coriaceous, lanceolate or oblong-lanceolate, rarely elliptic, abruptly and very shortly and bluntly apiculate; the base narrowed and equal-sided; both surfaces shining, the reticulations minute and distinct; main nerves 6 to 7 pairs, sub-erect; length 7 to 10 in., breadth 2 to 3·5 in., petiole ·75 to 1 in., thickened at the apex. *Cymes* axillary or extra-axillary, from the smaller branches, on thick peduncles, nearly as long as the petioles, with many very short branches at their apices, many-flowered. *Male flowers* ·3 in. in diam., on pedicels less than ·5 in. long. *Sepals* 4, imbricate, ovate. rotund, blunt, concave, pubescent externally. *Petals* 8, broadly ovate, blunt, with ciliate edges, each with a flat fleshy pubescent gland with white ciliate apex. *Stamens* 24 ; anthers elongate, deeply cordate ; the filaments short and with long white hairs. Female flowers unknown. *Fruit* globular, as large as an orange, on a thick peduncle ·25 in. long; the rind minutely granular, tawny-velvety, the outer layer thick and fibrous, the inner thin. *Seeds* numerous, irregularly oval, embedded in pulp. *Hydnocarpus heterophyllus,* Kurz (not of Blume) F. Flora B. Burmah i. 77. Wall. Cat. (*indeterminatae*) No. 7508.

Burmah; Griffith, (Kew Dist. 4363), Falconer, Brandis, Kurz,' Gallatly. Chittagong; Lister, Schlich. Sylhet, Wall. Cat., 7508.

This is the plant referred to in Hooker's Fl. B. Ind. i. 197 as "too immature for description." Since that remark was written, better material has been got from Burmah, on which Kurz described the species in his Forest Flora as *Hydnocarpus heterophylla,* Bl., with Blume's description of which it, however, manifestly disagrees. Kurz had modified the description of the genus *Hydnocarpus* to admit this plant. Female flowers of it I have never seen; but the males agree with those of *Taraktogenos.*

## 7. PANGIUM, Reinw.

A tree with entire, rarely 3-lobed, ovate-cordate, acuminate leaves. *Flowers* diœceous, axillary, solitary, large. *Calyx* globose, sepals 2-3, concave. *Petals* 5-6, each with a large sericeous scale at its base. *Male Fl.,* stamens 20 to 25 ; anthers ovate, innate ; *ovary* 0. *Female Fl.,* staminodes 5 or 6 ; *ovary* ovoid, 1-celled, with 2 parietal multiovulate placentas ; *stigma* sessile, obscurely 2-4-lobed. *Fruit* large, ovoid, indehiscent, many-seeded, pulpy. *Seeds* large, ovoid, angled, rugose, with a large elongate hilum, copious oily albumen, and broad foliaceous cotyledons.

P. EDULE, Reinw. in Syll. Pl. Soc. Ratisb., ii. p. 13. *Leaves* 6 to 8 in. long, by 3·75 to 5·5 in. broad. *Ripe fruit* with crustaceous pericarp, brown with white dots, 9 in. long by 6 in. in diam. ; seeds nearly 2 in. long. Miq. Fl. Ind. Bat. I, pt. 2, p. 109 ; Benn. Pl. Jav. Rar. 205, t. 43 ; Blume Rumphia iv, 20, t. 178 ; Mus. Bot. i, p. 14.

Perak ; King's Collector. Distrib. Malayan Archipelago.

## 8. RYPAROSA (RYPARIA), Blume.

Trees or shrubs with entire, alternate, elongate, petiolate leaves finely reticulate and more or less glaucescent beneath. *Flowers* rather small, diœceous; the males in long axillary racemes ; the females in shorter racemes, solitary, or in pairs. *Calyx* globose in bud, 3- to 5-cleft. *Petals* 5, imbricate, coriaceous; in the female flower each with a large sericeous scale at its base. *Male flower*; filaments united in a column with 5, ovate, 2-celled, extrorse anthers at its apex. *Female flower*; *staminodes* 5, alternate with the petals. *Ovary* 1-celled, with 1 to 3, biovulate, parietal placentas. *Stigmas* 2 to 3, sessile, broad, emarginate. *Fruit* baccate with little pulp; the pericarp coriaceous, tomentose. *Seeds* 1 or 2, sub-globular, smooth.

*Note.*—This genus was first published by Blume in his Bijdragen (p. 600) as *Ryparosa*, and in that work he published only the single species *R. cæsia*. In a footnote to the preface of his *Flora Javae* (p. viii), the same author referred to the genus (apparently by inadvertence) as *Ryparia* instead of *Ryparosa*; and the name *Ryparia* has been adopted by most subsequent authors. Blume regarded the genus as *Euphorbiaceous*, in which view he was followed by Endlicher (Gen. 5836), Hasskarl (Pl. Jav. Rar., p. 267), and Baillon (Etud. Euph., p. 339). Müll. Arg. (in DC. Prod. XV, ii., p. 1260) excluded the genus from *Euphorbiaceae;* and, in their Genera *Plantarum*, the late Mr. Bentham and Sir J. D. Hooker, (G. P. iii., 257), also exclude it; but, having seen no specimens either of it or of *Bergsmia*, they make no suggestion as to the true position of *Ryparosa* or of the relation of *Bergsmia* to it. Kurz (Journ. Bot. for 1873, p. 233, and For. Fl. Burm. I. 76) was the first to refer *Ryparosa* to *Bixineæ*. But Kurz made the mistake of describing in the latter work, as " *Ryparia caesia*," a plant which agrees neither with Blume's description nor with his specimens of *Ryparosa cæsia*. The name of Kurz's plant I have therefore altered to *R. Kurzii*. In 1848, Blume published, in Rumphia IV, p. 23, t. 178 C., fig. 2, a new genus called *Bergsmia* which, as Kurz also pointed out (Journ. of Bot. for 1873, p. 233), is nothing more or less than his older *Ryparosa*. Only one species ( *B. javanica*) was known to Blume. To this Miquel added (Fl. Ind. Bat. Suppl. 389) two species, namely, *B. Sumatrana* and *B.?* *acuminata*. I have seen neither of these ; but the cymose inflorescence of *B. Sumatrana* leads me to believe that it must be a *Hydnocarpus*, while the second ( *B. ?* *acuminata*) was referred doubtfully to *Bergsmia* by its author himself. The collections brought, within the past year or two, from Perak by the collectors of the Calcutta garden contain copious suites of specimens of *Ryparosa* and, from an examination of these, I have no doubt that *Ryparosa* belongs to *Bixineae*, and that *Bergsmia* must be reduced to it. Besides the seven species described below, there are in the Calcutta Herbarium imperfect materials belonging to several additional species from Perak, and to some from Sumatra. Wall. Cat. No. 7847 B. (from Penang), and Beccari's No. 702 (from Sumatra), are also clearly species of *Ryparosa*.

1. RYPAROSA KURZII, King. A tree or shrub. Young shoots ad-

pressed ferruginous-pubescent. *Leaves* elliptic to elliptic-oblong, shortly and bluntly acuminate, the base slightly narrowed ; upper surface shining, glabrous except the puberulous midrib; lower glaucous, the reticulations distinct ; nerves 7-8 pairs, spreading, prominent beneath ; length 8 to 12 in., breadth 4 to 5·5 in. ; petiole 1·5 in., thickened in its upper fourth, pubescent. *Male racemes* 5 to 10 in. long, ferruginous-tomentose, the petals reflexed ; *female racemes* shorter and subglabrous. *Fruit* globose, the size of a cherry, lenticellate, 2-seeded. *R. cæsia,* Kurz F. Fl. Burm., i, 78, not of Bl.

Andamans ; Kurz, King's Collector. Nicobars ; Kurz.

2. RYPAROSA WRAYI, King, n. sp. A tree 60 to 80 feet high, glabrous except the inflorescence. *Leaves* coriaceous, ovate-lanceolate to oblong-lanceolate or elliptic, the apex sub-acute ; slightly narrowed to the base ; upper surface shining; lower dull yellowish green when dry, the midrib and 4 pairs of sub-erect nerves prominent below as are the transverse veins; length 6 to 10 in., breadth 2·5 to 4·5 in. ; petiole 1 to 1·25 in., slightly winged at the apex. *Racemes* solitary or in pairs, axillary or from below the leaves, 6 to 9 in. long, longer in fruit. *Male flowers* pedicelled. *Calyx* with 3 broad ovate teeth, pubescent externally. *Petals* 5, oblong-ovate; pubescent externally, each with a triangular sericeous scale half as long as itself. *Staminal tube* pubescent; anthers 5, ovate, reflexed. *Female flower ; sepals* and *petals* as in the male ; disc annular, with 5 conical staminodes. Ovary rugulose, pubescent, globular; 1-celled. Stigmas 2, sub-bifid, spreading. *Fruit* globular, crowned by the stigmas, rugose, pubescent, ·5 to ·75 in , 1-seeded.

Perak ; King's Collector, Wray ; rather common.

3. RYPAROSA HULLETTII, King, n. sp. A small nearly glabrous tree. *Leaves* membranous, obovate-elliptic, with a very short abrupt acumen, the base narrowed ; both surfaces shining, the midrib and 3-4 pairs of spreading nerves prominent on the lower, as are the reticulations ; length 5 to 7 in., breadth 3 in. ; petiole 1·5 in., thickened in its upper fourth. *Male racemes* a foot or more long, puberulous. *Male flowers* ; calyx membranous, with 3 broadly ovate teeth. *Petals* 5, ovate ; scale small, sericeous. *Staminal tube* glabrous ; *anthers* 5, ovate, reflexed. Female flower and fruit unknown.

Singapore ; on Bukit Timah, R. H. Hullett.

Distinguished from the other species by its thin obovate leaves.

4. RYPAROSA SCORTECHINII, King, n. sp. A slender tree ; the branchlets and inflorescence rusty, otherwise glabrous. *Leaves* large, thinly coriaceous, oblong-lanceolate or oblanceolate, shortly and abruptly acuminate, gradually narrowed from the middle to the base ; both surfaces glabrous ; the upper shining, the lower dull, pale ; the midrib and 5 or 6 pairs of

nerves very prominent; length 10 to 15 in., breadth 4 to 6 in.; petiole 2 to 2·5 in., thickened and bent at the apex. *Racemes* in tufts from tubercles on the stem and large branches, the male 8 to 12 in. long. *Calyx* splitting into 3 ovate segments, tomentose. *Petals* 4, oblong; the gland large, rotund, sericeous. *Female racemes* shorter; sepals and petals as in the male; ovary tomentose, 4-angled; styles 2, discoid. *Fruit* angled when young; when ripe transversely oblong, 1·5 in., by 1 in., velvety rusty-tomentose with green or white dots; seeds two, plano-convex.

Perak; Scortechini, Kunstler, Wray; common.

The male flowers have been found only by Scortechini from whose field notes the above description of them has been taken.

5. RYPAROSA KUNSTLERI, King, n. sp. A glabrous tree, the branchlets smooth. *Leaves* coriaceous, ovate-oblong, obovate-oblong to oblong, shortly and abruptly acuminate, the base narrowed; upper surface shining; lower dull, pale, much reticulate, the midrib and 5-7 pairs of nerves very prominent; length 5 to 8 in., breadth 2·5 to 3·5 in.; petiole 1·25 to 1·75 in., swollen and bent towards the apex. *Racemes* axillary, solitary, rarely 2-3 from an axil, the male 6 to 8 in. long, the female half as long; flowers pedicelled. *Male flower; Calyx* thin, pubescent outside, with 3 ovate broad teeth. *Petals* 5, oblong-lanceolate, pubescent externally, each with a large sericeous gland at its base; *staminal tube* glabrous, the anthers ovate-oblong. *Female flower. Sepals* and *petals* as in the male; annular disc at base of ovary small; staminodes none. *Ovary* ovoid, angled, tomentose, 1-celled, with 4 parietal biovulate placentas; stigmas obovate, radiating. *Fruit* globular, yellowish, velvety, about 1·5 in. diam.; *seeds* 5 or 6, oblong, compressed, striate, about ·75 in. long.

Perak, at elevations up to 800 feet; common. A tree 40 to 100 feet in height, with shorter and (in proportion) broader leaves than *R. fasciculata*, 4 stigmas and more globular pedicellate fruit.

6. RYPAROSA FASCICULATA, King, n. sp. A glabrous tree 30 to 60 feet high. Young branches lenticellate. *Leaves* thinly coriaceous, narrowly oblong, acuminate, the base narrowed, shining above, pale beneath; the midrib, 5 to 7 pairs of lateral nerves, and the bold sub-erect transverse nerves and reticulations very distinct especially beneath; length 9 to 15 in., breadth 2·25 to 3·25 in.; petiole 1 to 1·5 in., grooved, thickened in its upper fourth. *Racemes* in fascicles of 4-7 from tubercles on the large branches and stem. *Petals* rotund, much imbricate and inflexed. *Female* flower with annular disc bearing 5 conical staminodes, the petals with hairy scales at their bases; stigmas 3, large, reniform. *Fruit* sessile, rusty-tomentose, pyriform, the apex mammillate and crowned for some time by the remains of the stigmas, about 6-seeded, 1·5 to 2 in. long.

Perak at elevations up to 800 feet; common.

7. RYPAROSA CAESIA, Bl. Bijdr. 600; *Ryparia*, Fl. Javac (praef. VIII). A small tree, the branchlets and inflorescence ferrugineous-silky. *Leaves* coriaceous, oblong, shortly acuminato, the base slightly narrowed; upper surface shining; lower pale, rather densely adpressed-sericeous; nerves 5 to 6 pairs, ascending; length 6 to 9 in.; breadth 2 to 3 in.; petiole 1·25 in., stout, thickened in its upper fourth. *Racemes* solitary, supra-axillary, the female longer than the leaves. *Male flowers; sepals* and *petals* 4, tomentose, the latter with a small basal hairy scale. *Staminal tube* short, glabrous; *anthers* 4, broadly ovate, reflexed. *Fruit* crowned by the 2 shortly-stalked fleshy radiating reniform emarginate stigmas, globose, ferruginous-tomentose, ·5 to ·7 in. diam. Hassk. Pl. Javan. Rar. 267; Baillon Euphorb. 339; Miq. Fl. Ind. Bat. i. pt. 2, p. 361; DC. Prod. XV, 2, p. 1260; Kurz in Journ. Bot. 1873, p. 233.

Java; Blume. Sumatra; Teysmann, Forbes, at an elevation of 3,500 feet.

Blume describes the lower surfaces of the leaves as "tenuiter strigosis;" but the hairs, although adpressed, are not stiff but silky. This is the only species in which the hairs on the lower surface of the leaves are at all conspicuous. The leaves of the Andaman plant referred to *R. caesia* by Kurz are nearly glabrous beneath.

ORDER X. PITTOSPOREÆ.

Trees or shrubs. *Leaves* alternate or subverticillate, quite entire (very rarely toothed); exstipulate. *Flowers* usually hermaphrodite, terminal or axillary. *Sepals* 5, imbricate. *Petals* 5, hypogynous, imbricate. *Torus* small. *Stamens* 5, opposite the sepals; anthers versatile. *Ovary* 1-celled, with 2-5 parietal placentas, or 2-5-celled by the projection of the placentas; style simple; stigma terminal, 2-5-lobed; ovules many, parietal or axile, anatropous. *Fruit* capsular or indehiscent. *Seeds* usually many, albumen copious; embryo small, radicle next the hilum.—Distrib. Genera 9; species about 90, chiefly Australian.

1. PITTOSPORUM, Banks.

Erect trees or shrubs. *Sepals* free or connate below. *Petals* erect, claws connivent or connate. *Stamens* 5, erect; anthers 2-celled, introrse, bursting by slits. *Ovary* sessile or shortly stalked, incompletely 2-3-celled; ovules 2 or more on each placenta. *Capsule* 1-celled, woody, 2-rarely 3-valved; valves placentiferous in the middle. *Seeds* smooth, imbedded in pulp. Distrib. Species about 50, subtropical Asiatic, Australian, and Oceanic.

PITTOSPORUM FERRUGINEUM, Ait., DC. Prod. I, 346. A tree 40 to 60
feet high. Young branches leaves and inflorescence softly ferruginous-pubescent. *Leaves* membranous, lanceolate or ovate-lanceolate,
acute or acuminate at base and apex, the edges minutely undulate ; when
adult glabrous except the midrib and larger nerves ; upper surface shining, the lower dull with the minute reticulations distinct ; nerves 7 to 8
pairs, not prominent, spreading ; length 2 to 3 in., breadth 1 to 1·5 in. ;
petiole slender, rusty-tomentose, ·5 in. long. *Flowers* ·25 in. long, greenish-white, in short terminal corymbs. *Sepals* lanceolate, pubescent.
*Petals* linear, the apices reflexed, pubescent, 3-nerved. *Ovary* cylindric,
rusty-tomentose ; style short, glabrous, excentric ; capsule globose, when
ripe compressed, rugose, with 6 to 8 black flat seeds. Hook. fil. Fl. Br.
Ind. i., 199 ; Putterl. Monogr. Pittosp. 7 ; Benth. Fl. Austral. i. 112 ;
Bot. Mag. 2075.

At elevations of from 800 to 1500 feet ; common. Distrib. Burmah,
the Malayan Archipelago, Philippines, Queensland.

There is some variability in leaf in different individuals of this
species, some having leaves narrowly lanceolate, others ovate-lanceolate.

ORDER XI. POLYGALEÆ.

Annual or perennial herbs, erect or scandent shrubs, or timber
trees. *Leaves* alternate (rarely whorled) or occasionally reduced to
scales or 0, simple, quite entire. *Stipules* 0. *Flowers* irregular, 2-sexual,
3-bracteate. *Sepals* 5, unequal, 2 inner often petaloid (*wing sepals*),
deciduous or persistent, imbricate in bud. *Petals* 5 or 3, distinct, unequal, the inferior usually keel-shaped. *Stamens* 8 (in *Salomonia* 4-5,
in *Trigoniastrum* 5) hypogynous, filaments united into a sheath, more rarely
distinct ; anthers opening by terminal pores, rarely by slits. *Ovary* free,
1-3-celled ; style generally curved, stigma capitate ; ovules 1 or more
in each cell, anatropous. *Fruit* generally a 2-celled, 2-seeded, loculicidal capsule ; or indehiscent and 1-seeded, or (in *Trigoniastrum*) of 3
indehiscent carpels. *Seed* usually strophiolate, albuminous, rarely exalbuminous. Distrib. The whole world except New Zealand, chiefly in
warm regions ; genera 16 ; species 450-500.

Herbs or (more rarely) erect shrubs. Capsule loculicidal, 2-celled :—
Stamens 8, united ; 2 interior sepals alæform  1. *Polygala.*
Stamens 4-5, united ; sepals petaloid, near-
ly equal  ...  ...  ...  2. *Salomonia.*
Climbing shrubs :—
Stamens 8, united ; fruit 1-celled, inde-
hiscent, samaroid  ...  ...  3. *Securidaca.*

Trees or erect shrubs :—
Stamens 5, united; fruit of 3 samaroid
  carpels      ...      ...      ...   4. *Trigoniastrum.*
Stamens 8, distinct : fruit 1-celled, not
  winged      ...      ...      ...   5. *Xanthophyllum.*

## 1. POLYGALA, Linn.

Herbs or more rarely shrubs. *Leaves* alternate. *Sepals* usually persistent; 2 inner larger, usually petaloid. *Petals* 3, united at the base with the staminal sheath, the inferior keel-shaped and generally crested. *Stamens* 8, filaments united for their lower half into a split sheath ; anthers opening by pores. *Ovary* 2-celled, ovules 1 in each cell, pendulous. *Capsule* 2-celled, loculicidal, 2-seeded. *Seeds* almost always strophiolate and albuminous. Distrib. conterminous with the order, except Tasmania. About 250 species.

Sect. I. CHAMÆBUXUS, (Tourn. genns). Shrubs with large handsome flowers. *Calyx* deciduous, the lower sepal large, concave-cucullate. *Keel* crested. *Seeds* with a large strophiole, exalbuminous.

1. POLYGALA VENENOSA, Juss. in Poir. Dict. V. 493. A glabrous shrub 4 to 10 feet high. *Leaves* membranous, lanceolate or oblanceolate to oblong-lanceolate, acuminate, entire, narrowed to the short petiole; primary nerves 7 or 8 pairs, the secondary nerves nearly as prominent, the reticulations open, rather prominent; length 5 to 8 in., breadth 1·5 to 2·5; petiole ·2 in. *Racemes* axillary, pendulous, 1 to 3 in. long, often much elongated in fruit. *Flowers* more than ·5 in. long. *Capsule* reniform, striate, more or less 4-winged, 4 in. in diam.; DC. Prod. I, 331; Bl. Bijdr. 59; Miq. Fl. Ind. Bat. I, pt. 3, p. 126; *Chamæbuxus venenosa*, Hassk. Pl. Jav. Rar., 294; Pl. Jungh., I, 126.

Var. *robusta.* Miq. l. c.; Hassk. Pl. Jungh. l. c. Leaves large, elliptic-oblong to oblong.

In all the Provinces at low elevations. Distrib. Malayan Archipelago.

A common shrub with handsome flowers ; the inner sepals white with pink veins ; the petals white, spotted with pink and the keel pink.

Sect. II. Herbs. *Flowers* small. *Calyx* deciduous after flowering. *Keel* not crested. *Seeds* albuminous.

2. POLYGALA TRIPHYLLA, Ham. in Don Prodr. 200; var. glaucescens, Hf. Fl. Br. Ind. I, 199. A glabrous, weak, erect or ascending herb. *Leaves* thinly membranous, lanceolate or ovate-lanceolate, sub-acute, contracted into the petiole ; main nerves about 7 pairs, thin ; length 1·5 to 2 in., breadth ·75 in., petiole ·5 to ·75 in. *Racemes* axillary, 2 to 4 in. long, (or more) slender. *Flowers* ·1 in. long. *Lateral sepals* petaloid,

as large as tho corolla. *Keel* hooded. *Capsule* sub-orbicular, entire,
narrowly 2-winged. Wall. Cat. 4182 (species).

**Perak.** At low elevations.

Sect. III. Herbs, sometimes woody at the base. *Calyx* persistent.
*Keel* crested. *Seeds* albuminous.

3. POLYGALA LEPTALEA, DC. Prod. I, 325. A perennial glabrous
herb, the root-stock woody. *Stems* erect, rigid, boldly striate, few-
leaved. *Leaves* sessile, linear-lanceolate, ·5 to ·75 in. long. *Racemes*
1 to 3 in. long, elongating with age, slender. *Flowers* ·2 to ·25
in. long. *Capsule* ovoid, emarginate at the apex, narrowly winged ;
Hook. fil. Fl. Br. Ind. I, 202 ; Benth. Fl. Austral. i. 139 ; Hassk. in Miq.
Ann. Mus. i. 173. *P. oligophylla*, DC. l. c. 325; Wall. Cat. 4188.
*P. discolor*, Ham. in Don Prodr. 199.

Nicobar Islands. Distrib. British India, Ceylon.

4. POLYGALA BRACHYSTACHYA, Bl. Bijdr. 69. A slender, prostrate
or sub-erect herb. Branches puberulous, terete below, angled above,
4 to 6 in. long. Leaves with very short petioles, linear-lanceolate, bristle-
pointed, glabrous, ·4 in. long, and ·05 in. broad. *Racemes* much longer
than the leaves, few-flowered, slender, axillary or extra-axillary; pedicels nearly as long as the flowers. *Flowers* ·15 in. long ; lateral sepals
obovate-oblong. Keel narrow below ; the apex suddenly dilated, 3-lobed.
*Capsule* sub-orbicular, the apex emarginate, the edges ciliolate. Hassk.
in Miq. Ann. Mus. Lngd. Bat. I, 157 ; Fl. Ind. Bat. I, pt. ii, p. 125.
*P. chinensis*, Linn., var. *brachystachya*.

Malacca, Griffith. Distrib. Java, Sumatra.

5. POLYGALA TELEPHIOIDES, Willd. Sp. Pl. iii, 876. A prostrate
annual with a woody root. *Stems* 2-4 in. long, pubescent or glabrous.
*Leaves* glabrous, often imbricate, fleshy, sessile, obovate or oblong,
obtuse or acute, the margins recurved, the base slightly narrowed, the
midrib prominent, nerves obsolete ; length ·5 to ·65 in. *Flowers* ·1 in.
long, in short, extra-axillary racemes. *Capsules* ·1 in. long, sub-orbicular,
notched at apex, not winged. Hook. fil. Br. Ind, I, 205 ; DC. Prod.
I, 332 ; W. & A. Prod. I, 36, *? P. serpyllifolia*, Poir. Dict. V, 499 ; DC.
l. c. 326. *P. buxiformis*, Hassk. in Miq. Mus. Lngd. Bat. I, 161.

Nicobar Islands. Distrib. Peninsular India, Ceylon, Malayan Archipelago, Philippines, China.

## 2. SALOMONIA, Lour.

Leafy diffuse annuals, or (Sect. *Epirhizanthes*) parasites with leaves
reduced to scales. *Flowers* minute, in dense terminal spikes. *Sepals*
nearly equal, the 2 interior somewhat larger. *Petals* 3, united at the base
with the staminal tube ; the inferior keel-shaped, galeate, not crested.

*Stamens* 4-5, filaments united for their lower half into a sheath ; anthers opening by pores. *Ovary* 2-celled, each cell with one pendulous ovule. *Capsule* much compressed laterally, 2-celled, loculicidal, margins toothed. *Seeds* albuminous, not or scarcely strophiolate. Distrib. Species about 8, natives of Eastern tropical Asia and tropical Australia.

Sect. I. SALOMONIA, DC. Stems leafy.

1. SALOMONIA CANTONIENSIS, Lour. Fl. Coch. Ch. 14. A diffuse, much-branched, glabrous annual ; stem and branches winged. *Leaves* shortly petiolate, ovate-cordate, 3-nerved, length ·25 to ·4 in. *Spikes* numerous, terminal, dense above but lax below, 1-3 in. long; bracts minute, fugacious. *Flowers* ·05 in. long. *Sepals* linear. *Capsule* flat, reniform, its edges with bold recurved triangular teeth. Seeds black, estrophiolate. Hook. fil. Fl. Br. Ind. I, 206; DC. Prod. I, 334; Benth. Fl. Hongk. 44; Miq. Flor. Ind. Bat. I, pt. ii, 127; Hassk. in Miq. Ann. Mus. Lugd Bat. I, 144. *S. subrotunda*, Hassk. l. c. 146.

In all the provinces except the Nicobars and Andamans; in swampy places. Distrib. Brit. India, Malayan Archipelago.

2. SALOMONIA OBLONGIFOLIA, DC. Prod. I, 354. An erect, simple or little-branched, glabrous annual, 3-6 in. high ; stem and branches very slightly winged. *Leaves* elliptic or ovate-lanceolate, sessile, ·15 to ·4 in. long. *Bracts* linear, often persistent. *Spikes* terminal, 1-3 in. long, naked below. *Flowers* crowded above, ·05 in. long. *Sepals* nearly equal, lanceolate. *Capsule* reniform, teeth pointed, spreading. *Seeds* black, estrophiolate. Hook. fil. Fl. Br. Ind. I, 207; Hassk. in Miq. Ann. Mus. Lugd. Bat. I, 147 ; Arn. Pug. Ind. IV ; Deless. Ic. Sel. III, t. 19. *S. sessiliflora*, Ham. in Don Prodr. 201. *S. obovata*, Wight Ill. i, t. 22 B. *S. camarana, rigida, ? Horneri, ? uncinata* and *? setosa-ciliata*, Hassk. l. c. 147, 148, 149. *? S. stricta*, Sieb. et Zucc. Abh. d. k. Baier. Akad. d. Wiss. IV, 2, 152.

In all the provinces except the Nicobars and Andamans, in swampy places. Distrib. Brit. India, Malayan Archipelago.

Sect. II. EPIRHIZANTHES, Blume (genus). Parasitic leaves none, or reduced to scales.

3. SALOMONIA APHYLLA, Griff. in Trans. Linn. Soc. xix, 342. A brownish-purple, erect, little-branching, parasitic herb, 3-6 in. high. *Leaves* reduced to a few distant, brown scales. *Spikes* terminal, dense, 1-3 in. long. *Bracts* minute, persistent. *Flowers* pale brown, ·04 in. long. *Sepals* ovate. *Capsule* transversely ovate, with a single apical tooth. *Seeds* black, strophiolate. Hook. fil. Fl. Br. Ind. I, 207. *S. parasitica*, Griff. Notul. IV, 538. *S. tenella*, Hook. fil. in Trans. Linn. Soc. xxiii. 158. *Epirhizanthes*, Bl. Cat. Hort. Buitenz. and in Flor. Bot. Zeit. 1825, p. 133 ; Reuter in DC. Prod. XI, p. 44.

Perak ; in dense Bamboo Forests. Distrib. Java, Borneo, Tenasserim.

### 3. SECURIDACA, Linn.

Shrubs, almost always scandent. *Flowers* in terminal or axillary, usually compound, racemes. *Sepals* deciduous, the 2 inner (*wings*) larger and petaloid. *Petals* 3, lateral nearly or quite distinct from the galeate crested keel, superior petals 0. *Stamens* 8, filaments united ; anthers 2-celled, dehiscing by oblique pores. *Ovary* 1-celled, 1-ovuled. *Fruit* a 1-celled samara, 1-seeded ; wing broad, coriaceous. *Seeds* exalbuminous, estrophiolate. Distrib. Species about 25 ; most numerous in tropical America, rarer in tropical Africa and Asia.

SECURIDACA BRACTEATA, Benn. in Hook. fil. Fl. Br. Ind. I, 208. A powerful climber ; branches terete, puberulons. *Leaves* elliptic, shortly and bluntly acuminate, the margins revolute when dry, base rounded or slightly narrowed ; upper surface shining, lower densely covered with minute pale pubescence ; nerves 5-6 pairs. *Flowers* in racemes or panicles ; bracts ovate, acuminate, pubescent, deciduous. *Outer sepals* nearly equal, small, ovate, very hairy, ciliate ; wings large, rotund, pubescent externally. *Lateral petals* truncate ; keel with a recurved, plaited crest. *Ovary* orbicular ; style curved. *Stigma* large. *Samara* 3-3½ in., the nucleus smooth, sub-globular, ·4 in. in diam. ; the wing obliquely oblanceolate, membranous, with prominent transverse curving arched nerves ; the upper edge thickened entire, the lower erose.

Malacca, Maingay ; Perak, Wray.

Not having seen any specimen with perfect flowers, I have copied the description of the sepals and petals from Bennet.

### 4. TRIGONIASTRUM, Miquel.

A shrub or small tree. *Leaves* hoary beneath. *Flowers* in slender terminal panicles. *Sepals* 5, nearly equal, the two outer larger. *Petals* 5, imbricate, unequal, the two lower partially united to form a keel ; the odd petal the largest, saccate at the base and with a large hairy gland in its concavity. Stamens 5 or 6, the filaments united into a group between the ovary and the keel. *Ovary* densely hairy, 3-locular ; ovules pendulous, solitary in each cell. *Fruit* of 3 samaroid, ultimately almost distinct, carpels. *Seeds* 1 in each carpel, not strophiolate.

TRIGONIASTRUM HYPOLEUCUM, Miq. Fl. Ind. Bat. Suppl. I, 395, A slender tree, 30 to 60 feet high ; young branches hoary-pubescent ; the older with dark brown, lenticellate bark. *Leaves* elliptic-lanceolate, shortly and bluntly acuminate, narrowed to the base ; both surfaces shining ; the upper glabrous ; the lower pale, very minutely scurfy-pubescent,

the reticulations and 6 pairs of nerves bold and prominent; length 4 to
5 in., breadth 1·25 to 1·75 in. ; petiole ·2 to ·3 in.  *Panicles* axillary and
terminal, slender, spreading.  *Flowers* ·25 in. in diam. ; shortly pedicellate.
*Sepals* 5, pubescent, slightly unequal.  *Ovary* hairy.  *Samaras* 2 in.
long ; the nucleus 1 in. to 1·25 long, triangular, flat ; the wing thinly
membranous, pale yellow, oblong, its apex blunt, oblique, venation
vertical, the areolae wide.  *Isopteris penangiana*, Wall. Cat. 7261.
    Penang.  Malacca.  Perak ; common.
    The pubescence on the under surfaces of the leaves is so minute
that, without a good lens, it is not seen.

## 5. XANTHOPHYLLUM, Roxb.

    Trees or shrubs.  *Leaves* coriaceous or sub-coriaceous, usually yellow-
ish green.  *Sepals* 5, nearly equal.  *Petals* 5 or 4 ; the inferior keeled,
not crested.  *Stamens* 8, distinct, 2 attached to the base of petals, the
others hypogynous.  *Ovary* often surrounded by a hypogynous disc,
stipitate, 1-celled ; style more or less filiform, ovules various in insertion
and number.  *Fruit* 1-celled, 1-seeded, indehiscent.  *Seeds* exalbuminous, estrophiolate.  Distrib.  Species about 27, mostly Malayan, a few
Indian and one in Queensland.
Ovules 4.
    Leaves membranous or sub-coriaceous (coriace-
      ous in No. 2)  small ; flowers less than ·4 in.
      long
        Fruit not verrucose.
          Ovary glabrous, fruit shining    ... 1. *X. Andamanicum.*
          Ovary tomentose
            Nerves of leaves 3 to 4, young
              fruit tomentose  ...        ... 2. *X. Griffithii.*
            Nerves of leaves 4 to 5, fruit tomen-
              tose, branches very slender  ... 3. *X. Maingayi.*
        Fruit verrucose.
          Fruit verrucose only when ripe, glab-
            rous ; leaf-nerves 9 to 10 pairs  ... 4. *X. glaucum.*
          Fruit verrucose from its youngest state.
            Nerves of leaves 4 to 5, ovary vil-
              lous ; fruit glabrous, vertically
              grooved    ...        ... 5. *X. Palembanicum.*
            Nerves of leaves 4 to 5 ; ovary vil-
              lous, fruit puberulous, not ver-
              tically grooved  ...        ... 6. *X. curhynchum.*

Nerves of leaves 10 to 13; ovary
villous; ripe fruit glabrous, not
vertically grooved ... 7. *X. Wrayi.*
Leaves coriaceous, shining on both surfaces;
flowers large, ·4 to ·6 in. long, (small in
No. 8).

    Leaves 3 to 5 in. long, nerves 5 to 6 pairs    8. *X. Curtisii.*
    ,,    4 to 9 in. ,,    ,,   8 to 10 ,,     9. *X. Kunstleri.*
    ,,    9 to 11 in. ,,   ,,   6 to 8 ,,    10. *X. Hookerianum.*
    ,,    9 to 14 in. ,,   ,,  14 to 16 pairs 11. *X. venosum.*

Ovules 6 to 14.

*Shrubs or trees with glabrous leaves.*
    Leaves shining on both surfaces, drying
    brown. Flowers large, in short racemes.
      Ovary cottony    ...    ... 12. *X. stipitatum.*
      Ovary glabrous.
      Leaves 2 to 2·25 in. long, nerves 3 pairs 13. *X. Scortechinii.*
      ,,    2·5 to 4 in., nerves 5 to 6 pairs... 14. *X. ellipticum.*
      ,,    4 to 6 in., nerves 5 to 6 pairs... 15. *X. obscurum.*
    Leaves dull white below, not shining, green-
    ish above when dry; flowers large, ovary
    tomentose    ...    ...    ... 16. *X. pulchrum.*
    Leaves shining on both surfaces, drying yel-
    lowish or greenish.
      Leaves not cordate; panicles small,
      not spreading    ...   .. 17. *X. affine.*
      Leaves minutely cordate at base; pa-
      nicles large, wide-spreading    ... 18. *X. bullatum.*
*Trees with leaves pubescent beneath, and tomentose*
*inflorescence.*
    Pubescence sulphureous; ovary glabrous... 19. *X. sulphuratum.*
    ,,    rufous; ovary tomentose    ... 20. *X. rufum.*
Doubtful species.
Fruit many-seeded, 3 in. in diam.    ...    ... 21. *X. insigne.*

    1. XANTHOPHYLLUM ANDAMANICUM, King, n. sp. A tree 20 to 30 feet
high, glabrous except the inflorescence, branches dark brown. *Leaves*
thinly coriaceous, drying a pale greenish passing into brown, elliptic to
elliptic-oblong, rarely sub-obovate, sub-acute, the base cuneate; upper surface smooth, shining; the lower dull, pale, minutely reticulate; main nerves
7 to 8 pairs, rather prominent; length 3 to 4 in., breadth 1·25 to 1·75
in.; petiole ·3 to ·4 in. *Flowers* ·3 in. long, their pedicels about as long.
*Panicles* extra-axillary or terminal, 1·5 to 3 in. long, lax, few-branched,

slender. *Sepals* rotund, pubescent, edges ciliolate. *Petals* broadly ovate, glabrous, keel pubescent. *Filaments* flat and pubescent at the base, otherwise glabrous. *Ovary* ovoid-elongate, glabrous, 4-ovuled; style rather flat, pubescent; disc small, glabrous. *Fruit* globose, ·5 in. in diam., smooth ; pericarp thin, crustaceous.

Andaman Islands ; Helfer, Kurz, King's Collector. Burmah, Kurz. This is not unlike *X. Griffithii* in its leaves ; but it differs in its glabrous ovary and fruit.

2. XANTHOPHYLLUM GRIFFITHII, Hook. fil. Fl. Br. Ind. I, 210. A tree 40 to 50 feet high ; glabrous, except the inflorescence and young fruit. Branchlets robust, dark brown, polished, terete. *Leaves* coriaceous, elliptic-lanceolate or lanceolate, acute or acuminate, the edges slightly revolute, the base acute ; upper surface dark (when dry) and shining ; the lower pale, sub-glaucous, the minute reticulations and 3-4 pairs of main nerves distinct; length 3 to 5 in., breadth 1 to 1·5 in., petiole thick, dark-coloured, ·35 in. long. *Flowers* about ·35 in. long, in tomentose axillary racemes shorter than the leaves, or in terminal few-branched panicles ; pedicels short, stout. *Sepals* broadly ovate, blunt, dark brown, tomentose externally. *Petals* oblong, blunt, glabrous except the pubescent keel. Lower half of *filaments* thickened, hairy. *Ovary* sessile, tomentose, with 4 ovules from its base. *Style* cylindric, sparsely pilose. *Fruit* (young), globular, tawny-tomentose.

Malacca and Perak. Distrib. Burmah (at Mergui). Ripe fruit of this is unknown.

3. XANTHOPHYLLUM MAINGAYI, Hook. fil. in Fl. Br. Ind. I, 210. A tree 20 to 40 feet high. Branches very slender with pale brown bark, the youngest puberulous. *Leaves* membranous, lanceolate or elliptic-lanceolate, cordate-acuminate, the base narrowed ; both surfaces smooth, the lower pale but not glaucous ; nerves 4-5 pairs, the reticulations fine, not prominent ; length 2 to 3 in., breadth ·65 in. to 1 in. ; petiole slender, about ·25 in. *Racemes* axillary and terminal, sometimes branched, the rachises tawny-tomentose, 2 to 3 in. long ; flowers white, ·35 in. long. *Sepals* rhomboid, unequal, pubescent. *Petals* much longer than the sepals, spreading, glabrous except the pubescent keel. *Filaments* much curved, with a hairy thickening above the base. *Ovary* shortly stipitate, ovoid, ridged, tawny-tomentose; ovules 4, parietal. *Fruit* globose, ·5 in. in diam., minutely tomentose ; pericarp moderately thick, puckering when dry.

Penang, Malacca and Perak ; at low elevations.

4. XANTHOPHYLLUM GLAUCUM, Wall. Cat. 4199. A tree 20 to 30 feet high. Young branches terete, smooth, pale, the very youngest brown and slightly angled, *Leaves* sub-coriaceous, oblong-lanceolate to

elliptic-lanceolate, sub-acute, the base narrowed into the petiole; upper surface shining; lower dull, sub-glaucous; nerves 8 to 10 pairs, not prominent, reticulations minute; length 3 to 4 in., breadth 1 in. to 1·4 in.; petiole rather thick, less than ·25 in. *Flowers* ·25 in. long, otherwise as in *X. Griffithii.* Fruit globose when quite ripe, slightly warted, and 1 in. in diam. Hook. fil. Fl. Br. Ind. I, 209; Hassk. in Miq. Ann. Mus. Lugd. Bat. I, 193.

Trang ; King's Collector No. 1427. Distrib. Chittagong, Burmah. Thsi differs from *X. Griffithii* chiefly by having many more nerves in its leaves.

5. XANTHOPHYLLUM PALEMBANICUM, Miq. Ann. Mus. Lugd. Bat. I, 317. A glabrous tree 30 to 40 feet high ; branchlets slender, pale. *Leaves* membranous, drying pale green, lanceolate, rarely elliptic, caudate-acuminate, the base narrowed ; acute or cuneate ; upper surface shining, the lower dull, pale but not glaucous, main nerves 4 to 5 pairs, slightly prominent, reticulations minute ; length 3 to 4·5 in., breadth 1·1 to 1·3 in. ; petiole ·2 in., slender. *Flowers* ·4 in. *Racemes* axillary, few-flowered, slender, shorter than the leaves. *Sepals* unequal, rhomboid, spreading, flat, puberulous externally. *Petals* spathulate, glabrous except the pubescent keel. *Filaments* thickened and pubescent in the lower half. *Ovary* shortly stipitate, villous ; the annular disc surrounding it small ; style sparsely villous. Ovules 4, from near base of ovary. *Fruit* globose, ·75 in. in diam., glabrous, boldly verrucose and with several irregular vertical grooves; pericarp ·1 in. thick, crustaceous.

Perak, rather common. Distrib. Sumatra.

This is not unlike *X. Maingayi,* but is at once distinguished from that by its deeply grooved fruit.

6. XANTHOPHYLLUM EURHYNCHUM, Miq. Ann. Mus. Lugd. Bat. I, 277. A glabrous tree 30 to 50 feet high ; branchlets terete, brown. *Leaves* drying pale green, sub-coriaceous, elliptic-lanceolate, tapering at both ends, to elliptic with rounded base and apex shortly acuminate ; both surfaces smooth, the upper shining, the lower dull slightly paler; main nerves 4 to 5 pairs, prominent beneath ; length 4 to 5·5 in., breadth 1·75 to 2·5 in., petiole ·3 in. *Flowers* ·25 in. long, the pedicels not longer than the calyx. *Racemes* shorter than the leaves, pubescent, axillary and solitary or in terminal few-branched panicles. *Sepals* unequal, ovate-oblong, blunt, spreading, minutely tomentose externally. *Petals* spathulate, glabrous except the pubescent keel. *Filaments* slightly flattened and pubescent in the lower half. *Ovary* villous, 4-ovuled. *Style* slightly curved, villous. *Fruit* globose, ·75 in. in diam., puberulous, rather minutely verrucose, not vertically ridged ; pericarp crustaceous, brittle, ·2 in. thick.

Perak, Pangkore. Distrib. Sumatra.

This comes very near *X. Palembanicum*, but is distinguished from that species by its more robust branches, longer leaves, and velvety fruit which is not vertically ridged. The two are, however, closely allied.

7. XANTHOPHYLLUM WRAYI, King, n. sp.   A shrub 3 to 8 feet high, the young branches puberulous. *Leaves* sub-coriaceous, drying a pale yellowish green, elliptic to oblong, more or less acuminate, the base rounded or narrowed ; upper surface shining ; lower slightly dull, pale and rather minutely reticulate ; the main nerves 10 to 13 pairs, prominent, forming arches ·2 in. within the edge ; length 6 to 10 in., breadth 2·25 to 4·5 in., petiole ·3 to ·6 in. *Flowers* ·35 in. long, their pedicels shorter than the calyx, in terminal or axillary racemes or small panicles less than a third of the length of the leaves. *Sepals* ovate, blunt, puberulous. *Petals* oblong, obtuse, puberulous in the upper, pubescent in the lower, half. *Filaments* flat, pubescent. *Ovary* on a glabrous stalk, densely villous, 4-ovuled, the disc glabrous. *Fruit* globose, ·75 in. in diam. ; when young sericeous, when ripe quite glabrous and boldly verrucose.

Penang, Curtis; No. 677.   Perak, King's Collector, Wray.

This is not unlike *X. affine*, Korth, but is distinguished from that species by its sericeous ovary and deeply warted fruit. In its fruit this resembles *X. Palembanicum* and *eurynchum ;* but it differs from both in its much larger and more numerously veined leaves.

8. XANTHOPHYLLUM CURTISII, King, n. sp.   A glabrous tree, 30 to 50 feet high. Young branches rather robust, dark brown, glabrous. *Leaves* coriaceous, drying brown, ovate-lanceolate, shortly acuminate, the base rounded or cuneate; both surfaces smooth, dull ; the lower paler, minutely reticulate ; main nerves 5 to 6 pairs, not much more prominent than the secondary nerves ; length 3 to 5 in., breadth 1 to 1·3 in., petiole ·4 in. *Flowers* ·4 in. long, the pedicels about as long as the calyx. *Panicles* axillary or terminal, few-branched, nearly as long as the leaves. *Sepals* nearly equal, rotund, tomentose. *Petals* oblong, obtuse, glabrous except the broadly obovate pubescent keel. *Filaments* with an ovoid pubescent swelling near the base. *Ovary* sessile, ridged, pubescent, 2- to 4-ovuled, thick-walled, surrounded by a fleshy glabrous slightly angled but not wavy disc. *Style* conical, pubescent. *Fruit* (very young) globose, tomentose.

Penang ; Curtis, No. 1591.   Singapore ; Murton.

The leaves of this dry of an olivaceous brown colour.

Ripe fruit is unknown.   Curtis' No. 1486 from Penang, of which I have seen no very complete specimen, is probably a variety of this with leaves more attenuated to both base and apex, and with longer more spreading panicles.

9. XANTHOPHYLUM KUNSTLERI, King, n. sp. A glabrous tree 50 to 80 feet high; the branchlets robust, dark brown, sub-glaucous. *Leaves* large, coriaceous (with a yellowish green tinge when dry) elliptic to elliptic-oblong, the apex very shortly and rather suddenly acuminate, the edges (when dry) undulate, the base rounded or slightly narrowed to the stout petiole; both surfaces shining; nerves sub-horizontal, 8 to 10 pairs, prominent beneath as are the secondary nerves and reticulations; length 4·5 to 9 in., breadth 2·25 to 4·25 in., petiole ·6 to ·75 in. *Flowers* ·6 in. long, shortly pedicellate, in axillary racemes less than half as long as, or in terminal few-branched spreading panicles longer than, the leaves. *Sepals* slightly unequal, ovate-rotund, fleshy, thickened along the midrib, minutely tomentose on both surfaces; the edges thin, ciliolate. *Petals* oblong-obtuse, glabrous except the pubescent keel. *Filaments* with a pubescent ovoid swelling above the base. *Ovary* almost sessile, surrounded by a shallow wavy fleshy disc, ovoid, grooved, tomentose, 4-ovulate; style conical, slightly curved, pubescent. *Fruit* globose, ·65 in. in diam. (young) deciduously tomentose; pericarp thick, spongy, the calyx persistent.

Perak. At low elevations, not common; King's Collector (Kunstler). Penang, Government Hill; Curtis, No. 1590.

10. XANTHOPHYLLUM HOOKERIANUM, King, n. sp. A glabrous large-leaved shrub; young branches rather stout, sub-glaucous. *Leaves* coriaceous, (drying yellowish) elliptic-oblong with a rather abrupt bluntish acumen 1 in. long, the edges slightly revolute when dry, the base slightly narrowed to the petiole; both surfaces smooth, the lower slightly paler; main nerves 6 to 8 pairs, thin but rather prominent beneath as are the reticulations; length 9 to 11 in., breadth 3 to 4 in.; petiole ·75 in. *Flowers* ·4 in. long, the pedicels about as long as the calyx. *Sepals* nearly equal, broadly ovate, minutely tomentose on both surfaces, the edges thin, ciliolate. *Petals* oblong, obtuse, glabrous except the pubescent keel. *Filaments* flat, pubernlons. *Ovary* sessile, ovoid, tomentose, 4-ovuled; style flat, grooved, pubescent, slightly curved. *Fruit* unknown.

Perak; King's Collector, No. 5997.

11. XANTHOPHYLLUM VENOSUM, King, n. sp. A glabrous long-leaved tree 20 to 30 feet high; young branches rather robust, with very dark brown bark. *Leaves* (drying pale yellowish-green), coriaceous, oblong, sub-acute, the edges recurved when dry, gradually but slightly narrowed in the lower fourth to the rounded or minutely cordate base; both surfaces shining, the lower slightly paler; main nerves 14 to 16 pairs, horizontal near the base, sub-ascending towards the apex, prominent on the lower surface and forming bold arches ·2 in. from the margin,

secondary nerves and minute reticulations distinct; length 9 to 14 in., breadth 2·5 to 3·25 in., petiole ·75 to 1 in. *Flowers* ·4 in. long, their pedicels longer than the calyx. *Panicles* axillary, few-branched, 3 to 6 in. long. *Sepals* nearly equal, broadly ovate, fleshy with thin edges, puberulous. *Petals* oblong, obtuse, glabrescent, the keel pubescent. *Filaments* flat, puberulous. *Ovary* surrounded by a shallow glabrous wavy disc, ovoid, minutely tomentose, grooved, 4-ovuled; style pubescent, slightly curved. *Fruit* globose with a conical apex, deeply rugose, verrucose; diam., ·35 in. (young).

Perak, King's Collector; Nos. 10614 and 10804.

Ripe fruit of this is unknown. This species, in leaf characters, approaches the Bornean *X. cordatum*, Korth.; but the fruit of that is smooth; of this the fruit is deeply corrugated-rugose as in. *X. Palembanicum* and *eurhynchum*.

12. XANTHOPHYLLUM STIPITATUM, A. W. Benn. in Hook. Fl. Br. Ind. I. 210. A tree with slender, smooth, brown branches. *Leaves* subcoriaceous, ovate or elliptic, shortly and obtusely caudate-acuminate; upper surface dull, the nerves obsolete; lower shining, the 3 to 4 pairs of nerves indistinct; length 1·75 to 2·25 in., breadth ·75 to 1 in., petiole ·15 in. *Racemes* slender, axillary, longer than the leaves, pubescent. *Flowers* ·4 in. long, shorter than their slender pedicels. *Sepals* subequal, oblong, obtuse, glabrescent. *Petals* twice as long as sepals, spathulate, oblong. *Filaments* thickened and hairy in their lower half. *Ovary* stipitate, cottony; style sparsely hairy; ovules 8 to 10.

Malacca.

Hitherto known only by specimens from Malacca; fruit not collected.

13. XANTHOPHYLLUM SCORTECHINII, King, n. sp. A tall glabrous tree. *Leaves* thinly coriaceous, drying brown, ovate, obtusely acuminate, the base slightly cuneate, shining on both surfaces; nerves 3 pairs, suberect, not prominent; length 2 to 2·25 in, breadth 1 in., petiole ·35 in. *Flowers* handsome, ·6 in. long, their pedicels ·35 in. *Racemes* axillary, solitary, ·2 in. long, few-flowered. *Sepals* fleshy, glabrous with ciliolate edges; the three outer ovate blunt, the two inner rotund. *Petals* broadly obovate, clawed, glabrescent, the keel pubescent. *Filaments* as long as the petals, flat, pubescent; anthers short, ovate. *Ovary* shortly stalked, elongated-ovoid, ridged, glabrous, 6-ovuled; style little curved, glabrous; stigma capitate; disc small, annular, glabrous.

Perak; Father Scortechini, No. 2079.

Of this distinct and handsome species fruit is as yet unknown.

14. XANTHOPHYLLUM ELLIPTICUM, Korth. in Miq. Ann. Mus. Lugd. Bat. I, 276. A glabrous tree 30 to 60 feet high; branchlets slender,

pale. *Leaves* drying pale brown, sub-coriaceous, elliptic-lanceolate to elliptic, shortly and bluntly acuminate, the base narrowed or rounded;; above shining; below dull, the reticulations distinct; nerves 5 or 6 pairs; length 2·5 to 4 in., breadth 1·2 to 1·75 in., petiole ·25 in. *Flowers* ·25 in. long. *Racemes* axillary, shorter than the leaves, the pedicels longer than the flowers. *Sepals* ovate, fleshy, glabrous, concave, unequal. *Petals* thin, spathulate, much longer than the sepals, glabrescent. *Ovary* glabrous, shortly stipitate; the stalk surrounded by a shallow entire, undulate, annular disc; 10-ovulate; style glabrous. *Fruit* globular, ·5 to ·75 in. in diam., when ripe smooth, pulpy; pericarp thin, leathery. Hoook. fil. Fl. Br. Ind. I, 211.

Malacca, Perak.

15. XANTHOPHYLLUM OBSCURUM, A. W. Benn. in Hook. fil. Fl. Br. Ind. I, 211. A large tree; branches stout, glabrous, lenticellate. *Leaves* coriaceous, elliptic, blunt, narrowed in the lower third to the stout petiole, drying to a dark brown; both surfaces shining; the lower slightly paler; main nerves 5 to 6 pairs, thin, rather prominent as are the intermediate nerves; length 4·25 to 4·75 in., breadth 2 to 2·25 in., petiole ·4 in. *Racemes* axillary, 1 to 1·15 in. long, few-flowered. *Flowers* ·6 in. long, the pedicels short. *Sepals* oblong, blunt, the edges ciliate, the 3 outer small. *Petals* glabrous, oblong, sub-spathulate. *Ovary* ovoid, glabrous, ovules 8 to 10; style glabrous.

Singapore; Maingay, Hullett.

16. XANTHOPHYLLUM PULCHRUM, King, n. sp. A glabrous shrub or small tree, the young branches rather robust, the bark very pale. *Leaves* coriaceous, shortly petiolate, elliptic, acute or very shortly and bluntly acuminate, the edges recurved when dry, base rounded; upper surface shining; lower dull, pale yellow, glaucous, the 5-6 pairs of nerves and the fine reticulations very prominent; length 4 to 8 in., breadth 2·25 to 4 in.; petiole stout, ·2 in. *Flowers* ·6 in. long, on short thick pedicels, in dense, solitary, axillary, rufous-tomentose racemes half as long as the leaves, or less; bracts broadly ovate, concave, deciduous, coloured. *Sepals* nearly equal, ovate-rotund, with fleshy tomentose midribs and thin minutely ciliate edges, coloured. *Petals* oblong, blunt, glabrous except the pubescent keel. *Filaments* rather short, broad, flat, puberulous. *Ovary* ovoid, pointed, tomentose, surrounded by a thin, rather deep, glabrous disc; style glabrescent, slightly curved; ovules 12. *Fruit* globose, ·75 in. in diam., minutely tomentose when ripe, pericarp thin.

Perak; rather common. A handsome bush or treelet with yellowish flowers tinged with pink.

In leaf characters this comes near to the Sumatran *X. vitellinum*, Blume ; but the two differ in flower and fruit.

17. XANTHOPHYLLUM AFFINE, Korth. in Miq. Ann. Lugd. Bat. I, 271. A shrub or tree; young branches glabrous, pale brown. *Leaves* thinly coriaceous (drying of a more or less yellowish pale green, especially beneath), elliptic to oblong-lanceolate, shortly and bluntly acuminate, the base cuneate ; upper surface smooth, shining, lower dull pale and yellowish ; main nerves 5 to 8 pairs, ascending, prominent beneath ; length 4 to 7 in., breadth 1·8 to 2·5 in., petiole ·3 to ·4 in. *Flowers* ·35 in. long, the pedicels nearly as long. *Panicles* axillary or terminal, few-branched, minutely tomentose, the axillary half as long, the terminal as long as, the leaves. *Sepals* unequal, ovate-rotund to rotund, blunt, tomentose externally. *Petals* oblong, obtuse, pubescent near the base or wholly glabrous, the keel always pubescent. *Filaments* flat, puberulous. *Ovary* shortly stipitate, glabrous, from 8- to 14-ovuled. *Style* short, flat, pubescent. *Disc* annular, fleshy, glabrous, often wavy. *Fruit* globose, ·5 to 1·25 in. in diam., smooth ; pericarp thin, crustaceous. Hook. fil. Fl. Br. Ind. I, 209.

In all the provinces ; common. Distrib. Malayan Archipelago generally. Tenasserim.

This occurs as a bush and also as a tree. It varies a little as to colour and shape of leaves, and as to the pubescence on the petals. But, when its commonness is considered, its characters are really remarkably constant, that of the size of the individual alone excepted.

18. XANTHOPHYLLUM BULLATUM, King, n. sp. A shrub or small tree with large, sub-sessile leaves ; young branches robust, pale, puberulous, lenticellate. *Leaves* coriaceous, drying a pale greenish-yellow, elliptic-oblong, sub-obovate, shortly and bluntly acuminate, the edges sub-recurved ; slightly narrowed to the cordate, sub-auriculate, slightly unequal base ; bullate, especially in the lower half, shining and glabrous on both surfaces, the lower a little paler ; main nerves 18 to 25, bold on lower surface, and sometimes puberulous as in the strong midrib ; length 11 to 18 in., breadth 4 to 6·5 in. ; petiole ·25 in., very stout, glandular. *Flowers* ·4 in. long, the pedicels twice as long as the calyx. *Panicles* terminal, many-branched, spreading, pubescent; bracts deciduous, ovate. *Sepals* unequal, rotund, fleshy, concave, tomentose, the edges of the inner two thin and ciliate. *Petals* ovate-rotund, glabrous, not much larger than the sepals. *Filaments* flat, fleshy, glabrous. *Ovary* glabrous, ovoid, 8-ovuled, surrounded by a glabrous fleshy annular wavy disc ; style glabrous ; stigma conical, pubescent. *Fruit* globose, glabrous, ·75 in. in diam. (? ripe ;) pericarp thick, crustaceous.

82

Perak, on low Hills.

This resembles *X. adenotus*, Miq., but differs in venation of leaves and in inflorescence.

19. XANTHOPHYLLUM SULPHUREUM, King, n. sp.  A tree 100 (or even 150) feet high; branches with very dark brown bark, the youngest minutely tomentose.  *Leaves* coriaceous, drying bright yellowish-green, elliptic-oblong, sometimes sub-obovate, acuminate, the edges recurved when dry, the base cuneate; upper surface glabrous, shining; lower sulphureous, softly but minutely pubescent especially on the midrib and 6 to 7 pairs of prominent ascending nerves; length 5·5 to 7·5 in., breadth 2 to 2·5 in., petiole ·5 to ·65 in.  *Flowers* ·4 in. long, the pedicels about as long as the calyx.  *Panicles* tomentose, with ovate deciduous bracts, compact, many-branched, axillary or terminal, less than half the length of the leaves.  *Sepals* unequal, ovate-rotund, minutely tomentose on both surfaces, the edges ciliolate.  *Petals* oblong, obtuse, glabrous except the pubescent tips; the keel obovate, vertically 9- to 10-ridged, tomentose.  *Filaments* flattened, glabrous.  *Ovary* shortly stipitate, glabrous, ovate, 8-ovuled, surrounded at the base by a fleshy glabrous much waved annular disc.  *Style* villous in its lower, glabrous in its upper, half.  *Fruit* (young) globose, sulphureous, glabrous, faintly rugose.

Perak, on low hills.

Not unlike *X. rufum*, A. W. Benn. in general aspect; but with smaller flowers, glabrous ovary and fruit, and leaves intensely sulphureous beneath.

20. XANTHOPHYLLUM RUFUM, A. W. Benn. in Hook. fil. Fl. Br. Ind. I 210.  A tree 40 to 50 feet high.  Branchlets stout, terete, pale scurfy-pubescent, the youngest rufous-tomentose.  *Leaves* coriaceous, elliptic to ovate or obovate-elliptic, very shortly and suddenly acuminate, narrowed in the lower third to the stout short petiole, the edges recurved when dry; upper surface glabrous, dull; lower paler, covered with short soft pubescence especially on the midrib and 7-8 pairs of bold semi-erect nerves; length 4·5 to 6·5 in. (acumen ·5 in.), breadth 2·25 to 2·75 in., petiole ·5 in.  *Panicles* terminal or from the axils of the uppermost leaves, lax, few-branched, 3 to 7 in. long, densely tomentose, the ends of the branches and sepals rufous.  *Flowers* ·6 in. long, on short pedicels in the axils of ovate sub-persistent bracts.  *Sepals* unequal, ovate to sub-reniform, densely tomentose on both surfaces, fleshy, concave.  *Petals* more than twice as long as the sepals, oblong, obtuse, glabrous except the tomentose keel.  *Filaments* glabrous, flat at the base.  *Ovary* surrounded by a shallow fleshy disc, sessile, ovate, pointed, ridged, tomentose as is also the conical style; ovules 12 to 16.  *Fruit* ·75 in. in diam., (? mature) globose, with 4 vertical rufous pubescent ridges; pericarp thick, crustaceous.

Malacca; Perak.   Distrib. Sumatra.   (Beccari, P. S. No. 646.)
21. XANTHOPHYLLUM INSIGNE, A. W. Benn. in Hook. fil. Fl. Br. Ind.
I, 211.   A glabrous tree with pale branchlets. *Leaves* drying brown,
coriaceous, elliptic, obtuse, the base slightly narrowed; upper surface
shining; lower dull, pale, the 6-8 pairs of nerves and rather wide reticulations prominent; length 4·5 to 6·5 in., breadth 3 to 3·5 in.; petiole stout,
·6 in.   Racemes 3 to 4 in. long, axillary, sometimes terminal and panicled.
*Flowers* ·6 to ·75 in. long; *sepals* unequal, sub-orbicular, puberulous.
*Petals* spathulate, glabrous except the pubescent claw; keel adpressed-
sericeous, its claw pubescent.   *Stamens* 8, the filaments flat at the base
and rising from an annular entire undulate disc which surrounds the
ovary.   *Ovary* ovoid, ridged, glabrous, ovules 16; style little curved,
glabrous.   *Fruit* globose, 3 in. in diam., minutely rugose, pericarp ·5 in.,
thick; seeds oblong, 1 in. long, embedded in pulp.
Malacca; Maingay, No. 348, Miller.

## ORDER XII.   PORTULACEÆ.

Herbs, rarely undershrubs.   *Leaves* opposite or alternate, entire;
nodes with scarious or hairy appendages, rarely naked.   *Inflorescence*
various.   *Sepals* 2, imbricate.   *Petals* 4-5, hypogynous or perigynous,
free (or united below), fugaceous.   *Stamens* 4-∞, inserted with (rarely
upon) the petals, filaments slender; anthers 2-celled.   *Ovary* free, or
½-inferior, 1-celled; style 2-8-fid, divisions stigmatose; ovules 2-∞, on
basal funicles or a central column, amphitropal.   *Capsule* with transverse or 2-3-valvular dehiscence.   *Seeds* 1-∞, compressed; embryo
curved round a mealy albumen.   Distrib. Cosmopolitan, chiefly American; genera 15, species about 125.

### 1.   PORTULACA, Linn.

Diffuse, usually succulent, annual or perennial herbs.   *Leaves* with
scaly or hairy nodal appendages.   *Flowers* terminal, surrounded by a
whorl of leaves, solitary or clustered.   *Sepals* connate below, the free
part deciduous.   *Petals* 4-6, perigynous or epipetalous.   *Ovary* ½-inferior;
style 3-8-fid; ovules ∞.   *Capsule* crustaceous, dehiscing transversely.
*Seeds* ∞, reniform.   Distrib. Tropical regions, chiefly American;
one or two are cosmopolitan weeds extending to temperate regions;
species 16.

1.   PORTULACA OLERACEA, Linn.   An annual glabrous, sub-succulent,
prostrate herb, 6 to 12 inches long; sometimes with minute scarious
appendages at the nodes.   *Leaves* flat, cuneate-oblong, rounded or
truncate at the apex, ·25 to 1·25 in. long; petiole very short.   *Flowers* in
few-flowered terminal heads or in dichotomous cymes, sessile, surrounded

by a few ovate, pointed scarious bracts; *petals* 5, equal to the sepals, yellow; *stamens* 8 to 12; *style* 3-8-cleft; *seeds* punctate; Roxb. Fl. Ind. II, 463; W. & A. Prodr. 356. *P. lævis*, Ham. in Wall. Cat. 6841 ; Hook. fil. Fl. Br. Ind. I, 246. *P. suffruticosa*, Thw. Enum. 24 (not of Wight).

In the Andamans, and probably in all the Provinces, in waste places. Distrib. All warm climates.

2. Portulaca quadrifida, Linn. An annual with diffuse filiform stems, rooting at the nodes; nodal appendages copious, pilose. *Leaves* flat, opposite, ovate or ovate-lanceolate, acute, almost sessile; length ·2 to ·35 in. *Flowers* solitary, terminal; calyx-tube partly immersed in the extremity of the axis, surrounded by long silky hairs and by about 4 bracteoles; *petals* 4, yellow; *stamens* 8 to 12; *style* filiform, deeply 4-fid. *Seeds* minutely tuberculate. DC. Prod. III, 354 ; Wight Ill. ii, t. 109 ; Hook. fil. Fl. Br. Ind. I, 247; Oliver Fl. Trop. Africa, I, 149. *P. meridiana*, L. Roxb. Fl. Ind. II, 463. *P. geniculata*, Royle Ill. 221. *P. anceps*, Rich. Fl. Abyssin., I, 301.

Penang, in the Fort; Curtis. Distrib. Throughout the Tropics of Asia, S. Africa.

ORDER XIII. HYPERICINEÆ.

Herbs or shrubs, rarely trees. *Leaves* opposite, often punctate with pellucid glands or dark glandular dots, entire or glandular-toothed ; stipules 0. *Flowers* solitary or cymose, terminal, rarely axillary. *Sepals* and *petals* each 5, rarely 4 ; petals contorted in bud. *Stamens* indefinite, or rarely definite, 3- or 5-delphous, rarely free or all connate; anthers versatile. *Ovary* 3- 5-carpellary, 1- or 3-5-celled ; styles as many, filiform, free or united ; ovules few or numerous, on parietal or axile placentas, anatropous, raphe lateral or superior. *Fruit* capsular or baccate. *Seeds* exalbuminous, sometimes winged ; embryo straight or curved. Distrib. Temp. countries and mountains of warm regions; genera 8, species about 210.

1. CRATOXYLON, Blume.

Shrubs or trees. *Leaves* entire, usually papery. *Inflorescence* axillary or terminal, cymose. *Sepals* and *petals* each 5. *Stamens* 3- or 5-delphous, with fleshy hypogynous glands alternating with the bundles. *Ovary* 3-celled ; styles distinct ; ovules 4- 8 in each cell. *Capsule* 3-valved, seeds winged. Distrib. Tropical Asia ; species about 12.

Sect. I. ANCISTROLOBUS, Spach. *Petals* sub-persistent, inappendiculate. *Stamens* 3-delphous ; glands more or less cucullate.

1. CRATOXYLON POLYANTHUM, Korth. Verhand. Nat. Gesch. Bot. 175, t. 36. A large shrub, or tree 30 to 40 feet high, all parts glabrous ;

young branches palo brown, compressed. *Leaves* membranous, minutely pellucid-punctato, elliptic-oblong, almost equally acute at base and apex; abovo shining, below rather dull; nerves about 7 to 10 pairs, pale; reticulations minute; length 1·5 to 3·5 in., breadth ·75 in. to 1·25 in., petiole ·1 in. *Flowers* slightly supra-axillary, solitary or in 1- to 3-flowered cymes, ·5 in. in diam. *Sepals* elliptic, obtuse, as long as the petals. *Petals* oblanceolate, veined. *Hypogynous glands* large, fleshy. *Capsule* slightly exceeding the persistent sepals. *Seeds* obliquely winged. Hook. fil. Fl. Br. Ind. I, 257; Miq. Fl. Ind. Bat. I, pt. ii, p. 516.

All tho Provinces. Distrib. British India, China, Philippines.

Var. I. LIGUSTRINUM, Blume Mus. Bot. II, 16 (sp.); leaves narrowed at both ends, acuto. *C. lanceolatum*, Miq. Fl. Ind. Bat. Supp. I, 500. *Ancistrolobus ligustrinus*, Spach Suit. Buff. V, 361. *A. brevipes*, Turcz. Bull. Mosc. 1858, I, 383. *Hypericum pulchellum*, Wall. Cat. 4821. *H. carneum*, Wall. Cat. 4820.

Andamans, Malacca, Penang.

Var. 2. WIGHTII, Bl. l. c. 18. (sp.) Leaves broadly oval, mostly obtuse. *Ancistrolobus* sp. Wight. Ill. I, 111. *Hypericum horridum*, Wall. Cat. 4822. *Elodea* sp. Griff. Notul. IV, 569.

Perak, King's Collector.

Sect. II. TRIDESMIS, Spach. *Petals* not persistent, with a basal squamule. *Stamens* 3- or 5-adelphous.

2. CRATOXYLON ARBORESCENS, Blume Mus. Bot. II, 17. A tree 15 to 50 feet high, all parts glabrous. Young branches robust, the bark palo brown, often ridged. *Leaves* coriaceous, broadly oblanceolate, obovate-elliptic or oblong-obovate, very shortly acuminate; dull on both surfaces, the lower pale, yellowish-brown with black dots; nerves numerous, obsolete; length 3 to 4·5 in., breadth 1·25 to 2 in.; petiole ·25 in., stout. *Cymes* in large terminal panicles usually longer than the leaves. *Flowers* ·3 in. in diam. *Sepals* unequal, ovate-rotund, veined. *Petals* about as long as the sepals, broadly cuneate, conspicuously veined and with a laciniate basal scale. *Capsule* longer than tho persistent sepals. *Seeds* winged all round. Hook. fil. Fl. Br. Ind. I, 258; Kurz Fl. Burm. I, 84. *C. coccineum*, Planch. *Hypericum arborescens*, Vahl Symb. II. 86, t. 43. *H. coccineum*, Wall. Cat. 4823. *Ancistrolobus glaucescens*, Turcz. Bull. Mosc. 1858, I, 383. *Vismia? arborescens*, Choisy Prod. Hyp. 36.

In all tho Provinces. Distrib. Malayan Archipelago; Burmah.

Var. *Miquelii*, branches more slender than in tho typical form; leaves thinner in texture, oblanceolate, acuminate. A small tree 15 to 20 feet. *C. cuneatum*, Miq. Fl. Ind. Bat. I, pt. ii, 517.

Penang, Perak. Distrib. Sumatra.

3. CRATOXYLON FORMOSUM, Benth. and Hook. fil. Gen. Pl. I, 166. A
shrub or tree 20 to 6 feet high, all parts quite glabrous, young branches
pale. *Leaves* membranous, broadly elliptic to elliptic-oblong, acute or
rounded, the base slightly narrowed; upper surface shining; lower dull,
pale, glaucescent and with numerous minute black dots; main nerves 6
to 8 pairs, little more prominent than the secondary; length 3 to 4 in.,
breadth 1·75 to 2·25 in., petiole ·25 in., thin. *Cymes* axillary, or from
above the scars of fallen leaves, 2-3-flowered. *Flowers* ·75 in. long, their
pedicels ·5 in. *Sepals* elliptic, pointed, faintly veined, ·2 in. long, nearly
equal, not accrescent. *Petals* thin, prominently veined, elliptic, with a
narrow scale above the slender claw; *hypogynous glands* small, oblong
or quadrate, crimson. Tubes of staminal bundles long, slender, exserted.
*Capsules* cylindric, acute, ·6 in. long. *Seeds* ·3 in. long, with an obtuse
obovate unilateral wing. Hook. fil. Br. Ind. I, 258; Kurz Fl.
Burm. I, 84. *Tridesmis formosa,* Korth. Verh. Nat. Gesch. Bot. 179, t.
37; Miq. Fl. Ind. Bat. I, pt. ii, p. 517. *T. ochnoides,* Spach Suit. Buff.
V, 359. *Elodea formosa,* Jack in Hook. Journ. Bot. I, 374.

In all the Provinces. Distrib. Siam, Philippines, Malayan Archipelago.

4. CRATOXYLON MAINGAYI, Dyer in Hook. fil. Fl. Br. Ind. I, 258. A
tree 30 feet high; all parts glabrous; young branches with pale brown,
ridged bark. *Leaves* coriaceous, elliptic, acuminate at apex and base;
both surfaces shining, the lower paler; main nerves obscure, about 6
pairs; length 2 to 3·5 in., breadth 1 to 1·5 in., petiole ·25 in. *Cymes*
axillary, or from above the scars of fallen leaves, few-flowered. *Flowers*
about ·5 in. long, their pedicels ·25 in. *Sepals* oblong, unequal. *Petals*
elliptic, clawed, veined, the basal scale acute. *Capsule* ·5 in. long,
narrowly cylindric. *Seeds* ·25 in. long, with oblong unilateral wing.

Penang; Maingay.

Apparently an uncommon tree, since only Maingay has as yet
collected it.

ORDER XIV. GUTTIFERÆ.

Trees or shrubs with yellow or greenish juice. *Leaves* opposite,
coriaceous or membranous, rarely whorled or stipulate. Flowers axillary or terminal, solitary, fascicled, subracemose or panicled, white, yellow
or red, regular, diœcious, polygamous or hermaphrodite. *Sepals* 2-6,
imbricate or in decussate pairs. *Petals* 2-6 (rarely more, or 0), usually
much imbricated or contorted. MALE FL.; *Stamens* usually indefinite,
hypogynous; filaments free or variously connate, monadelphous or in as
many bundles as there are petals; anthers various. FEMALE FL.;
*Staminodes* various. *Ovary* 1-2-∞ -celled; style slender, short or 0; stigmas

as many as the cells, free or connate, sometimes peltate; ovules 1-2 or ∞, axile, or erect from the base of the cell. *Fruit* usually baccate and indehiscent. *Seeds* large, albumen 0; embryo consisting of a large radicle (*tigellus*) with small or obsolete cotyledons, or of thick free or consolidated cotyledons with a very short inferior radicle. A large tropical family, common in Asia and America, rare in Africa, of 24 genera and 320 species.

Tribe I. *Garcineæ.* Cells of ovary 1-ovuled; stigma sessile, or sub-sessile, peltate, entire or with radiating lobes. Fruit baccate, indehiscent; embryo with cotyledons minute or undistinguishable.

    Calyx of 4 or 5 sepals     ...     ...    1. *Garcinia.*

Tribe II. *Calophylleae.* Ovary with 1 to 4 erect ovules; style 1, slender; stigma peltate, 4-fid. Fruit fleshy, usually indehiscent. Embryo with 2 distinct cotyledons.

    Ovary 1-celled.

      Ovules solitary, style 1, stigma peltate  ...   2. *Calophyllum.*

      Ovules 4; style 1, 4-fid. with a stigma
        above each segment     ...     ...   3. *Kayea.*

    Ovary 2-celled, 4-ovuled.

      Style 1, stigma peltate     ...     ...   4. *Mesua.*

### 1. GARCINIA, Linn.

Trees, usually with yellow juice. *Leaves* evergreen, coriaceous, very rarely stipulate. *Flowers* solitary, fascicled, or panicled; axillary or terminal; polygamous. *Sepals* 4-5, decussate. *Petals* 4-5, imbricate. MALE FL.; *Stamens* ∞, free, or collected into a ring, or an entire globose or conical 4-5-lobed mass, often surrounding a rudimentary ovary; anthers sessile, or on short thick filaments, 2-rarely 4-celled, adnate or peltate, dehiscing by slits or pores, or circumsciss. FEMALE OR HERMAPHRODITE FL.; *Staminodes* 8-∞, free or connate. *Ovary* 2-12-celled; stigma sessile or sub-sessile, peltate, entire or lobed, smooth or tubercled; ovules solitary in each cell, attached to the inner angle of the cell. *Berry* with a coriaceous rind. *Seeds* with a pulpy aril. Distrib. Tropical Asia, Africa, and Polynesia; species about 100.

Sub-genus I. GARCINIA proper. Sepals 4, decussate; petals 4, imbricate.

Sect. 1. Stamens of male flower occupying both sides of 4 pedicelled fleshy processes; anthers sessile, 2-celled, the cells more or less orbicular

(4-celled in *cuspidata*) dehiscing longitudinally, the connective thick ; rudimentary stigma hemispheric, entire, discoid and flat, or concave (in *Merguensis* and *rostrata*) ; the style long, cylindric      *species* 1 to 7.

Sect. 2. Stamens of male flower in a 4-lobed mass surrounding the rudimentary ovary ; anthers 2-celled, oblong, dehiscing longitudinally.

     Rudimentary stigma 6 to 8-lobed      ...      *species* 8 to 10.

     Rudimentary stigma none      ...      ...      *species* 11 to 13.

Sect. 3. Stamens in a single un-lobed mass; anthers 2-celled.

     Stamens of male flower in a cone ; rudy. stigma large, convex ...      ...      ...      14. *Malaccensis.*

     Staminal receptacle stipitate ; anthers broadly oblong, curved, dehiscence longitudinal ; rudy. stigma broad, discoid ...      ...      15. *Maingayi.*

     Stamens in whorls in a thin annular fleshy receptacle      ...      ...      ...      16. *atro-viridis.*

     Stamens of male flower on a flat or convex sessile receptacle.

         Anthers bent round the apex of the connective (horse-shoe-shaped) dehiscing along the convexity ; rudy. stigma 0.

             Leaves white beneath      ...      17. *opaca.*

             „      green      „      ...      18. *calycina.*

         Anthers thick, cuneate, with flat broad tops, the connective large, cells dehiscing longitudinally.

             Rudy. stigma large, discoid      ...      19. *costata.*

             „         „      none      ...      20. *Griffithii.*

         Anthers with small connective, cells suborbicular, dehiscing longitudinally ; rudy. stigma 0      ...      ...      21. *Forbesii.*

         Anthers with the connective lengthened transversely and bearing at its extremities the small oval anther cells ; rudy. stigma 0      ...      ...      ...      22. *Bancana.*

Sect. 4. Anthers 4-celled ...      ...      ...      *species* 23 to 26.

Sect. 5. Anther-cells surrounding the central connective, often confluent, their dehiscence circumscissile ; rudy. stigma 0      ...      ...      *species* 27 to 30.

Sub-genus II.   XANTHOCHYMUS.

Sepals and petals 5, all imbricate      ...   *species 31 to 36.*
Sub-genus I.   GARCINIA proper, sepals 4, decussate; petals 4, im-
cate.

1.  GARCINIA EUGENIÆFOLIA, Wall. Cat. 4873.   A small tree; the
young branches thin, 4-angled, rather pale when dry.   *Leaves* sub-coriaceous, elliptic, tapering to each end, the apex with a short blunt tail;
upper surface shining; the lower dull, pale, opaque; nerves thin, spreading, less than ·1 in. apart, very indistinct on either surface; length 2 to
3·5 in., breadth ·9 to 1·35 in., petiole ·2 to ·25 in.   *Male flowers* ·2 in. in
diam., in axillary or terminal, minutely bracteate, 3-to 6-flowered fascicles; pedicels ·3 in. long.   *Sepals* 4, orbicular, the outer pair small,
the inner pair as large as the petals.   *Petals* 4, orbicular, thin with a
circular thickened coloured fleshy spot near the base; *Stamens* numerous, forming with the rudy. stigma a dense convex mass; *anthers* numerous, on both sides of 4 fleshy processes, orbicular-oblong, 2-celled,
the dehiscence vertical; rudy. stigma large, hemispheric, the style
cylindric.   *Female flower*; ·25 in. in diam., in pedunculate 3-flowered
cymes, sometimes several from same axil, pedicles ·25 to ·35 in.   *Sepals* 4;
the outer pair small, fleshy, ovate-orbicular; the inner pair thin, nearly
as large as the petals, slightly keeled at the base; *petals* as in the male;
Staminodes and disc absent.   *Stigma* large, hemispheric, sub-papillose,
entire, covering nearly the whole of the ovary.   *Fruit* in fascicles of 2 to
4, globular, ·75 in. in diam., smooth, brown, crowned by the papillose
stigma; calyx not persistent. Hook. fil. Fl. Br. Ind. I, 268; Pierre Fl.
Forest. Coch-Chine, fasc. VI, p. vi, in part; *G. brevirostris*, Scheff. Obs.
Phyt. II, 41.

Penang; Wallich, Curtis, No. 669.   Tenasserim and Andamans;
Helfer, 855.   Perak; King's Collector Nos. 8604, 5954, Wray No. 461.

There are two specimens in the Calcutta Herbarium of *G. brevirostris,* Scheffer, named by the author himself; and they agree absolutely
with Wallich's No. 4873. This species is quite distinct from Griffith's
No. 858 (Kew Dist.) from Malacca, which Pierre not only reduces here,
but of which he figures (tab. 90 E. F.) the flowers as the flowers of this.
This species does not appear to be a common one. Specimens of other
things appear to have been so much confounded with it, that I forbear
to quote more synonyms than *G. brevirostris.*

2.  GARCINIA MERGUENSIS, Wight Ill. 122, 1c. 116.   A tree 30 to 40
40 feet high; young branches thin, terete, dark brown when dry.
*Leaves* ovate-elliptic to lanceolate, bluntly caudate-acuminate, the base
cuneate; upper surface when dry shining, dark brown; the lower dull

pale brown, the midrib distinct on both ; nerves indistinct, thin, spreading, about ·075 in. apart ; length 3 to 3·5 in., breadth 1·1 to 1·4 in.,
petiole ·25 in. *Male flowers* ·15 in. in diam., in rather dense axillary
minutely bracteolate 3- to 6-flowered cymes longer than the petioles ;
pedicels ·2 in., buds globose ; *sepals* 4, fleshy, the outer pair small, ovate-
orbicular, sub-acute ; the inner pair orbicular, all concave ; petals 4,
orbicular, fleshy, concave, covering the stigma in bud ; *anthers* numerous
on both sides of 4 fleshy processes, sessile, oblong, dehiscing suturally ;
rudy. style long, cylindric, thick ; stigma discoid, smooth, flat. *Hermaphrodite flowers* ; *sepals* 4, the outer pair as in the male ovate-orbicular,
thin ; *petals* 4, orbicular-reniform, fleshy, not covering the stigma ;
*stamens* numerous on both sides of 4 triangular fleshy processes ;
*anthers* sessile, sub-orbicular, dehiscing vertically by the sutures ;
*stigma* sessile, very large, hemispheric, convex, smooth, covering the
anthers when young. *Female flowers* ; *sepals* 4, the outer pair much
smaller than the inner, all thin and concave ; *petals* 4, orbicular, about
the same size as the inner sepals, concave, thinly coriaceous, with a
thickened coloured patch at the base ; staminodes and disc 0 ; *stigma*
semi-hemispheric, almost covering the whole ovary. *Fruit* pedicelled,
globular, ·75 in. in diam., smooth, covered by the concave smooth
stigma. Hook. fil. Fl. Br. Ind. I, 267 ; Kurz Fl. Burm., I, 89 ; Pierre
Flora Forest. Coch-Chin. fasc. VI, p. vi, tab. 68, 69, 91, D.

Malacca ; Griffith, Maingay, No. 155, Kew Distrib. Perak ; Scortechini Nos. 244*a* and 812, King's Collector, No. 2660, Wray, 1075.
Penang ; Curtis, No. 900.

Maingay No. 155 is the type of Pierre's species *G. fulva*, but, in
spite of very careful dissection of many of the flowers of this most
puzzling plant, I cannot see my way to adopting that as a species
separable from *G. merguensis*, Wight.

3. GARCINIA ROSTRATA, Benth. and Hook fil. Gen. Plantar. I, 174.
A tree 30 to 40 feet high. Young branches terete, pale, slender. *Leaves*
thinly coriaceous, elliptic-oblong, with a rather short blunt acumen,
the base much narrowed ; both surfaces shining, the lower rather pale,
midrib prominent ; nerves very numerous, thin, sub-horizontal, rather
distinct when dry, especially on the lower surface ; length 2·5 to 3·5 in.,
breadth 1·25 to 1·75 in., petiole ·2 in. *Male flowers* ·15. in. in diam., in
slender, pedunculate, lax, often dichotomous, 3- to 9-flowered cymes which
are in fascicles of 2 to 4 in the axils of the leaves ; buds depressed-
globose, ·1 in. in diam. ; the pedicels ·25 to ·35 in., slender ; *sepals* 4,
orbicular, concave, the outer pair small, fleshy, the inner thin as large as
the petals ; *petals* 4, orbicular, concave, thin, each with a fleshy coloured
circular patch near its base ; *stamens* numerous ; *anthers* sessile, on both

sides of 4 thick fleshy processes; cells 2, orbicular-oblong seated on the apex of the thick connective, dehiscing along the convexity; stigma very large, discoid, smooth, depressed in the centre, covering the stamens. *Hermaphrodite flowers* in 3-flowered, axillary, solitary, sessile, bracteolate cymes; sepals as in the male; *petals* 4, orbicular-reniform, not covering the stigma, otherwise as in the male; *anthers* in 4 masses as in the male, 4-celled; *stigma* sub-sessile, covering the whole ovary, large, discoid, smooth, entire. *Fruit* solitary, or 2 or 3 from an axil, ovoid, sub-orbicular, ·5 to ·65 in. long and rather less in diam., smooth, crowned by the discoid sub-concave stigma. Pierre Fl. Forest. Coch-Chine, fasc. VI, p. v, tab. 91, B. *Discostigma rostratum*, Hassk. Cat. Pl. Hat. Bogor. 213. Hook. fil. Journ. Linn. Soc. XIV, 486.

Malacca; Griffith, No. 855, Maingay 156. Perak; Scortechini 1962, King's Collector Nos. 8486, 10762.   Distrib. Java.

This is readily distinguished by its small flowers on slender pedicels, and by its flattened buds; also by the sub-horizontal, close, rather distinct, venation of the leaves.

4.   GARCINIA CUSPIDATA, King, n. sp.   A tree 60 to 70 feet high; the young branches terete, dark-coloured. *Leaves* elliptic-ovate, shortly sub-spathulate, cuspidate, the base narrowed; upper surface shining, the lower dull; nerves sub-horizontal, distinct beneath when dry, ·1 in. apart, anastomosing with an intramarginal nerve; length 2·5 to 3·25 in., breadth 1·1 to 1·5 in., petiole ·3 in. *Male flowers* ·15 in. in diam., in shortly pedunculate, axillary, 6 to 9-flowered, spreading cymes; buds pyriform; pedicels slender, ·3 to ·6 long; *sepals* 4, equal, reflexed, orbicular, thin, concave; *petals* 4, reflexed, covering the stamens and stigma in bud, thin, orbicular, concave, a little larger than the sepals; *stamens* numerous, on both sides of 4 fleshy processes, filaments very short and thick; *anthers* with 4 globular cells, each dehiscing by a long vertical suture; *style* short, cylindric; stigma capitate, small, quite concealed by the staminal masses. *Female flowers* and *fruit* unknown.

Perak, at low elevations; King's Collector, No. 10865.

Collected only once by the late Mr. Kunstler. The leaves a good deal resemble those of *G. rostrata*, Hassk.; but the nerves are slightly more distinct, and the flowers have a different androecium, although externally they much resemble those of *G. rostrata*, Hassk.

5.   GARCINIA WRAYI, King, n. sp.   A small spreading tree; young branches very slender, terete, dirty yellow. *Leaves* thinly coriaceous, ovate or elliptic, the apex produced into a long sub-spathulate point, the base cuneate; upper surface shining, the lower dull, pale; nerves slender, sub-horizontal, ·05 in. apart, invisible on the upper and faint on the lower surface even when dry; length 2 to 2·5 in., of which

the apical tail is sometimes as much as ·75 in., breadth ·8 to 1·2 in. ; petiole ·2 in. *Male flowers* ·15 in. in diam., in axillary fascicles of 2 or 3 ; buds globose, pedicels ·15 in. ; perianth reflexed, *sepals* 4, the outer less than half as large as the inner pair, all orbicular, sub-coriaceous and concave ; *petals* 4, ovate-orbicular, blunt, with a thickened spot near the base, covering the stamens in bud ; *stamens* numerous, on both surfaces of 4 fleshy processes ; anthers sessile, globular-oblong, the connective rather thick, 2-celled, dehiscence vertical ; rudy. style cylindric, convex, smooth. *Male flowers* axillary, solitary, pedicels as in the male ; *sepals* 4, orbicular, thin, concave, about the same size as the petals and neither sepals nor petals quite covering the stigma in bud : *petals* 4, orbicular, concave, with a coloured thickened spot near the base ; disc lobed, shallow, fleshy ; staminodes none ; *ovary* cylindric ; *stigma* hemispheric, smooth, entire, enveloping the whole of the ovary, ultimately becoming discoid and slightly depressed in the middle.

Perak ; on Ulu Batang Padang and on Gunong Batu Patch, at elevations of 4,500 feet and upwards. Wray, Nos. 267, 362, 1527 ; Scortechini, No. 323b.

I have not seen ripe fruit of this, but (from the appearance of a young one) it is probably ovoid. In its leaves, this species rather resembles *G. merguensis* and *rostrata.*

6. GARCINIA DIVERSIFOLIA, King, n. sp. A tree 40 to 60 feet high ; young branches rather thick, 4-angled, yellowish. *Leaves* lanceolate and sub-acute, to elliptic and shortly and bluntly cuspidate, the base always cuneate ; upper surface shining, the lower dull, slightly pale when dry ; midrib prominent on both surfaces ; the nerves numerous, about ·05 in. apart. spreading, straight, visible on the upper, invisible on the lower surface ; length of the lanceolate form 3·5 to 4·5 in., breadth 1·5 to 1·75 in.; length of the elliptic form 2·5 to 4·25 in. ; breadth 1·4 to 2·75 in., petiole ·2 to ·4 in. *Male flowers* ·75 in. in diam., in 3- to 6-flowered, bracteolate, axillary cymes ; buds globular-ovoid ; pedicels unequal, from ·2 to ·4 in.; bracteoles orbicular, fleshy ; *sepals* 4, orbicular, fleshy, concave, the outer pair small, united by their bases and sometimes irregularly denticulate, inner pair as large as the petals ; *petals* 4, ovate-orbicular to orbicular, fleshy, concave ; *stamens* very numerous, occupying both sides of 4 fleshy processes ; anthers sessile, 2-celled, the connective thick, bifid, bearing at its upper part the two sub-orbicular suturally-dehiscent cells ; rudy. pistil with cylindric style thickened upwards ; the stigma large, hemispheric, entire, sub-papillose. *Female flowers* in cymes like the males but fewer-flowered and often terminal, perianth as in the male ; staminodes apparently none ; *ovary* depressed-globose, smooth ; *stigma* sessile, discoid, entire, its surface

minutely lobulose.   *Fruit* (unripe) sub-globular, ·6 in. in diam., crowned by the stigma.

Perak; at elevations of 3,000 to 3,500 feet, King's Collector, No. 6220, Wray, No. 1209.

7.  GARCINIA CADELLIANA, King, n. sp.   A tree about 30 feet high ; the young branches, slender, terete, brownish yellow.  *Leaves* thinly coriaceous, elliptic to elliptic-oblong, sub-acute, the base very cuneate ; both surfaces shining ; main nerves 10 to 12 pairs, inter-arching very near the margin, thin, slightly prominent on both surfaces when dry ; length 3·5 to 5·5 in., breadth 1·6 to 2·5 in.; petiole ·25 to ·35 in., stout.  *Male flowers* ·25 in. in diam., in dense 3- to 8-flowered axillary fascicles, buds globose. pedicels ·1 in., bracteolate at the base, stout ; *sepals* 4, slightly unequal, small, orbicular, fleshy, the edges thin ; *petals* 4, obovate-orbicular, fleshy, concave ; *stamens* numerous, on both sides but especially on the inner sides of 4 fleshy processes opposite the petals; anthers oblong, sessile, 2-celled, the dehiscence longitudinal ; style cylindric, as long as the staminal bundles ; stigma large, hemispheric, papillose, entire.  *Female flowers* and fruit unknown.

Andamans ; King's Collector, No 371.

8  GARCINIA SPECIOSA, Wall. Pl. As. Rar. iii. t. 258.  A tree 40 to 60 feet high ; young branches slightly 4-angled, yellowish when dry.  *Leaves* thinly coriaceous, oblong or elliptic-oblong, sometimes oblanceolate, shortly acuminate, the base cuneate ; both surfaces shining, the midrib, main and intermediate nerves all rather prominent; length, 5 to 8 in., breadth 2 to 3·75 in.; petiole ·5 to ·6 in.  *Male flowers* 1·5 to 2 in. in diam., terminal in fascicles of 4 or 5, or solitary ; peduncles longer than the petioles.  *Sepals* 4, fleshy, concave, slightly unequal, 1 pair ovate and 1 pair reniform.  *Petals* 4, yellow, larger than the sepals, rotund, slightly clawed.  *Stamens* numerous, in 4 short, thick, diverging, oval masses confluent at the base ; filaments short; anthers oblong with longitudinal dehiscence.  *Style* short, thick, columnar ; rudy. stigma large, convex, with 6 shallow, broad, blunt lobes.  *Female flowers* solitary, terminal, on short thick pedicels ; perianth larger than in the male ; ovary sub-globular, the stigma large, convex, the margin 6- to 8-lobed.  *Flower* unknown.  Unripe *fruit* ovoid, sub-globose, apiculate, the hardened stigma and the thickened sepals persistent.  Wall. Cat. 4855, 4852 E.  *Garcinia affinis*, Wall. Cat. 4854; Choisy Guttif., Ind. 23 ; Planch. and Triana Mem. Guttif., 171 ; Kurz For. Fl. Burma. I, 88 ; Pierre Fl. Forest. Coch.-Chine, fasc. IV, p. xiv, tab. 79, excl. figs. H. and I.

Andaman Islands ; Kurz.  Distrib. Tenasserim.

This arboreous species is no doubt very closely allied to the shrubby

94

*G. Kurzii*, Pierre. And it appears probable that, although its head-quarters are Burmah and Sylhet, *G. speciosa* does occur on the Anda-mans. M. Pierre relies, as a diagnostic mark of his *G. Kurzii*, on its having solitary male flowers, whereas those of *G. speciosa* are fascicu-late. But in Calcutta Herbarium specimens of the same set which M. Pierre would refer to his *G. Kurzii*, the flowers are sometimes solitary and sometimes clustered. Another mark which M. Pierre relies on is that the peduncles of the flowers of *G. speciosa* are described by Wal-lich as two or three times as long as the leaf petioles, whereas in *G. Kurzii*, the peduncles are shorter than the petioles. The female flower of *G. speciosa* is described in the Flora of British India as unknown, and M. Pierre says the same of the female flower of his *G. Kurzii*. Wallich's specimens of *G. affinis* from Sylhet have advanced female flowers, and it is from these that I have described the female flower (figured by Pierre, tab. 79, fig. G.) ; for *affinis* appears to me in no way distinct from *speciosa*. Wallich was no doubt mislead by the size of the rudimentary stigma in the male flowers of *speciosa* into considering these as hermaphrodite, and it is probable that he never saw true female flowers. This view is supported by the fact that he does not describe either ovary or fruit. Pierre (l. c. t. 79, figs. H. and I.) gives drawings of what he believes to be the male and female flowers of *G. speciosa*. But in his text (fasc. VI, p. xiv), he states that the flowers thus figured were, in the specimen from which he took them, unattached to any leaf-twig and were mixed with flowers of other species. They are therefore altogether doubtful even for M. Pierre.

9. GARCINIA KURZII, Pierre Flor. Forest. Coch.-Chine, fasc. VI, p. xiv, t. 78 B, A shrub with the branchlets and leaves of *G. speciosa*, but the leaves less acuminate and with longer petioles. Flowers as in *speciosa*, but the stamens less numerous and the rudimentary stigma discoid and flat. Ripe fruit unknown.

Andaman Islands ; Kurz, King's Collector.

This differs from *G. speciosa* chiefly in being a shrub, and in its rudimentary stigma being flat and discoid, instead of convex. Both this and *speciosa* differ but little from *G. cornea*, Roxb., a species indi-genous to Amboina.

10. GARCINIA HOMBRONIANA, Pierre Fl. Forest. Cochin-Chine, fasc. VI, p. xii, t. 79, figs. D. E. F. J. A tree, with rather stout, 4-angled branches, yellowish when dry. *Leaves* elliptic, to oblong-elliptic, slight-ly inequilateral, sub-acute or very shortly and abruptly blunt-acumiuate ; the base cuneate, slightly unequal ; upper surface slightly glossy, the under rather dull ; nerves numerous, ascending, not prominent on either surface ; the midrib bold on both ; length 3·5 to 5 in., breadth 1 to 2·75

in., petiole ·5 in. *Male flowers* about 1 in. in diam., terminal, in fascicles of 3 to 6, pedicels ·2 to ·4 in. *Sepal* thinly coriaceous, concave, the outer pair orbicular; the inner ovate-oblong, blunt. *Petals* ovate-orbicular, twice as long as the sepals. *Stamens* numerous; the filaments united in a fleshy, slightly 4-lobed, annulus on which the broad, oblong, vertically dehiscing anthers are inserted; rudimentary stigma slightly protruding above the mass of stamens, flat, 8-lobed. *Female flower* terminal, solitary, with sepals and petals like the male; staminodes absent. *Ovary* globose; the stigma large, convex, recurved at the edge when young, when adult with 8 shallow crenations. *Fruit* sub-globular, not · mammillate, about 1 in. in diam.; the pericarp rather thin, sub-crustaceous; sepals persistent. *Seeds* about 6, oblong, with soft juicy arillus.

Malacca; Griffith, No. 857 (Kew Dist.), Perak; Scortechini (1 specimen). Nicobar Islands; Kurz, Jelinek.

This species, which has been established by M. Pierre, comes (as his own description and figures show) very near to *G. cornea*, Linn. It differs chiefly from *cornea* by its broader leaves, stouter branchlets and 8-lobed stigma. Curtis's Penang specimen No. 690 probably belongs to this species.

11. GARCINIA MANGOSTANA, Linn. A glabrous tree 20 to 30 feet high; young branches cylindric, slightly grooved, the bark smooth, green. *Leaves* thickly coriaceous; shining on both surfaces, elliptic-oblong, acute or shortly acuminate, the base cuneate; nerves sub-horizontal, numerous, interarching with a double intra-marginal nerve, rather prominent beneath when dry; length 6 to 10 in., breadth 2·5 to 4·25 in., petiole ·75 to 1 in. *Male flowers* 1·5 in. in diam., in terminal fascicles of 3 to 9; pedicels ·5 to ·75 in., with several orbicular, concave, scarious bracts. *Sepals* 4, unequal, coriaceous, rotund, concave. *Petals* 4, large than the sepals, ovate, fleshy, yellowish tinged with greenish red. *Stamens* indefinite, in a 4-lobed mass; filaments short; anthers oblong, ovate, recurved, dehiscence longitudinal. *Pistil* 0. *Disc* fleshy, as long as the stamens, its apex conical. *Hermaphrodite flowers* 2 in. in diam., solitary or in pairs at the apices of the young branches, and usually on different trees from the male flowers; pedicel ·5 in. long, stout, woody. *Calyx* and *corolla* as in the male, but larger. *Stamens* many; filaments slender, connate below; anthers irregular and mostly abortive. *Ovary* globular, 4- to 8-celled; stigma sessile, 8-rayed; ovules solitary. *Fruit* as large as a small orange, smooth, dark purplish brown; pericarp thick; seeds oblong, flattened, with large white juicy arillus. Bl. Bijdr. 213; DC. Prod. i, 560; Roxb. Fl. Ind. ii, 618; Bot. Mag. t. 4847; Choisy Guttif. Ind. 33; Planch. and Triana Mem. Guttif. 170; Miq. Fl. Ind. Bat. I, pt. ii, p. 506; Hook. fil. Fl. Br. Ind. i, 26); Kurz For. Fl.

Burm. I, 87; Lanessan Mom. Garcin. 15; Pierre Flor. Forest. Cochin-Chine t. 54.

Wild and cultivated in the Malayan Peninsula and Archipelago; cultivated also in Burma, Ceylon and a few places in the S. of India.

12. GARCINIA MICROSTIGMA, Kurz, Journ. Bot. 1875, p. 324; For. Flora Burmah, I, 91. A shrub 4 to 6 feet high; young branches obscurely 4-angled, the bark dark-coloured. *Leaves* elliptic to elliptic-oblong or lanceolate, sub-acute, the base cuneate; rather dull on both surfaces when dry, the midrib distinct beneath; main nerves 7 to 8 pairs, thin, interarching ·1 to ·2 in. from the margin; length 3 to 4 in., breadth 1·5 to 2·3 in., petiole ·5 to ·75 in. *Male flowers* ·3 in. in diam., in 2 to 3-flowered, few bracteoled, axillary cymes; buds globose; pedicels ·2 to ·25 in. long; *sepals* 4, the outer pair ovate-acute, fleshy, keeled, the edges thin, longer than the inner obovate-orbicular, very concave, thinner pair; *petals* 4, obovate-orbicular, fleshy, concave, about the same size as the inner sepals and barely covering the stamens; *stamens* about 20, on a single convex receptacle, filaments short; anthers red, broadly ovate, 2-celled, the dehiscence longitudinal; rudy. stigma 0. *Female flowers* (fide Kurz) on shorter pedicels than the male and probably solitary, terminal. *Fruit* globose, 1·5 to 2 in. in diam. the pericarp smooth, thin, red, the sepals persistent at its base, and its apex bearing the very minute discoid sessile entire stigma; *seeds* 2 or more. Pierre Fl. Forest. Coch-Chine, fasc. VI, p. xix.

South Andaman; Kurz.

13. GARCINIA PENANGIANA, Pierre Fl. Forest. Cochin-Chine, fasc. VI. p. xxxvii, No. 46a. A slender tree 20 to 30 feet high; the young branches glossy, pale brown when dry, slightly 4-angled. *Leaves* oblong-lanceolate, shortly and rather bluntly acuminate, the base cuneate; upper surface shining, the lower slightly dull and paler, both, (but especially the lower) with a reddish tint when dry; the midrib stout; nerves close, straight, sub-horizontal, faintly visible; length 4·5 to 7 in., breadth 1·5 to 2·5, or even 3 in.; petiole ·5 in. or less. *Male flowers* 1 in. in diam., in terminal fascicles of 3 to 6, pedicels about ·5 in. *Sepals* 4; the outer pair rotund, fleshy, very concave; the inner pair larger, thinner, elliptic, obtuse. *Petals* 4, rather longer than the inner sepals, oblong, blunt, creamy-white. *Stamens* indefinite, the filaments united in a slightly 4-lobed short fleshy mass; anthers short, broad, with longitudinal dehiscence; pistil 0. *Female flowers* terminal, solitary, larger than the male and on shorter stouter peduncles. *Style* short, thick; *ovary* globular; the stigma large, convex, hemispherical, corrugated and deeply 4-lobed; *stamens* none. *Ripe fruit* globular, more than 1·75 in. in diam., crowned by the persistent stigma, the thickened

sepals persistent at its base; pericarp thin, crustaceous.  *Seeds* few, ovate.
*G. cornea*, Wall. Cat. 4852 D.; Hook. fil. Fl. Br. Ind. I, 260 (in part).
*G. fascicularis*, Wall. Cat. 4853, Pierre l. c., p. xvi.

Penang; Porter (Wallich's Collector), Curtis. Perak; King's Collector, Scortechini.

This plant, first distinguished as a species by M. Pierre, seems to be rather common in Penang and Perak. Ripe fruits are as yet unknown; those in Mr. Kunstler's specimens No. 3583 (noted by him as unripe) measure 1·25 in in diam. A fruit on one of Scortechini's specimens measures half an inch more. Mr. Kunstler notes the tree as occurring at elevations of 300 up to 3,000 feet. The foregoing description of the flower does not quite agree with that of M. Pierre which was drawn up from specimens without female flowers and with buds only of the male flowers.

14. GARCINIA MALACCENSIS, Hook. fil. Fl. Br. Ind. I, 251. A tree; the branchlets rather stout, 4-angled. *Leaves* brown when dry, elliptic, shortly and abruptly acuminate, the base much narrowed, shining above, the lower surface rather dull; midrib bold, prominent on both surfaces; nerves numerous, faint, sub-horizontal, connected by oblique secondary nerves; length 4 to 8 in., breadth 1·5 to 2·5 in.; petiole ·4 to ·6 in., channelled. *Male flowers* 1 in. in diam., in terminal fascicles of 4 to 6, pedicels ·35 to ·65 in. long. *Sepals* orbicular, concave, fleshy. *Petals* twice as long as the sepals, dull red, broadly ovate, shortly clawed. *Stamens* very numerous, densely imbricated in a sub-cylindric or conical truncate mass formed of the fleshy conjoined filaments; *anthers* adnate, broadly ovoid, 2-celled, the connective broad; stigma large, convex. *Ovary* abortive. *Female flowers* 1·5 to 2 in. in diam. terminal, solitary, red. *Staminodes* few or 0. *Ovary* globose, 8-celled; stigma sessile, large, convex, enveloping half the ovary, much corrugated and deeply 8-lobed. *Fruit* unknown. Pierre Flore Forest. Coch.-Chine, fasc. VI, p. xi, t. 78, fig. D.

Malacca; Maingay (Kew Distrib. No. 149). Of this I have seen only Maingay's specimens. In its leaves, in the colour of its flowers, and in its 8-lobed stigma, this resembles *G. mangostana*.

15. GARCINIA MAINGAYI, Hook. fil. Fl. Br. Ind. I, 267. A tree 40 to 60 feet high; young branches thick, 4-angled and dark-coloured when dry. *Leaves* oblong-elliptic, obtuse with short blunt apiculus, the base narrowed; both surfaces shining, the lower pale brown when dry; nerves 9 to 13 pairs, bold, spreading, prominent beneath as is the midrib; length 4·5 to 7 in., breadth 2·25 to 3·25 in., petiole ·75 in. *Male flowers* 1 to 1·25 in. in diam., waxy white, in terminal or axillary, 3 to 6-flowered, shortly peduncled umbels; pedicels ·25 to ·5 in. long. *Sepals*

4, orbicular, fleshy, concave, the outer pair rather smaller than the inner. *Petals* 4, larger than the sepals, ovate-orbicular, fleshy, concave. *Stamens* very numerous, forming with the rudimentary pistil a dense globular mass; the filaments slender, in several rows from a stipitate fleshy receptacle; anthers oblong, curved, 2-celled, with extrorse longitudinal dehiscence; rudy. pistil cylindric, with a broad discoid stigma. *Female flowers* solitary, terminal, sub-sessile; *sepals* and *petals* as in the male but smaller; *staminodes* few, slender; *ovary* globose 4- to 6-celled; the stigma large, convex papillose, entire or very slightly 4-lobed. *Fruit* globular, 2 to 2·5 in. in diam. when quite ripe; crowned by the large, flat, discoid, papillose, slightly 4-lobed stigma. Pierre Flore Coch.-Chine, fasc. VI, p. xvii.

Perak; common. Malacca; Maingay (Kew Dist. 160 and 161). Pangkore; Curtis No. 1610.

In Scortechini's field-note on this species, the young branches are described as terete; but, in the dried state, they are distinctly 4-angled.

Var. *stylosa*; stigma on a thick style ·3 in. long.

Perak. King's Collector, No. 5359.

Only specimens with immature fruit are known, but these differ from the typical form in no respect except the stout style.

16. GARCINIA ATROVIRIDIS, Griff. MSS. A graceful tree 40 to 60 feet high; the young branches rather thick, sub-terete, yellowish-grey when dry. *Leaves* coriaceous, both surfaces shining; narrowly oblong, very shortly but sharply acuminate, the base cuneate; nerves numerous, spreading, straight, indistinct when fresh, but rather distinct when dry, anastomosing ·05 to ·1 in. from the edge with a fine intra-marginal nerve; length 4·5 to 8 in., breadth 1·25 to 2 in., petiole, ·6 to ·75 in. *Male flowers* 1·25 in. in diam., in terminal clusters of few-flowered cymes, pedicels unequal, from ·25 in. to ·75 in., long. *Sepals* 4, fleshy, concave; the outer pair orbicular or transversely oblong; the inner pair broadly oblong or orbicular, fleshy with thin edges, larger than the outer pair, streaked with red inside. *Petals* 4, orbicular-obovate, concave, fleshy, larger than the sepals, red. *Stamens* very numerous, forming with the large convex rudimentary stigma a globose mass; filaments slender, nearly as long as the anthers, inserted in whorls on a thin annular fleshy receptacle; *anthers* narrowly oblong, 2-celled, extrorse, the dehiscence longitudinal. Rudy. style cylindric. *Female flowers* terminal, solitary, rarely geminate; *sepals* and *petals* as in the male, but the petals smaller; *staminodes* small, attached to a thin fleshy wavy annulus which surrounds the ribbed, sub-cylindric, 12- to 16-celled ovary. *Stigma* thick, fleshy, very convex, pileate, deep red, the edges undulate. *Fruit* ( fide

Scortechini) globular, 3 in. in diam. yellowish-green, crowned by the sessile, concave, ribbed stigma.    Hook. fil. Fl. Br. Ind. I, 266; Pierre Fl. Coch.-Chine, fasc. VI, p, xxiv, tab. 80, fig. C.

Malacca ; Maingay (Kew Dist. No. 154.) Perak ; Scortechini, Wray. Wellesley Province, King's Collector.    Penang, Curtis, No. 855.

According to Mr. Curtis, the fruit is eaten by the Malays in curries, and the tree is a very handsome one with pendulous branches.    I have not seen fully ripe fruit, and the above description of it is taken from Fr. Scortechini's field notes.

17. GARCINIA OPACA, King.    A tree 40 to 60 feet high; the branchlets when dry dull dirty yellow, striate.    *Leaves* oblong, narrowed at each end, the apex sometimes shortly acuminate, the base cuneate ; upper surface slightly shining, lower surface opaque, whitish ; the rather numerous nearly horizontal nerves thin, little visible on either surface, the midrib bold and prominent on both ; length 4 to 5 in., breadth 1·75 to 2·25 in., petiole ·6 in.    *Male flowers* ·75 in. in diam., in shortly pedicelled, 2- to 3-flowered, ebracteolate, terminal or axillary cymes; pedicels ·25 in., annulated ; *sepals* 4, obovate, concave, thin, veined ; *petals* 4, similar to the sepals but a little larger ; *stamens* numerous, on a single, convex, fleshy receptacle ; *anthers* sessile, depressed-globular, with circumscissile dehiscence ; rudy. *stigma* 0.    *Female flowers* solitary, terminal ; *sepals* more coriaceous than those of the male flower ; staminodes 0 ; *ovary* cylindric ; stigma convex, smooth, the edge irregularly subcrenate.    *Fruit* solitary, terminal, ovate-globose, slightly mammillate, crowned by the broad flat stigma which has 4 broad shallow rounded lobes ; the sepals rounded, cartilaginous, persistent; pericarp brown when dry, thin, crustaceous.    Seeds several, ovoid, flattened on one side. *G. cornea*, Wall. Cat. 4852 E.

Perak ; King's Collector, Scortechini.

Distinguished by its opaque leaves whitish beneath and with faint sub-horizontal nerves.    In fruit this resembles *G. Penangiana ;* but it has a very different stigma.

18. GARCINIA CALYCINA, Kurz, Journ. Bot. 1875, p. 324.    A shrub 15 feet high ; young branches slender, slightly angled, pale brown when dry. *Leaves* thinly coriaceous, elliptic-oblong to elliptic, abruptly and shortly caudate-acuminate or sub-acute, the base cuneate ; upper surface shining, the lower rather dull and pale ; main nerves 7 or 8 pairs forming bold intra-marginal arches, the intermediate nerves very numerous, all slightly prominent beneath ; length 3 to 5 in., breadth 1·25 to 2 in., petiole ·3 to ·5 in.    *Male flowers* ·15 in. in diam., axillary, solitary or in 2- to 3-flowered fascicles ; buds globular, pedicels ·15 in. long.    *Sepals* and *petals* each 4, equal, orbicular, concave, the petals veined ; *stamens*

under 20, in a single convex group, the filaments very short, the connective rather thick, the elongate 2-celled anthers bent like a horse shoe over the apex of the connective and dehiscing along the convexity ; rudy. stigma 0. *Female flowers* larger than the male, subsessile, solitary, axillary ; *sepals* broadly ovate, the outer pair larger than the inner; staminodes about 12, distinct, short, square ; ovary hidden by the large hemispheric, lacunose, deeply 4-lobed stigma. *Fruit* (immature) ovoid-oblong, smooth, the sepals persistent at its base and the apex crowned by the sessile stigma. Pierre Flore Forest. Coch.-Chine, fasc. VI, p. xxxiii, tab. 87 D.

Nicobar Islands ; Kurz.

19. GARCINIA COSTATA, Hemsley MSS. in Herb. Kew. A tree 50 to 70 feet high ; young branches pale, flattened. *Leaves* thinly coriaceous, elliptic, acute, the base cuneate; both surfaces rather dull, the lower paler; nerves bold, spreading, 13 to 18 pairs, very distinct on the lower surface when dry ; length 6 to 14 in., breadth 3·5 to 6 in. ; petiole 1 to 1·5 in., stout. *Male flowers* 1 to 1·25 in. in diam., in shortly peduncled, 3- to 5-flowered, terminal cymes ; pedicels ·25 to ·5 in. *Sepals* 4, equal, orbicular, fleshy, concave. *Petals* larger than the sepals, pale yellow with a reddish tinge, orbicular-ovate, fleshy, concave. *Stamens* numerous, forming with the discoid stigma an oblong 4-angled mass ; filaments short, thick, inserted on a fleshy receptacle ; anthers thick, cuneate with flattish tops, 2-celled ; the cells large, curved, with extrorse longitudinal dehiscence ; rudimentary stigma large, discoid. *Female flowers* solitary, terminal, on short thick pedicels ; *sepals* and *petals* as in the male ; staminodes about 12 ; *ovary* with many vertical grooves; *stigma* large, discoid, with radiating grooves corresponding to those of the ovary, the edge wavy. *Fruit* depressed-spheroidal, 3 in. in diam. by 2 in. high, with many deep vertical grooves, pale rose-coloured to crimson.

Perak ; on Gunong Bubo at elevations of 2,500 to 3,000 feet, King's Collector ; Maxwell's hill, Wray.

A remarkably fine species, at once known by its large deeply grooved eatable fruit.

20. GARCINIA GRIFFITHII, T. Anders. in Hook. Fl. Ind. 1, 266. A tree 60 to 100 feet high, the young branches sub-tetragonous, yellowish-green. *Leaves* large, coriaceous, bullate, oval to ovate-elliptic, sub-acute or rather blunt ; the base slightly narrowed, sometimes slightly cordate ; both surfaces shining, the lower paler; midrib stout; nerves 16 to 24 pairs, bold, sub-horizontal ; length 9 to 16 in., breadth 4 to 8 in., petiole ·6 in. *Male flowers* ·75 in. in diam., in dense 3- to 10-flowered cymes from tubercles in the axils of leaves or of fallen leaves ; pedicels ·25 in. *Sepals* 4, equal, orbicular, fleshy, concave. *Petals* 4, oblong, blunt

101

fleshy, red. *Stamens* from 25 to 40, in a square flat-topped mass; anthers nearly sessile, broadly oblong, the connective wide; the cells 2, lateral, slightly curved, their dehiscence longitudinal; rudy. ovary 0. *Female flowers* in few-flowered axillary cymes; pedicels thick, ·2 in long. *Sepals* and *petals* as in the male; staminodes in 4 bundles of unequal length; ovary ribbed, 12-celled; style very short; stigma with many conical papillae, peltate, slightly depressed in the middle, its margins creuate. *Fruit* sub-globular, 2 or 3 in. in diam. when ripe, greenish yellow, crowned by the concave papillate stigma, very glutinous. Pierre Flore Coch.-Chine, fasc. VI, p. xxvi, tab. 80, fig. B.

Malacca; Griffith 861, Maingay 153 (Kew Distrib). Perak, common. Pangkore, Curtis 1609. Distrib. Sumatra; Forbes, No. 2994.

21. GARCINIA FORBESII, King, n. sp. A small tree, young branches sub-tetragonous, yellowish. *Leaves* thinly coriaceous, oblanceolate to ovate-lanceolate, shortly acuminate, the base cuneate; both surfaces slightly dull when dry, the lower slightly pale; nerves spreading, anastomosing with an intra-marginal nerve, ·15 in. apart, the intermediate rather bold, all distinct below when dry; length 3·5 to 5 in. breadth 1·5 to 2·5 in., petiole ·3 to ·4 in. *Male flowers* ·25 in. diam. in 3- or 4-flowered clusters from small axillary tubercles, buds sub-globose, pedicels ·1 to ·15 in. *Sepals* 4, equal, rather thin, pale-coloured, orbicular, concave. *Petals* 4, fleshy, orbicular, dark-coloured, concave. *Stamens* numerous, in a single convex mass, the connective small; anthers sessile, sub-orbicular, 2-celled with longitudinal dehiscence; rudy. ovary 0. *Female flowers* axillary, solitary, sessile; sepals 4, broadly ovate, blunt, fleshy, concave; petals 4, orbicular, fleshy, concave, red to orange; *stigma* sessile, convex, completely covering the ovary, entire, its surface with prominent glandular papillæ. *Fruit* (young) ovoid, crowned by the stigma.

Perak; Wray 3396.  Sumatra; Forbes Nos. 2936 and 3152.

22. GARCINIA BANCANA, Miq. Fl. Ind. Bat. Suppl., 494. A tree 60 to 80 feet high; young branches stout, nodular, not angled, black and shining when dry. *Leaves* coriaceous, large, broadly obovate-lanceolate; the apex rounded, often slightly and bluntly mucronate; much narrowed in the lower third into the stout winged petiole; upper surface shining, the numerous and very oblique nerves distinct; lower surface dull, opaque, pale brown, the nerves obsolete; midrib prominent in both; length 5 to 7 in., breadth 2 to 3 in., petiole ·75 to 1·25 in. *Male flowers* ·15 in. in diam., in crowded fascicles of 6 to 12, from short densely bracteolate tubercles in the axils of leaves or of fallen leaves; pedicels unequal, ·25 to ·5 in. long; bracteoles ovate, coloured, ·1 in. or less. *Sepals* 4, orbicular, concave, fleshy, the outer pair larger than the inner. *Petals* 4, ovate, blunt, fleshy concave. *Stamens* numerous, in a

102

convex sub-cylindric mass; the anthers sub-sessile, broad, with 2 small oval cells at the extremities of the transversely lengthened connective; rudimentary pistil 0. *Female flower* solitary ?, sub-sessile; staminodes 6 to 10, solitary or in two or three groups. *Ovary* sessile, globular, slightly grooved vertically ; stigma hemispheric, with 8 triangular rays. *Fruit* ovoid, 1·25 in. long, and 1·1 in. diam.; about 8-seeded. Miq. Ann. Mus. Lugd Bat. I, 408 ; Hook. Fl. Br. Ind. I. 263 ; Scheff. Obs. Phyt. pt. ii, 41 ; Pierre Flora Forest. Cochin-Chine, fasc. VI, pp. xxvi and xxxviii. *Garcinia Lamponga*, Miq. Fl. Ind. Bat. Suppl. 494 ; Ann. Mus. Lugd. Bat. I, 208 ; Pierre l. c. *G. Hookeri*, Pierre l. c. p. xxvii. *G. leucandra*, Pierre, l. c. xxvii.

Perak ; King's Collector, Scortechini. Malacca ; Maingay (No. 158, Kew Dist). Distrib. Banka, Sumatra.

I have examined the type specimens of Miquel's *G. bancana* and *Lamponga*, and I believe them to belong to one and the same species. One of his Sumatran specimens of *G. Lamponga* bears, however, besides leaves of the shape described above, some that are broadly elliptic. Pierre reduces to this two more of Miquel's Sumatran species, namely, *G. oxyedra* and *G. ? oxyphylla* (Fl. Ind. Bat. Suppl. 494, 495) ; but of these I have not seen Miquel's types.

23. GARCINIA COWA, Roxb. Fl. Ind. II, 622. A dioecious tree 30 to 60 feet high ; young branches slender, not angled, dark-coloured when dry. *Leaves* broadly lanceolate, acute at both ends, the apex sometimes acuminate, both surfaces rather dull when dry ; the nerves thin but rather distinct when dry, numerous, rather straight, oblique ; length 3·5 to 5 in., breadth 1 to 1·75 in., petiole ·3 to ·5 in. *Male flowers* ·4 in. in diam., axillary or terminal, in fascicles of 3 to 8 ; pedicels ·25 in. *Sepals* broadly ovate, fleshy, yellow. *Petals* twice as long as the sepals, obovate or oblong, blunt, yellow. *Stamens* numerous, on a convex fleshy receptacle, anthers 4-celled, stigma rudimentary. *Female flowers* ·8 in. in diam., terminal, in fascicles of 2 or 3, pedicellate like the males ; *ovary* sub-globose, 6- to 8-celled ; *stigma* sessile, flat, deeply divided into 6 or 8 papillose wedge-shaped rays ; *staminodes* in 4 clusters of 3 to 8, unequal. *Fruit* globular-depressed, not mammillate, with 4 to 8 vertical grooves, smooth, yellow, ·8 to 1·5 in. in diam.; pericarp thin ; *seeds* ·5 to ·75 in. long, oblong, with a soft arillus. DC. Prodr. i, 561 ; W. and A. Prodr. i, 101 ; Chois. Guttif. Ind. 34 ; Planch. and Triana Mem. Guttif. 186 ; Wall. Cat. 4863 ; Lanessan Mem. Garcin. 54 ; *G. Roxburghii*, Wight Ic. 104 ; Kurz For. Fl. Burm. I, 90. *Oxycarpus Gangetica*, Ham. in Mem. Wern. Soc. V, 344.

Andaman Islands ? Distrib. Assam and base of the Khasia Hills, Chittagong, Burmah ; in tropical forests.

103

This is very near *G. Kydiana* but differ in the points noted under that species.

24. GARCINIA KYDIANA, Roxb. Fl. Ind. II, 623. A diœceous tree, 25 to 40 feet high ; the branchlets dark-coloured when dry, not angled. *Leaves* thinly coriaceous, lanceolate, acuminate, the base acute, both surfaces shining ; nerves thin but distinct when dry, rather few for this genus ; length 3 to 5 in., breadth ·75 to 1·5 in., petiole ·35 to ·5 in. *Male flowers* ·75 in. in diam., in small axillary or terminal pedunculate umbels of 3 to 5, or solitary ; pedicels ·25 in. long ; peduncles of the umbels ·4 to ·6 in. *Sepals* 4, equal, ovate, obtuse, fleshy, yellow. *Petals* twice as large as the sepals, broadly ovate, blunt, pale yellow. *Anthers* numerous, inserted into the slightly 4-lobed fleshy mass of conjoined filaments, square, 5-celled (a cell at each angle) pistil 0. *Female flowers* axillary and terminal, solitary, sessile. *Sepals* and *petals* as in the male ; staminodes 4, small, 3- or 4-fid. *Ovary* globular, sessile, 6- to 8-lobed ; stigma sub-sessile, with 6 to 8 spreading glandular rays. *Fruit* 1 to 1·5 in. in diam., smooth, yellow, globular, depressed, with 6 to 8 deep vertical grooves near the apex, and with a nipple-like protuberance from the depressed apex on which is inserted the persistent stigma. *Seeds* 6 to 8, oblong, ·85 in. long ; the arillus soft, acid, juicy. Kurz For. Fl. Burm. I, 90, *in part ;* Pierre Fl. Forest. Coch.-Chine, fasc. VI, p. xxix. Lanessan Mem. Garcin. 59, *in part ; G. Roxburghii,* Wight Ic. 113 ; *G. Cowa* Roxb. Hook. fil. Fl. B. Ind. I, *in part.*

Andaman Islands.

Of the true Roxburghian *G. Kydiana,* the only specimens that I have seen are from the Andamans. The Burmese specimens referred to this species by Pierre and others belong mostly to *G. Cowa* as Roxburgh described and figured it. But the two species are very closely allied. The chief points that separate *Kydiana* from *Cowa* are its larger flowers, the arrangement of the males in distinct pedunculate umbels, the females always solitary and sessile ; and, in the fruit, the curious nipple rising from the depressed apex, and the restriction of the vertical grooves to the neighbourhood of the apex. In the Flora of Br. India the two are united under *G. Cowa.* Griffith's Nos. 865 and 867, referred to *Kydiana* by Pierre, belong in my opinion to *G. nigro-lineata,* Planch.

25. GARCINIA NIGRO-LINEATA, Planch. MSS. A tree 20 to 50 feet high ; young branches not angled, their bark rather dark. *Leaves* thinly coriaceous, lanceolate and acuminate, or ovate-lanceolate and shortly caudate-acuminate, the base acute ; both surfaces shining, the lower ferruginous in some stages ; midrib rather stout ; main nerves rather distinct when dry, about ·1 to ·15 in. apart, the intermediate nerves almost as prominent ; length 3 to 4·5 in., breadth 1 to 1·5 in., petiole

·2 to ·4 in. *Male flowers* ·25 in. in diam., in umbels of 3 to 8, on the apices of the branches, or from the axils of leaves or of fallen leaves; pedicels ·25 to ·5 in., slender. *Sepals* orbicular, fleshy, concave. *Petals* longer than the sepals, oblong, obtuse, concave. *Stamens* about 20, forming a tetragonal mass inserted on a convex receptacle, the filaments very short; anthers broad, cuneate with flat tops, 4-celled with vertical dehiscence, the connective thick; pistil 0. *Female flowers* apetalous, solitary, or in clusters of 2 to 5, axillary; *ovary* ovoid, 5- or 7-celled; stigma large, convex with a central smooth depression, bearing many black papillae, and obscurely 5- to 7-lobed; staminodes about 8 to 10, not branched, their heads flat. *Fruit* ovoid-globose, ·5 to ·75 in. in diam., orange-coloured, pulpy, with a thick fleshy apiculous crowned by the persistent stigma. Hook. fil. Fl. Br. Ind. I, 263. Pierre Flore Forest. Coch.-Chine, fasc. VI, p. xxix, (excl. t. 81. fig. F.) *G. parvifolia,* Miq. Ann. Mus. Lugd. Bat. I, 208. *Rhinostigma parvifolium,* Miq. Fl. Ind. Bat. Supp. 495.

In all the provinces; in tropical forests. Distrib. Sumatra.

This is one of the commonest species of the genus. The pulpy fruit is eaten by the aborigines. Griff. No. 854 and Maingay Nos. 152 and 162 are the specimens on which Planchon founded the species. Griff. Nos. 865 and 867 (referred by Planchon and Triana and also by Pierre to *G. Kydiana,* Roxb.) in my opinion fall here, as also does *G. umbellifera,* Wall. Cat. 4864; but Anderson reduces the latter to *G. Cowa,* Linn. Pierre's figure, (t. 81, fig. F.), which he names *S. nigro-lineata,* does not represent the flowers of the type specimens in the Calcutta Herbarium which bear the numbers which Pierre quotes. I fear therefore that there must have been some confusion in the distribution of the Griffithian collections.

It is quite possible that the description which I have given above may cover two species. The specimens with lanceolate-acuminate leaves have rather more erect and fainter nerves than those with ovate-lanceolate caudate-acuminate leaves. But, although I have dissected a large number of the male flowers of each, I cannot detect any tangible difference. Unfortunately I have been able to find very few female flowers. An examination of Miquel's type specimen of his *Rhinostigma parvifolium* leaves no doubt whatever that it is identical with Planchon's *G. nigro-lineata.*

I never find the petals reflexed; but Anderson, in Hook. fil. Fl. Br. Ind. (l. c.), describes them, and Pierre (l. c.), figures them, as reflexed from about the middle.

26. GARCINIA KUNSTLERI, King, n. sp. A shrub, 6 to 8 (rarely 15) feet high; the young branches dark-coloured, not-angled. *Leaves* mem-

branous, oblanceolate to narrowly ovate-lanceolale, shortly but sharply acuminate, the base much narrowed ; smooth on both surfaces, the lower rather pale ; nerves indistinct, spreading, 9 to 13 pairs, some of the intermediate almost as distinct as the primary ; length 3·5 to 5 in., breadth 1·15 to 1·8 in., petiole ·35 to ·6 in. *Male flowers* about ·15 in. in diam., in small axillary or terminal fascicles of 3 to 6. *Sepals* 4, orbicular, fleshy, concave, their edges thin. *Petals* 4, broadly ovate, blunt, fleshy, concave. *Stamens* about 15, forming a convex mass ; the anthers transversely flattened, 4-celled, the connective broad; rudimentary pistil 0. *Female flowers* solitary ; *sepals* 4, orbicular, membranous, veined. *Petals* 0. Staminodes 4, with filaments half as long as the ovary, and flat square heads. *Ovary* thick, cylindric, vertically grooved ; the stigma convex, with large prominent black-tipped conical papillæ, and with about 10 inconspicuous lobes. *Fruit* orange-yellow, depressed, sub-globose, nearly 1·5 in. in diam., smooth, the sepals persistent at its base for some time.

Perak ; at low elevations, common ; King's Collector, Scortechini, Wray.

This is allied to the Burmese *G. linoceroides*, T. Anders.; but has smaller flowers, fewer stamens and more acuminate leaves.

27. GARCINIA SCORTECHINII, King, n. sp. A tree 20 to 40 feet high ; branchlets yellowish, slightly angled. *Leaves* thinly coriaceous, ovate-elliptic, occasionally ovate-lanceolate, shortly and rather bluntly acuminate, the base acute ; both surfaces shining, the lower rather paler ; main nerves 5 or 6 pairs, spreading, anastomosing ·2 in. from the margin. very distinct on the under surface when dry, reticulations indistinct ; length 2·75 to 4·75 in., breadth 1·5 to 2·5 in., petiole ·25 in. *Male flowers* ·2 in. in diam., sessile or shortly pedicellate, in clusters of 3 to 6 from small axillary tubercles ; *sepals* 4, orbicular, concave, thin, veined ; *petals* 4, broadly ovate, fleshy, concave ; *stamens* varying from 10 to 20, inserted on a 4-angled receptacle ; *anthers* with circular peltate tops, the connective in the centre the cells circumferential, dehiscing along the edge; filaments slender, shorter than the anthers ; rudimentary pistil 0. *Female flowers* nearly ·5 in. in diam., axillary, solitary, sessile or shortly pedicellate. Ovary globose ; stigma sessile, with large lobules, obscurely 4-lobed. *Fruit* globular, ·6 to ·75 in. in diam. ; the pericarp thick, leathery ; seeds about 4.

Perak, common. Malacca ; Griffith (Kew Distrib. 859). Penang, Curtis, 1249.

This is not very different from *G. Choisyana*, Wall. to which indeed Pierre refers the Griffithian specimen 859. But Wallich's specimens of *G. Choisyana* have leaves of so much thinner texture that, on the strength of this character alone, the two must be kept distinct. This species is readily known by its boldly 5- to 7-nerved leaves and hard globular, small fruit.

28. GARCINIA UROPHYLLA, Scortechini MSS. A tree; the branchlets very slender, terete, yellowish. *Leaves* thinly coriaceous, ovate-lanceolate, caudate-acuminate, the base cuneate; both surfaces shining, the lower pale; nerves 4 to 5 pairs, ascending, distinct below when dry; length 2·25 to ·3 in., breadth ·7 to 1·2 in., petiole ·15. *Male flowers* ·2 in. in diam., solitary or in pairs, from small bracteolate axillary tubercles; buds globose, pedicels 1 in. or less; sepals 4, obtuse, sub-coriaceous, concave, subequal, ovate-orbicular; *petals* 4, orbicular, almost flat, very fleshy, much thickened near the base; *stamens* about 12, in a single group, with flat circular tops, the connective in the middle, and the anther round the edge dehiscing circumferentially; filaments thick, fleshy; rudy. stigma 0. *Female flowers* axillary, solitary, sub-sessile; *sepals and petals* as in the male; staminodes about 6, free; *ovary* cylindric; *stigma* convex, boldly lobulate and deeply 4-cleft. *Fruit* ovoid-orbicular, ·4 in. long by ·35 in. in diam., smooth, crowned by the sessile lobulate stigma.

Perak; Scortechini Nos. 32ᵃ, 723. Distrib. Sumatra; Beccari, No. 963.

In the size and shape of the leaves, this has a superficial resemblance to *G. rostrata*, *eugeniæfolia* and *merguensis;* but the nerves are only 4 or 5, while in these the nerves are numerous. Moreover the androecium of this is totally different.

29. GARCINIA UNIFLORA, King, n. sp. A small tree; the young branches rather stout, terete, of a dirty yellow when dry. *Leaves* ovate-oblong to elliptic-oblong, the apex abruptly shortly and sharply acuminate, the base cuneate; both surfaces dull, the lower pale and opaque; main nerves 12 to 20 pairs, thin, but rather prominent, the secondary nerves almost as distinct; length 5·5 to 7·5 in., breadth 2·5 to 4·25 in.; petiole ·75 to 1 in. thick, channelled. *Male flowers* ·75 in. in diam., solitary, sessile in the axils of fallen leaves, buds globular; *sepals* 4, sub-equal, obovate-orbicular, membranous, veined, concave; *petals* 4, orbicular, concave, fleshy, smaller than the sepals; *stamens* rather numerous, in an undivided globose mass; anthers sessile, peltate, or sub-globose with flat tops, dehiscing by a circular infra-marginal slit. *Female flowers* solitary and axillary like the males, and with a similar perianth; staminodes 0; stigma convex, deeply papillose; the ovary short, cylindric. *Fruit* unknown.

Perak; on Gunong Batu Putch, at elevations of 3000 to 4000 feet, King's Collector 8081, Scortechini 364ᵇ.

30. GARCINIA DUMOSA, King, n. sp. A shrub 3 or 4 feet high; young

branches 4-angled, yellow. *Leaves* thinly coriaceous, ovate-lanceolate to oblong-lanceolate, shortly caudate-acuminato, the base acute; both surfaces shining, the lower pale, opaque, nerves 7 to 9 pairs, bold and prominent beneath as are some of the secondary nerves; length 4·5 to 5·5 in., breadth 1·5 to 2 in., petiole ·25 in. *Male flowers* about ·2 in. in diam. in dense 6- to 10-flowered fascicles from small minutely bracteolate axillary tubercles; buds turbinate; pedicels slender, ·1 in. or less in length; *sepals* 4, membranous, orbicular, concave; the outer pair much smaller and thicker, keeled; *petals* 4, smaller than the sepals, fleshy, orbicular, concave; *stamens* from a small receptacle; filaments short; anthers broadly reniform, the connective expanded transversely, the cells sometimes confluent, bent round it, and dehiscing along the convexity; rudy. stigma 0. *Female flowers* on axillary tubercles like the males, but fewer-flowered; *sepals* as in the male but subequal; *petals* as in the male; *staminodes* about 10, distinct, the filaments broad, the pseud-anthers flat, ovate; *stigma* convex with 8 radiating ridges, its margin 8-angled; *ovary* thick, cylindric, nearly as wide as the stigma. *Fruit* (*fide* Kunstler) ovoid, pointed.

Perak; at low elevations, Wray No. 2162, King's Collector, No. 2531.

Subgenus II. XANTHOCHYMUS, Roxb. (Gen.). *Sepals* and *petals* 5, very rarely 4. *Filaments* connate in 5, rarely in 4, erect distant pedicelled spathulate bodies, antheriferous at the top, free portions very short, incurved; anthers small, didymous.

31. GARCINIA XANTHOCHYMUS, Hook. fil. Fl. Br. Ind. I, 269. A medium-sized tree; the branches glabrous, angled. *Leaves* glabrous, shining; narrowly oblong or oblong-lanceolate, acute, the base cuneate; nerves numerous, not prominent; length 8 to 15 in., breadth 1·75 to 3·75 in., petiole about ·75 in. *Male flowers* ·5 to ·75 in. in diam., in 4-10-flowered fascicles, axillary or from the axils of fallen leaves, greenish-white; pedicels stout, ·5 to 1 in. long. *Sepals* ·25 in. in diam., orbicular, unequal, fleshy, concave. *Petals* ·35 in., orbicular, spreading, thin. *Stamens* in 5 broad bundles of 3 to 5, alternating with 5 fleshy glands; anthers 2-celled. *Hermaph. flowers* like the male, the pedicels 2 or 3 times as long. *Ovary* ovoid, pointed, usually 4-celled; *stigma* with 5, spreading, oblong blunt lobes. *Ripe fruit* globose, pointed, 2·5 in. in diam., dark yellow. *Seeds* 1 to 4, oblong. Kurz For. Flora Burma i, 93; Pierre Flore Forest. Cochin-Chine, fasc. VI, p. iii. t. 21 A. *Xanthochymus pictorius,* Roxb. Corom. Pl. ii, 51, t. 196; Fl. Ind. ii, 633. *X. tinctorius,* DC. Prodr. i, 562; Chois. Guttif. Ind, 32; Planch. and Triana Mem. Guttif. 159; W. and A. Prodr. 102; Wall. Cat. 4837, except C.

Andamans, Penang. Distrib. Burmah and Chittagong, base of E. Himalaya and Assam, S. India up to 1500 feet.

Sheet C of No. 4837 of Wall. Cat. (said to have been collected in Penang) does not in my opinion belong to this species. Its leaves have too few nerves.

32. GARCINIA NERVOSA, Miq. Ann. Mus. Lugd. Bat. I, 208. A tree 40 to 80 feet high; young branches stout, compressed, 4-angled, 2 of the angles winged. *Leaves* large, glabrous, very coriaceous, oblong-oblanceolate or oblong-ovate, sub-acute or obtuse, slightly narrowed below to the rounded or minutely cordate base; upper surface shining; the lower dull, pale; main nerves bold, numerous, anastomosing ·1 in. within the margin with the bold intra-marginal nerve; secondary nerves and reticulations rather prominent; length 9 to 20 in., breadth 3·5 to 7 in., petiole 1·25 in. *Male flowers* unknown. *Female flowers* ·75 in. in diam., in axillary fascicles of 8 to 10; pedicels thickened upwards, 1 to 1·25 in. long, (longer in the fruit). *Sepals* 5, unequal, orbicular, much imbricate and very concave, very coriaceous, pubescent externally. *Petals* 5, much larger than the sepals, orbicular, concave, thin. Disc of 5 thick, fleshy, pitted glands with 5 minute staminodes between them each bearing 4-5 minute anthers. *Ovary* ovoid, narrowed into a distinct 5-rayed style, 5-celled. *Ripe fruit* ovoid or obovoid, yellow with red blotches, 2 in. long and 1·5 in. in diam., with a large eccentric mammilla crowned by the persistent 5-lobed stigma. *Seeds* about 2, elongate-ovoid. *G. Andersoni*, Hook. fil. Fl. Br. Ind. I, 270, 715; *Stalagmites ? nervosa*, Miq. Fl. Ind. Bat. Suppl. 496.

Perak; King's Collector 10491, Scortechini. Malacca; Maingay (Kew Distrib. 157). Distrib. Sumatra.

Var. *pubescens.* *Leaves* densely and minutely pubescent below, cordate and slightly unequal at the base, 15 to 24 in. long, the edges re-curved when dry; petiole triquetrous, very stout. Fruit bright yellow, 3 in. long, 2 in. in diam., the mammilla about ·75 in.

Perak; King's Collector, No. 3197.

This may be separable as a species when further material shall be forthcoming. Male flowers are unknown.

33. GARCINIA DULCIS, Kurz For. Flora Burmah I, 92. A tree 30 to 40 feet high; the young branches 4-angled, pale yellow. *Leaves* oblong to ovate-oblong, with an abrupt short sharp point, the base rounded or slightly narrowed; upper surface shining, the lower slightly pale and dull when dry, the midrib rather prominent on both; main nerves about 10 pairs, interarching near the margin, not much more prominent than the intermediate nerves; length 5 to 10 in., breadth 1·75 to 4·5 in.; petiole ·4 to ·6 in., stout. *Flowers* globular, hardly expanding, about ·25 in. in diam., male and hermaphrodite mixed in dense many-flowered fascicles from small tubercles in the axils of the leaves or of the fallen

leaves; pedicels ·25 to ·35 in., *sepals* 4 to 6, usually 5, orbicular, unequal, fleshy, concave, the 3 outer smaller. *Petals* usually 5, larger than
the sepals, orbicular, fleshy, concave; *Stamens* about 40, in 5, pedicelled,
fan-shaped groups; filaments short, thick; anthers sub-globular,
2-celled, with sutural dehiscence; disc in the male depressed, truncate,
corrugated, fleshy; in the female with 5 lobes which alternate with the
staminal groups. *Stigma* in the male absent; in the hermaphrodite
5-rayed, the ovary ovoid-globular. *Fruit* 2·5 in. long, from globular to
pear-shaped, pedunculate, smooth, yellow, with much sweet pulp; seeds 1
to 5, oblong, pointed; pedicels 1 in. Pierre Flor. Forest. Cochin-Chine,
fasc. VI, p. iv. *Xanthochymus dulcis*, Roxb. Cor. Pl. t. 270; Wight Ic.
270; Bot. Mag. 2088; Choisy Gutt. Ind. 32; Planch. and Triana Mem.
Guttif. 149. *Garcinia elliptica*, Choisy in DC. Prod. i, 561 (not of Wall.
Cat.) *X. Javensis*, Blume Bijdr. 216; *Stalagmites dulcis*, Cambess. Mem.
Mus. xvi. 392, 425; Miq. Fl. Ind. Bat. I, Pt. 2, 508; Hassk. Pl. Jav.
Rar. 275.

   Perak; King's Collector No. 5750. Distrib. Malayan Archipelago.

   34. GARCINIA ANDAMANICA, King, n. sp. A tree from 20 to 40 feet
high; young branches 4-angled, pubescent. *Leaves* elongate-ovate, often
inequilateral, sub-acute; the base broad, rounded or slightly cordate;
both surfaces glabrous, shining; main nerves 14 to 16 pairs, rather prominent; length 8 to 11 in., breadth 4 to 5·5; petiole ·5 in., stout. *Male
flowers* about 3 in. in diam., in short dense axillary fascicles from short
wart-like branches. *Sepals* 5, coriaceous, ovate-rotund, imbricate,
pubescent externally. *Petals* 5, larger than the sepals, thin, rotund,
clawed, imbricate, glabrous. *Stamens* indefinite, in 5 thick fleshy
bundles; anthers minute, sub-globular, introrse. *Disc* of 5 broad
corrugated glands much shorter than the bundles of stamens and alternating with them. *Pistil* 0. *Female flowers* unknown. *Fruit* globular
or oval, smooth, bright yellow, 1 to 1·5 in. long, and ·75 to 1·25 in. in
diam., shortly apiculate; the 5-lobed stigma persistent. *G. dulcis*, Kurz
(not of Roxb.) For. Flora Burma i, 92; Pierre Fl. Forest. Cochin.-Chine,
fasc. VI, p. vi, *in part*.

   Andaman Islands; Helfer No. 872, Kurz, King's Collector, No.
224.

   Var. *pubescens*, leaves shortly pubescent beneath, the base cuneate.
Andamans; King's Collector, No. 136.

   This species was considered by Kurz to be identical with *Xanthochymus dulcis*, Roxb., a native of the Moluccas cultivated in the Botanical
Gardens, Calcutta. It does not, however, agree with specimens still
in cultivation there, nor with Roxburgh's description. Pierre (l. c.)
expresses his doubt as to the identity of the Andaman and Molucca plants;
110

but he adopts Kurz's name for the latter. The variety named above *pubescens* may turn out to be a distinct species. At present only fruiting specimens of it are known. This species is closely allied to *G. Villersiana*, Pierre, a common Cambodian plant. The leaves of the latter, as shown in Pierre's figure (Fl. Forest Cochin-Chine, t. 21) have however more nerves; the flowers have longer pedicles, and the staminal bundles are longer and more slender than in this species; the lobes of the disc are also narrower and longer.

This is the Helferian plant referred to under *G. Xanthochymus* in Fl. Br. Ind. i, 269, as allied to, but differing from that species.

35. GARCINIA DENSIFLORA, King, n. sp. A tree 60 to 80 feet high; young branches stout, 4-angled, brown when dry. *Leaves* thickly coriaceous, broadly elliptic to elliptic-oblong, sub-acute or rather blunt, the base cuneate; both surfaces shining; the midrib stout; nerves 10 to 12 pairs, sub-horizontal, anastomising by arches ·1 in. from the margin; length 4·5 to 6 in., breadth 2·5 to 3·75 in.; petiole ·6 in., thick, deeply channelled. *Male flowers* ·35 in. in diam., in large dense many-flowered clusters 1 to 1·5 in. in diam., on bracteolate tubercles from the axils of the fallen leaves; buds globose; pedicels unequal, from ·2 to ·35 in long; *sepals* 5, the 3 outer smaller than the inner 2, orbicular, fleshy, concave, puberulous externally, the margins ciliolate; the inner 2 as large as the petals, glabrous, the margins ciliolate; *petals* 5, fleshy, orbicular, concave, yellow; *stamens* in 5 pedicelled, fan-shaped branches of about 12; filaments thick, fleshy; anthers with 2 orbicular, suturally dehiscent cells; *Disc* large, fleshy, much corrugated, with 5 radiating lobes which alternate with the staminal groups; *rudy. style* cylindric, corrugated; the stigma oblong, smooth, small. *Female flowers* and *fruit* unknown.

Perak at elevations under 1000 feet, King's Collector, No. 5933. A very distinct species collected only once.

36. GARCINIA PRAINIANA, King n, sp. A small tree; young branches terete or compressed, not angled, pale yellowish. *Leaves* more or less broadly elliptic-oblong, narrowed to the rounded or slightly cordate base, shining on both surfaces; nerves 12 to 15 pairs, spreading, inter-arching submarginally, rather prominent beneath when dry; intermediate nerves prominent, bifurcating; the midrib stout; length 4·5 to 9 in., breadth 1·75 to 4 in.; petiole ·25 in., stout. *Male flowers* ·4 in. in diam., in dense, 6- to 12-flowered, bracteate, terminal cymes; bracts numerous, lanceolate, fleshy, keeled; pedicels thick, flat; *sepals* 5, fleshy, concave, orbicular; *petals* 5, darker in colour than the sepals, fleshy, concave, sub-orbicular; *stamens* numerous, in a 5-lobed annulus round the globose rudimentary ovary, 2-celled, with sutural dehiscence. *Female flowers* unknown. *Fruit* young 1·3 in diam. globular, pulpy, smooth, crowned

by the sessile smooth concave stigma ; the sepals persistent at its base, coriaceous, concave, about ·5 in. long.

Perak ; at Kwala Dynong, Scortechini, No. 1796.

### DOUBTFUL SPECIES.

GARCINIA JELINEKII, Kurz MSS. in Herb. Hort. Calc. A specimen with leaves like a *Garcinia* and detached fruit of a true *Garcinia* (No. 169 Exped. *Novara*), collected in the Nicobars by Dr. Jelinek, has been thus named in the Calcutta Herbarium. The material is too imperfect to be dealt with.

### 2.  CALOPHYLLUM, Linn.

Trees. *Leaves* opposite, shining, coriaceous, with innumerable parallel slender veins at right angles to the midrib. *Flowers* polygamous, in numerous axillary or terminal panicles. *Sepals* and *petals* 4-12; imbricate in 2-3 series. *Stamens* very many, filaments filiform, often flexuous, free or connate below; anthers erect, 2-celled, dehiscence vertical. *Ovary* 1-celled ; style slender, stigma peltate; ovule solitary, erect. *Drupe* with a crustaceous putamen. *Seed* erect, ovoid or globose; testa thin, or thick and spongy. Distrib. About 35 species, chiefly tropical Asiatic with a few American.

### SERIES A.  SEPALS 4  PETALS 0.  (*Apoterium*, Bl.).

| | |
|---|---|
| Flowers axillary, solitary or in pairs | ... 1. *C. microphyllum.* |
| Flowers in axillary fascicles     ...  | ... 2. *C. Kunstleri.* |
| Flowers in axillary racemes. | |
|   Glabrous everywhere. | |
|     Leaves less than 5 in. long. | |
|       Leaves ovate or obovate-lanceo-<br>late, pedicels 2 or more times as<br>long as the flowers; fruit<br>ovoid ...        ...        ... | 3. *pulcherrimum.* |
|       Leaves elliptic-oblong, pedicels as<br>long as flowers, fruit yellowish;<br>young branches yellowish     ... | 4. *Prainianum.* |
|       Leaves elliptic-lanceolate, pedicels<br>not exceeding flowers, racemes<br>very numerous; young branches<br>brown     ...        ...        ... | 5. *floribundum.* |
|     Leaves more than 5 in. long     ... | 6. *spectabile.* |
|   Apices of young branches, petioles and<br>inflorescence    ferruginous-pubescent. | |

Leaves thinly coriaceous, lanceolate  
or oblong-lanceolate ; outer sepals  
oblong ; fruit globose or sub-  
ovoid ; racemes not bracteate...    7. *amœnum.*  
Leaves coriaceous, ovate to ovate-  
elliptic ; outer sepals obovate,  
clawed ; fruit globose; racemes  
not bracteate    ...    ...    8. *retusum.*  
Leaves coriaceous, narrowly elliptic,  
blunt or retuse; racemes brac-  
teate at base ; fruit globose    ...    9. *Curtisii.*  
Young parts and leaves, except when  
very old, softly ferruginous-tomentose 10. *molle.*  
Flowers in terminal panicles    ...    ...    11. *canum.*  

Series B.    Sepals 4.    Petals 4 or more.

Leaves elliptic, rarely obovate, blunt or emargi-  
nate, thinly coriaceous, fruit spherical    ...    12. *Inophyllum.*  
Leaves obovate, retuse or emarginate, thickly  
coriaceous, fruit ovoid    ...    ...    13. *Inopylloide.*  
Leaves oblong, acuminate.  
    Young leaves and inflorescence rufous...    14. *Wallichianum.*  
    All parts glabrous.  
        Leaves 4 to 6 in. long, flowers ·35  
        in. in diam.    Petals 4    ...    15. *Griffithii.*  
        Leaves 5 to 10 in. long; flowers 1  
        in. in diam.    Petals 4.    Fruit 5  
        in. long    ...    ...    16. *macrocarpum.*  
        Leaves 1·75 to 3 in. long ; flowers  
        ·5 in. in diam.    Petals 4 to 6  ...  17. *venustum.*  

1.    Calophyllum microphyllum, T. Anders. in Hook. Fl. Br. Ind. i,  
272.    A glabrous, much branched, very leafy shrub; youngest branches  
4-angled.    *Leaves* rigidly coriaceous, obovate-cuneate or obovate, obtuse  
or retuse, much narrowed to the base ; nerves slightly and equally pro-  
minent on both surfaces ; length ·75 to 1·5 in., breadth ·35 to ·75 in.,  
petiole ·1 to ·2 in.    *Flowers* solitary or in pairs in the axils of the  
younger leaves, minute ; pedicels ·15 to ·25 in., slender, recurved, with  
2 bracts at the base.    *Sepals* sub-orbicular.    *Fruit* pisiform, topped by  
remains of style.  
    Mount Ophir, near the summit.  
    The leaves are not unlike those of *C. floribundum*, but the inflores-  
cence is quite different.

2. CALOPHYLLUM KUNSTLERI, King, n. sp. A tree 40 to 60 feet high, all parts glabrous except the buds, the 4-angled young branches, and the petioles and lower part of rachis of inflorescence with its bracts which are ferruginous-pubescent. *Leaves* thinly coriaceous, narrowly elliptic-oblong, the apex obtusely acuminate, tapering in the lower third to the short stout petiole; both surfaces shining; the nerves very close together and like the midrib most distinct on the lower; length 3·5 to 5 in., breadth 1·25 to 1·4 in., petiole ·3 to ·4 in. Flowers in solitary fascicles from the axils of the older or of fallen leaves, about 1·5 in. long, 3-4-flowered; bracts at base of pedicel 4, ovate, boat-shaped. *Flowers* ·25 in. in diam.; the pedicels often very unequal, the uppermost 1 in. and about twice as long as the lower. *Sepals* 4, the outer pair obovate-oblong, the inner oblong, all obtuse. *Petals* 0. *Fruit*, (not ripe) ovoid or globular, glabrous; pericarp thick, crustaceous.

Perak; King's Collector, Nos. 5328, 5374, 5459.

A common species; varying a little as to the amount of pubescence on the branchlets and buds, and in the form of the fruit. Ripe fruit has not, however, yet been collected; and it may prove that when ripe the fruit is uniformly globular. The nervation is closer than in any other species that I have seen, and the surfaces of the leaves have a peculiarly lustrous sheen.

3. CALOPHYLLUM PULCHERRIMUM, Wall. Cat. 4848. A glabrous tree, 20 to 60 feet high; the young branches as thick as a crow quill, 4-angled. *Leaves* thinly coriaceous, ovate or obovate-lanceolate, shortly and obtusely acuminate, much narrowed to the base; both surfaces shining; the edge a little thickened and undulate, the midrib stout; length 1·75 to 2·5 in., breadth ·8 to 1·2 in., petiole ·3 in. *Racemes* solitary, about half as long as the leaves, from the axils of the older leaves, lax, spreading, few-flowered. *Flowers* ·25 in. in diam.; pedicels very slender, about ·5 in. long. *Sepals* broadly ovate, the inner pair slightly larger and more membranous. *Ovary* globose. *Fruit* ovoid with a very short beak, ·65 in. long. Chois. Guttif. Ind. 14; Planch. and Triana Mem. Guttif. 246; Hook. fil. Fl. Br. Ind. i, 271; Pierre Fl. Coch.-Chine, t. 104.

Singapore, Malacca, Perak. Distrib. Cochin-China.

Miquel's three species *bancanum*, *plicipes* and *gracile* are reduced to this in Hooker's Fl. Br. Ind. Miquel ascribes 4 petals to *gracile*, which would throw it into another section. Pierre (l. c.) expresses doubts as to *bancanum* and *gracile* falling here, and considers *C. plicipes* as totally distinct both as to leaves and flowers. Of *C. mesuaefolium*, (Wall. Cat. 4850), only fragmentary specimens exist. In the Fl. Br. Ind. it is reduced here; but Planchon and Triana consider it quite different.

Var. *oblongifolium*, T. Anderson (in Hook fil. Fl. Br. Ind. l. c.); leaves oblong, tip rounded.

4. CALOPHYLLUM PRAINIANUM, King, n. sp. A glabrous tree 40 to 60 feet high; the youngest branchlets polished, terete, yellowish. *Leaves* thinly coriaceous, elliptic-oblong, shortly sub-abruptly and obtusely acuminate, narrowed in the lower third to the short petiole; the nerves rather distinct on both surfaces; lower surface paler than upper, both shining; the edge pale yellow, very slightly thickened; length 2·5 to 4 in., breadth 1 to 1·5 in., petiole 2·5 to 4 in. *Racemes* solitary, axillary, rarely supra-axillary, about 1·5 in. long, ebracteate, lax, few-flowered. *Flowers* ·25 in. in diam.; pedicels slender, ·25 in., the upper rather longer. *Sepals* 4; the outer pair orbicular, concave, puberulous externally; the inner pair larger, imbricate, orbicular-oblong, glabrous. *Ovary* ovoid, stigma very broad. *Fruit* spherical, ·4 in. in diam., crowned by the thin style, pericarp thin.

Perak; King's Collector, Nos. 5366 and 7243.

Very like *C. pulcherrimum*, but with globular fruit; also like *C. Teysmannii*, but the nervation of the leaves in that species is unusually oblique for the genus, whereas in this the nerves are almost horizontal.

5. CALOPHYLLUM FLORIBUNDUM, Hook. fil. Fl. Br. Ind. I, 272. A tree? much branched and everywhere glabrous; branchlets glaucous, 4-angled, as thick as a crow-quill. *Leaves* coriaceous, elliptic-lanceolate, obtusely acuminate, the edges thickened and pale, the base acuminate, the numerous nerves and midrib most distinct on the under sub-glaucescent surface, upper surface shining; length 1·24 to 1·5 in., breadth ·5 to ·6 in., petiole ·25 in. *Racemes* from most of the leaf-axils erecto-patent, more than half as long as the leaves; pedicels opposite, spreading, not much longer than the diameter of the flowers. Flowers ·25 in. in diam. The outer pair of *sepals* broadly ovate, sub-acute, the inner broadly obovate, blunt, membranous. *Stamens* numerous, style not longer than the sepals.

Malacca; Maingay, Nos. 170, 171.

This is closely allied to *C. pulcherrimum*, Wall., but has smaller leaves, the racemes are more numerous and longer in proportion to the leaves, while the pedicels of individual flowers are much shorter.

6. CALOPHYLLUM SPECTABILE, Willd. A tall tree; when adult all parts glabrous, the buds and young parts ferruginous-pubescent. *Leaves* thinly coriaceous, narrowly or broadly oblong, rarely elliptic, sub-acute or obtuse, undulate, the base cuneate; both surfaces shining, the nerves very numerous, the midrib strong; length 6 to 12 in., breadth 1·5 to 3 in., petiole ·5 to ·75 in. *Racemes* umbelliform, axillary, solitary, lax, few-flowered, ·5 in. in diam.; pedicels slender, ·5 in. *Sepals* orbicular, glabrous. *Ripe fruit* spherical, ·75 in. in diam. DC. Prod. i, 562; Choisy Guttif. Ind. 43, in part; Planch. and Triana Mem. Guttif. 238; Wight Ill. i, 129; Miq. Fl. Ind. Bat. i, Pt. 2, 510; Pierre Fl. Coch-Chine, t.

107; Kurz Fl. Burm. i, 94. *C. tetrapetalum*, Roxb. Fl. Ind. ii, 608;
*C. Moonii*, Wight Ill. i, 129, Ic. t, 111; Thw. Enum. 52; Beddome Flor.
Sylvat. Gen. xxii. *C. cymosum*, Miquel Fl. Ind. Bat. Suppl. i. 497;
*C. Diepenhorstii*, Miq. l. c. 497. *C. hirtellum*, Miq. Pl. Jungh. i, 291; Fl.
Ind. Bat. I, Pt. 2, 511. *Apoterium Sulatri*, Bl. Bjdr. 218.

Penang, Singapore, Andamans, Nicobars. Distrib. Malayan
Archipelago, Cochin-China, Fiji, Society Islands.

7. CALOPHYLLUM AMOENUM, Wall. Cat. 4849. A tree 20 to 40 feet
high; the apices of the youngest branches, the buds, the leaf-petioles,
and the rachises of the racemes minutely ferruginous or griseous-pubes-
cent. *Leaves* thinly coriaceous, lanceolate or oblong-lanceolate, rarely
ovate-lanceolate or obovate-elliptic, acute or very shortly and obtusely
acuminate; the base cuneate; nerves very close, about equally prominent
on both surfaces; length 2·5 to 3·5 in., breadth 1 to 1·5 in., petiole ·4 in.
*Racemes* stout, sub-erect, shorter than the leaves, few-flowered. *Flowers*
·25 in. in diam., pedicels ·2 in. *Sepals* reflexed, the outer pair oblong,
ferruginous-tomentose externally; the inner pair longer, sub-glabrous.
*Fruit* globose or sub-ovoid, ·3 in. long, the pericarp pulpy. Choisy
Guttif. de l'Inde, 41; Planch. and Triana Mem. Guttif. 235; Kurz Fl.
Burm. i, 95.

Andamans; King's Collector. Tenasserim, Helfer, No. 881;
Amherst, Wallich 4849.

None of the Andaman specimens which I have seen are in fruit;
and none of the Burmese are in flower. But in leaf and other characters
the specimens are alike. The species seems to me a good one and to be
distinct from *C. retusum*, Wall., with which it has however been united
in Fl. Br. Ind., and this is also the opinion of Planchon and Triana.

8. CALOPHYLLUM RETUSUM, Wall. Cat. 4846. A much-branched, very
leafy shrub; the young branches 4-angled, softly ferruginous-pubescent,
as are the petioles and inflorescence. *Leaves* coriaceous, ovate to ovate-
elliptic, obtuse, the base rounded or slightly narrowed; nerves rather distant for the genus, more visible on the upper than on the lower surface;
length 1·75 to 2·25 in., breadth ·8 to 1·1 in.; petiole ·2 in. stout, pubescent,
when old glabrous. *Racemes* solitary, axillary, sub-erect, ferruginous-
pubescent, especially at the base, 1 in. long. *Flowers* ·25 in. in diam.,
pedicels ·2 in. *Sepals;* the outer obovate, clawed; the inner ovate-oblong.
*Fruit* pisiform. Pierre Fl. Coch-Chine, t. 102. *C pisiferum*, Planch.
and Triana Mem. Guttif. 266. *C. retusum*, Hook. fil. Fl. Br. Ind. i, 272,
(excl. syn. *C amœnum*, Wall.).

Malacca; Griffith, Maingay (Kew Distrib. No. 166). Singapore;
Wallich, No. 4846.

9. CALOPHYLLUM CURTISII, King, n. sp. A tree; the young branches,

116

buds, petioles and inflorescence ferruginous-pubescent, sub-pulverulent. *Leaves* coriaceous, narrowly elliptic, blunt or retuse, the base narrowed; upper surface glabrous, shining; the lower rather dull, pubescent on the prominent midrib; the nerves rather distinct on both surfaces; length 2 to 2·75 in., breadth 1 to 1·4 in., petiole ·4 in. *Racemes* solitary or two together, axillary, umbellate, compact, 3- to 5-flowered, ferruginous-tomentose, much shorter than the leaves and with several navicular ferruginoustomentose bracts at their base. *Flowers* ·25 in. in diam., the pedicels ·2 long, more than twice as long in fruit, and the uppermost the longest. Sepals 4; the outer oblong, sub-obovate, ferruginous-tomentose; the inner smaller, oblong, sub-glabrous. Petals 0. Fruit ovoid.

Penang; on Government Hill, at 500 feet, Curtis, No. 523.

A very distinct species ripe fruit of which is unknown.

10. CALOPHYLLUM MOLLE, King, n. sp. A tree 40 to 80 feet high; the young shoots, buds, under surfaces of adult leaves, and young fruit softly ferruginous-tomentose. *Leaves* coriaceous, narrowly oblong, gradually narrowed in the upper fourth to the sub-obtuse apex, the edges thickened and slightly recurved, the base rounded or slightly narrowed; upper surface when adult sub-glabrous, the nerves close, slightly visible, the midrib sparsely and coarsely pubescent; lower surface pale and, except when very old, more or less softly tomentose especially on the very stout midrib; length 5 to 8 in., breadth 1·25 to 2·25 in., petiole ·4 in. to ·6 in. *Racemes* axillary, solitary, about 1 in. long, 1- to 2-flowered, densely ferruginous-tomentose as are the ovary and young fruit. *Sepals* 4, the outer oblong, ferruginous-tomentose externally. *Petals* 0. *Fruit* globular, slightly apiculate, 1 in. long, sub-glabrous when ripe.

Penang; Curtis, No. 1426. Perak; King's Collector, many numbers.

A species collected by Sig. Beccari in Sumatra (P. S. 953) comes very near this; but the leaves are broader and more inclined to be oblanceolate, the thickening of the edge is greater and is pale in colour, while the young fruit is ovoid and not tomentose. Judging from Pierre's figure (he gives no description) of his *C. Dongnaiense*, Fl. Coch.-Chine, t. 108, that species and this must be near allies.

11. CALOPHYLLUM CANUM, Hook. fil. Fl. Br. Ind. i, 271. A tree 40 to 80 feet high; young branchlets as thick as a goose-quill, smooth. *Leaves* coriaceous, glabrous, narrowly elliptic-oblong, bluntly and shortly acuminate, slightly undulate, the base acute, upper surface shining, the lower less so; midrib very strong, nerves very thin and numerous; length 5 to 7 in., breadth 1·75 to 2·25 in., petiole ·5 to ·75 in. *Flowers* ·75 in. in diam., in terminal hoary-pubescent panicles less than half as long as the leaves; or in axillary racemes, pedicels ·15 in. *Sepals* hoary-puberulous, orbicular; the outer pair coriaceous, concave; the inner pair larger and

thinner, imbricate, the upper edge incurved, ciliate. *Petals* 0. Stamens very numerous. *Stigma* discoid. *Ovary* depressed-spherical, glabrous. *Fruit* ovoid, smooth, ·75 in. long.

Malacca; Maingay. Perak; King's Collector, No. 5420; Scortechini, No. 2044. Penang; Curtis, No. 1543. Distrib. Cochin-China, British India.

Not unlike *C. Wallichianum*, Planch. and Triana; but apetalous and the leaves never tomentose.

12. CALOPHYLLUM INOPHYLLUM, Linn. sp. 732. A glabrous tree 20 to 30 feet high; young branches stout. *Leaves* thinly coriaceous, elliptic, rarely obovate-oblong, apex rounded or emarginate, the base acute, shining on both surfaces; length 4 to 6 in., breadth 2·5 to 3·5 in., petiole ·75 in. broad. *Racemes* in the upper axils, lax, 3 to 4 in. long, few-flowered. *Flowers* ·75 in. in diam.; pedicels slender, 1 to 1·75 in. *Sepals* 4, the 2 inner petaloid. *Petals* 4, longer than the sepals. *Filaments* 4-delphous. *Ovary* stipitate, globose. *Style* longer than the stamens; stigma peltate, lobed. *Fruit* globular; the pericarp smooth, fleshy, 1 in. in diam. or more. DC. Prod. I, 562; Bl. Bijdr. 217; Chois. Guttif. Ind. 42; Planch. and Triana Mem. Guttif. 254; Roxb. Fl. Ind. ii, 606; W. and A. Prod. 103; Miq. Fl. Ind. Bat. I, pt. 2, p. 510; Wight Ill. i, 128; Ic. 77; Hook. Fl. B. Ind. i, 273; Kurz Fl. Burm. i, 95. *C. Blumei*, Wight Ill. i, 128. *C. Bintagor*, Roxb. Fl. Ind. ii, 607. (?)

On the Coasts, in all the Provinces. Distrib. Burmah, S. India and Ceylon, E. African Islands, Australia, Polynesia.

The pure white flowers are delightfully fragrant, the seeds yield a beautiful mild oil, and the wood is useful for spars of boats and ships.

13. CALOPHYLLUM INOPHYLLOIDE, King, n. sp. A glabrous tree, 60 to 80 feet high; the young branches about as thick as a goose-quill, dark brown. *Leaves* thickly coriaceous, obovate or obovate-oblong, the apex retuse or emarginate, the edges thickened, recurved (when dry), gradually narrowed from about the middle to the stout petiole; both surfaces shining, the lower less so and paler; nerves very numerous, little prominent, the midrib stout; length 3·25 to 4·5 in., breadth 1·75 to 2·75 in.; petiole ·6 to 1 in. long, broad at the apex. *Racemes* from the axils of the upper leaves, 2 to 3 in. long, lax, few-flowered. *Flowers* globular in bud, about ·75 in. in diam. when expanded. *Outer sepals* rotund, concave, reflexed, 4 in. long; the inner petaloid, larger than the outer. *Petals* narrower than the sepals; pedicels slender, ·65 to 1 in. long. *Style* stout; stigma broad, discoid. *Fruit* (not quite ripe) ovoid, ·75 in. long, the pericarp not pulpy.

Perak; on low Hills, elevation 300 to 500 feet.

The leaves of this much resemble those of *C. Inophyllum*, but they are thicker, smaller, and invariably obovate and retuse. The flowers are smaller than those of *C. Inophyllum ;* the fruit also differs in being smaller, ovoid and not pulpy. This species also resembles the British Indian *C. Wightianum,* Wall. The existence of petals is certain, but the condition of the flowers on the only specimens hitherto collected is such that their number cannot be made out with certainty.

14. CALOPHYLLUM WALLICHIANUM, Planch. and Triana Mem. Gutt. 249. A tree; the branchlets pale yellowish, the youngest 4-angled and, with the buds under surface of young leaves and inflorescence, minutely ferruginous-tomentose. *Leaves* thinly coriaceous, narrowly elliptic-oblong, the apex shortly and obtusely acuminate, the base acute ; upper surface shining, the midrib narrow ; lower surface dull, the midrib prominent, at first minutely ferruginous-tomentose, when adult glabrous ; length 4·5 to 6 in., breadth 1·5 in. ; petiole ·75 in., rusty. *Racemes* axillary and terminal, less than half as long as the leaves, ferruginous-tomentose, erecto-patent. *Flowers* ·5 in. in diam., pedicels ·2 in. *Sepals* 4, orbicular, ferruginous-tomentose on both surfaces. *Petals* 4, cuneate-oblong, glabrous internally. Fruit ( *fide* F. B. Ind.) globose, the size of a cherry. Wall. Cat. No. 4843, in part; Hook. fil. Fl. Br. Ind. i, 273.

Malacca ; Maingay.

This species was founded by Planchon and Triana on a specimen mixed with Wall. Cat. No. 4843, (the bulk of which is *C. spectabile,* Willd.) This does not appear to be a common. species, and its fruit I have not seen. It may be readily distinguished by its yellow branches, the pale ferruginous, almost cinnamoneous, colour of its leaves when dry, and its darkly rusty racemes.

15. CALOPHYLLUM GRIFFITHII, T. Anders. in Hook. Fl. Br. Ind. i, 273. A glabrous tree, the youngest shoots 4-sided. *Leaves* thinly coriaceous, oblong or elliptic-oblong, acute or obtuse, the base shortly cuneate, shining on both surfaces, the rather distant nerves equally distinct on both, the midrib more distinct and pale-coloured on the lower; the edges with a pale thickening ; length 4 to 6 in., breadth 1·75 to 2 in., petiole ·4 to ·6 in. *Racemes* solitary, axillary, from 1·5 to 2·5 in. long, few-flowered. *Flowers* ·35 in. in diam., glabrous ; pedicels unequal, ·2 to ·5 in. long, slender, each with a small deciduous bract at its base. *Sepals* 4, outer pair orbicular, inner pair longer but narrower. *Petals* 4, oblong, obtuse. *Fruit* (young) ovoid, smooth.

Malacca ; Griffith. Distrib. Sumatra, Forbes, No. 322a.

16. CALOPHYLLUM MACROCARPUM, Hook. fil. Fl. Br. Ind. I, 273. A glabrous tree ; branchlets polished, sharply 4-angled. *Leaves* coriaceous, narrowly oblong or elliptic-lanceolate, shortly and obtusely acuminate,

much narrowed at the base, edge slightly thickened, upper surface shining, the midrib prominent on the rather dull lower surface, nerves rather bold and equally prominent on both; length 5 to 10 in., breadth 2 to 3 in., petiole 1 to 1·25 in. *Racemes* not half the length of the leaves, axillary, solitary, 6-10-flowered, minutely ferruginous-puberulous. *Flowers* 1 in. in diam.; pedicels 1 to 1·25 in. *Sepals* 4, the outer pair puberulous externally; the inner pair larger, imbricate, oblong-rotund, orbicular, concave, rusty, obtuse, petaloid. *Petals* 4, smaller than the inner sepals, oblanceolate, clawed. *Stamens* short. *Fruit* (*fide* Maingay) ellipsoid, ·5 in. long.

Malacca; Maingay (Kew Distrib. 174). Perak; King's Collector, No. 8851.

17. CALOPHYLLUM VENUSTUM, King, n. sp. A glabrous tree, 20 to 30 feet high. *Leaves* thinly coriaceous, shining, rigid, elliptic-ovate, retuse, the base cuneate, the margin thickened; nerves rather distant for the genus and equally distinct on both surfaces; length 1·75 to 3 in., breadth 1 to 1·5 in., petiole ·3 to ·4 in. *Racemes* 3-5-flowered, solitary, axillary, half as long as the leaves, very lax, spreading. *Flowers* large (·5 in. in diam.), on long (·75 in.) pedicels; buds ovoid. *Sepals* 4, more or less orbicular. *Petals* 4 to 6, narrower than the sepals, the inner oblong, veined, all obtuse. *Fruit* unknown.

Perak; King's Collector, No. 7763.

A very handsome species of which the fruit is unknown. In leaf it resembles *C. amœnum*, but differs greatly in the flowers.

## 3. KAYEA, Wall.

Trees. *Leaves* opposite; veins rather distant, arched. *Flowers* hermaphrodite, either large and solitary, or small and collected in terminal panicles. *Sepals* and *petals* 4 each, imbricate. *Stamens* numerous; filaments slender, free or connate at the base; anthers small, subglobose, 2-celled, dehiscence vertical. *Ovary* 1-celled; style slender, stigma acutely 4-fid; ovules 4, erect. *Fruit* subdrupaceous, fleshy, indehiscent, 1-5-seeded. *Seeds* thick, testa thin and crustaceous.—Distrib. Tropical Asia, 7 species.

Flowers in racemes.
    Racemes 2 to 3 in. long; flowers 1 in.
        or more in diam.        ...        ...    1. *K. Wrayi.*
    Racemes less than 1 in. long; flowers
        less than 1 in. in diam.        ...    2. *K. racemosa.*
Flowers solitary, axillary.
    Nervation of leaves bold, distinct.

Fruit turbinate, quite enveloped by
the outer sepals when ripe ...   3. *K. grandis.*
Fruit ovoid, pointed, only partly
covered by sepals.
Leaves tapering to the mo-
derately long petiole   ...   4. *K. Kunstleri.*
Leaves rounded or cordate at
base, sub-sessile   ...   5. *K. nervosa.*
Fruit ovoid with a much elongate
hooked apex, leaves caudate-
acuminate   ...   ...   6. *K. caudata.*
Nervation of leaves indistinct.
Young branches slender, smooth,
flowers axillary ...   ...   7. *K. elegans.*

1. KAYEA WRAYI, King, n. sp. A small glabrous tree; the young branches pale, polished, terete, often whorled. *Leaves* very thickly coriaceous, broadly elliptic, acute or acuminate, the edges much recurved when dry, the base rounded; nerves about 15 pairs, unequal, only slightly prominent on the lower and less so on the upper surface; both surfaces smooth, the upper shining, the lower dull; length 3 to 4·5 in., breadth 1·75 to 2·25 in.; petiole ·4 in., thick. *Racemes* axillary or terminal, 2 to 3 in. long, sub-erect, stout, with minute subulate bracts at the base, 3- to 5-flowered. *Flowers* 1 to 1·25 in. in diam., pedicels ·5 to 1·25 in. long. *Sepals* orbicular, nearly equal, glabrous, the outer pair coriaceous, the inner thinner. *Petals* much larger than the sepals, broadly-obovate or orbicular, clawed. *Stamens* much shorter than the petals. *Fruit* unknown.

Pahang; on Gunong Brumber, elevation 7000 feet, L. Wray, junior. A remarkable species quite unlike any hitherto described.

2. KAYEA RACEMOSA, Planch. and Triana Mem. Guttif. 269. A glabrous tree, 40 to 60 feet high. *Leaves* sub-coriaceous, elliptic-oblong, with a very short blunt acumen, slightly narrowed to the petiole; upper surface rather dull, the lower paler and shining; main nerves 18 to 25 pairs, bold, spreading; length 6 to 9 in., breadth 1·5 to 2·5 in.; petiole ·75 in., stout. *Racemes* less than 1 in. long, few-flowered, bracteolate, crowded at the apices, or in the axils near the apices, of the rather long naked often whorled branchlets; pedicels stout, ·15 in. long. *Flowers* ·5 in. in diam. *Sepals* rotund, thickly coriaceous. *Petals* longer than the sepals, thin. *Stamens* numerous, in one series, monodelphous at the base; Wall. Cat. without number or locality; Hook. fil. Fl. Br. Ind. i, 276, (excl. syn. *Mesua Singaporiana*, Wall. Cat. 4836.)

Malacca; Maingay (Kew Distrib. 177). Perak; Scortechini, 97.

The foregoing description has been drawn up from Maingay's Malacca specimens above quoted, which have been accepted by Sir Joseph Hooker as of the species described by Planchon and Triana as *K. racemosa*. These authors founded the species on a Wallichian specimen in M. de Candolle's Herbarium, without number or indication of locality, which had been separated from some other Wallichian number, and which bears the following note by Choisy " *Mesua speciosa ? specimen imperfectum sine notula in herb. Wallichiano repertum.*" This specimen I have not seen. Of Wall. Cat. No. 4836, (*Mesua Singaporiana*), there is a specimen in Herb. Calc. ; and it is certainly different from Maingay's 177, being more like a *Mesua* than a *Kayea*.

3.   KAYEA GRANDIS, King, n. sp.   A glabrous tree, 40 to 80 feet high. *Leaves* large, coriaceous, oblong to elliptic-oblong, sub-acute, the edges revolute (when dry), slightly narrowed towards the rounded or sub-acute base ; both surfaces rather dull (when dry), the 20 to 25 pairs of main nerves sub-horizontal, prominent, the secondary nerves also prominent ; length 9 to 18 in., breadth 3 to 4·5 in. ; petiole ·4 to ·75 in., smooth. *Flowers* sub-globose in bud, pedicelled, in short few-flowered axillary cymes crowded in the axils of the leaves, rarely solitary, about 1·25 in. in diam. when expanded ; pedicels ·5 in.   *Sepals* rotund, the outer concave, very coriaceous ; the inner thin, not larger than the outer. *Petals* elliptic-oblong, acute, larger than the sepals (·5 in. long or more). *Ripe fruit* turbinate, 2 to 2·5 in. in diam. and 1·25 in. thick, leathery, completely enveloped by the persistent, thickened, outer sepals.

Malacca ; Maingay (Kew Distrib. 178), Cantley No. 2354.   Perak ; King's Collector.

A very fine species ; at once distinguished by its large leaves and depressed turbinate fruit.   The fruit, and probably the whole plant, abounds in yellow juice.   According to M. Cantley the wood sinks in water.

4.   KAYEA KUNSTLERI, King, n. sp.   A glabrous tree, 30 to 50 feet high ; the branchlets brownish, sub-striate, not tuberculate. *Leaves* thinly coriaceous, elliptic-lanceolate, acuminate, sub-undulate, the base much narrowed to petiole ; both surfaces rather dull (when dry) with a few scattered opaque black dots ; the lower pale, sub-glaucescent (when dry) ; length 4 to 6 in., breadth 1 to 2·25 in. ; petiole ·25 in. to 4 in., rugose ; nerves 20 to 24 pairs, unequal, prominent ; the lower horizontal, the upper slightly curving upwards.   *Flower* solitary, axillary or terminal, 1·5 to 2 in. diam., on a very short smooth pedicel, bracts at its base linear-subulate ; bud globose, smooth.   *Sepals* unequal as in *K. nervosa*.   *Petals* oblong-acuminate, 1 in. long.   Ripe fruit ovoid, gradually narrowing into a short subulate apical beak.

Perak ; King's Collector, Nos. 3301, 6850; Penang, Curtis, No. 1419 ; Malacca, Maingay, No. 176.

This is allied to *K. nervosa*, T. Anders. ; but it is readily distinguished from that by its smooth branchlets, by the leaves much and gradually narrowed to both base and apex, and by the oblong-acuminate petals. A shrubby form of this occurs in Penang (Curtis, Nos. 805, 1418), and in Perak (King's Collector, No. 1345) in which all the parts are smaller and the leaves are less acuminate at the apex, and rounded instead of much attenuated at the base.

5. KAYEA NERVOSA, T. Anders. in Hook. fil. Fl. Br. Ind. i, 277. A glabrous tree ; the branchlets minutely tubercled, 4-angled. *Leaves* sub-sessile, membranous, elliptic-oblong, shortly and bluntly acuminate, the base rounded or emarginate ; both surfaces (when dry) dull coppery brown, the lower paler; nerves 16 to 20, unequal, rather prominent beneath ; length 3 to 5 in., breadth 1·25 in. to 2 in. ; petiole ·15 in. long, rugose as is the base of the midrib. *Flowers* axillary, usually solitary (sometimes 2 or 3 from an axil), or terminal, 1·75 in. in diam. ; pedicels ·75 in. or less, tubercled, each with several linear lanceolate bracts at its base. *Sepals* unequal, the outer very coriaceous, sub-orbicular ; the inner nearly twice as large but thinner. *Petals* obovate ; filaments about as long as the slender pistil. *Ripe fruit* sub-globular, beaked, leathery, ·75 in. in diam. or more, the calyx marcescent. Kurz Flora Burm. i, 96 ; *Mesua nervosa*, Pl. and Triana Mém. Guttif. 279.

Malacca, Perak. Distrib. Burmah.

6. KAYEA CAUDATA, King, n. sp. A slender glabrous tree, 20 to 30 feet high, with drooping habit ; the branchlets slender, pale brown, striate. *Leaves* membranous, obovate-elliptic, caudate-acuminate, mucronulate, edges undulate, slightly narrowed to the rounded base; both surfaces dull, the lower pale ; nerves 12 to 14 pairs, prominent, sub-horizontal ; length 3 to 3·5 in. of which the acumen forms ·7 in. ; breadth 1 to 1·15 in., petiole ·15 in. *Flowers* unknown. *Fruit* solitary, terminal, shortly pedicellate, narrowly ovoid-cylindric, tapering very much to the apex and often curved, less narrowed to the base, 2 to 2·5 in. long, and 1 to 1·25 in. in diam. at the middle; bracts at base of pedicel subulate, 1-nerved. *Sepals* persistent, the outer pair orbicular, the inner oblong; pedicel about ·2 in. long, rather shorter than the subulate bracts.

Perak ; King's Collector, No. 7937.

Only once collected and only in fruit. Easily recognisable by its caudate-acuminate leaves and fruit.

7. KAYEA ELEGANS, King, n. sp. A glabrous tree 40 to 60 feet high, with slender drooping branches ; branchlets very thin, pale grey. *Leaves* thinly but rigidly coriaceous, lanceolate, acuminate, the base acute, the

edges undulate (when dry); both surfaces rather dull, the nerves numerous but indistinct, the midrib slightly prominent; length 2·25 to 3 in., breadth ·5 to ·75 in., petiole ·25 to ·35 in. *Flower* solitary, axillary or terminal, ·4 in. in diam.; pedicel ·1 in. long with several ovate acute bracts at its base. *Sepals* nearly equal, the outer coriaceous. *Petals* oblong, acute, smaller than the sepals. *Ovary* narrowly ovoid, attenuate above, and passing into the long filiform curving exserted style. *Fruit* unknown.

Perak; on Gunong Bubu, elevation from 1500 to 2000 feet. King's Collector.

A very distinct and elegant species, distinguished by its thin rigid lanceolate leaves and very slender branches.

## 4  MESUA, Linn.

Trees. *Leaves* opposite, rigidly coriaceous, often pellucid-dotted; veins very numerous, very slender, at right angles to the midrib. *Flowers* polygamous or hermaphrodite, large, axillary or terminal, solitary. *Sepals* and *petals* 4 each, imbricate. *Stamens* very numerous, filaments filiform, free or connate at the base; anthers erect, oblong, 2-celled, dehiscence vertical. *Ovary* 2-celled; style long, stigma peltate; ovules 2 in each cell, erect. *Fruit* between fleshy and woody, 1-celled by the absorption of the septum, at length 4-valved, 1-4-seeded. *Seeds* without an aril, testa fragile. Distrib. Tropical Asia; 3 species.

1.  MESUA FERREA, Linn. sp. 734. A medium sized tree with spreading head; branches faintly 4-angled, glaucous. *Leaves* coriaceous, linear-lanceolate to oblong-lanceolate, acute or acuminate, the base acute or rounded; above shining; below pruinose, glaucous or glaucescent; nerves numerous, close, inconspicuous; length 3 to 6 in., breadth ·75 to 1·25 in., petiole ·25 to ·35 in. *Flowers* ·75 to 3 in. in diam., in pairs or solitary, usually terminal. *Sepals* orbicular, fleshy, the margins thin. *Petals* 4, obovate, white; anthers large, elongate. *Fruit* ovoid-conic to sub-globose, from 1 to 2 in. long, the sepals persistent. Choisy in DC. Prod. i, 562; Choisy Guttif. Ind. 40; Planch. and Triana Mem. Guttif. 271; Roxb. Fl. Ind. ii, 635; W. and A. Prod. 102; Wall. Cat. 4834; Wight Ill. 127, Ic. t. 118; Beddome Flor. Sylvat. Gen. xxiii; Hook. fil. Fl. Ind. i, 277; Bl. Bijdr. 216; Miq. Fl. Ind. Bat. i, Pt. 1, 509; Kurz For. Fl. Burm. i, 97. *M. speciosa*, Chois. in DC. l. c.; Guttif. Ind. 40; Wight Ic. t. 118 and 961; Wall. Cat. 4835; Pl. and Trian. l. c. 375; Beddome l. c. xxiii. *M. pedunculata*, Wight Ill. 127; Ic. t. 119. *M. coromandeliana*, Wight Ill. 129; Ic. t. 117; Pl. and Trian. l. c. 378; Beddome Flor. Sylvat. t. 64. *M. Roxburghii*, Wight Ill. 127; Beddome l. c. xxiii. *M. salicina*, *M. Walkeriana* and *M. pulchella*, Planch. and Trian. l. c. 373,

124

374 and 379. *M. sclerophylla,* Thwaites Enum. 407; Beddome l. c., xxiii. *M. Nagana,* Gard. in Calc. Journ. Nat. Hist. vii, 4.

In all the Provinces. Distrib. Eastern and Southern provinces of British India ; Ceylon ; often cultivated.

A variable species to which many names have been given. A form with narrow leaves (·5 in. broad) and small flowers is found in Ceylon, and was distinguished by Thwaites as var. *angustifolia* ( *M. salicina,* Pl. and Tri.). In other forms from Ceylon and the South of India, the leaves have very little of the characteristic white waxy powder on their under surfaces ; and these formed the bases of Planchon and Triana's species *M. pulchella,* and of Wight's *M. Coromandeliana.*

2. MESUA LEPIDOTA, T. Anders. in Hook. fil. Fl. Br. Ind. I, 288. A slender glabrous tree, 60 to 80 feet high ; the branches pale brown, the youngest minutely rugose when dry. *Leaves* coriaceous, shining, narrowly elliptic or oblong-lanceolate, the apex shortly acuminate, the base acute ; lower surface pale, nerves indistinguishable but the midrib prominent on both surfaces ; length 2 to 3 in., breadth ·75 to 1·2 in., petiole ·15 in. *Flowers* unknown. *Fruit* solitary, terminal, pedicellate, broadly ovoid or depressed-globular when young, slightly pointed when mature, apiculate, 1 in. or more in diam., subtended at the base by the 4 lignified sub-rotund spreading sepals ; pericarp thick, woody, rugulose, dehiscing vertically by 2 (rarely 3) pointed valves. *Seeds* two, plano-convex, or one depressed-globose ; the testa brown, brittle ; pedicels 1 to 1·5 in. long, thickened upward, and with several minute subulate deciduous bracts at their bases.

Malacca ; Griffith (Kew Distrib. No. 845). Perak ; Scortechini, No. 183ᵇ, King's Collector, Nos. 4551 and 5881.

It is suggested in Fl. Br. Ind. (I, 278) that Griffith's No. 845, although now put with *Mesua,* is probably the type of a new genus between *Kayea* and *Mesua.* Griffith's specimens have no flowers, and unfortunately neither have those of the Perak collectors. The latter appear to belong to the same plant as Griffith's ; but their leaves are rather smaller, the branchlets more slender, and the pericarp slightly thinner. It may therefore be found, when fuller material is forthcoming, that there are two species here, and that neither belongs to *Mesua.*

ORDER XV. TERNSTRŒMIACEÆ.

Shrubs rarely climbing, or trees. *Leaves* alternate, simple (in Indian species) entire or often serrate, usually coriaceous, exstipulate. *Flowers* handsome, seldom small, usually subtended by 2 sepal-like bracts, rarely diclinous, axillary, 1 or more together, rarely in lateral or terminal racemes or panicles. *Sepals* 5, rarely 4-7, free or slightly con-mate, the innermost often larger. *Petals* 5, rarely 4-9, free or connate below, imbricate or contorted. *Stamens* numerous (definite in *Sladenia* and *Stachyurus*) free or connate, usually adnate to the base of the deciduous corolla ; anthers basifixed or versatile, dehiscing by slits or rarely by terminal pores. *Ovary* free (½-inferior in *Anneslea*), sessile, 3-5-celled, (many-celled in *Actinidia*) ; styles as many, free or connate, stigmas usually small ; ovules 2-8 in each cell, rarely solitary, never orthotropous. *Fruit* baccate or capsular. *Seeds* few or numerous, placentas axile ; albumen scanty or 0, rarely copious ; embryo straight or hippocrepiform, cotyledons various. Distrib. Rare in temperate, abundant in tropical Asia and America, almost wanting in Africa and entirely in Australasia ; species about 270.

Tribe I. TERNSTRŒMIEÆ. *Peduncles* 1-flowered. *Petals* imbricate. *Stamens* adherent to the base of the corolla ; anthers basifixed. *Fruit* (in Indian genera) indehiscent. *Seeds* usually few ; albumen fleshy, usually scanty. *Embryo* curved ; cotyledons shorter than the radicle and about as broad.

&ast; Fruit inferior.

                                      1. *Anneslea.*

&ast;&ast; Fruit superior.

    Flowers hermaphrodite.

        Anthers usually pilose, stamens and
           seeds numerous, ovary 3-5-celled   2. *Adinandra.*

    Flowers diœcious.

        Flowers large, on long pedicels      ...   3. *Ternstrœmia.*
        Flowers small, sessile or sub-sessile  ...   4. *Eurya.*

Tribe II. SAURAUJEÆ. *Peduncles* many-flowered. *Petals* imbricate. *Anthers* versatile. *Fruit* usually pulpy, rarely sub-dehiscent. *Seeds* numerous, minute, albumen abundant. *Radicle* straight or slightly curved and longer than the cotyledons.

    Climbers, diœcious       ...       ...   5. *Actinidia.*
    Trees or shrubs ; usually hermaphrodite  ...   6. *Saurauja.*

Tribe III. GORDONIEÆ. *Peduncles* 1-flowered, very often short. *Petals* imbricate. *Anthers* versatile. *Fruit* indehiscent or loculicidal. *Albumen* scanty or 0. *Cotyledons* various; radicle short, straight or curved.

&ast; Fruit indehiscent ...     ...     ...   7. *Pyrenaria.*

&ast;&ast; Fruit dehiscent.

    Ovules lateral, seeds winged, radicle in-
        ferior       ...       ...       ...   8. *Schima.*
    Ovules pendulous, seeds winged, radicle
        superior       ...       ...       ...   9. *Gordonia.*

Tribe IV. BONNATIEÆ. *Flowers* in lateral
panicles crowded near the apices of the branches.
*Anthers* versatile ; *capsule* dehiscing from base ...    10. *Archytæa.*

## 1. ANNESLEA, Wall.

Evergreen glabrous trees or shrubs. *Flowers* in terminal corymbs,
large, white, 2-bracteolate. *Sepals* 5, their lower part fleshy, connate, and
adherent to the ovary, their upper part coriaceous and crowning the
fruit. *Petals* 5, connate by their bases. *Stamens* numerous ; the filaments
short, inserted on the torus ; anthers narrow, elongate, 2-celled, introrse,
with a long apiculus from the connective. *Ovary* half immersed in
the torus, 3-celled ; style cylindric, 3-fid ; ovules many, pendulous.
*Fruit* a leathery inferior berry crowned by the sepals. *Seeds* oblong,
flattened, emarginate at one end, with a hard testa and fleshy albumen.
Distrib. Burmah and Malayan Peninsula ; species 2.

ANNESLEA CRASSIPES, Hook. in Choisy Mem. Ternst. 41. A bush or
small tree ; young branches stout, the bark rough, rather pale. *Leaves*
coriaceous, obovate or oblanceolate with short abrupt blunt acumen, or
oblong-lanceolate and acute, much narrowed at the base ; the edge thickened and obscurely glandular-serrate ; nerves 6 to 8 pairs, invisible in
the fresh, faint in the dry state, the midrib prominent in both ; length
2·5 to 6 in., breadth 1·5 to 2·25 in., petiole ·6 to 1 in. *Flowers* 1 to 1·25 in.
in diam., in corymbs of 3 to 6 ; pedicels ·5 to 1·25 in., recurved ; bracteoles
fleshy, square, keeled. Free portion of *sepals* fleshy, ·65 in. long, yellow,
rounded or emarginate. *Petals* smaller than the sepals, membranous,
ovate-acuminate. *Stamens* about 30. *Fruit* ovoid, 1 to 1·5 in. long
(excluding the free part of the sepals), rough ; style persistent. Hook.
fil. Fl. Br. Ind. i, 280.

Mount Ophir in Malacca ; Griffith, &c.—Perak ; on Gunong Batu
Puteh at 3,400 ft. Wray, Scortechini.

Var. *obovata.* A bush. *Leaves* obovate, minutely and bluntly
mucronate ; *fruit* conspicuously verrucose.

Perak, Gunong Bateh, at an elevation of 6,700 feet.
Anneslea is practically a *Ternstrœmia* with half-inferior fruit.

## 2. ADINANDRA, Jack.

Small evergreen trees with the habit of *Ternstrœmia* or *Gordonia.*
*Peduncles* axillary, solitary, recurved, 2-bracteate at the apex. *Flowers*
often silky outside. *Sepals* 5. *Petals* 5, connate at the base. *Stamens*
many, often 1-4-delphous, the inner smaller ; anther cells lateral, narrow, elongate, the connective apiculate, usually hairy. *Ovary* 3-5-
celled ; style ultimately elongate, entire or shortly 3-5-fid ; ovules many

127

in each cell. *Fruit* globose, subtended by the persistent calyx and
crowned by the style. *Seeds* many, small, albumen fleshy. Distrib.
Confined (except the W. African *A. Mannii*) to the Malay Peninsula and
Indian Archipelago ; species 12.

1. ADINANDRA DUMOSA, Jack in Malay Misc. ii, No. 7, p. 50. A large
shrub or small tree, glabrous everywhere except the stamens ; young
branches slender, terete, dark brown. *Leaves* coriaceous, glabrous, reddish
beneath, oblong-lanceolate to elliptic, more or less acute or obtusely acuminate, the base narrowed ; edges entire or obsoletely serrate ; midrib
prominent especially beneath, nerves invisible ; length 2 to 4 in., breadth
1·25 to 1·75 in., petiole ·1 to ·2 in. *Flowers* ·65 in. in diam., peduncles
·4 to ·75 in. long, not thickened after flowering ; bracteoles leathery,
broadly ovate, opposite, close to the calyx. *Sepals* sub-erect, glabrous,
leathery, ovate-rotund, blunt, sometimes emarginate. *Petals* longer
than the sepals, membranous, oblong-lanceolate with broad bases, the
apex minutely apiculate, erect, conniving. *Stamens* about 30, the inner
shorter ; filaments united by their bases, pilose ; *anthers* with 2 narrow
lateral cells ; the connective broad and pilose behind, its apex mucronate.
*Ovary* 5-celled, the placentas incurved, multi-ovulate ; style subulate ;
stigma small, simple. *Fruit* ·4 to ·5 in. in diam., baccate, dry, with coriaceous pericarp, imperfectly 4-5-celled. *Seeds* numerous, reniform.
Wall. Cat. 3664, (corrected at p. 215 to 3666) and 7071. Dyer in Hook.
fil. Fl. Ind. i, 282; Miq. Fl. Ind. Bat. i, Pt. 2, p. 477 ; Choisy Mem.
Ternst. 24. *A. Jackiana* and *trichocoryna*, Korth. Verh. Nat. Gesch.
Bot. 106, 107. *A. cyrtopoda*, *stylosa* and *glabra*, Miq. Flor. Ind. Bat.
Suppl. i, 478, 479. *Ternstræmia ? dumosa*, Wall. Cat. 2245. *Camellia ?
Scottiana*, Choisy l. c. (not of Wall. Herb.).

In all the provinces except the Andamans and Nicobars, at low
elevations, common. Distrib. Malay Archipelago.

2. ADINANDRA ACUMINATA, Korth. Verh. Nat. Gesch. Bot. 109. A
tree 40 to 60 feet high ; all parts except the stamens glabrous ; young
branches slender, smooth, dark-coloured ; the older pale and rough.
*Leaves* coriaceous, oblong-lanceolate, acuminate, the base acute, both surfaces shining ; midrib prominent below ; the 9 to 11 pairs of nerves rather
prominent below when dry, forming a double series of arches inside the
margin ; length 3·5 to 6 in., breadth 1 to 2·5 in., petiole ·25 in. *Flowers*
·9 in. in diam.; peduncles ·75 to 1 in. long, thickened and verrucose after
flowering ; bracteoles leathery, lanceolate, at some distance from the calyx,
alternate. *Sepals* leathery, glabrous ; the two outer small, ovate ; the
three inner much larger, spreading, rotund, the edges serrulate. *Petals*
larger than the inner sepals, rotund, spreading, fleshy, the edges thin.
*Stamens* about 40, the inner smaller ; filaments united by their bases,

short, coarsely pilose as are the narrow elongate apiculate anthers. *Ovary* 5-celled, depressed, ribbed, pubescent; style filiform, pilose; stigma small, conical. *Fruit* ·5 to ·75 in. in diam., baccate, dry, with coriaceous, pubescent, but ultimately glabrous, pericarp, 2-celled, many seeded; *Seeds* oval, flat, furrowed on both sides. Hook. fil. Fl. Br. Ind. I, 282. Miq. Fl. Ind. Bat. I, Pt. 2, p. 478. *Gordonia acuminata*, Wall. Cat. 3664. *Ternstræmia? coriacea*, Wall. Cat. 1453. *Camellia axillaris*, Wall. Cat. 1453, p. 158 (not of Roxb. ex Bot. Reg. 349, see Journ. Linn. Soc. xii, 330). *Polyspora axillaris*, Chois. Mem. Ternstr. 91 (not of Don.)

In all the provinces, from 1000 feet to (in Perak) 4000 feet; common. Distrib. Sumatra.

In Journ. Linn. Soc. xiii, 330, there is a note by Mr. W. T. Thiselton Dyer (who elaborated this family of *Ternstræmiaceæ* in the Flora of Brit. India) on the plant issued by Wallich as *Ternstræmia coriacea*, and identified by him (in an appendix to his Catalogue), with *Camellia axillaris*, Roxb. Mr. Dyer shows that, under the name *C. axillaris*, Roxb., a totally different plant (=*Gordonia anomala*, Spreng) was figured in the Bot. Register (t. 349), and that Wallich's *T. coriacea* was neither Roxburgh's plant nor that figured in the Bot. Register, but really *A. acuminata*, Korth.

3. ADINANDRA MACULOSA, T. Anders. Hook. fil. Fl. Br. Ind. I, 282. A tree 40 to 60 feet high; young branches dark-coloured, pubescent near the apex, not silky. *Leaves* coriaceous, elliptic to sub-rotund, shortly bluntly and abruptly acuminate, entire, the base acute; upper surface smooth, shining; the lower pale brown, dull, opaque, minutely rugulose when dry; main nerves 8 to 10 pairs, spreading, very indistinct; midrib distinct; length 3·5 to 5·5 in., breadth 1·5 to 2·25 in.; petiole ·25 to ·4 in., glabrous. *Flowers* ·75 in. in diam.; peduncles little longer than the petioles, pubescent; bracteoles sub-rotund, opposite, close to the calyx. *Sepals* unequal, the two inner smaller, rotund to broadly ovate, very fleshy, puberulous externally, the edges thin and glabrous as is the whole internal surface. *Petals* membranous, ovate, acute, glabrous, connivent. *Stamens* about 30; filaments attached to the petals, short, glabrous. *Anthers* narrow, the cells elongate, lateral; the connective sericeous with short glabrous apiculus. *Ovary* depressed-hemispheric, 5-ribbed, 5-celled. *Style* cylindric, glabrous; stigma small, conical. *Fruit* ·5 in. in diam., globular, baccate, 4-celled, the leathery pericarp at first pubescent but finally glabrous; *seeds* numerous, black, shining, horse-shoe-shaped, small. *Ternstræmia integerrima*, Wall. Cat. 1452 (in part).

Penang; Wallich, Curtis. Perak; Scortechini, Wray, King's Collector; at elevations of from 1800 to 4000 feet.

4. ADINANDRA INTEGERRIMA, T. Anders. Hook. Fl. Br. Ind I, 282.
A small tree; young branches with dark-coloured bark, the extremities
fulvous-pubescent, the leaf-buds sericeous. *Leaves* sub-coriaceous, ovate
to ovate-oblong, shortly acuminate, the base acute or rounded, the margin
minutely glandular-serrulate; upper surface glabrous, shining, greenish
when dry; the lower pale brown, sparsely pubescent and with many
minute black glands; main nerves 10 to 14 pairs, thin, interarching
·2 in. from the margin, slightly prominent on both surfaces (when dry)
as are the reticulations; length 3·5 to 5 in., breadth 1·4 to 2·25 in.; petiole
·2 to ·25 in., pubescent. *Flowers* ·6 in. in diam.; peduncles not much
longer than the petioles, strigose; bracteoles ovate, acute, opposite, close
to the calyx. *Sepals* spreading, broadly ovate, acute; the two outer
larger, very thick, the edges thin, serrulate-denticulate; externally
adpressed-sericeous, internally smooth and shining. *Petals* smaller than
the sepals, coherent by their bases, connivent, ovate, acute, membranous, glabrous except an adpressed sericeous patch on the back, connivent.
*Stamens* about 30, adnate to the base of the petals; filaments short,
glabrous; anthers elongate, fusiform, the cells lateral; connective sericeous with a long glabrous apiculus. *Ovary* depressed-hemispheric,
adpressed-sericeous, 5-celled, multi-ovulate; style cylindric, expanded
below, sericeous; stigma small, sub-capitate. *Fruit* (*fide* Dyer) baccate,
adpressed-pubescent, ·7 in. in diam. *Seeds* small, shining. Dyer in
Hook. fil. Fl. Br. Ind. I, 282; Pierre Fl. Forest Coch.-Chine, t. 125,
(excl. syn. *T. villosa*, Choisy). *Ternstrœmia dasyantha*, Choisy (not of
Korth.). *Ternstrœmia? integerrima*, Wall. Cat. 1452 (in part) and
2246. *Gordonia reticulata*, Wall. Cat 3663 and 7070.

Penang; Wallich. Perak, Scortechini; at low elevations.

The specific name is unfortunate, as in all the specimens I have
seen the leaves are as described above and not entire.

5. ADINANDRA VILLOSA, Choisy Mem. Ternstr. 24. A pubescent tree,
40 to 50 feet high; young branches pilose, pale brown, leaf-buds sericeous.
*Leaves* coriaceous, oblong-lanceolate, shortly acuminate, entire or faintly
glandular-crenate; the base rounded, rarely acute; upper surface shining, glabrous except the pubescent midrib; under surface yellowish,
sparsely pubescent; nerves 7 to 9 pairs, ascending, interaching within
the margin, not prominent; midrib bold; reticulations rather distinct;
length 4 to 5·5 in., breadth 1·5 to 1·75 in.; petiole ·2 in., pilose. *Flowers*
·75 in. in diam., from ·4 to nearly 1 in. long, pilose; bracteoles ovate,
from the middle of the peduncle, fugaceous. *Sepals* spreading, sub-
equal, rotund, fleshy, pilose externally, smooth internally. *Petals* ovate,
blunt, membranous, adpressed-sericeous externally, the edges glabrous.
*Stamens* about 30, attached to the bases of the petals, sericeous, the

filaments short ; cells narrow, elongate, connective with a glabrous apiculus. *Ovary* depressed-hemispheric, adpressed-sericeous, 5-ridged, 5-celled ; *style* cylindric, glabrous except at the base ; stigma minute. *Fruit* ·5 in. in diam., baccate, adpressed-sericeous, 4-celled. *Seeds* numerous, reniform, brown, small. Hook. fil. Fl. Br. Ind. I, 283 ; Kurz Fl. Burm. i, 100. *Ternstrœmia ? sericea,* Wall. Cat. 1454. *Schima Wallichii,* Choisy Mem. Ternst, 91 (not of Choisy in Zoll. Cat.)

Perak ; King's Collector, Wray, at elevations from 3000 to 4000 feet. Distrib. Tavoy.

6. ADINANDRA HULLETTII, King, n. sp. A tree ; young branches densely and minutely rusty-tomentose. *Leaves* coriaceous, elliptic-oblong, shortly acuminate, the base acute ; edges glandular, denticulate, slightly recurved when dry ; upper surface smooth, shining ; lower brown, densely and minutely tomentose, the midrib prominent ; the nerves spreading, obscure, about 10 pairs ; length 3·5 to 5·5 in., breadth 1·5 to 2·5 in., petiole ·25 in. *Flowers* ·75 in. in diam. ; peduncles ·4 to ·5 in., tomentose ; bracteoles broadly ovate, acute, opposite, close to the calyx. *Sepals* spreading, fleshy, all glabrous internally ; the outer 2 rotund, tomentose externally, larger than the others ; the inner 3 ovate-rotund, tomentose externally, the edges glabrous. *Petals* longer than the sepals, membranous, oblong, blunt, glabrous, the tips reflexed. · *Stamens* from 40 to 50, epipetalous ; anthers densely pubescent, the connective with a long blunt glabrous apiculus ; filaments short, geniculate. *Ovary* conic-hemispheric, adpressed-sericeous ; style glabrous ; stigma small, conic. *Fruit* unknown.

Singapore ; Murton, No. 144 ; Hullett, No. 103. Penang ; Curtis, No. 275, in part.

A very distinct species of which fruit is as yet unknown.

7. ADINANDRA MACRANTHA, Teysm. and Binn. Nat. Tijd. Ned. Ind. xxv, 421. A tree 20 to 50 feet high ; young branches with pale glabrous bark, the apices and buds sericeous. *Leaves* coriaceous, elliptic-oblong, with a broad apex suddenly contracted to a short blunt acumen, narrowed in the lower third to the sub-acute base ; the edges entire or faintly crenate ; both surfaces glabrous ; the upper greenish, the lower pale yellowish when dry ; midrib bold, sometimes puberulous ; main nerves 15 to 20 pairs forming a double series of arches inside the margin, rather prominent as are the reticulations ; length 5 to 7 in., breadth 2 to 3·25 in. ; petiole ·25 in., stout. *Flowers* 1·4 in. in diam. ; peduncles 1 to 1·5 in. long, smooth ; bracteoles rotund-reniform, opposite, close to the calyx. *Sepals* spreading, fleshy with thin ciliolate edges, smooth, rotund, the two outer smaller. *Petals* larger than the sepals, sub-coriaceous with thin edges, rotund, spreading. *Stamens* 40 to 50, adpressed-

sericeous everywhere, the apiculus of the connective with a terminal tuft ; filaments short. *Ovary* depressed-hemispheric, 3-4-celled, smooth as is the cylindric style ; stigma small, conical. *Fruit* ·75 in. in diam., imperfectly 3- to 4-celled, pericarp smooth. *Seeds* few, large, brown, horse-shoe-shaped, punctate, shining.

Perak ; from 500 to 1500 feet, King's Collector, Scortechini. Distrib. Sumatra.

The Perak specimens agree perfectly with Teysmann's types collected in Sumatra.

8. ADINANDRA MIQUELII, King, n. sp. A medium-sized tree ; young branches stout, the bark white and polished. *Leaves* thickly coriaceous, oblanceolate, apex with a short blunt abrupt point, gradually narrowed in the lower half to the petiole, entire ; midrib prominent ; nerves 5 to 7 pairs, ascending, anastomosing ·2 in. from the margin, invisible in the fresh, inconspicuous in the dry state ; length 4 to 6 in., breadth 1·5 to 2·25 in. ; petiole ·6 to ·75 in., stout. *Flowers* about ·75 in. in diam., scattered below the apices of the branches, axillary and extra-axillary, polygamous ; peduncles spreading, solitary, compressed, pale, ·75 to 1 in. long; the bracteoles just below the flower, small, fugaceous. *Sepals* fleshy, rotund, the 2 outer much smaller. *Petals* larger than the sepals, rotund, clawed, fleshy. *Stamens* numerous, chiefly from the torus, pubescent, the connective with a long apiculus, filaments short. *Ovary* depressed-globose, 2-celled, narrowed above into the short cylindric style ; *stigma* shortly bifid; the lobes narrow, acute, spreading. *Fruit* unknown. *Ternstrœmia bancana*, Miq. Fl. Ind. Bat. Suppl. 477.

Penang ; Curtis, No. 1612. Distrib. Bangka.

The stigma shows that this does not belong to the genus *Ternstrœmia* into which Miquel put it. It is evidently a rare plant in Penang, as Curtis's specimen (which agrees perfectly with Miquel's type-specimens from Bangka) is the only one which I have seen from that island.

### 3. TERNSTRŒMIA, Linn.

Evergreen glabrous trees or shrubs. *Leaves* more or less coriaceous, entire or crenate-serrate. *Peduncles* axillary, solitary or sub-fasciculate, recurved, 2-bracteolate, flowers usually diœcious. *Sepals* 5, imbricate. *Petals* 5, imbricate, connate by their bases. *Stamens* many, mostly adherent to the base of the corolla, anthers glabrous. *Ovary* 2- to 3-celled, style simple or absent. *Stigma* broadly 2- to 3-lobed or sub-entire ; ovules usually 2 in each cell, pendulous. *Fruit* indehiscent, sub-baccate. *Seeds* rather large, the albumen copious or scanty. Distrib. Tropical Asia and America ; species about 30.

132

1. TERNSTRŒMIA PENANGIANA, Choisy. Mém. Ternst. 20. A tree 40
to 60 feet high ; young branches rough, stout, pale brown. *Leaves* coriaceous, oblanceolate to obovate, sub-acute or bluntly mucronate, rarely
blunt or emarginate, entire, the base narrowed to the petiole; nerves 5 to
7 pairs, spreading, invisible when fresh and inconspicuous when dry, the
midrib prominent; length 3·5 to 6 in., breadth 1·5 to 2·5 and (in
Wallich's specimen) to 4 in., petiole ·6 to ·75 in. *Flowers* ·8 to 1·25 in.
in diam., diœcious, solitary, axillary ; pedicels ·75 in. long, recurved or
straight. *Sepals* rotund, fleshy with thin edges. *Petals* much larger
than the sepals but similar in texture, rotund with a broad claw, the edges
sub-denticulate. *Stamens* in the male very numerous, crowded, short, (reduced to filaments in the female) ; connective slightly produced beyond
the anther cells, truncate ; ovary globular, its cells biovulate. *Stigmas* 2,
large, reniform, with erose glandular edges. Berry dry with coriaceous
epicarp, globular, 1 to 1·5 in. in diam., subtended by the thickened rugulose connate sepals ; *Seeds* about 4, oblong. Dyer in Hook. fil. Fl. Br.
Ind. I, 281 ; Kurz For. Fl. Burmah i, 99 ; Miq. Fl. Ind. Bat. J, Pt. 2, p.
469 ; Pierre Fl. For. Coch.-Chine, t. 123. *T. macrocarpa*, Scheff. Obs.
Phyt. i, p. 5. *Erythrochiton Wallichianum*, Griff. Notul. iv. 565, t. 585
A, fig. 7. *Fagraea dubia*, Wall. Cat. 4456. *Garcinia acuminata*, Wall.
Cat. 4871 A, in part, (*fide* Hooker in Journ. Linn. Soc. xiv. 486.)

Penang ; Wallich, Griffith, Curtis. Andamans and Nicobars ; Kurz,
King's Collector. Distrib. Java.

This species was founded by Choisy on the imperfect Wallichian
specimens from Penang issued by Wallich as his No. 4456. These
specimens consist of leaves and fruit with some imperfect flowers. The
leaves are obovate, almost rotund, and broader than those of any *Ternstrœmia* which has been collected since. It is therefore not quite certain that the Andaman and Nicobar plant is really the same as Wallich's,
although in stigma and fruit it agrees. The plant described and figured
as *T. Penangiana* by Pierre (1. c.) is obviously the same as the Andaman
and Nicobar species, but whether it is the same as Wallich's No. 4456, I
am not prepared to say.

2. TERNSTRŒMIA SCORTECHINII, King, n. sp. A tree, 20 to 40 feet
high ; young branches with pale brownish-grey bark, striate when dry.
*Leaves* coriaceous, verticellate, drying of a pale green, oblanceolate, the
apex shortly abruptly and rather bluntly acuminate, narrowed from above
the middle to the rather stout short petiole ; edges entire ; under surface
rather pale ; midrib distinct on both surfaces ; nerves visible on neither ;
length 3 to 5 in., breadth 1·25 to nearly 2 in. ; petiole ·4 to ·5. *Flowers*
diœcious, ·6 to ·7 in. in diam., pedunculate, axillary, solitary or in fascicles of 2 to 6 ; peduncles slender, compressed, 1 to 1·5 in. long ; the 2

bracteoles about ·2 in. below the calyx, minute, fugacious. *Sepals* sub-equal, fleshy with thin edges, rotund. *Petals* much larger than the sepals, orbicular to reniform, clawed. *Stamens* in the male very nu-merous; anthers sub-sessile, the connective broad, bearing the 2 cells on its edges and produced above them into a broad short truncate pro-cess; rudimentary ovary flattish, without stigma. *Female flowers* like the males but with fewer stamens; *ovary* hemispheric, imperfectly 2-celled; stigmas 2, sub-sessile, flat, foliaceous, each divided into 3 or 4 lobes with thick corrugated edges. *Fruit* a dry ovoid berry with coriaceous dark-coloured epicarp, ·75 in. long, and ·2 in. in diam., subtended by the leathery calyx. *Seeds* 2, large, broad, horse-shoe-shaped, flattish, ·5 in. long.

Perak, at low elevations; Scortechini, King's Collector.

A very distinct species with leaves curiously like those of *Illicium evenium*, and with smaller flowers than the other species.

3.  TERNSTRŒMIA CORIACEA, Scheff. Obs. Phyt, ii, p. 16, (not of Wall.).

A tree 50 to 70 feet high ; young branches light brown, smooth. *Leaves* coriaceous, usually oblong-oblanceolate with an abrupt short blunt api-culus, sometimes oblong-lanceolate and acute ; attenuate in the lower third to the stout petiole; midrib bold ; main nerves 5 to 9 pairs, spreading, anastomosing ·2 in. from the entire margin, rather inconspicuous even when dry ; length 4 to 6 in., breadth 1·75 to 2·5 in., petiole ·75 to 1 in. *Flowers* 1·25 to 1·5 in. in diam., diœcious, solitary, axillary or from the axils of fallen leaves ; peduncles flattened, deep brown, 1·5 to 2 in. long, slender ; bracteoles alternate, minute, about ·25 in. below the calyx. *Sepals* fleshy with thin edges, rotund ; the 2 outer rather smaller than the inner 3. *Petals* larger than the sepals, much imbricate, rotund, fleshy, not clawed. *Stamens* in the male numerous, from the torus, the connective with a broad rounded apical appendage ; quite absent in the female flower. Ovary globular, 2-celled ; *stigmas* 2, sub-sessile, each deeply divided into 6 to 8 sub-spathulate lobes. *Fruit* baccate, globu-lar-ovoid, ·75 in. in diam. and nearly 1 in. long, dry, with a coriaceous rind, subtended by the slightly enlarged hardened calyx and crowned by the remains of the stigma. Seeds about 4, oblong, the testa rugulose.

Malacca; Griffith (Kew Distrib.) 183. Penang ; Curtis, No. 1055. Perak ; King's Collector. Distrib. Bangka.

Distinguished from the preceding by its anthers, by the venation of its leaves, and by its young branches. Teysmann's specimens from Bangka in no way differ from those from Perak, Malacca and Penang. Wallich's fragmentary specimens, (Cat. No. 7430,) probably fall here. The plant issued by Wallich as *Ternstrœmia coriacea (Cat. No. 1453)* is, as suggested by the late Dr. Anderson and Mr. Dyer (Journ. Linn. Soc. xiii, 331), *Adinandra acuminata*, Korth.

## 4. EURYA, Thunb.

Shrubs. *Leaves* narrow, usually crenate-serrate. *Flowers* small, unisexual, sessile or shortly pedicelled, in axillary fascicles, rarely solitary, with persistent bracteoles. *Sepals* 5. *Petals* 5, united at the base. *Stamens* 15 or less, rarely 5; anthers glabrous. *Ovary* 3- (rarely 2-5)-celled; styles 3 (rarely 2-5) free or united; ovules many in the inner angle of each cell. *Fruit* baccate. *Albumen* fleshy. Distrib. S. E. Asia, Indian Archipelago and Pacific Islands; described species more than 30, probably reducible to 10.

1. EURYA ACUMINATA, DC. Mém. Ternst. 29. A tree 30 to 40 feet high; young branches slender, pubescent to minutely tomentose. *Leaves* thinly coriaceous, narrowly oblong-lanceolate or oblanceolate, acuminate, serrulate, the base acute; upper surface glabrous, shining; the lower paler, pubescent especially on the midrib, or sub-glabrous; length 2·5 to 3·5 in., breadth ·5 to ·75 in., petiole ·1 in. or none. *Flowers* ·25 in. in diam., in 2- to 6-flowered clusters; pedicels short, pubescent, bracteolate. Buds globose. *Sepals* unequal, the outer smaller, rotund with a thickened wrinkled patch near the base, pubescent externally. *Petals* larger and thinner than the sepals, oblong, blunt, veined, glabrous. *Male flowers;* stamens about 12, glabrous; filaments slender, anthers oblong, blunt; rudimentary ovary conic, without styles, or absent. *Female flower* as in the male, but the sepals and petals smaller and narrower; stamens 0; ovary ovoid-conic; smooth, 3 or (by abortion) 2-celled, multi-ovulate; styles 3, united or free in the lower two-thirds, cylindric, about as long as the ovary. *Stigmas* on the inner surfaces of the upper part of the styles. *Fruit* globular, ·15 in. in diam., smooth, subtended by the persistent calyx and crowned by the styles. *Seeds* small, angled, pitted, shining, brown. *Diospyros serrata*, Ham. in Don Prod. Fl. Nep. 143.

In all the provinces at low elevations, common. Distrib. Subtropical Himalaya, Assam, Chittagong and Burmese Ranges, Malay Archipelago, Fiji Islands.

In a plant with such a wide distribution, variations in form are only to be expected. Many of these have been treated as species which, in Sir J. D. Hooker's Flora of British India, Mr. Thiselton Dyer has reduced to varieties as follows:

Var. 1. *euprista*, Korths. Verh. Nat. Gesch. Bot. 113 (sp.); styles distinct. Griff. Ic. 604, f. 3. *E. multiflora*, DC. 1. c. 25. *E. serrata*, Blume Fl. Jav. præf. vii. *E. angustifolia*, Wall. Cat. 1465.

*E. acuminata*, Royle Ill. 127, t. 25.   *E. salicifolia*, Blume Mus. Bot. II, 118.   *E. chinensis*, Hook. f. and Thoms. Herb. Ind. Or. (not of Brown).

Var. 2. *Wallichiana*, Stend. in Blume Mus. Bot. ii, 118 (sp.) ; styles united. *E. lucida*, Wall. Cat. 1462. *E. fasciculata*, Ham. in Wall. Cat. 1463. *E. acuminata*, Wall. Cat. 1464. *E. bifaria*, Wall. Cat. 3721 ? *E. membranacea*, Gardn. in. Calc. Journ. Nat. Hist. vii, 444. *E. japonica*, β *acuminata*, Thw. Euum. Pl. Cey. 41.

2.   EURYA WRAYI, King, n. sp.   A small tree ; young branches slender, purplish-brown, laxly pubescent towards the apex. *Leaves* drying greenish-yellow, thinly coriaceous, oblong-lanceolate, bluntly acuminate, minutely serrulate, the base rounded ; upper surface glabrous, shining ; lower paler, dull, sparsely pubescent ; length 2 to 2·75 in., breadth ·5 to ·7 in., petiole ·1 in. *Flowers* narrowly ovate, pointed, scarcely expand-ing, ·1 in. in diam., and ·2 in. long, axillary, solitary, or in 2-6-flowered sessile umbels, quite glabrous ; pedicels slender, glabrous, ·1 to ·15 in. long, bi-bracteolate. *Sepals* unequal, erect, fleshy, ovate, acute, much imbricate. *Petals* sub-equal, erect, membranous, ovate, acute, connate in the lower third. *Stamens* 15, glabrous ; anthers narrow, elongate, shortly apiculate ; filaments short. *Ovary* ovoid, gradually narrowing into the thick style, imperfectly 3-celled ; stigmas short. *Fruit* un-known.

Perak ; at Tapa, Wray.

Distinguished by its narrowly ovate pointed flower-buds and flowers, and by the rounded bases of its leaves.

4.   ACTINIDIA, Lindl.

Glabrous, strigose, or tomentose shrubs ; usually climbers. *Leaves* entire or serrate, usually membranous, feather-veined. *Flowers* polyga-mous or diœcious, in axillary cymes, rarely solitary. *Sepals* 5, slightly imbricate, subconnate at the base. *Petals* 5, somewhat contorted-im-bricate. *Stamens* many ; anthers dehiscing by slits. *Ovary* many-celled ; the styles as numerous, divergent and elongated after flowering. *Fruit* baccate.   Distrib. Himalaya, China and Japan ; species about 8.

1.   ACTINIDIA MIQUELII, King, n. sp.   Slender, scandent, 30 to 60 feet long ; young branches cylindric, striate, glabrous, dark-coloured. *Leaves* membranous, ovate-acuminate to sub-rotund, mucronate, minutely glan-dular-dentate, the base rounded or slightly cordate ; upper surface glabrous, rigid, the nerves and midrib minutely pubescent, lower sur-face pale browb when dry, minutely but densely tomentose ; nerves about 5 pairs, the lower spreading, the upper sub-erect, prominent be-neath as are the midrib and transverse veins ; length 3 to 4 in., breadth

2·5 to 3·5 in.; petiole 1·25 to 1·5 in., slender. *Cymes* axillary, dichotomous. spreading, rusty-tomentose, on slender ebracteate peduncles 1·5 in. long which lengthen to 3 in. in fruit. *Flowers* numerous, diœcious, ·5 in. in diam.; pedicels ·3 to ·4 in. long. *Sepals* thick, ovate, blunt, densely rusty-tomentose externally. *Petals* larger than the sepals, membranous, oblong-obovate, blunt. *Stamens* in males very numerous, glabrous; the anthers broadly oblong, blunt, deeply cordate at the base; filaments slender. *Ovary* in the males absent or rudimentary, densely pilose, and with several rudimentary styles. *Female flowers* unknown. *Fruit* ovoid, ·75 in. long, and ·4 in. in diam., baccate, smooth, pulpy, subtended by the persistent calyx and crowned by the remains of 15 to 20 filiform styles. *Seeds* numerous, shining, brown, less than ·1 in. long, ovoid, sub-compressed, pitted and with several longitudinal grooves. *Kadsura pubescens*, Miq. Fl. Ind. Bat. Suppl. 620.

Perak ; on trees, at elevations of 3,500 to 4000 feet, King's Collector, Nos. 5437 and 8789. Distrib. Eastern Sumatra.

I have carefully examined a type specimen of Miquel's *Kadsura pubescens* from Sumatra named by the author's own hand ; and there is no doubt whatever that it is an *Actinidia* and not a *Kadsura* ; nor is there any that it is. identical with the above quoted numbers of the Calcutta Collector from Perak. Miquel is quite wrong in describing his plant as having 3 sepals and 6 petals ; there being 5 in each whorl.

### 6. SAURAUJA, Willd.

Trees or shrubs. *Branches* usually brown with whitish tubercular dots, both branches and leaves more or less strigose-pilose or scaly when young. *Leaves* approximate at the ends of the branches, usually serrate, with parallel veins diverging from the midrib. *Inflorescence* lateral, often from the axils of fallen leaves, cymose, subpaniculate, rarely few-flowered. *Bracts* usually small, remote from the calyx. *Flowers* usually hermaphrodite. *Sepals* 5, strongly imbricate. *Petals* 5, usually connate at the base. *Stamens* many ; anthers dehiscing by pores. *Ovary* 3-5-celled ; styles as many, distinct or connate, rarely dry and sub-dehiscent. Distrib. Tropical and sub-tropical Asia and America. Species about 60.

1. SAURAUJA TRISTYLA, DC. Mém. Ternstr. 31, t. 7. A shrub or tree 2 to 3 feet high ; young branches with grey, faintly striate bark, deciduously scurfy and strigose towards the apices. *Leaves* membranous, oblanceolate, abruptly and shortly acuminate, minutely and remotely serrulate or sub-entire, the base acute ; both surfaces glabrous, except the midrib and main nerves which have a few scale-like hairs, the lower pale brown when dry ; nerves 10 to 12 pairs, erecto-patent, rather prominent be-

137

neath; length 5 to 8 in., breadth 1·5 to 3 in., petiole ·5 to 1 in. *Flowers*
·2 to ·3 in. in diam., narrowly ovate in bud, in fascicles of 2 to 5 from
small axillary tubercles, but mostly from the axils of fallen leaves; the
pedicels slender, minutely bracteolate, ·75 in. long, scurfy. *Male flower;*
*sepals* erect, unequal, the two outer smaller, more or less broadly ovate,
blunt; *petals* larger than the sepals, sub-erect, membranous, veined,
oblong, blunt; stamens about 25, glabrous; the anthers broadly ovate,
blunt, with sutural dehiscence; rudimentary ovary none. *Female flowers;*
*sepals* and *petals* as in the male; stamens absent. *Ovary* ovoid, glabrous;
*styles* 3, distinct to the base, or united half way. *Fruit* globular, sub-
dehiscent, scarcely exceeding the calyx. *Seeds* broadly ovate, angled,
deeply pitted. Dyer in Hook. fil. Fl. Br. Ind. i, 287; Miq. Fl. Ind. Bat.
i, Pt. 2, p. 483; Kurz For. Fl. Burm. i, 104. *Scaphu Candollei* and
*S. Pinangiana,* Choisy. Mém, Ternst. 31. *Ternstroemia pentapetala,* Jack in
Malay Misc. i, No. 5, 40. *T. trilocularis,* Roxb. ex Wall. Pl. As. Rar.
ii, 40. *T. bilocularis,* Roxb. Fl. Ind. ii, 522?

In all the provinces (except the Andamans and Nicobars from
which it has not as yet been sent); at low elevations, common.

The plant figured under this name by Pierre (Fl. Forest Coch.-
Chine) is obviously a different species; for it has 5 styles, and it differs
also in other respects.

2.  SAURAUJA NUDIFLORA, DC. Mém. Soc. Genève i, 422. A tree 20 to
30 feet high; youngest branchlets dark-coloured, squamulose towards the
apex; the older esquamulose, pale, faintly striate. *Leaves* membranous,
oblanceolate, shortly and sharply acuminate, minutely glandular-serrate,
narrowed in the lower half to the acute base; both surfaces glabrous;
the midrib and 12 to 13 pairs of bold spreading nerves puberulous on
the upper, sparsely covered with flattened hairs on the lower surface;
length 6 to 10 in., breadth 2·25 to 3·75 in., petiole ·5 to 1·25 in. *Flowers*
·25 to ·4 in. in diam., white, glabrous, solitary or in 2- to 3-flowered fascicles from tubercles in the axils of leaves or of fallen leaves; pedicels
·5 to 1 in. long, slender, sparsely scurfy, and with several acute bracteoles. *Sepals* rotund, fleshy with thin edges, united at the base. *Petals*
oblong-obovate, emarginate, united below, larger than the sepals, *Sta-*
*mens* 25 to 30, attached to the base of the corolla; anthers oblong-ovate,
curved, the dehiscence sutural, not apiculate; filaments short. *Ovary*
hemispheric, pubescent. *Styles* 3 to 5, united in the lower half *Fruit*
covered by the accrescent calyx. *Seeds* ovate, deeply foveolate, pale
brown, shining. Miq. Fl. Ind. Bat. I, Pt. ii, p. 484.? *S. Noronhiana,*
Bl. Bijdr. 126.

Perak 800 to 3,500 feet, common. Distrib. Sumatra and Java.

This differs from *S. tristyla* in its rotund sepals, larger flowers, pu-
138

bescent ovary, sub-globular seeds, and in its often having 5 styles. There may be two species covered by the foregoing description ; but I cannot find a constant character to separate them. I believe this to be Blume's *S. Noronhiana* and De Candolle's *S. nudiflora ;* but, not having been able to consult any authentic specimen of the former and only moderately good ones of the latter, I am not quite satisfied of the identity with them of this common Perak tree. The genus *Sauranja* is a very puzzling one. The species come very close together, and Miquel's descriptions of the numerous species which he named are so incomplete that it is almost impossible to recognise them with any certainty.

3. SAURAUJA CAULIFLORA, Bl. Bijdr. 128, var. *calycina*, King. A tree ; young branches and petioles densely covered with long paleaceous yellowish hairs. *Leaves* elliptic-oblong, shortly and sharply acuminate, the edges faintly aristate-serrate, the base acute; upper surface glabrous ; lower pale brown when dry, strigose on the midrib nerves and veins ; main nerves 12 to 14 pairs, spreading, prominent beneath ; length 6 to 9 in., breadth 2·25 to 2·75 in., petiole about 1 in. *Flowers* ·4 in in diam., on long pedicels, crowded in large fascicles from flat tubercles on the larger branches and stem ; pedicels from ·75 to 1·5 in. long, tomentose-squamulose, rufous. *Sepals* rotund, the outer densely tomentose-squamulose ; the inner almost glabrous, veined. *Petals* obovate-oblong, blunt, united in their lower third, membranous, nerved, scarcely so large as the sepals. Stamens about 25, adherent to the corolla, elongate-ovate, adnate, dehiscing by two large apical pores. *Ovary* scaly, 3-celled, multiovulate. *Styles* 3, united by their bases only. *Fruit* enveloped by the slightly accrescent calyx, sub-glabrous, 3-celled. *Seeds* small, ovate-rotund, compressed, foveolate, pale brown. DC. Mém. Soc. Geneve I, 425 ; Korth. Verh. Nat. Gesch. Bot. 126 ; Hassk. Pl. Jav. Rar. 273 ; Miq. Fl. Ind. Bat. I, Pt. ii, p. 486 ; Ann. Mus. Ludg. Bat. IV, 106.

Perak ; Batu Kurau. Scortechini, No. 1614.

This differs in no respect from the plant described by Blume, of which I have seen good specimens, except in its larger sepals which are densely tomentose-squamulose externally.

## 7. PYRENARIA, Blume.

Shrubs or trees. *Leaves* serrate, large and sub-membranous. *Flowers* sub-sessile, axillary, erect or nodding. *Sepals* usually 5, unequal, graduating from the bracts to the petals. *Petals* connate at the base, *Stamens* very numerous, mostly connate, adnate to the base of the petals, *Ovary* 5-celled ; styles 5, free, or partially united ; ovules 2 in each cell, attached laterally. *Fruit* drupaceous, indehiscent. *Seeds* oblong, stout, with a thick woody testa, wingless.; albumen 0 ; cotyledons large, crumpled-pled or conduplicate ; radicle inferior, inflexed.   Distrib. Malay Peninsula and Indian Archipelago.   Species about 7.

1.  PYRENARIA ACUMINATA, Planch. ex Choisy Mém. Ternstr. 84.   A shrubby tree, 15 to 30 feet high ; young branches densely tawny- or fulvous-tomentose.   *Leaves* elongate-oblanceolate, sometimes oblong-elliptic, acuminate, minutely serrulate, the base attenuate ; upper surface glabrous, shining, the midrib and nerves puberulous, greenish when dry ; the lower softly pubescent, minutely papillose ; the midrib stout, tomentose ; main nerves about 10 pairs, sometimes forking and always interarching about ·25 in. from the margin ; length 6 to 12 in., breadth 2 to 3·5 in. ; petiole ·4 in., tomentose.   *Flowers* 1·5 in. in diam., shortly pedicellate, solitary, crowded towards the ends of the branches in the axils of leaves or of abortive leaves ; pedicels recurved, tomentose ; bracteoles lanceolate, close to the calyx, tawny-silky externally as are sepals and petals. *Sepals* and *petals* graduated in size from the bracts inwards, broadly ovate, acuminate, glabrous and brownish internally ; *anthers* ovate, adnate, only about one-fourth the length of the slender filaments.   *Ovary* ovoid, sericeous ; styles united in the lower half, free above ; stigmas small.   *Fruit* depressed-globose, 1·5 in. in diam., and 1 in. long ; the pericarp sericeous, becoming glabrescent, leathery, sub-succulent.   *Seeds* few, large, sub-reniform, compressed.   Miq. Fl. Ind. Bat. I, Pt. ii, p. 493 ; Dyer in Hook. fil. Fl. Br. Ind. i, 290.   *Ternstrœmia ? macrophylla,* Wall. Cat. 3663.   *Gordonia (Camellia ?) acuminata,* Wall. Cat. 3664.

Singapore, Malacca, Penang and Perak ; at low elevations.

2.  PYRENARIA KUNSTLERI, King, n. sp.   A tree 15 to 30 feet high ; all parts glabrous except the very apices of the branches, the youngest leaf-buds, and the flowers.   *Leaves* elliptic-oblong to oblong-oblanceolate, acuminate, faintly serrate in the upper three-fourths ; the base entire, acute ; both surfaces, but especially the lower, much pustulate when dry ; the lower brown, the upper greenish ; midrib and 6 to 8 pairs of erecto-patent main nerves rather prominent below, the latter interarching ·3 to ·4 in. from the edge ; secondary nerves prominent ; length 5·5 to 7 in., breadth 1·8 to 2·5 in., petiole ·3 to ·4 in.   *Flowers* ·75 in. in diam., on peduncles ·1 in. long ; bracteoles 2, opposite, broad, close to the calyx. *Sepals* rotund, coriaceous, pubescent externally.   *Petals* larger than the sepals, rotund, glabrous, fleshy with thin edges, white.   *Stamens* numerous ; anthers broadly ovate, apiculate, 4 or 5 times as long as the slightly flattened filaments.   *Ovary* ovoid-conic, ridged, adpressed-pubescent, 5-celled.   *Style* short, conic, glabrous, 5-ridged.   *Stigmas* small, acute, connivent.   *Fruit* 1·25 in. long, and ·9 in. in diam., ovoid, bluntly 5-ridged, pubescent.   *Seeds* few, ovate, sub-compressed, ·6 in. long.

Perak ; at elevations of 500 to 2000 feet ;   King's Collector.

3. PYRENARIA WRAYI, King, n. sp. A bush; the young branches pale, minutely adpressed-pubescent towards the apices as are the leaf-buds. *Leaves* thinly coriaceous, oblong-oblanceolate, shortly acuminate, obscurely crenate-serrate to sub-entire; the base attenuate, entire; both surfaces glabrous, the lower yellowish-green, pustulate when dry, the upper greenish; midrib prominent especially beneath; main nerves 10 to 12 pairs, interarching ·25 in. from the margin, rather prominent beneath; length 6 to 8 in., breadth 1·75 to 2·25 in.; petiole ·3 or ·4 in., stout. *Flowers* ·5 in. in diam., buds globose; peduncle very short, glabrous; bracteoles 3, broadly ovate, connate just below the calyx. *Sepals* 6, increasing in size inwards, rotund, minutely pubescent externally. *Petals* 6, rotund, concave, thinner than the sepals, puberulous externally with broad glabrous edges. *Stamens* numerous; anthers broadly ovate, about one-fourth as long as the filaments. *Ovary* shortly ovate-conic, with many lines of white hair, 5- or 6-celled. *Styles* 3, united for half their length; stigmas vertically flattened. Ovules 2 in each cell. *Fruit* sub-globular, bluntly 5-ridged, deciduously pubescent, 1 in. in diam. *Seeds* ovoid, sub-compressed, smooth, ·6 in. long, the hilum very large.

Perak; at low elevations, Wray, Scortechini.

Closely allied to *P. Kunstleri;* but the leaves have many more nerves, the flowers are 6-merous with only 3 styles, and the fruit is more globular than in that species.

## 8. SCHIMA, Reinw.

Trees with evergreen leaves. *Peduncles* usually erect, axillary or solitary, or the uppermost shortly racemed. *Flowers* handsome, 2-bracteolate. *Sepals* 5, subequal, united below. *Petals* 5, much larger, connate at the base, the outermost concave and sub-cucullate. *Stamens* many, adnate to the base of the petals. *Ovary* 5- (rarely 4-6) celled; styles united, or partially free at the apex with broad spreading stigmas; ovules 2-6 in each cell, attached laterally, sub-pendulous. *Capsule* woody, depressed-globose, loculicidal, with a persistent axis. *Seeds* flat, kidney-shaped, dorsally ridged, hilum central, albumen scanty; cotyledons foliaceous, flat or crumpled, accumbent; radicle inferior, curved upwards. Distrib. Tropical Asia. Species about 3.

1. SCHIMA NORONHAE, Reinw. in Bl. Bijdr. 130. A tree 40 to 80 feet high; young branches with pale brown bark, deciduously pubescent, lenticellate. *Leaves* sub-coriaceous, narrowly elliptic to elliptic-lanceolate, acuminate, faintly crenate-serrate, often sub-entire, the base narrowed or rounded; both surfaces glabrous, the lower pale, dull; main nerves 9 or 10 pairs, spreading, slender, rather distinct below when dry, the minor nerves obsolete; length 4·5 to 6 in., breadth 1·4 to 2·5 in., petiole

·75 to 1·25 in., flat, more or less winged. *Flowers* 1·25 to 1·5 in. in diam., axillary, crowded at the apices of the branches and forming lax terminal pseudo-corymbs; peduncles 1 to 1·5 in. long, slender, thickened towards the apex, glabrous or pubescent, bracteoles minute. *Sepals* rounded or sub-acute, glabrous or glabrescent, the margins minutely ciliate, about ·15 in. long. *Petals* thin, veined, obovate, clawed, their bases pubescent and their edges ciliate in the lower half, white or pale pink. *Stamens* 5-delphous; anthers sub-rotund, small, the filaments 4 or 5 times as long. *Ovary* depressed-hemispheric, pubescent, 5-celled. *Style* thick; stigma discoid, with 5 blunt lobes. *Fruit* ·75 in. in diam., adpressed-pubescent when young, glabrous or sub-glabrous when old; upper part of columella expanded, 5-angled. Korth. Verh. Nat. Gesch. Bot. 143, t. 29, figs. 21 to 27; Choisy. Mém. Ternst. 54; Miq. Fl. Ind. Bat. I, Pt. i, p. 492; Ann. Mus. Lugd. Bat. IV, 112; Kurz For. Fl. Burm. i, 107. *S. crenata,* Korth. l. c. t. 29, figs. 1 to 20; Miq. Flora l. c. 491; Ann. l. c. 113; Kurz l. c. 107; Hook. fil. Fl. Br. Ind. i, 289; Pierre Fl. Forest. Coch.-Chine, t. 121. *Gordonia floribunda,* Wall. Cat. 1456; Griff. Not. iv, 563. *G. oblata,* Roxb. Fl. Ind. ii, 572.

In all the provinces except the Andamans and Nicobars. Distrib. The Malayan Archipelago, Burmah, at elevations of 1000 to 3000 feet.

This rather widely distributed species varies remarkably little. In spite, however, of this, Korthal, carved out of it his species *S. crenata,* which he states to have the same calyx, corolla, stamens, ovary, style and stigma as Reinwardt's *S. Noronhae,* but to differ in the leaves and capsule. His own descriptions and figures of leaves and capsule, however, of both species are practically identical. The only other really distinct species of the genus appear to me to be *S. Khasiana,* Dyer, *S. bancana,* Miq. and perhaps *S. Wallichii,* Choisy.

## 9. GORDONIA, Ellis.

Trees with evergreen entire or crenate leaves. *Flowers* usually large, often subsessile, solitary in the axils of the leaves or collected at the ends of the branches, 2-4-bracteolate. *Sepals* usually 5, unequal, graduating from the bracts to the petals. *Petals* free or united at the base, imbricate, the inner larger. *Stamens* indefinite, 5-delphous or 1-delphous, adnate to the petals; anthers versatile. *Ovary* 3-5-celled; style single; the stigma flat, rotund, rather thick, sometimes lobed; ovules pendulous, 4 to 8 in each cell. *Capsule* oblong, woody, loculicidal, with a persistent column. *Seeds* flat or compressed, the apex often winged, albumen none; embryo usually straight, the cotyledons ovate, flat or plicate. Distrib. Tropical, Asia N. America. Species about 15.

1. GORDONIA EXCELSA, Bl. Bijdr. 130. A tree 30 to 40 feet high; young branches slender, smooth, pale brown, pubescent towards the apex. *Leaves* thinly coriaceous, glabrous, elliptic-lanceolate, acuminate, the edge slightly recurved, sub-serrulate, base acute; midrib bold, puberulous near the base beneath; main nerves 5 to 7 pairs, indistinct, bifurcating ·3 in. from the edge and forming wide intra-marginal areolae; length 2·5 to 5 in., breadth 1 to 1·5 in.; petiole ·3 in., slender. *Flowers* 1·5 in. in diam., subsessile, solitary, in the upper axils only; pedicel about ·1 in.; bracteoles lanceolate, small, fugaceous. *Sepals* spreading, free, orbicular, pubescent externally, fleshy. *Petals* white, much larger than the sepals, orbicular, minutely pubescent externally, fleshy with broad membranous glabrous margins. *Anthers* ovoid, only a quarter of the length of the flattened filaments. *Ovary* hemispheric-conic, vertically ridged, densely sericeous, 5-celled. *Styles* single, slender, 5-angled; stigma small, with 5 blunt radiating lobes. Capsule 1·5 in. long, ·75 in. in diam., deciduously adpressed-pubescent. *Seeds* 1 in. or more long, three-fourths being wing. Dyer in Hook. fil. Fl. Br. Ind." i; 291; Miq. Fl. Ind. Bat. I, pt. ii, p. 489. *G. singaporiana,* Wall. Cat. 1457 (in part). *Antheeischima excelsa,* Korth. Verh. Nat. Gesch. Bot. 138, t. 27. *Dipterospermæ, sp.* Griff. Notul. iv, 564.

Malacca. Penang; Curtis, No. 834, King's Collector. Perak; King's Collector, Wray; at elevations of 1200 to 2,500 feet. Distrib. Outer ranges of Eastern Himalaya.

Allied to *G. Maingayi,* but with much larger flowers and fruit and differently shaped leaves.

2. GORDONIA GRANDIS, King, n. sp. A tree 80 to 120 feet high; all parts except the flowers glabrous; young branches as thick as a goose-quill, dark purplish-brown when dry. *Leaves* coriaceous, oblong-oblanceolate, shortly acuminate, faintly serrate-crenate in the upper two-thirds, entire in the lower third and prolonged along the petiole; upper surface greenish when dry, shining; the lower dull, brown; nerves 10 to 12 pairs, indistinct, interarching ·15 in. from the margin; length 4:5 to 6 in., breadth 1·1 to 1·5 in., petiole proper ·15 in. *Flowers* 1·5 to 2 in. in diam., solitary, axillary, about ·3 in. long, puberulous; buds globose; bracteoles few, small, fugaceous. *Sepals* and *petals* greenish, rotund, minutely adpressed-sericeous externally, coriaceous, the edges thin and glabrous; the petals much the larger, spreading. *Stamens* very numerous; anthers narrowly oblong, about a fifth of the length of the slender slightly flattened filaments. *Ovary* narrowly ovoid, vertically ridged, minutely adpressed-sericeous. *Style* longer than the ovary, vertically ridged and sericeous like the ovary. *Stigma* with 5 small roundish lobes. *Fruit* unknown.

Perak, at elevations of 500 to 1000 feet, King's Collector.

3. GORDONIA MAINGAYI, Dyer in Hook. fil. Fl. Br. Ind. I, 291. A tree 30 to 40 feet high; young branches slender, with glabrous pale roughish bark, pubescent towards the apices. *Leaves* coriaceous, broadly oblanceolate, shortly and bluntly acuminate, obscurely serrulate in the upper two-thirds, the lower third gradually attenuate, entire; both surfaces glabrous, the upper greenish, the lower brownish when dry, the midrib bold and sparsely pubescent beneath; lateral nerves 6 pairs, indistinct; length 2·5 to 3 in., breadth 1 to 1·4 in., petiole ·25 in. *Flowers* sub-sessile, ·8 to 1 in. in diam., buds sub-globular; bracts, sepals and petals forming a cone, all adpressed-sericeous externally except the glabrous edges; pedicels about ·15 in. long. *Sepals* and *petals* orbicular, blunt or retuse. *Stamens* numerous; anthers elongate-ovoid; filaments much longer, slender. *Ovary* ovoid-conic, vertically ridged, adpressed-sericeous, 4- or 5-celled. *Style* single, angled. Stigmas 4 or 5, acute, connivent. *Capsules* 4- to 5-angled; woody, 1 to 1·25 long, ·5 to ·6 in diam., 4- to 5-celled, backs of valves flat. *Seeds* ·9 in. long of which three-fourths are wing.

Malacca; Maingay, No. 192. Perak, Scortechini, Wray; at about 1000 feet.

4. GORDONIA SCORTECHINII, King, n. sp. A tree; young branches slender, dark brown, glabrous, the apices and leaf-buds minutely puberulous. *Leaves* coriaceous, narrowly elliptic, blunt, or sub-emarginate, slightly narrowed to the sub-acute or rounded base; both surfaces glabrous, the lower dull, pale; the upper shining, green, when dry; midrib bold; nerves about 8 pairs, faint on the upper, invisible on the lower surface; length 2 to 3 in., breadth ·8 to 1·4 in., petiole ·25 in. *Flowers* ·6 in. in diam., solitary, axillary, only towards the apices of the branches, on very short curved pubescent peduncles. Buds ovoid. *Sepals* orbicular, fleshy, unequal, pubescent externally. *Petals* twice as large as the sepals, membranous, puberulous externally. *Stamens* few, (only about 30); anthers broadly ovate, about a fourth as long as the flattened filaments. *Ovary* narrowly ovoid, pubescent, 3-celled. *Styles* 3, thick, shorter than the ovary, pubescent; stigmas on the inner surface only, slightly spreading. *Fruit* unknown.

Perak; Scortechini, No. 362b.

This has a superficial resemblance to *G. Maingayi*, to which the late Father Scortechini referred it. But it has smaller flowers with fewer stamens, and very different ovary and styles; the leaves moreover are thicker than those of *G. Maingayi*, and are not oblanceolate.

5. GORDONIA IMBRICATA, King, n. sp. A tree? Young branches rather stout, glabrous, dark purplish-brown when dry. *Leaves* coria-

144

ceous, oval-oblong, sometimes slightly oblanceolate, the apex obtuse, very slightly emarginate, the edges thickened and slightly recurved, quite entire or very faintly sub-serrulate; the base slightly narrowed, roundish; both surfaces shining, the upper greenish; the lower dull, tinged with brown when dry, midrib bold; nerves about 12 pairs, thick but inconspicuous; length 1·75 to 2·25 in., breadth ·9 to 1·1 in.; petiole ·15 in., thick. *Flowers* about 1 in. in diam., axillary, solitary, sub-sessile, only in the upper axils; the buds elongate-obovoid; the bracts numerous, closely imbricate, passing into the sepals, all orbicular and pubescent externally with broad scarious glabrous edges. *Petals* much larger than the sepals, orbicular, densely and minutely pubescent externally, fleshy with thin glabrous edges. *Stamens* numerous; anthers ovate, about one-fourth of the length of the slender cylindric filaments *Ovary* ovoid-conic, ridged, adpressed-pubescent, 5-celled. *Style* single, boldly 5-ridged; stigmas distinct, small. *Fruit* slightly under 1 in. long, ·4 in. in diam., 5-angled, adpressed-pubescent, subtended by the elongate imbricate cup formed by the sepals and bracts. *Seeds* ·75 in. in length, of which one half is wing.

Perak. Scortechini, No. 402b.

Father Scortechini's scanty specimens are accompanied by no notes; but, from the species of *Hymenophyllum* growing on the branches of some of them, I conclude that they were collected probably at elevations of 4000 or 5000 feet. The remarkable imbricate buds at once distinguish this species.

6. GORDONIA MULTINERVIS, King, n. sp. A tree 40 to 50 feet high; young branches smooth, greenish, sub-compressed, all parts glabrous except the flowers. *Leaves* thinly coriaceous, obovate, apex rounded or mucronate, faintly crenate-serrate or subentire, attenuate below the middle and passing into the short petiole; upper surface greenish when dry, the lower brown, midrib bold; main nerves 12 to 18 pairs, spreading, rather faint, interarching ·2 in. from the edge, length 5·5 to 8 in., breadth 2·5 to 3·25 in.; petiole ·2 to ·25 in., stout. *Flowers* 1·25 in. in diam., on stout curved peduncles ·5 to ·6 in. long; bracts small, few, fugaceous. *Sepals* rotund, fleshy, spreading, adpressed-sericeous externally, the edges glabrous. *Petals* like the sepals but larger and thinner, spreading. *Anthers* short, broadly ovate, only a quarter of the length of the slender slightly flattened filaments. *Ovary* ovoid-conic, adpressed-sericeous, 5-celled. *Style* single, thick, sub-glabrous. *Stigma* discoid, with 5 blunt lobes. *Fruit* unknown.

Perak; Scortechini, No. 1968.

The style and stigmas are quite those of a *Gordonia*. The leaves, however, are more those of a *Pyrenaria* and are very like those of the Burmese *P. attenuata*, Seem.

### 10. Archytæa, Martius.

Glabrous shrubs or trees with semiamplexicaul leaves. *Flowers* on a lateral, compressed, 1- to 4-flowered peduncle. *Bracts* large, leaf-like. *Sepals* and *petals* each 5. *Stamens* numerous, 5-adelphous; anthers versatile. *Ovary* 5-celled; styles distinct, or wholly united; ovules numerous, in many imbricating rows. *Capsule* acuminate, septicidal from below, with a persistent axis. *Seeds* linear-subcylindric, albumen scanty. Distrib. Trop. Amer. and Indian Archipelago. Species 3.

1. Archytæa Vahlii, Choisy Mém. Ternstr. 73. A glabrous shrub (sometimes epiphytic) or small tree; the young branches, pale, smooth. *Leaves* thinly coriaceous, sessile, narrowly oblanceolate, acute, entire, slightly narrowed to the truncate or slightly amplexicaul base; nerves about 15 pairs, straight, erect, interarching with an intra-marginal nerve; length 3 to 4·5 in., breadth ·5 to ·75 in. *Flowers* 1 to 1·25 in. in diam.; peduncles crowded towards the end of the branches, coloured; bracts close to the flowers, oblong, sub-serrulate, ·5 to ·75 in. long. *Sepals* ovate-rotund, coriaceous. *Petals* obovate, much larger than the sepals, membranous, veined, pink. *Fruit* ·75 in. long, narrowly ovoid, acuminate, crowned by the persistent styles. Hook. fil. Fl. Br. Ind. i, 294; Pierre Fl. For. Coch.-Chine, t. 129. *Ploiarium elegans*, Korth. Verh. Nat. Gesch. Bot. 135, t. 25; Miq. Fl. Ind. Bat. I. Pt. ii, 491. *Hypericum alternifolium*, Vahl. Symb. ii, t. 42; DC. Prodr. i, 445; Wall. Cat. 4806.

In all the provinces except the Andaman and Nicobar Islands. Distrib. The Malayan Archipelago.

** Note on the fruit of *Xanthophyllum Scortechinii*, King.

Since the pages describing the genus *Xanthophyllum* were printed off, I have received from Mr. Curtis, of the Forest Department, Penang, complete specimens of this species; and I am therefore now able to add to the account of it given on p. 140 the following description of the young fruit.

*Fruit* globular or ovoid-globular, ·75 to 1 in. in diam., shortly apiculate, smooth, shining; the pericarp very thick.

Ripe fruit is still a desideratum.

# MATERIALS

## FOR A

# FLORA OF THE MALAYAN PENINSULA.

BY

GEORGE KING, M. B., LL. D., F. R. S., C. I. E.

SUPERINTENDENT OF THE ROYAL BOTANIC GARDEN, CALCUTTA.

[*Reprinted from the Journal of the Asiatic Society of Bengal, Vol.* LX, *Part* II, *No.* 1, 1891.]

## NO. 3.

~~~~~~~~~~~~~~~~~~~

CALCUTTA:

PRINTED AT THE BAPTIST MISSION PRESS.

1891.

*Materials for a Flora of the Malayan Peninsula.—By* GEORGE KING, M. B., LL. D., F. R. S., C. I. E., *Superintendent of the Royal Botanic Garden, Calcutta.* No, 3.

(Continued from page 206 of Vol. LIX of 1890.)

[Received 2nd March 1891. Read April 1st 1891.]

In the arrangement of the Natural families which is being followed in these papers (that of DeCandolle as modified by the late Mr. Bentham and Sir Joseph Hooker), the family *Dipterocarpeae* should have preceded *Malvaceae.* Delays have, however, occurred in the elaboration of that family ; and, rather than postpone the publication of the remaining three *Thalamifloral* orders, I have decided to submit my account of these to the Society now, deferring my paper on the *Dipterocarpeae* and on the previously omitted *Anonaceae* to a future occasion.

## ORDER XVII. MALVACEÆ.

Herbs, shrubs or trees ; herbaceous portions often stellate-hairy or scaly. *Leaves* alternate, palminerved, simple, lobed, or rarely compound. *Stipules* free, sometimes caducous. *Bracteoles* 3 or more, free or combined, often forming an epicalyx. *Flowers* axillary or terminal, solitary, fascicled or cymose-paniculate, regular, hermaphrodite or 1-sexual. *Sepals* 5, valvate, free or connate. *Petals* 5, twisted-imbricate. *Stamens* ∞, rarely definite, adnate to the base of the petals ; filaments monadelphous, forming a tube ; anthers oblong or reniform, cells sinuous or twisted, linear or annular, ultimately 1-celled bursting longitudinally. *Ovary* 2-many-celled, entire, or lobed, of 2-5 or usually more carpels whorled round a central axis ; styles connate below or throughout their length ; ovules 1 or more, curved, attached to the inner angle of each carpel. *Fruit* of dry cocci, or capsular and loculicidal, often large and woody. *Seeds* reniform or obovid, sometimes arillate ; albumen scanty, often mucilaginous or 0 ; embryo curved ; cotyledons leafy, usually

147

folded or crumpled.—Distrib. Abundant in warm regions, common in temperate, absent from arctic. Genera 57 ; known species about 700.

A. Staminal tube entire, or but slightly divided at the apex.

Tribe I. *Malveœ*. Herbs or shrubs. *Ripe carpels* separating from the axis. *Styles* as many as the carpels.

Ovules solitary ; carpels with convergent, often beaked, apices ... ... ... 1. *Sida.*

Ovules 2 or more ; carpels with divergent, not beaked, apices ... ... ... 2. *Abutilon.*

Tribe II. *Ureneœ. Styles or stigmatic branches* twice as many as the carpels.

Fruit of indehiscent cocci ... ... 3. *Urena.*

Tribe III. *Hibisceœ.* Herbs or shrubs. *Fruit* capsular. *Sepals* leafy. *Staminal-tube* truncate or 5-toothed at the apex.

Calyx toothed : stigmas distinct, spreading ... 4. *Hibiscus.*

„ truncate : stigmas united ... ... 5. *Thespesia.*

B. Staminal tube short or divided into single filaments to its base.

Tribe IV. *Bombaciae.* Trees. *Sepals* leathery : styles connate or free. *Fruit* capsular.

Leaves digitately compound, calyx truncate or irregularly 3 to 5-lobed ; seed silky outside.

Anthers solitary ... ... ... 6. *Bombax.*

„ in groups of 2 or 3 ... ... 7. *Eriodendron.*

Leaves simple, usually scaly ; fruit woody, muricate ; seeds arillate.

Calyx tubular or bell-shaped.

Anthers linear, cells sinuous ... 8. *Durio.*

Anthers globose, opening by a pore ... 9. *Boschia.*

Calyx dilated at the base.

Calyx finally forming a cushion-shaped annulus ... ... ... ... 10. *Neesia.*

Calyx 5-pouched at the base, petals inserted on the calyx ... ... ... 11. *Cœlostegia.*

## 1. SIDA, Linn.

Herbs or undershrubs. *Leaves* entire or lobed. *Bracteoles* 0. *Calyx* of 5 valvate sepals, tubular below. *Corolla* of 5 petals, free above, connate below and adnate to the tube of the stamens. *Staminal-tube* dividing at the summit into numerous anther-bearing filaments. *Carpels* 5 or more, whorled ; styles as many as the carpels, stigmas terminal.

148

*Ripe carpels* separating from the axis, generally 2-awned at the summit, and dehiscing irregularly or by a small slit. *Seed* solitary, pendulous or horizontal; radicle superior.—Distrib. A genus of about 80 species, most of them being tropical weeds.

1. S. MYSORENSIS, W. & A. Prod. I, 59. A sub-erect, sometimes decumbent, herb 1 to 2 feet high, covered with more or less glutinous hairs. *Leaves* cordate-ovate, acuminate, coarsely serrate-crenate, 1·5 to 2·5 in. long and 1 to 1·5 in. broad; petiole about half as long as the blade. *Stipules* linear, less than half as long as the petiole. *Flowers* less than ·5 in. in diam., in few-flowered axillary racemose cymes, corolla yellow; pedicels shorter than the petioles, jointed near the middle. *Carpels* shorter than the calyx, sub-glabrous, each with a short awn, or awnless. Mast. in Hook. fil. Fl. Br. Ind. I, 322; Thwaites Enum. 28. *S. hirta*, Wall. Cat. 1855, not of Lam. *S. urticæfolia*, W. & A., l. c. *S. nervosa*, Wall. Cat. 1853 E. *S. olens*, Ham. in Wall. Cat. 1874. *S. glutinosa*, Roxb. Hort. Beng. 97; Fl. Ind. iii, 172; Wall. Cat. 1855, not of Cav. *S. tenax*, Ham. in W. & A. Prodr. i, l. c.; Wall. Cat. 1855, E. F. *S. fasciculiflora*, Miq. Fl. Ind. Bat. i, Pt. 2, 140. *S. radicans* Cav. Diss. i, 8 : W. & A. Prod. i, 59.

A weed by roadsides; in Perak and probably in the other provinces. Distrib. India, Java.

2. S. CARPINIFOLIA, L. An undershrub 2 to 3 feet high; glabrous or sub-glabrous; a few minute stellate hairs on the stems and petioles. *Leaves* linear-lanceolate, acute, serrate, 2 to 3 in. long and ·25 to ·35 in. broad; petioles ·1 to ·2 in. *Stipules* subulate, nerved, much longer than the petiole. *Flowers* ·5 in. in diam., solitary, axillary; corolla yellow, peduncles as long as the petiole, jointed, minutely bracteolate. *Carpels* shorter than the sub-globose ribbed calyx, glabrous, rugulose, each with 2 short awns. DC. Prod. i. 460. Mast. in Hook. fil. Fl. Br. Ind. i. 323; Wall. Cat. 1871. *S. acuta*, Burm.; Cav. Diss. i p. 15, t. 2, f. 3; DC. Prodr. i. 461; Wall. Cat. 1868, 1, 2. 3, 4, 5; Roxb. Fl. Ind. iii. 171; W. & A. Prodr. i. 57; Dalz. & Gibs. Bomb. Fl. 17; Thwaites Enum. 27; Miq. Fl. Ind. Bat. i. Pt. 2. p. 143; Wight Ic. t. 95; Bl. Bijdr. 55; Wall. Cat. 1868 G. *S. lanceolata*, Roxb. l.c. 175; Wall. Cat. 1868 F. *S. stipulata*, Cav. Diss. i. t. 3, f. 10; DC. Prodr. i. 460; W. & A. Prodr. l.c. *S. Stauntoniana*, DC. l.c.; *S. scoparia*, Lour. ex W. & A. lc.

In all the provinces as a weed. Distrib. India and Tropics generally.

3. S. RHOMBIFOLIA, Linn. sp. 961. An erect under shrub 2 to 3 feet high, from glabrous to hoary, stellate-pubescent. *Leaves* varying

from ob-lanceolate or obovate to rhomboid, but always with tapering
bases, serrate to crenate; under surface hoary, rarely green; length
·5 to 2·5 in., petiole ·1 to ·2 in. *Stipules* setaceous, longer than the
petioles. *Flowers* ·5 in. in diam., axillary, solitary; corolla yellow,
rarely white; peduncles much longer (sometimes six times) than the
petioles, variously and sometimes indistinctly jointed, ebracteolate.
*Carpels* smooth or pubescent, or reticulate, each usually with 1 or 2
rather long awns, sometimes awnless, generally longer than the calyx.
Mast. in Hook. fil. Fl. Br. Ind. i. 323; Miq. Fl. Ind. Bat. i. pt. 2. p. 142;
DC. Prodr. i. 462; Roxb. Fl. Ind. iii. 176; Wall. Cat 1862, 2; Thwaites
Enum. 28. *S. canariensis*, Willd.; DC. Prodr. i. 462. *S. compressa*,
Wall. Cat. 1866; DC. Prodr. i. 462.

This very polymorphic species has been divided into varieties by
Dr. Masters in Hooker's Fl. Br. Ind. l.c. as follows:—

"Var. 1. *scabrida*, W. & A. Prodr. i. 57 (sp.); sprinkled with rigid
hairs, leaves concolorous, peduncles joined at the base, carpels awned.

"Var. 2. *retusa*, Linn. (sp.); leaves obovate retuse hoary underneath,
peduncles equalling the leaves jointed above the middle, carpellary
awns short.—Cav. Diss. i. t. 3, f. 4, and Diss. v. t. 131, f. 2; Bl. Bijdr.
75; W. & A. Prodr. i. 38; Wall. Cat. 1870; DC. Prodr. i. 462; Roxb.
Fl. Ind. iii. 175; Dalz. & Gibs. Bomb. Fl. 17; Miq. Fl. Ind. Bat. i. pt.
2, 142. *S. chinensis*, Retz ex Roxb. Hort. Beng. 97; Fl. Ind. iii. 174.
*S. philippica*, DC. Prodr. i. 462; W. & A. Prodr. l.c.; Wall. Cat. 1869;
Rheede Hort. Mal. x. 18; Rumph. Amb. v. t. 19.—The *S corynocarpa*,
Wall. Cat. 1870, seems to be a form of this variety, with densely intricate
woody branches, and long carpellary awns.

"Var. 3. *rhomboidea*, Roxb. Hort. Beng 50; Fl. Ind. iii. 176 (sp.);
leaves rhomboid hoary beneath, peduncles jointed at the base, carpellary
awns very short inflected. DC. Prodr. i. 462; W. & A. Prodr. i. 57,
Wall. Cat. 1862 E., 1863; Thwaites Enum. 28. *S. rhombifolia*, Wall.
Cat. 1862 F.? *S. orientalis*, Cav. Diss. i. t. 12.—The flowers expand at
noon (Roxb.).

"Var. 4. *obovata*, Wall. Cat. 1864 (sp.); leaves 1½ by 2 in., broadly
obovate, hoary beneath, apex coarsely toothed, base cuneate, petiole ¼
in., peduncle longer than the petiole shorter than the blade.

"Var. 5. *microphylla*, Cav. Diss. i. t. 12, f. 2 (sp.); leaves small,
elliptic dentate hoary beneath, peduncle slightly exceeding the petiole,
carpels 5-7 awned.—Roxb. Fl. Ind. iii. 170; DC. Prodr. i. 461."

In all the provinces—a common weed. Distrib. The Tropics
generally.

4. S. CORDIFOLIA, Linn. spec. 961. An erect softly hairy undershrub

2 to 3 feet high, the hairs on the branches and petioles long and spreading. *Leaves* oblong-cordate, obtuse, rarely acute, crenate; both surfaces, but especially the pale lower surface, softly hairy; length 1·25 to 2 in., breadth ·8 to 1·25 : petiole slightly longer than the blade. *Stipules* linear, less than half the length of the petiole. *Flowers* ·6 in. in diam., axillary, solitary; corolla yellow; peduncles jointed near the apex, varying in length, the lower longer, the upper shorter, than the petioles. *Carpels* boldly 3-angled, reticulate, sub-glabrous, crowned by 2 strong, divergent, retro-hispid awns. DC. Prod i. 464, Roxb. Fl. Ind., iii. 177; Wall. Cat 1819; W. & A. Prod. i. 58; Thwaites Enum. 28. Dalz. & Gibs. Fl. Bombay, 17; Mast. in Hook. fil. Fl. Br. Ind. i. 324, and in Oliver's Fl. Trop. Afr i. 181; Miq. Fl. Ind. Bat. i. pt. 2, 140. *S. herbacea,* Cav. Diss. i. 19, t. 13, f. 1; DC. Prodr. i. 463. *S. micans,* Cav. Diss. i. 19, t. 3. f. 1. *S. rotundifolia,* Cav. Diss. i. 20, t. 3, f. 6, and Diss. vi. t. 194, f. 2; Wall. Cat. 1849, D; DC. Prodr. i. 464. *S. althæifolia,* Swartz, Guill. & Per. Fl. Seneg. i. 73.—Rheede Hort. Mal. x. t. 54.

In Malacca: and probably in all the Provinces as a weed. Distrib. The Tropics generally.

## 2. ABUTILON, Gærtn.

Herbs or undershrubs more or less covered with down. *Leaves* angled or palmately-lobed. *Inflorescence* axillary or terminal. *Bracteoles* 0. *Calyx* of 5 valvate sepals, tubular below. *Corolla* of 5 petals, free above, connate below and adnate to the tube of the stamens. *Staminal-tube* divided at the apex into numerous filaments. *Carpels* 5-8. Styles as many as the carpels. Ripe *carpels* separating from the axis, awned or not, 1- or more-seeded. *Seeds* reniform. Distrib. About 70 species, all tropical or subtropical.

A. INDICUM, G. Don. Gen. Syst. i. 504. An annual or perennial undershrub. *Leaves* broadly cordate, irregularly and coarsely toothed or sub-entire, pale and minutely pubescent on both surfaces, often with a few longer hairs intermixed, length 1 to 2 in., breadth. 1 to 2 in.; petiole usually longer than the blade. *Flowers* 1 in. in diam, axillary, solitary, the peduncles longer than the petioles, jointed near the top; corolla yellow. *Sepals* ovate, acute, shorter than the spreading petals. *Carpels* 15 to 20, longer than the calyx, truncate or with short spreading awns, tomentose at first, ultimately sub-glabrous. *Seeds* dark brown, minutely stellate-hairy. Mast. in Hook. fil. Fl. Br. Ind. i. 326; *A. asiaticum,* W. & A. Prodr. i. 56, not *Sida asiatica,* Linn.; W. & A. Prodr. i. 56; Wight Ic. t. 12; Dalz. & Gibs. Bomb. Fl. 18; Thwaites Enum. 27; Mast. in Oliv. Fl. Trop, Afr. i. 186; Miq. Fl. Ind.

Bat. i. pt. 2, 146. *Sida indica*, L.; DC. Prodr. i. 471; Cav. Diss. i. p. 33, t. 7, f. 10; Roxb. Fl. Ind. iii. 179; Wall. Cat. 1859, 1, 2, D. F. *Sida populifolia*, W. & A. l.c. *A. populifolia*, G. Don. l.c. *Sida populifolia*, DC. Prod. i. 470; Cav. Diss. i. t. 7, fig. 9; Roxb. Fl. Ind. iii. 179; Bl. Bijdr. 79. *S. Beloere*, L'Her. Stirp. i. 130. *S. Eteroomischos*, Cav. Diss. ii. 55 and v. p. 275, t. 128.

Singapore, Selangore and probably in all the other provinces. A weed.

## 3. URENA, Linn.

Herbs or undershrubs, more or less covered with rigid stellate hairs. *Leaves* angled or lobed. *Flowers* clustered. *Bracteoles* 5, adnate to the 5-cleft calyx, sometimes coherent at the base into a cup. *Petals* 5, often tomentose at the back, free above, connate below and united to the base of the tube of the stamens. *Staminal-tube* truncate or minutely toothed. *Anthers* nearly sessile. *Ovary* 5-celled, cells 1-ovuled, opposite the petals; stigmatic branches 10; stigmas capitate. *Ripe carpels* covered with hooked bristles or smooth, indehiscent, separating from the axis when ripe. *Seed* ascending; cotyledons bent and folded; radicle inferior. Distrib. Species 4-5, natives of tropical and subtropical countries, 2 only being confined to Asia.

U. LOBATA, Linn. Spec. 974. A herbaceous undershrub 1 to 3 feet high, more or less hairy. *Leaves* very variable; the lower rotund to reniform, more or less cordate at the base, the apex usually acute, edges with 5 to 7 shallow lobes or sub-entire, 5 to 7-nerved; length 1 to 2 in., breadth 1 to 2·5 in.; upper leaves smaller and sometimes ovate to linear-lanceolate, 3-nerved. *Petiole* shorter than the blade; bracteoles oblong-lanceolate, as long as the sepals. *Corolla* pink, ·5 to 1 in. in diam. *Carpels* tomentose, and with many smooth hooked spines. Mast. in Hook. fil. Fl. Br. Ind. i. 329; Miq. Fl. Ind. Bat. i. pt. 2, p. 149; Cav. Diss. iv. p. 336, t. 185, fig. 1; Miq. Pl. Jungh. 283; DC. Prodr. i. 441; Roxb. Fl. Ind. iii. 182; W. & A. Prodr. i. 56; Wall. Cat. 1928; Dalz. & Gibs. Bomb. Fl. 18; Thwaites Enum. 25; Miq. Fl. Ind. Bat. i. pt. 2, 148. *U. cana*, Wall. Cat. 1930 B. *U. palmata*, Roxb. Fl. Ind. iii. 182. *U. tomentosa*, Bl. Bijdr. 65.

All the Provinces: a weed. Distrib. The tropics generally.

Var. 1. *sinuata*, Miq. Fl. Ind. Bat. l.c.; leaves deeply 5-lobed, the lobes narrowed at the base, serrate, often pinnatifid, bracteoles linear; flowers often smaller than in the typical plant. *U. sinuata*, Linn.; DC. Prodr. i. 441; Roxb. Hort. Beng. 50; Fl. Ind. iii. 182; Wall. Cat. 1933 E.; W. & A. Prodr. i. 46; Hook. Fl. Br. Ind. i. 329; Thwaites Enum.

44

Pl. Coy. 25 ; Dalz. & Gibs. Bomb. Fl. 18. *U. muricata*, DC. Prodr. i. 442. *U. Lappago*, DC. Prodr. i. 441. *U. morifolia*, DC. Prodr. i. 442 ? *U. heterophylla*, Smith in Rees' Cycl. 37 ; Wall. Cat. 1933 E, F. G, H, K. *U. tomentosa*, Wall. Cat. 1933 H. ;—Burm. Zeyl. t. 69, f. 2. Distributed like the last.

Var. 2. *scabriuscula*, DC. Prod. i. 441 (sp.) ; herbaceous ; leaves roundish, scarcely lobed, with 1-3 glands beneath ; bracteoles linear, longer than the sepals. *U. scabriuscula*, Wall. Cat. 1928 F ; W. & A. Prodr. i. 46 ; Dalz. & Gibs. Bomb. Fl. 18.

## 4. Hibiscus, Linn.

Herbs, shrubs, or trees. *Leaves* stipulate, usually more or less palmately-lobed. *Inflorescence* axillary, rarely terminal. *Bracteoles* 5 or more, free, or connate at the base. *Calyx* 5-toothed or 5-fid, valvate, sometimes spathaceous. *Petals* 5, connate at the base with the staminal-tube. *Staminal-tube* truncate or 5-toothed at the summit; filaments many; anthers reniform, 1-celled. *Ovary* 5-celled, cells opposite the sepals, each with 3 or more ovules ; styles 5, connate below ; stigmas capitate or sub-spathulate. *Capsule* loculicidally 5-valved, sometimes with a separate endocarp, or with false dissepiments forming a spuriously 10-celled fruit. *Seeds* glabrous, hairy or woolly. About 150 species ; distributed chiefly in the tropical regions of both hemispheres.

Calyx spathaceous, deciduous ... ... 1. *H. Abesmoschus.*
Calyx persistent, 5-cleft.
  Bracteoles of involucre distinct, their apices spathulate ... ... 2. *H. Surattensis.*
  Bracteoles united at the base, nearly as long as the calyx ... ... 3. *H. macrophyllus.*
  Bracteoles united into a cup much shorter than the calyx.
    Involucre and calyx softly pubescent 4. *H. tiliaceous.*
    „ „ rugulose 5. *H. floccosus.*

1. H. Abelmoschus, Linn. Spec. 980. A stout annual under-shrub 2 to 3 feet high : young branches and peduncles retro-hispid, all other parts hispid or stellate-hispid. *Leaves* variable, usually with 3 to 5, deep, oblong-lanceolate or linear, serrate-crenate, acute lobes, sometimes hastate or sagittate, the base always rounded ; length and breadth 3 to 5 in.; petiole longer than the blade : stipules minute, subulate, fugaceous. *Flowers* 3 in. in diam., axillary, solitary ; peduncles shorter than the petioles, ebracteate. *Involucres* 8 to 12, linear, ·5 to ·75 in. long. *Calyx* 1·25 in. long, toothed at the apex. *Corolla* yellow with a crimson

153

45

centre, glabrous. *Capsule* oblong, pointed, hispid, becoming sub-
glabrous, 1 to 3 in. long. *Seeds* reniform, striate, glabrous, musky.
Mast. in Hook. fil. Fl. Br. Ind. i. 342 (excl. syn. *H. sagittifolius*, Kurz.);
DC. Prod. i. 452; Roxb. Fl. Ind. iii. 202; Griff. Not. iv. 521. *Abelmoschus moschatus*, Mœnch; W. & A. Prod. i. 53; Wight Ic. t. 399; Wall.
Cat. 1915, F, G, H, I, K, L; Thwaites Enum. 27; Miq. Fl. Ind. Bat.
i. pt. 2, 151. *H. flavescens*, Cav. Diss. iii. t. 70, f. 2; DC. l.c. 454.
*H. spathaceus*, Wall. Cat. K. *H. ricinifolius*, Wall. Cat. 1915. *Bamia
chinensis*, Wall. Cat. 1616? *Hibiscus pseudo-abelmoschus*, Bl. Bijdr.
70. *H. longifolius*, Willd. Spec. iii. 827; DC. Prod. i. 450. *Bamia
multiformis* and *betulifolia*, Wall. Cat. 1917 and 1918.

In all the Provinces; cultivated or naturalised. Distrib. the
tropics generally.


2. H. SURATTENSIS, Linn. Spec. 979. A weak straggling undershrub; the branches, petioles and peduncles with small recurved prickles and a few soft spreading pale hairs. *Leaves* palmately 3 to 5-partite,
rarely ovate, sub-entire, serrate, sparsely pilose; length and breadth
1·5 to 3 in.; petiole slightly longer than the blade. *Stipules* broadly
ear-shaped. *Flowers* 2 to 2·5 in. long, solitary, axillary, corolla yellow
with dark centre; bracts of involucre 10 to 12, linear with spathulate
apices. *Capsules* membranous, the individual carpels with 3 bold
aculeate nerves and a long terminal point. *Seeds* with long straight
brittle yellowish hairs. Mast. in Hook. fil. Fl. Br. Ind. i. 334; Miq.
Fl. Ind. Bat. i. pt. 2, 161; Bl. Bijdr. 68; DC. Prodr. i. 449; W. & A.
Prodr. i. 48; Roxb. Fl Ind. iii. 205; Wight Ic. t. 197; Cav. Diss. iii.
t. 53, f. 1; Thwaites Enum. 26; Wall. Cat. 1893, 1, 2, 3, D, E, F, G;
Dalz. & Gibs. Bomb. Fl. 20; Mast. in Oliv. Fl. Trop. Afr. i. 201; Miq.
Fl. Ind. Bat. i. pt. 2, 161. *H. furcatus*, Wall. Cat. 1896 C, not of Roxb.

Malacca, Perak, and probably in the other Provinces. Distrib.
The tropics generally.

This has a decumbent or even climbing habit.


3. H. MACROPHYLLUS, Roxb. Hort. Beng. 51. A large shrub or
small tree, all parts more or less covered with pale soft minute velvetty
tomentum; the young branches, petioles, pedicels, bracteoles and calyx
bearing, in addition, numerous more or less deciduous tufts of long
spreading stiff tawny hairs. *Leaves* large, on long petioles, cordate-
orbicular to reniform, the apex shortly sharply and abruptly acuminate,
the edges entire; palmately 7 to 9-nerved; length and breadth 7 to 12
in.; petiole usually longer than the blade. *Stipules* oblong, convolute,
hispid-tomentose, 3 to 4 in. long. *Flowers* in terminal cymes, pedicels

1·5 to 2 in. long, articulate near the apex and bearing two large broadly ovate deciduous bracts. *Involucres* of the individual flower 10 to 12, linear-lanceolate, connate at the base, as long as the calyx, hispid-tomentose like the calyx. *Calyx* with 5 deep linear teeth; the tube 10-ribbed, 1 to 1·25 in. long. *Corolla* 4 in. in diam., purple. *Fruit* pointed, hispid, as long as the persistent calyx. *Seeds* reniform, their edges densely fulvous-sericeous. Mast. in Hook. fil. Fl. Br. Ind. i. 337 ; Kurz For. Fl. Br. Burm. i. 126 ; DC. Prod. i. 455 ; Wall. Pl. As. Rar. i. 44, t. 51 ; Wall. Cat. 1903. *H. setosus*, Roxb. Fl. Ind. iii. 194. *H. vestitus*, Griff. Notul. iv. 519.

Penang, Perak. Distrib. Java, India.

4. H. TILIACEUS, Linn. Spec. 976. A small much branched tree ; young branches minutely pubescent. *Leaves* sub-coriaceous, broadly cordate to reniform, minutely crenulate or entire, rarely lobed, acute ; upper surface scaly, minutely pubescent, glabrescent or glabrous ; lower densely and minutely hoary-pubescent; nerves 7 to 9 pairs, palmate ; length and breadth 3·5 to 6·5 in., petioles ·5 to 2 in., stipules oblong, oblique, shorter than the petiole. *Flowers* solitary ; or in pedunculate, solitary, 2 to 3-flowered, axillary cymes ; the peduncles 2 or 3 times as long as the petioles, with 2 obliquely oblong, opposite, pubescent, caducous bracts. *Involucres* 7 to 10, acute, united above the middle. *Sepals* 5, like the involucres but twice as long, with an elongated gland externally. *Corolla* campanulate, 4 in. in diam., yellow with crimson centre. *Fruit* as long as the calyx or shorter, ovate-acute, stellate-pubescent, spuriously 10-celled. *Seeds* few, obovate-reniform, faintly striate, sparsely scaly, pubescent, or glabrous. Mast. in Hook. fil. Fl. Br. Ind. i. 343 ; Kurz For. Fl. Burm. i. 126 ; DC. Prod. i. 454 ; Cav. Diss. iii, p. 151, t. 55, f. 1 ; Bl. Bijdr. 72 ; Roxb. Fl. Ind. iii. 182 ; Miq. Fl. Ind. Bat. i. pt. 2, 153 ; Beddome Fl. Sylvat. Anal. Gen. t. 4. *Paritium tiliaceum*, A. Juss. in St. Hil. Fl. Bras. Med. i. p. 156 ; (excl. syn. *H. elatum*) W. & A. Prodr. i. 52 ; Wight Ic. t. 7 ; Wall. Cat. 1912 ; Thwaites Enum. 26 ; Dalz. & Gibs. Bomb. Fl. 17 ; Griff. Notul. iv. 523. *H. tortuosus*, Roxb. Fl. Br. Ind. iii. 192 ; Wall. Cat. 1912 G, 1913 B.

All the provinces ; near water. Distrib. The tropics generally near the coasts.

5. H. FLOCCOSUS, Mast. in Hook. fil. Fl. Br. Ind. i. 343. A tree 30 to 40 feet high ; young branches, petioles, peduncles and outer surfaces of involucres and calyx rngulose and minutely rusty-puberulous. *Leaves* sub-coriaceous, cordate-reniform, 5-angled, acute, irregularly and

distantly sub-crenate ; both surfaces minutely and sparsely stellate-pubescent, glabrescent when old, harsh ; length and breadth 2 to 6 in., petiole less than half as long as the blade. *Flowers* in stout few-flowered terminal racemes longer than the leaves ; peduncles stout, very rugulose, ebracteate, ·75 to 1·5 in. long. *Involucres* combined into a bluntly-lobed cup much shorter than the calyx. *Sepals* oblong-lanceolate, 1·5 in, long, coriaceous, united for half their length or more. *Petals* membranous, spathulate, 4 in. long, glabrous inside, boldly striate and hispid-pubescent externally. *Staminal-tube* stellate-pubescent. *Capsule* obovoid, truncate, shorter than the persistent closely adherent calyx, densely stellate-pubescent and very rugulose, 5-valved, dehiscing only at the apex. *Seeds* numerous, obovate, sub-compressed, with shortly pilose angles, the rest of the surface scaly.

Mount Ophir, Malacca ; Maingay (Kew Distrib.) 216. Perak ; King's Collector 7024.

I have not been able to detect stipules on any of the specimens I have seen. They are probably fugacious.

## 5. THESPESIA, Corr.

Trees or shrubs. *Leaves* entire. *Inflorescence* axillary. *Bracteoles* 5-8, arising from the thickened end of the peduncle, deciduous. *Calyx* cup-shaped, truncate, minutely 5-toothed. *Corolla* convolute. *Staminal-tube* 5-toothed at the apex. *Ovary* 4-5-celled ; style club-shaped, 5-furrowed, entire or 5-toothed ; ovules few in each cell. *Capsule* loculicidal or scarcely dehiscent. *Seeds* tomentose ; cotyledons conduplicate, black-dotted.—Natives of tropical Asia, Madagascar, and Australasia ; species about 6.

T. POPULNEA, Corr. in Ann. Mus. ix. p. 290. A tree 20 to 30 feet high, young shoots scaly. *Leaves* on long petioles, sub-coriaceous, broadly cordate, acuminate, entire, glabrous above, sparsely scaly on lower surface ; the base 5 to 7-nerved with a glandular pore between the nerves ; length 4·5 to 6 in., breadth 3 to 4 in. petiole 2·5 in. *Flowers* 2 to 3 in. in diam., solitary, axillary, on peduncles shorter than the petioles ; *petals* bright yellow with a brown spot at the base ; bracteoles close to the calyx, lanceolate, often abortive. *Capsule* 1 to 1·5 in. in diam., depressed-spheroidal, scaly, becoming glabrescent ; pericarp of 2 layers. *Seeds* 1 to 3 in each cell, reniform, minutely tomentose or mealy. Mast. in Hook. fil. Fl. Br. Ind. i. 345 ; Kurz For. Fl. Burm. i. 128 ; Miq. Fl. Ind. Bat. i. pt. 2, 150 ; Pierre Fl. For. Coch-Chine x. 173 ; Bl. Bijdr. 73 ; Cav. Diss. iii. 152, t. 56, f. 1 ; DC. Prodr. i. 456 ; W. & A. Prodr. i. 54 ; Wight Ic. t. 8 ; Thwaites Enum. 27 ; Beddome Fl. Sylvat. t. 63 ;

Dalz. & Gibs. Bomb. Fl. 18; Wall. Cat. 1888, 1, 2, & C to H. Miq. Fl. Ind. Bat. i. pt. 2, 150. *Hibiscus populneus*, L.; Roxb. Hort. Beng. 51; Flor. Ind. iii. 190. *H. populneoides*, Roxb. l.c. *Malvaviscus populneus*, Gærtn. Fruct. ii. 253, t. 135. *Azanza acuminata*, Alefeld Bot. Zeit. 1861, 299.

In all the provinces, on the sea-shore. Distrib. Tropics generally.

## 6. BOMBAX, Linn.

Trees. *Leaves* digitate, deciduous. *Peduncles* axillary or subter-minal, solitary or clustered, 1-flowered. *Flowers* appearing before the leaves. *Bracteoles* 0. *Calyx* coriaceous, cup-shaped, truncate or lobed. *Petals* obovate or oblong. *Stamens* in 5 bundles opposite the petals : filaments numerous ; anthers reniform, 1-celled. *Ovary* 5-celled, multi-ovulate ; style clavate, stigmas 5. *Capsule* loculicidally 5-valved, valves coriaceous, wooly within. *Seeds* silky, the testa thin, albumen small ; cotyledons contortuplicate. About 10 species, all tropical and mostly American ; 1 in Africa.

1.  B. INSIGNE, Wall. Pl. As. Rar. i. 71, t. 79, 80 ; Cat. 1841. A tall tree ; trunk without prickles ; branchlets armed or not ; all parts glabrous. *Leaves* 7-9-foliolate ; leaflets sub-coriaceous, obovate or ob-lanceolate, shortly acuminate, attenuate at the base, glaucous beneath ; length 5 to 8 in., breadth 2·5 to 3 in. ; petiolules ·5 to ·75 in. : petioles longer than the leaflets. *Flowers* 5 or 6 in. long, solitary towards the end of the leafless branches ; peduncles ·75 in. long, stout, clavate. *Calyx* 1·5 in. long, thickly coriaceous, urceolate-globose, obscurely and irregularly lobed, ultimately 2-cleft, sub-glabrous outside, silky inside. *Petals* fleshy, oblong, obtuse, recurved, internally glabrous, externally shortly sericeous, red to orange or yellowish. *Stamens* many ; filaments fleshy, united for ·5 in. above the base into 4 or 5 bundles. *Capsule* oblong, 10 in. long by 1·5 in thick, curved, glabrous. Mast. in Hook. fil. Fl. Br. Ind. i. 349 ; Kurz For. Fl. Burm. i. 130 ; Journ. As. Soc. Beng. 1873, ii. p. 61. *B. festivum*, Wall. Cat. 1841.

Andamans. Distrib. Burmah.

The earliest name of this is *B. festivum* (1828). But at p. 89 of his Catalogue, Wallich changed this to *B. insigne*, under which name he figured and described it. It comes very near to *B. malabaricum*, DC. ; but Wallich says it is a much smaller tree, and Kurz says it has many more stamens, than the former. 1 include it as an Anda-man plant solely on the authority of the late Mr. Kurz, but 1 have seen no specimen collected by him or by any other person in the Andamans. And I have a strong suspicion that what Kurz regarded as *B. insigne* is really an undescribed species which Wallich issued as 1840-2 B of his Catalogue under the name *B. malabaricum*, var. *albiflora*. His No. 3 of the same name I have not seen. A tree with leaves exactly like Wallich's 1840-4 and with unarmed trunk and branches has recently been collected in the little Coco Island by Dr. D. Prain for the Calcutta Herbarium.

2. B. MALABARICUM, DC. Prod. i. 479. A tree with the general characters of the last, but much larger ; and with the trunk and branches prickly, the leaflets much narrower (lanceolate not obovate) and the flowers and fruit smaller. Mast. in Hook. fil. Fl. Br. Ind. i. 349 ; Kurz For. Fl. Burm. i. 136 ; Bl. Bijdr. 81 ; Wight Ill. t. 29 ; W. & A. Prodr. i. 61 ; Wall. Cat. 1840 (exclude No. 4 and possibly No. 2 B) ; Beddome Fl. Sylvat. t. 82. *Salmalia malabarica*, Schott Meletem, 35 ; Thwaites Enum. 28 ; Dalz. & Gibs. Bomb. Fl. 22 ; Miq. Fl. Ind. Bat. i. pt. 2, 166. *Bombax heptaphylla*, Cav. Diss. v. p. 296 ; Roxb. Hort. Beug. 50 ; Cor. Pl. iii. t. 247 ; Fl. Ind. iii. 167. *B. Ceiba*, Burm. Fl. Ind. 145, excl. syn. *Gossampinus rubra*, Ham. in Trans. Linn. Soc. xv.

Andaman Islands ; common.

## 7. ERIODENDRON, DC.

Trees. *Leaves* digitate, deciduous. *Flowers* appearing before the leaves, tufted at the ends of the branches, or axillary, large white or rose-coloured. *Bracteoles* 0. *Calyx* cup-shaped, truncate, or 3-5-fid. *Petals* oblong. *Staminal* bundles 5, opposite the petals, connate at the base, each bearing 2-3 sinuous or linear anthers. *Ovary* ovoid, 5-celled ; style cylindrical, dilated, stigma obscurely 5-lobed. *Capsule* oblong, coriaceous or woody, 5-celled, 5-valved, valves densely silky within. *Seeds* globose or obovoid ; testa crustaceous, smooth with silky hairs, albumen scanty ; cotyledons contortuplicate.—About eight species—1 Asiatic and African, the others American.

1. E. ANFRACTUOSUM, DC. Prod. i. 479. A tall tree, the trunk prickly when young ; branchlets stout, smooth, glaucous. *Leaflets* 8 or 9, lanceolate, acuminate, entire or serrulate towards the apex, the base acute ; glaucous beneath ; length 3 to 4 in., breadth ·75 to 1 in., petiolule ·25 in. broad ; petioles usually longer than the leaflets. *Flowers* pedunculate, in fascicles of 3 to 8 below the apices of the branches ; peduncles 1 to 2 in. long, minutely bracteate : involucre none. *Calyx* cup-shaped, with 5 rounded lobes, glabrous externally, sericeous internally. *Petals* oblanceolate, tomentose externally, glabrous within, 1 to 1·5 in. long, whitish. *Filaments* shorter than the petals. *Capsule* oblong, 3 to 5 in. long, smooth. *Seeds* numerous, sub-ovoid, black. Mast. in Hook. fil.

Fl. Br. Ind. i. 350; Bl. Bijdr. 81; W. & A. Prodr. i. 61; Wight Ic. t.
400; Griff. Not. iv. 533; Dalz. & Gibs. Bomb. Fl. 22; Miq. Fl. Ind.
Bat. i. pt. 2, 166; Beddome Fl. Sylvat. Anal. Gen. t. 4. Wall. Cat.
1839. *Bombax pentandrum*, Linn. Sp. Pl. 989; Cav. Diss. v. 293, t. 151;
Roxb. Fl. Ind. iii. 165. *B. orientale*, Spreng. Syst. iii. 124. *Ceiba
pentandra*, Gærtn. Frnct. ii. 244, t. 133; Ham. in Trans. Linn. Soc.
xv. 126. *Eriodendron orientale*, Steud. Nomencl. 587; Thwaites Enum.
28; Kurz For. Fl. Br. Burm. i. 131.

In all the provinces. Distrib. Malayan Archipelago, British
India, West Indies. Often planted.

## 8. DURIO, Linn.

Trees, with entire coriaceous penni-nerved leaves, scaly beneath
(except in *D. Oxleyanus*). *Flowers* in lateral cymes: peduncles angular.
Bracts 2 or 1, connate into a cup, or distinct below, tips free, deciduous.
*Calyx* bell-shaped, leathery, like the bracteoles densely scaly, the sepals
distinct, or 5-fid, lobes valvate oblong or rounded. *Petals* 5, contorted-
imbricate, spathulate, longer than the sepals. *Staminal-tube* divided
into 4-5 phalanges opposite the petals; filaments many, bearing a
globose head of sinuous 1-celled anthers, or (in *D. Oxleyanus*) a single
annular 1-celled anther. *Ovary* usually scaly externally, 4-5-celled;
styles connate, stigmas capitate; ovules many and 2-seriate in each cell.
*Fruit* very large, subglobose or oblong, spiny, indehiscent or loculicidally
5-valved. *Seeds* arillate; cotyledons fleshy, often connate. Distrib.
Malay Peninsula and islands; species 3.

1. D. ZIBETHINUS, Linn. Syst. Nat. edit. xiii. p. 581. A tall tree;
young branches thin and, like all the soft parts except the upper
surfaces of the leaves, minutely scaly. *Leaves* elliptic-oblong, rarely
obovate-oblong, shortly and abruptly acuminate, the base rounded;
both surfaces shining, the upper glabrous, the lower adpressed-lepidote;
main nerves 10 to 12 pairs, thin, slightly ascending; length 4·5 to 6 in.,
breadth 1·5 to 1·8 in., petiole ·4 to ·5 in. *Flowers* 2 in. long, 2 to 3 in.
in diam., on long slender pendulous dichotomus peduncles in fascicles
from the stem and larger branches, globose in bud: peduncles lepidote,
3 in. long, the bracts embracing the calyx and shorter than it. *Calyx*
tubular, ventricose at the base, the limb with 5 or 6 short broad teeth.
*Petals* twice as long as the calyx, spathulate. *Stamens* in 5 bundles
united only at the very base; the filaments in each bundle united for
one-fourth of their length: anthers glomerulate, reniform, compressed.
*Ovary* elongate-ovoid, scaly; style pubescent, as long as the stamens.
*Fruit* ovoid-globose, 8 to 12 in. long, woody, densely covered with strong

smooth pyramidal spines, 5-valved. *Seeds* few, large, with copious suc-
culent arillus. Mast. in Hook. fil. Fl. Br. Ind. i. 351, and Journ. Linn.
Soc. xiv. 501; Beccari Malesia, iii. 230, t. xii. f. 1 to 5, xxxvi. f. 1 to 12;
Kurz For. Fl. Burm. i. 131; DC. Prod. i. 480; Bl. Bijdr. 81; Koen. in
Trans. Linn. Soc. vii. 266, t. 14—16; Roxb. Fl. Ind iii. 399. Miq. Fl.
Ind. Bat. i. pt. 2, 167. Griff. Not. iv. 528; Ic. t. 596. Wall. Cat. 1842.
—Rumph. Amb. i. 99, t. 29.

In all the provinces except probably the Nicobars, cultivated.
Distrib. Malayan Archipelago.

2. D. Lowianus, Scortechini MSS. A tree 50 to 60 feet high;
young branchlets and petioles and lower surface of midrib with rather
large loose scales. *Leaves* narrowly elliptic-oblong, shortly acuminate;
the base rounded, not attenuate; upper surface glabrous, the midrib
puberulous, lower quite covered with adpressed scales, mostly minute,
but a few larger and loose; main nerves 14 to 18 pairs, faint, sub-horizontal; length 4·5 to 5·5 in., breadth 1·5 to 2 in.; petiole ·5 in. stout.
*Cymes* crowded on small tubercles on branches several years old, tricho-
tomous, 3 in. in diam. and about as long. *Flower-pedicels* ·5 to ·75 in.
long, angled, covered with loose coppery scales. *Flowers* 2 in. in diam.;
bracts 2 or 3, ·5 in. long, broadly ovate, connate, deciduous. *Calyx* campanulate, its base sub-inflated, ·75 in. long, its mouth with 3 broad
blunt, shallow teeth, glabrous inside, covered with large silvery scales
outside. *Petals* 5, oblanceolate, glabrous inside, pubescent outside,
1·25 in. long. *Stamens* in 5 phalanges, dividing shortly above the base
into about 8 processes each dividing at its apex into several short filaments, each bearing a single reniform anther with marginal dehiscence.
*Ovary* broadly ovoid, densely covered with large loose scales, 5-celled
with 4 ovules in each, biseriate. *Style* cylindric, tapering, pubescent:
stigma capitate. Fruit unknown.

Perak. Scortechini No. 1969.

A species collected only once and named by the late lamented
Father Scortechini in honour of Sir Hugh Low, representative of the
British Government at Perak, and to whose enlightened help Malayan
Botany owes very much. The species approaches *D. Zibethinus* in many
respects.

3. D. malaccensis, Planch. MSS. Mast. in Hook. fil. Fl. Br. Ind.
i. 351. A tree; the young branches thin, very minutely adpressed-
scaly. *Leaves* elliptic-lanceolate with acute apices; the base acute, sometimes slightly rounded; main nerves about 20 pairs, thin, almost
horizontal; both surfaces shining, the upper glabrous, the lower very

minutely adpressed-scaly; length 5 to 6·5 in., breadth 1·5 to 1·8 in.; petiole ·5 in., scaly like the branches. *Peduncles* ·5 to 1 in. long, in fascicles from tubercles on the stem, angled, bifurcating at the apex and bearing two pedicellate flowers, sometimes bearing one or two pedicels below the apex: pedicels two or three times as long as the common peduncles, angled, loosely scaly. *Flowers* 2·5 to 3 in. long. *Bracts* 2, broadly ovate, acute, embracing the buds. *Sepals* 5, ovate-oblong, blunt, valvate, 1·25 in. long, glabrous internally but with numerous very loose scales externally. *Petals* nearly twice as long as the sepals, narrowly oblong, pubescent on both surfaces, the outer with a few loose scales. *Anthers* narrowly oblong, 1-celled, sessile in groups on the apices of groups of combined filaments which are again united into 5 phalanges which, for more than half their length, form a tube round the ovary and style. *Ovary* oblong, angled, densely covered with scales with long cylindric stalks and flat heads. *Style* shorter than the staminal tube, pubescent, slightly scaly. *Stigma* capitate. *Young fruit* globular, densely covered with subulate pubescent spines. *Ripe fruit* unknown. Mast. in Journ. Linn. Soc. xiv. p. 501, t. xiv. fig. 17 to 20: *Beccari Malesia*, iii. 237, t. xii. fig. 6 to 8.

Malacca; Griffith; Maingay (No. 212, Kew Distrib.) Distrib. Burmah.

This is known only from Malacca and Burmah. It is distinguished from *D. Perakensis*, which in other respects it much resembles, by the stalked scales on the ovary, and by the larger and looser scales on the leaves. Doubtless when ripe fruit of both is found, better characters will be yielded by it. Beccari's specimen No. 852, and the same distinguished botanist's Nos. 2190 and 2590 from Borneo, have been referred by Masters (Journ. Linn. Soc. l. c.) to this species. But Beccari (in *Malesia* iii. 238, 244) founded his species *D. affinis* on the former, and his *D. testitudinarum* on the two latter.

4. D. TESTITUDINARUM, Becc. Malesia, iii. p. 244, t. xiii and xiv. A tall tree bearing flowers only near the base of the trunk; young branches rather slender, minutely sub-adpressed scaly. *Leaves* narrowly elliptic-oblong or oblanceolate-oblong, acute or shortly acuminate, the margins (in var. 2) sometimes with a single wide shallow indentation, the base rounded; upper surface glabrous, the lower densely covered with sub-adpressed scales: main nerves 18 to 22 pairs, rather bold, subhorizontal: length 4·6 to 8·5 in. (only 2·5 to 3·5 in. in var. 1 and much longer and broader in var. 2); breadth 1·4 to 2·2 in.; petiole 6 to ·25 in., thickened at apex. *Flowers* 3 to 3·5 in. long, in short condensed bracteolate racemes from tubercles near the base of the trunk;

the axes, pedicels, bracteoles and bracts densely covered with large loose scales: bracts enveloping the buds 2, broadly ovate, blunt. *Sepals* 5, valvate, wide and saccate at the base, the apices narrowed, glabrous inside, densely covered outside with loose large scales. *Petals* narrowly oblong, obtuse, more than twice as long as the sepals. *Stamens* as in *D. Malaccensis. Ovary* oblong, densely covered with loose, flat, sessile scales. *Style* shorter than the stamens, pubescent, sparsely scaly. *Stigma* capitate. *Fruit* (according to Beccari) on long peduncles, globose, 4 in. in diam., with 4 or 5 slight superficial grooves, densely covered with short broad pyramidal spines. *Seeds* sub-ovate, obtuse, angled ; the arillus short, thin, cup-shaped.

Perak ; at low elevations, Kunstler, Wray. Distrib. Borneo.

Var. 1. *Pinangiana*, Becc. l. c. 246. Leaves narrowly lanceolate, acuminate, 2·5 to 3·5 in. long by ·6 to ·9 in. broad. *Flowers* smaller than in the typical form : fruit unknown.

Penang, at 2,500 feet; Curtis No. 293. This variety, of which only imperfect specimens have as yet been obtained, will probably, when full material shall be forthcoming, prove to be a distinct species.

Var. 2. *macrophylla*, King. *Leaves* 10 to 17 in. long, 2·5 to 5·5 in. broad, the edge sometimes with a single shallow indentation. Racemes 3 in. long, many-flowered, with numerous bracteoles.

Perak ; Kunstler 7497, Wray 3397. No fruit of this variety has as yet been collected. Like the last, it may prove to be a distinct species.

5. D. WRAYII, King, n. sp. A large tree ; young branches very slender and, like the petioles and under surface of midrib, covered with rather large adpressed pale brown scales. *Leaves* narrowly elliptic-oblong with caudate acuminate apex and rounded base; upper surface quite glabrous, lower closely covered with thin adpressed silvery scales smaller than these on the midrib ; main nerves 10 to 12 pairs, sub-horizontal, faint: length 5·5 to 8·5 in., breadth 2 to 2·5 in., petiole ·75 in. *Flowers* nearly 2 in. long, from the branches ; pedicels of individual flowers rather more than 1 in. long, with many large loose scales. *Bracts* 3, broadly ovate, connate. *Calyx* cup-shaped, the mouth with 5 broad, rather deep, sub-acute teeth ; inside glabrous, outside covered with large adpressed silvery scales as are also the bracts. *Petals* 1·5 in. long, oblanceolate, or spathulate-clawed, the claw very narrow, pubescent on both surfaces but especially on the outer. *Stamens* in 5 phalanges united at the bases only, each phalange dividing into 5 or 6 processes at the apices of which are born about 8 narrow reniform anthers dehiscing by their edges. *Ovary* broadly ovoid, loosely scaly.

*Style* longer than the stamens, cylindric, pubescent, not scaly : stigma capitate. *Fruit* unknown.

Upper Perak at 300 feet ; Wray.

The fruit of this is unknown. Mr. Wray describes the petals as pink. The caudate-lanceolate leaves of this are different from those of any other *Durio* of the Malayan Peninsula.

6. D. OXLEYANUS, Griff. Notul. iv. 531. A tree, the young branches, petioles and under surfaces of the midrib adpressed-lepidote. *Leaves* elliptic-oblong, rounded at base and apex ; upper surface glabrous ; the lower softly pubescent, not scaly except on the midrib, the 15 to 18 pairs of main nerves stout, sub-horizontal, prominent beneath; length 3·5 to 5 in., breadth 1·5 to 2 in., petiole ·5 in. *Flowers* about 1 in. in diam., in few-flowered scaly cymes from the smaller branches. *Involucral bracts* 2, broadly ovate, pubescent, sparsely and minutely scaly. *Calyx* cup-shaped, the mouth with 4 broad shallow rather blunt teeth, inside glabrous, outside with many large loose scales. *Petals* 4, oblanceolate or spathulate, little longer than the calyx, pubescent on both surfaces, not scaly. *Stamens* 20, shorter than the petals ; 5 free and alternating with 5 phalanges of 3 each which are slightly united by the bases of their filaments : *anthers* solitary, drum-shaped, the dehiscence circular. *Ovary* depressed-globular, 4-celled, densely stellate-hairy. *Style* cylindric, pilose ; stigma capitate. *Fruit* unknown. Mast. in Hook. fil. Fl. Br. Ind. i. 351 and Journ. Linn. Soc. xiv. 501, t. xvi. fig. 13 to 16. Beccari Malesia, III, 252. *Neesia Griffithii*, Planch. MSS.

Malacca, Griffith No. 545. Maingay, No. 220, (Kew Distrib.)

This differs, as Beccari has well pointed out (Malesia l. c.), from the other species of *Durio* by the absence of scales from every part of the leaf except the petiole and midrib; by the single, not glomerulate, anthers ; by the hairy, not squamose, ovary. Should the fruit when found also present differences, it may be desirable to create a new genus for this species.

9. BOSCHIA, Korth.

Trees. *Leaves* oblong, entire, scaly beneath. *Flowers* small, axillary. *Bracteoles* 2-3, connate at the base, deciduous. *Calyx* deeply 4-5 parted. *Petals* linear-ligulate, entire or laciniate. *Stamens* many, some free, others irregularly coherent, outermost without anthers ; anthers globose, 1-celled, opening by a terminal pore, solitary, or in groups of 2-6. *Ovary* 3-5-celled, style elongate ; ovules one or more in each cell. *Fruit* oblong, 3 to 5-celled, 3 to 5-valved, muricate. *Seeds* few, ovoid, half-covered by a fleshy, coloured, cup-shaped arillus ; cotyledons foliaceous. Species 4: all Malayan.

1. B. Griffithii, Masters in Hook. fil. Fl. Br. Ind. i, 352. A tree 40 to 60 feet high; young branches rather slender, pale, minutely furfuraceous. *Leaves* oblong, or elliptic-oblong, or obovate-oblong, shortly and abruptly acuminate, slightly narrowed towards the rounded base; upper surface quite glabrous; the lower pale, very minutely pubescent, the midrib and nerves slightly scaly; main nerves 8 to 11 pairs, spreading, prominent beneath and dark coloured; length 5 to 6·5 in., breadth 1·5 to 2·25. in., petiole ·4 to ·6 in.: stipules linear, deciduous. *Flowers* ·75 in. in diam., solitary, or in 2 to 3-flowered cymes from the axils of leaves or of fallen leaves; pedicels shorter than the petioles, bracteolate. *Involucral bracts* 2, broadly-ovate, blunt, connate at the base, closely enveloping the buds; scaly externally, glabrous within. *Sepals* 4, ovate, spreading, pubescent on both surfaces, scaly also on the outer. *Petals* 4 to 8, nearly twice as long as sepals, linear or linear-spathulate, ·1 in. broad. *Stamens* very numerous, unequal, slightly united by the bases of the filaments: the outer without anthers, some flat resembling the petals, a few of the inner longer and bearing at their apices 1 to 4 oblong obovoid anthers which dehisce by an apical pore. *Ovary* ovoid, 3-celled, densely covered with peltate, fimbriate, long-stalked scales. *Style* as long as the longest stamens. *Stigma* subcapitate. *Fruit* oblong, pointed at each end, 1·5 to 2 in. long, densely covered with sharp stout conical spines, 3-celled, dehiscent. *Seeds* 3 to 6, or fewer. Mast. in Journ. Linn. Soc. xiv. t. xv, fig. 29 to 39, t. xvi., fig. 40 to 42. *Beccari Malesia* III, p. 256. *Heteropyxis*, Griff. Not. iv. 524; Ic. Pl. As. t. 594.

Malacca; Griffith, Maingay. Perak, very common. Distrib. Sumatra, Forbes, No. 3068.

## 10. Neesia, Blume.

Trees. *Branches* marked with large leaf-scars. *Leaves* entire, pinnate-veined. *Stipules* leafy. *Cymes* from the stem in the axils of the fallen leaves. *Bracteoles* 3, connate into a cup, deciduous, covered, like the sepals, with peltate scales. *Calyx* ventricose, conical above, opening by a circular irregularly crenulate orifice at the top, ultimately dilated and cushion-shaped at the base. *Petals* 5, free, imbricate. *Stamens* numerous, the filaments more or less united; anthers 2-celled, opening lengthwise, connective thick; staminodes 0. *Ovary* oblong, 5-celled; style short; stigma capitate; ovules numerous, 2-seriate, horizontal, anatropous. *Fruit* ovoid, woody, muricate, loculicidally 5-valved. *Seeds* albuminous; aril 0; cotyledons flat, leafy. Distrib. Seven species, all Malayan.

56

N. SYNANDRA, Mast. in Hook. fil. Fl. Br. Ind. i. 352. A tree 70 to
100 feet high; young branches stout, their bark dark lenticellate and with
large cicatrices. *Leaves* large, crowded near the apices of the branches,
coriaceous, oblong-elliptic to obovate-elliptic: the apex rounded, emarginate; the edges sub-undulate, slightly narrowed in the lower third to
the sub-cordate base; upper surface glabrous, lower puberulous; nerves
13 to 22 pairs, spreading, stout and distinct on both surfaces, the reticulations also distinct; length 7 to 16 in., breadth 3·5 to 8 in.; petiole
1·5 to 3 in., thickened at base and apex; stipules foliaceous, with very
stout midribs, 1·5 to 2·5 in. long. *Cymes* short (1·5 in. long), crowded,
dichotomous, 8 to 12-flowered, from the axils of sub-apical fallen leaves;
the pedicels short, scaly. *Flowers* about ·6 in. long. *Bracts* connate into
a 3-lobed cup surrounding the base of the flowers. *Calyx* ventricose
with a contracted irregularly and minutely toothed mouth, densely
pubescent inside, scaly outside as are the bracts, ultimately involute so
as to form an annular cushion ·5 in. or more in diam. *Petals* 5, free,
much imbricate, ovate-lanceolate, glabrous. *Stamens* numerous, the
filaments more or less connate at the base, unequal; anthers sub-globular, 2-celled. *Ovary* conical, sessile, densely pilose, not scaly: style
slightly longer than the ovary; stigma capitate, 5-angled. *Fruit* 6 to
8 in. long and 4 to 5 in. in diam., ovoid-conic, pedunculate, with 5 bold
rounded vertical angles: the pericarp very thick, woody, externally
covered with stout pyramidal sharp spines, internally lined with a
dense layer of stiff yellow hair; 5-celled, dehiscent. Mast. in Journ.
Linn. Soc. xiv. p. 504. *Beccari Malesia*, iii. 263.

Malacca, Maingay. Perak; Scortechini, Wray, King's Collector.

I have seen no specimens of the plant (*N. altissima*) on which
Blume founded this genus. But, judging from his admirable description
and fine coloured figure (Nov. Act. Acad. Caes. xvii. 83, t. vi), this species
must be very closely allied to that. I find the stamens of this agree
both with Blume's description above referred to, and with Sig. Beccari's,
in his admirable and splendidly illustrated monograph in *Malesia* iii.
pp. 258 to 268. Ripe fruit and seeds of this are as yet unknown.

11.  CŒLOSTEGIA, Benth.

Tall trees. *Leaves* simple, entire, scaly beneath. *Flowers* small
(scarcely ·25 in. in diam.), cymose; the inflorescence, bracts and calyx
scaly. *Bracts* connate into a toothed cup. *Calyx* with constricted tube,
pouched above and constricted at the apex into 5 connivent lobes.
*Petals* 5, free, inserted near the apex of the calyx tube, connivent.
*Stamens* numerous; the filaments short, thick, slightly connate at the
base, the apex constricted; the anthers globose, 3 to 4-celled. *Ovary*

57

partly immersed in the calyx-tube, globular or sub-globular, 5-celled ; the ovules few, erect. *Style* short; stigma peltate, discoid, large. *Fruit* large, woody, muricate externally, hairy within, 5-celled, few-seeded, dehiscent. Three species; all Malayan.

C. GRIFFITHII, Benth. in Benth. & Hook. fil. Gen. Plant. i. 213. A tree; the young branches rather slender, dark-coloured, striate, minutely and deciduously scaly. *Leaves* coriaceous, oval, shortly and bluntly acuminate, the base rounded; upper surface glabrous, lower sparsely adpressed-scaly ; main nerves about 8 pairs, spreading, faint ; length 2·4 to 3·75 in., breadth 1·25 to 1·6 in.; petiole ·5 to ·75 in , minutely adpressed-scaly. *Inflorescence* of fasciculate cymose racemes about 2 in. long, from the axils of fallen leaves, many-flowered ; pedicels longer than the flowers. *Flowers* ·25 in. in diam., scaly. *Bracts* connate into a 3-lobed cup less than half as long as the calyx. *Calyx* constricted at the base, then dilated into a 5-pouched sac which is contracted and 5-toothed at its apex. *Petals* 5, distinct, inserted on the calyx at the apex of its tube, triangular, acute, connivent, fleshy, glabrous. *Stamens* numerous, attached to the petals ; the anthers small, globose, 3 or 4-celled. *Ovary* globular-obovate, densely covered with large loose scales. *Style* shorter than the ovary ; stigma peltate, thick, its edges wavy. *Fruit* unknown. Mast. in Hook. fil. Fl Br. Ind. i. 353 and Journ. Linn. Soc. xiv. 505, t. xvi, figs. 43 to 50. *Beccari Malesia*, iii. 270.

Malacca, Griffith ; Perak, Scortechini, King's Collector.

*Fruit* was not known when this genus was first established by the late Mr. Bentham ; and, of this species, fruit is still unknown. Sig. Beccari has, however, discovered two species in Sumatra and Borneo (*C. Sumatrana* and *Bornensis*) the fruit of which he describes and figures (*Malesia*, iii. 271, t. xxvii. to xxix) ; and from his description the generic description has been completed.

## Order XVIII. STERCULIACEÆ.

Herbs, shrubs or trees ; herbaceous portions usually more or less stellate-pubescent. *Bark* usually abounding in mucilage, inner fibrous. *Leaves* alternate, simple, often lobed, stipulate. *Inflorescence* axillary, rarely terminal, usually cymose. *Flowers* regular, uni- or bi-sexual. *Sepals* 5. often connate. *Petals* 5 or 0. *Andrœcium* columnar or tubular, of many stamens ; or stamens rarely few, free ; anthers in heads, or in a single ring at the apex of the column, or dispersed on the outside of the tube, or arranged along the edge of a cup or tube, with intervening staminodes or sterile stamens ; anther-cells always 2,

166

parallel or divergent. *Ovaries* 2 to 5, free, rarely 1, sessile or stalked ; styles slightly united and becoming free or slightly coherent, as many as the ovaries. *Ovules* many or few, attached to the inner angles of the ovaries, anatropous, ascending or horizontal, raphe ventral or lateral. *Fruit* dry or fleshy, dehiscent or indehiscent. *Seeds* sometimes arillate, albuminous or exalbuminous : cotyledons leafy, flat, folded or convolute ; radicle short, inferior, pointing towards, or remote from the hilum. Distrib. Abundant in the tropics of either hemisphere and in subtropical Africa and Australia. Genera 40—45 ; species from 500 to 600.

Tribe I. *Sterculieæ. Flowers* unisexual or polygamous. *Petals* 0. *Andrœcium* columnar ; the anthers clustered at its apex ; or in a 1-seriate ring.

Anthers numerous.

Ovary with 2 or more ovules in each cell ; fruit dehiscent ... ... 1. *Sterculia.*

Ovarian cells 1-ovuled ; fruit indehiscent 2. *Tarrietia.*

Anthers 5, whorled ; fruit indehiscent.

3. *Heritiera.*

Tribe II. *Helictereæ. Flowers* hermaphrodite. *Petals* deciduous. *Andrœcium* columnar below, dilated above into a cup, margin bearing on it the anthers usually alternating with staminodes.

Capsule membranous, inflated... ... 4. *Kleinhovia.*

Capsule more or less woody, not inflated.

Anther-cells divaricate; seeds not winged 5. *Helicteres.*

Anther-cells parallel ; seeds winged ... 6. *Pterospermum.*

Tribe III. *Hermannieæ. Flowers* hermaphrodite. *Petals* marcescent, flat. *Andrœcium* tubular at the base only ; stamens 5, staminodes 0.

Ovary 5-celled ... ... ... 7. *Melochia.*

Ovary 1-celled, 1-seeded ... ... 8. *Waltheria.*

Tribe IV. *Buettneriæ. Petals* concave or unguiculate at the base ; filaments in a tube with the anthers at its apex, solitary or in groups between staminodes.

Stamens in a single series.

Stamens in groups between the staminodes ;

Petals unguiculate ... ... 9. *Abroma.*

167

Stamens solitary between the staminodes.
Petals unguiculate, with 2 lateral
lobes and a long subterminal appendage ...  ...  ... 10. *Buettneria.*
Petals linear not lobed, concave not
unguiculate at the base  ... 11. *Commersonia.*
Stamens in several series  ...  ... 12. *Leptonychia.*

### 1. STERCULIA, Linn.

Trees or shrubs. *Leaves* simple, entire or palmately lobed, sometimes digitately compound. *Inflorescence* panicled or racemose, usually
axillary and crowded towards the apices of the branches. *Flowers* male
and hermaphrodite. *Calyx* campanulate or rotate, 4-5 lobed, often
coloured. *Petals* 0. *Staminal column* bearing a head or ring of usually
sessile, 2-celled, anthers at its apex, the cells often divergent. *Carpels*
5, distinct or slightly cohering, 2 to many-ovuled, borne on the apex of a
more or less elongated gynophore ; styles more or less connate : stigmas
free or united so as to form a peltate lobed disc. *Ripe carpels* distinct,
spreading, sessile or stalked, follicular, from membranous to woody, with
several (rarely many) seeds ; or navicular with a single seed. *Seeds* 1 to
many, sometimes winged, rarely arillate ; albumen bipartite, flat or lobed :
cotyledons thin flat and adherent to the albumen, or fleshy ; radicle near
to or remote from the hilum. Distrib. About 70 species tropical and
chiefly Asiatic.

Sect. I. *Eusterculia,* Endl. *Follicle* coriaceous or
 woody. *Seeds* two or more.
 Leaves simple, orbicular or reniform.
  Leaves lobed.
   Follicles glabrous within, the edges
    only ciliate ; gynophore and stami-
    nal tube glabrous  ...  ... 1. *S. villosa.*
   Follicles hispid-pilose within ; gyno-
    phore and staminal tube hairy ... 2. *S. ornata.*
  Leaves not lobed  ...  ... 3. *S. macrophylla.*
 Leaves simple, longer than broad ; not or-
  bicular or reniform.
  Leaves quite glabrous.
   Calyx-lobes not cohering by their
    apices  ...  ...  ... 4. *S. laevis.*
   Calyx-lobes cohering by their apices.
    Flowers in racemes : nerves of
     leaves 6 pairs or fewer  ... 5. *S. hyposticta.*

Flowers in panicles: nerves of
leaves more than 6 pairs
Leaves narrowly oblong-lanceo-
late; follicles 1 to 1·25 in.
long ... ... ... 6. *S. parvifolia.*
Leaves ovate or obovate-oblong
to narrowly elliptic.
    Ovaries 3, villous: stamens 7   7. *S. Kunstleri.*
    Ovaries 5, scaly; stamens 10   8. *S. parvifolia.*
Leaves more or less hairy.
  Calyx-lobes not cohering by their
    apices: leaves glandular-dotted
    beneath ... ... ... 9. *S. Scortechinii.*
  Calyx-lobes slightly cohering by their
    apices: leaves white beneath ... 10. *S. bicolor.*
  Calyx-lobes spreading, connivent and
    cohering by their apices.
    Stigmas free, long, recurved ... 11. *S. augustifolia.*
    Stigmas united into a lobed disc.
      Leaves more or less obovate 12. *S. rubiginosa.*
      „   lanceolate ... ... 13. *S. ensifolia.*
Species of uncertain position... ... 14. *S. pubescens.*

Sect. II. *Firmiana,* Marsili; Br. in Benn. Fl. Jav.
Rar. 235 (gen.). *Follicles* membranous,
opening long before maturity. *Seeds* two or
more.
  Calyx ·75 in. long: staminal tube about
    the same length; adult leaves glabrous... 15. *S. colorata.*
  Calyx 1·25 in. long, staminal tube ·5 in.
    longer: adult leaves minutely stellate-
    pubescent ... ... ... 16. *S. fulgens.*

Sect. III. *Pterygota,* Endl. (gen.). *Follicles* woody.
*Seeds* many, winged at the apex ... 17. *S. alata.*

Sect. IV. *Scaphium,* Endl. *Anthers* 15, (sometimes 10). *Stigmas* lobed. *Follicles* large,
membranous, boat-shaped, often gibbous,
opening long before maturity, containing
only 1 seed near the base.
  Leaves ovate-rotund, deeply cordate ... 18. *S. linearicarpa.*
  Leaves ovate to ovate-oblong: main
    nerves 2 to 4 pairs ... ... 19. *S. scaphigera.*

61

Leaves elliptic-oblong : main nerves 6 to
  7 pairs ... ... ... 20. *S. affinis.*
Sect. V. *Pterocymbium*, Br. in Benn. Pl. Jav. Rar.
  219 (gen.). *Flowers* sub-hermaphrodite.
  *Anthers* 10. *Styles* coherent, stigmas recurved. *Follicles* 4—6, membranous, opening long before maturity. *Seed* solitary.
    Leaves broadly ovate, acuminate, the
      base deeply cordate ... ... 21. *S. campanulata.*
    Leaves elliptic-oblong ; the base broadly
      rounded or sub-truncate, not cordate... 22. *S. tubulata.*

1. S. VILLOSA, Roxb. Hort. Beng. 50. A tree 30 to 60 feet high :
young branches thick, their apices tawny-tomentose and enveloped
by the large sub-caducous stipules, the bark pale with large leaf-
cicatrices. *Leaves* thickly membranous, rotund or reniform, with 5 to 7
broad abruptly acuminate often toothed lobes, the sinuses between the
lobes acute ; the base deeply cordate, the basal lobes rounded : upper
surface at first minutely stellate-pubescent, ultimately glabrous, except
the 5 to 7 radiating tomentose nerves : under surface uniformly and
minutely tomentose ; length and breadth from 12 to 18 inches : petiole
deciduously densely pubescent, about as long as the blade : stipules
ovate-lanceolate, acuminate, with cordate bases, pubescent, sub-caducous.
*Panicles* from the axils of the previous year's leaves, solitary, from
6 to 12 in. long : branches short, many-flowered, tomentose. *Calyx*
campanulate, ·4 in. in diam., with 5 ovate acute spreading lobes as long
as the tube, yellowish with purple fundus, veined, puberulous outside
especially towards the base, almost glabrous inside. *Male flower ;*
staminal column longer than the calyx-tube, slightly curved, quite glabrous, bearing at its apex 10 sub-sessile anthers with thick connective
and 2 divergent cells. *Female flower ;* gynophore glabrous, thickened
above ; ovaries 5, conjoined, tomentose ; styles conjoined, puberulous,
curved ; stigma small, lobed. *Follicles* 3 to 5, coriaceous, sessile, bright
red when ripe, oblong, tapering to both ends ; 2 to 2·5 in. long by 1 in.
broad ; shortly hispid-pubescent externally, smooth and shining internally and glabrous except along the placental edges which are strongly
ciliate. *Seeds* 6 or more, oval, smooth. Roxb. Fl. Ind. i. 153 ; Kurz
For. Fl. Burm. i. 136 ; Mast. in Hook. fil. Fl. Br. Ind. i. 355 ; Pierre Fl.
Forest. Coch-Chine, t. 185, fig. D. ; Wall. Cat. 1136, 2, 3, D. ; W. & A.
Prodr. i. 63 ; Dalz. & Gibs. Bomb. Fl. 22 ; Br. in Benn. Pl. Jav. Rar.
227.
    Andamans, Prain. Distrib. British India.
170

2. S. ORNATA, Wall. in Herb. Calcutta. A tree 20 to 30 feet high : young branches thick, glabrous, pale, the leaf-cicatrices very large, the apices decidnously pilose, coccineous drying into brown. *Leaves* thickly membranous, reniform, more or less deeply divided into 5 or 7 acuminate lobes, the sinuses between the lobes wide, the base deeply cordate; upper surface minutely strigose, often stellate, minutely pitted ; lower surface yellowish-brown, minutely and uniformly tawny-tomentose, minutely glandular-dotted under the hair ; the 5 to 7 radiating main nerves and the ascending secondary nerves bold and distinct; length about 12 in., breadth about 15 in. ; petiole 15 to 18 in. long, thickened at the base, minutely tomentose. *Panicles* from the axils of the previous year's leaves, solitary, 8 to 15 in. long, shortly branched, many-flowered, pulverulent reddish-tomentose. *Calyx* ochre-coloured with red fundus, veined, widely campanulate, sub-rotate, with 5 ovate acute spreading lobes longer than the tube, stellate-pubescent externally, puberulous internally ; ·75  in diam. *Male flower ;* gynophore about as long as the tube, curved,  ly glandular-hairy, bearing at its apex 10 small anthers with thick connective. *Female flower ;* gynophore thickened above, densely tawny-tomentose as are the conjoined ovaries and curved style ; the ovaries with a ring of about 10 sessile anthers at their base ; stigma discoid, rugulose, 5-lobed. *Follicles* about 5, sessile, coriaceous, narrowly oblong, very shortly beaked, brilliant orange scarlet when ripe, outside glabrescent, inside densely coccineous-pilose ; length 4 in., breadth 1·25 in. *Seeds* about 6, oval, smooth. Wall. in Voigt Hort. Calc. Suburb. 105 (*name only*) ; Kurz Journ. As. Soc. Beng. Vol. xlii. pt. 2, p. 258 ; Vol. xliii. pt. 2, p. 116 ; For. Fl. Burm. i, 136. *Sterculia armata*, Mast. in Hook. fil. Fl. Br. Ind. i. 357, *in part.* Pierre Fl. Forest. Coch-Chine, t. 185, fig. C.

Burmah ; Wallich, Brandis, Kurz. Andamans, Kurz.

I include this species because, although the evidence of its having been collected in the Andamans is not very good, I think it extremely likely that it does occur there, and that good unmistakeable specimens will soon be forthcoming. The species in many respects resembles *S. villosa*, with which it appears to have often been confused. The distinctive marks to separate it from *S. villosa* are that the leaves are minutely dotted and pitted ; that the apices of the young branches have red hairs (becoming brown on drying) ; that after the hairs have fallen the young branches have pale polished bark with very large leaf-cicatrices and some warts, but no sub-persistent stipules ; that the flowers are larger (·75 in. in diam. as against ·4 in) ; that the staminal column and gynophore are hairy ; that the follicles are larger and paler ; and that the whole of their inner surface is densely hispid-pilose.

3. S. MACROPHYLLA, Vent. Hort. Malm. ii. No. 91 (in note). A tree 80 to 120 feet high; young branches very thick, rough from the leaf cicatrices, the apices deciduously rufous or tawny-pilose. *Leaves* sub-coriaceous, broadly ovate to ovate-rotund or obovate-rotund, entire, narrowing to the slightly cordate 7-nerved base; upper surface sparsely and rather minutely pubescent, some of the hairs 2-branched, becoming glabrescent with age, the midrib and nerves always pubescent; under surface sub-tomentose, tawny, the midrib and 6 to 8 pairs of lateral nerves prominent, rufous-villose; transverse venation distinct, rather straight; length 8 to 16 in., breadth 6 to 12 in., petiole 3·6 to 6 in., softly hairy, tawny. *Panicles* solitary, axillary, nearly as long as the leaves, much-branched, many-flowered, hispidulous-pubescent, capillary, shorter than the flowers. *Flower-buds* minute, sub-globose. *Calyx* ·15 in. long, campanulate, stellate-hairy, 5-lobed; the lobes triangular, erect, shorter than the tube. *Follicles* 3 to 5, shortly stalked, woody, sub-rotund, about 2·25 in. each way, crimson w̶██████ripe, outside pubescent and longitudinally rugose; inside smooth ████ls oblong, black, smooth, ·75 in. long. Mast. in Hook. fil. Fl. Brit. Ind. i. 356; R. Brown in Benn. Pl. Jav. Rar. 230.

Malacca; Maingay No. 233 (Kew Dist.). Perak; at elevations of 200 to 500 feet; King's Collector Nos. 6052 and 7923; Scortechini, No. 230. Distrib. Java, Brit. North Borneo.

4. S. LAEVIS, Wall. Cat. 1138. A shrub or small tree; young branches rather thin, with pale striate bark, the apices deciduously rusty-puberulous. *Leaves* membranous, narrowly ovate-oblong, sometimes slightly obovate, the apex shortly and bluntly acuminate; the base tapering, acute, rarely rounded, faintly 3-nerved : both surfaces glabrous, shining, the midrib and 6 to 9 pairs of spreading nerves prominent on the lower : length 4·5 to 9 in., breadth 2 to 3 in.; petiole 1·1 to 2·5 in., smooth, thickened at the apex. *Panicles* meagre, solitary, axillary, slender, puberulous, shorter than the leaves, few-flowered; pedicels about as long as the flowers. *Flower-buds* oblong. *Calyx* ·5 in. long or more, pubescent on both surfaces but especially on the inner; the tube urceolate, divided at its apex into 5 linear-oblong sub-acute ascending lobes, longer than the tube, slightly connivent but not cohering by their apices, hispidulous on their inner surface. *Male flower;* staminal column shorter than the tube, glabrous; anthers 10, sessile at its apex, elongate-ovate. *Hermaph. flower:* gynophore very short; ovaries 5, boat-shaped, rusty-pubescent, sub-sessile, with a ring of 10 sessile anthers at their base outside : styles almost obsolete; stigmas 5, cylindric, free, radiating, recurved, pubescent beneath. *Follicles* 3 to 5, coriaceous, narrowly

oblong, with short straight beaks, bright red when ripe, puberulous externally, slightly curved, glabrous, shining and ridged internally, 2 in. long and about ·5 in. broad. *Seeds* 3 or 4 oblong, black, shining. Mast. in Hook. fil. Fl. Br. Ind. i. 357. Pierre Fl. Forest. Coch-Chine t. 192, figs. 1 to 7; Br. in Benn. Pl. Jav. Rar. 230; Miq. Fl. Ind. Bat. i. pt. 2, 174. *S. coccinea*, Jack Mal. Misc. i. 286, not of Roxb.

Penang, Perak, Malacca, Singapore : at low elevations : but not common.

5. S. HYPOSTICTA, Miq. Fl. Ind. Bat. Suppl. 399. A shrub or small tree, all parts glabrous except the inflorescence : young branches slender, dark and smooth becoming (by the falling off of the bark) pale and striate. *Leaves* membranous, oblong, to oblong-lanceolate, sometimes slightly obovate, abruptly acuminate or even caudate-acuminate, entire, the base slightly narrowed and rounded, or not narrowed and truncate, emarginate, rarely acute, 3-nerved ; both surfaces glabrous, shining : lateral main nerves 3 to 5 pairs, spreading, curved, inarching far from the margin, prominent beneath : length 3·5 to 5·5 in., breadth 1·5 to 2·25 in., petiole 1 to 1·5 in., thickened at base and apex. *Racemes* axillary, solitary, drooping, longer than the leaves, minutely whitish pubescent, with superficial brown stellate hairs: bracteoles linear, longer than the pedicels. *Calyx* with narrowly campanulate tube ·25 in. long, densely rufous-pubescent externally and glabrous inside: lobes 5, not quite so long as the tube, linear, spreading, connivent, cohering from some time by their tips, the edges recurved, glandular-pilose inside, sub-pubescent outside. *Male flower;* staminal column short, glabrous, with 8 sessile oblong 2-celled anthers at its apex. *Female flower :* gynophore short ; ovaries 4, ovoid, conjoined, shortly tomentose, with ring of 8 sessile anthers at their base. *Style* simple, curved, sparsely villous ; stigma large, glabrous, with 4 fleshy oblong-obovoid curved lobes. *Follicles* 2 or 3, coriaceous, bright red when ripe, narrowly oblong, tapering to each end, 2 to 2·25 in. long and ·65 in. broad; externally minutely rusty-pubescent; internally glabrous, wrinkled. *Seeds* 4, oblong, pointed, black. Kurz in Journ. As. Soc. Beng. Vol. xlv. pt. 2, p. 120.

Perak ; King's Collector, Wray. Nicobars, Kurz.

6. S. PARVIFOLIA, Wall. Cat. 1123. A tree 20 to 30 feet high : young branches slender, striate, the older pale, the younger dark-coloured, glabrous. *Leaves* membranous, drying of a pale green, oblong-lanceolate, rarely ovate-oblong, bluntly acuminate, entire ; the base acute or round-ed, faintly 3-nerved ; both surfaces glabrous : main nerves 6 to 8 pairs,

spreading, rather prominent on both surfaces as is the midrib : length 4 to 6·5 in., breadth 1 to 1·75 in.; petiole 1 to 1·75 in., smooth, slender, thickened at the apex. *Racemes* solitary, axillary, much shorter than the leaves, few-flowered, glabrous; flower-pedicels shorter than the flowers, capillary. *Flower-buds* oblong. *Calyx* less than ·5 in. long, glabrescent externally, puberulous internally especially on the lobes ; tube wide, cylindric, with 5 linear-lanceolate lobes about as long as itself, spreading, incurving and joined for some time by their tips. *Male flower :* staminal column shorter than the tube and bearing at its apex about 12 small oblong anthers with thick connective and diverging cells. *Herm. flower :* gynophore very short, glabrous ; ovaries 5, broadly ovate, rusty-pubescent ; styles united, recurved, with many white spreading hairs : stigmas clavate, flattened, recurved, spreading. *Follicles* 3 to 5, broadly oblong, with a straight beak, 1 to 1·25 in long, ·6 in. broad. *Seeds* 2, broadly ovoid, black, shining. Mast. in Hook. fil. Fl. Br. Ind. i. 356 ; R. Brown in Benn. Pl. Javan. Rar. 229 ; Miq. Fl. Ind. Bat. Vol. i. pt. 2, p. 173.

Penang, Perak, Malacca.

Closely allied to *S. laevis*, Wall.: but with smaller flowers and follicles, and with calyx lobes coherent at their tips.

7. S. KUNSTLERI, King, n. sp. A tree 30 to 60 feet high; all parts (except the inflorescence and the tips of the young branches) glabrous ; branches with pale smooth striate bark. *Leaves* thinly coriaceous, broadly ovate (or slightly obovate) to oblong or narrowly elliptic, the apex rounded, blunt, sub-acute or very shortly and sub-abruptly acuminate ; slightly narrowed to the rounded or sub-truncate, rarely acute, 3 to 5-nerved, base ; both surfaces shining ; lateral nerves about 7 to 9 pairs, spreading, slightly prominent beneath : length 4 to 9 in., breadth 2 to 4·5 in. ; petiole ·75 to 2·75 in., slender, glabrous. *Panicles* solitary, narrow, in the axils of (and shorter than) the mature leaves, or supra-axillary, slender ; the lateral branches short, 1-to 3-flowered, flocculent-tomentose, rusty ; bracteoles lanceolate to ovate, caducous. *Calyx* ·3 to ·35 in. long, the tube urceolate, densely stellate-tomentose outside, sub-glabrescent inside ; lobes 5, shorter than the tube, linear-lanceolate, villous on the inner surface, tomentose on the outer, spreading, conni-vent and slightly coherent by their tips. *Male flower :* staminal column slender, shorter than the calyx-tube, curved, bearing at its apex 5 to 7 sessile broad anthers. *Hermaph. flower :* gynophore short ; ovaries 3, ovoid, villous, with a ring of adpressed sessile oblong anthers at their base : styles distinct, short, thin, sparsely villous ; stigmas thick, fleshy, clavate, bent (outwards) on themselves. *Follicles* 2 or 3,

174

woody, from peach-coloured to carmine when ripe, oblong, rounded at
at the base, the apex acute and slightly curved; externally rugose (the
rugae mostly longitudinal), minutely tomentose, inside smooth; length
3 to 3·5 in., breadth 1·25 to 1·5 in. *Seeds* narrowly ovoid, nearly 1 in.
long, black.

Perak; King's Collector Nos. 3259, 7211, 7245, Scortechini No. 1805;
at 100 to 300 feet elevation. Distrib. Sumatra; Forbes, No. 2679.

In externals this species closely resembles *S. parviflora*, Roxb.
But, after numerous dissections, I conclude that the two species are
quite distinct. The ovaries of this are never more than 3, and they are
always densely villous; those of *parviflora* are invariably 5, and they
are scaly, not villous. The stigmas of this are long and are bent outwards on themselves; these of *parviflora* are short and recurved outwards from their junction with the styles: they are not bent on themselves. The follicles of this are thicker and more woody and the seeds
are larger than those of *S. parviflora*. Moreover this has never more
than 7 stamens, while *S. parviflora* has 10. The leaves of this are rather
thicker in texture and the young branches are thinner and paler than
those of *S. parviflora*.

8. S. PARVIFLORA, Roxb. Hort. Beng. 50. A tree 20 to 50 feet
high; young branches rather thick; the tips ferruginous-tomentose; the
bark pale, rough, glabrous. *Leaves* membranous, oval, ovate or obovate-
oblong, the apex rather abruptly shortly and bluntly acuminate, entire;
the base rounded and slightly cordate, or sub-truncate and emarginate,
5-nerved; both surfaces glabrous, but not shining; the midrib and 7 or
8 pairs of spreading rather prominent lateral nerves sparsely stellate-
pubescent on the lower when young; length 4 to 10 in., breadth 2 to
5·5 in.; petiole 1 to 4 in., deciduously rufous-tomentose. *Panicles* about
as long as the leaves, slender, the lateral branches short and the flower-
pedicels capillary, everywhere covered with rusty stellate tomentum,
ebracteolate. *Calyx* ·2 in. long with an urceolate tube, the mouth with
5 linear-lanceolate lobes almost as long as the tube, incurved and united
by their apices, stellate-tomentose externally, glabrous within. *Male
flower:* staminal column shorter than the calyx-tube, bearing at its apex
10 sessile short narrowly ovate anthers with thick connective. *Herm.
flower:* ovaries 5, ovoid, scaly, with a ring of anthers at their base: ovules
4 or 5. *Styles* slightly united, slender, sparsely villous, short; stigmas
united into a fleshy boldly 5-lobed disc, but easily separable into 5
fleshy flattish recurved stigmas. *Follicles* 1 to 5, thickly coriaceous,
brilliant red to orange, pubescent to glabrescent, oblong, shortly beaked,
2·5 to 3·5 in. long and 1·25 to 1·5 in. broad; inside glabrous, shining,

boldly ridged. *Seeds* broadly ovoid, black, ·6 in. long, smooth. Roxb. Fl. Ind. iii. 147; Brown in Bennett Pl. Jav. Rar. 232 : Wall. Cat. 1121. Kurz For. Fl. Burm. i. 138. Pierre Fl. Forest. Coch-Chine, t. 195 F. *S. Maingayi*, Mast. in Hook. fil. Fl. Br. Ind. i. 359; Pierre Fl. Forest. Coch-Chine, t. 188 A.

Penang, Malacca, Perak; at low elevations, common. Distrib. Burmah and Sylhet in British India ; Cochin China.

After careful dissection of the flowers of the types of the two species *S. parviflora*, Roxb. and *S. Maingayi*, Masters, and of flowers of many other specimens, I can come to no other conclusion than that they are one and the same. There is a curious tendency to inequality in size in the leaves, some being twice as large as others rising from the same twig within the distance of an inch. And the panicles usually follow the leaves in the matter of length.

9. S. Scortechinii, King, n. sp. A tall tree ; young branches rather thick, their bark pale, rough, the youngest parts deciduously rusty-pubescent. *Leaves* thinly coriaceous, oblong, slightly obovate, the apex rounded, with an abrupt short blunt point, entire ; the base slightly narrowed, rounded or minutely cordate, 3-nerved ; upper surface glabrous, shining; the lower slightly paler, dull, thickly dotted with minute reddish flat shining glands, the midrib and 4 to 5 pairs of prominent ascending lateral nerves stellate-pubescent: length 2·5 to 3·5 in., breadth 1·25 to 1·65 in. ; petiole ·65 to 1 in., deciduously pulverulent-tomentose. *Panicles* racemes-like, axillary, solitary, shorter than the leaves, densely pulverulent-tomentose, rusty ; pedicels as long as the buds : bracteoles ovate, ·25 in. long, imbricate, caducous. *Calyx* campanulate, divided almost to its base into 5 broadly ovate spreading not connivent lobes, pubescent-tomentose both internally and externally *Male flower :* staminal column shorter than the calyx, crowned by about 10 short anthers with thick sub-cuneate connective and short divergent cells. *Herm. flower :* Ovary 3-celled, obliquely ovoid, pubescent-scaly; ovules 3 or 4 in. each cell. *Styles* connate, pubescent. *Stigmas* 3, large, ovoid, spreading, glabrous, dark-coloured. *Follicles* not seen.

Perak ; Scortechinii, No. 2068.

Collected only once, and without fruit.

10. S. bicolor, Mast. in Hook. fil. Fl. Br. Ind. i. 359. A tree 40 to 60 feet high : young branches rather thin, cinereous, striate, glabrous, rufous-pubescent at the very tips. *Leaves* small, membranous, obovate-oblong, acute or shortly mucronate, entire, slightly narrowed to the minutely 2 to 3-nerved rounded base ; upper surface glabrous

when adult, with a few small scattered white stellate hairs when young ; under surface pale from a layer of minute whitish hairs, the midrib and 16 to 18 pairs of sub-horizontal lateral nerves rufous-tomentose; length 2·5 to 3 in., breadth 1·2 to 1·4 ; petiole about 1 in., slender, scaly-tomentose. *Panicles* about as long as the leaves, slender, in the axils of young leaves, pulverulent-tomentose, sub-ferruginous; branches short, spreading. *Calyx* pedicellate, ovoid-oblong, pointed in bud, when adult ·3 in. long, widely campanulate, with 5 linear incurved pubescent lobes as long as the tube. *Staminal column* shorter than the tube, glabrous; anthers about 12, sessile at the apex of the column, their connective thick, cuneate, the cells divergent. *Follicles* unknown.

Malacca; Maingay, No. 230 (Kew Distrib.) Perak. Wray, No. 2378.

Recognisable at once by its small leaves, white beneath. The figure named *S. bicolor*, Mast. by Pierre (Fl. Forest. Coch-Chine t. 187) agrees neither with M. Pierre's own description of it ; nor with Masters' type-specimen. There may probably have been some printer's blunder in the matter.

11. S. ANGUSTIFOLIA, Roxb. Hort. Beng. 50. A small tree : young branches densely velvetty rusty-tomentose; ultimately rather pale, glabrous, warted and striate. *Leaves* membranous, oblong-lanceolate rarely ovate-lanceolate, acuminate or acute, entire, slightly narrowed to the rounded 3-nerved base : upper surface glabrous, the lower more or less densely and softly rusty-tomentose : length 4 to 7 in., breadth 1·25 to 2·25 ; petiole ·6 to 1·1 in., rusty-tomentose. *Panicles* solitary, axillary, crowded at the apices of the branches, lax, drooping, longer than the leaves, everywhere densely rusty-tomentose ; pedicels much longer than the ovate pointed buds. *Calyx* ·2 in. in diam., hispidul-ous-pubescent everywhere except the tube which inside is glabrous, deeply divided into 5 linear-lanceolate lobes ; the lobes longer than the tube, spreading, connivent, cohering by their tips, their edges recurved. *Male flower* : staminal column as long as the tube, glabrous, recurved, bearing at its apex 10 oblong sessile anthers with large connective, the cells slightly divergent. *Herm. flower* : gynophore short, glabrous : ovaries 5, ovoid, rusty-tomentose with a ring of 10 sessile anthers at their base : style short, sparsely pilose : stigmas much longer than style, fleshy, spreading, recurved. *Follicles* 4 or 5, ovate-oblong with a short curved beak, 2·75 in. long and 1·35 in. broad, densely but minutely velvetty rusty tomentose outside, smooth shining and rugose inside and with a few small scattered whitish hairs. Roxb. Fl. Ind. iii. 148. Pierre Fl. For. Coch-Chine, t. 190; Wall. Cat. 1133; R. Brown in Benn. Pl. Jav.

177

Rar. 231. Kurz For. Fl. Burm. i. 138, in part. *S. mollis*, Wall. Cat.
1131; R. Brown in Benn. Pl. Jav. Rar. 231. *S. Balanghas*, L. var.
*mollis*, Mast. in Hook. fil. Fl. Br. Ind. i. 358.

Burmah; Griffith No. 578 (Kew Dist.); Helfer Nos. 579, 580;
Falconer. Perak, King's Collector, No. 8360.

Roxburgh left in the Calcutta Herbarim an excellent coloured
drawing of his S. *angustifolia*. In his Flora Indica he gives a very brief
account of the species, drawn up from specimens flowering in the Botanic
Garden and which he states came from Nepal. His description is too brief
to be of any use: but his figure is so good that I have no hesitation in
saying that no species of *Sterculia* collected since Roxburgh's time in
any part of the outer Himalaya, or from the plain at its base, is in the
least like this plant. I have little doubt that Roxburgh was deceived
as to its origin by some changing of labels of the native gardeners at
Calcutta (a sublimely inaccurate race!); and that the plant was really
received, like so many others during the early years of the garden, from
the Straits. Wallich, no doubt deceived by the alleged Himalayan
origin of the plant, distributed (as No. 1133 of his list) specimens from
the trees of it which were still in his time cultivated in the Calcutta
Garden under Roxburgh's name, while specimens collected in Burmah he
issued as No. 1131, under the name *S. mollis*, Wall. Pierre's figure
above quoted does not agree very well with Roxburgh's, the panicles
being by far too short and not nearly hairy enough.

12. S. RUBIGINOSA, Vent. Hort. Malmaison, ii. 91. A tree 20 to
50 feet high: young branches rather thick, their apices deciduously rufous-tomentose; the bark pale or brown, striate, glabrous. *Leaves* membranous, obovate-oblong, sometimes ovate-oblong, shortly and abruptly
acuminate, entire; narrowed to the acute, rounded or minutely cordate,
3-nerved base: upper surface glabrous, or sparsely stellate-pubescent; the
lower stellate-pubescent, most of the hairs pale and minute but these
on the midrib and 7 to 10 pairs of spreading stout nerves larger and
darker coloured: length 4·5 to 7·5 or rarely 12 in., breadth 2 to 3 in.,
rarely 4 in.; petiole varying with age from ·3 to 1·5 in, rufous tomentose as are the linear caducous ·5 in. long stipules. *Panicles* solitary
in the axils of the crowded young leaves, many-flowered, shorter than,
or as long as the leaves, rufous-tomentose like the outer surfaces of
the flowers; flower-pedicels spreading, capillary. *Flower buds* broadly
ovate. *Calyx* less than ·5 in. long, widely campanulate, divided for half
its length or more into 5 lanceolate spreading incurved lobes cohering
by their tips, the lobes densely covered inside with white hispidulous
hairs. *Male flower;* staminal column longer than the tube or about as
178

long, glabrous; *anthers* about 10, sessile at the apex of the column, 2-celled, the cells distinct. *Female flower;* gynophore very short; ovaries ovoid, villous (as are the united styles); with 10 sessile anthers at their base; stigma discoid, deeply 5-lobed. *Follicles* 5, coriaceous, crimson when ripe, oblong, shortly beaked, about 2 in. long and 1 in. broad; pubescent externally, glabrous shining and boldly ridged inside. *Seeds* oblong, ovoid, black. Mast. in Hook. fil Fl. Br. Ind. i. 358 : Kurz For. Fl. Burm. i. 138; Pierre Fl. Forest. Coch-Chine, t. 194 B; Blume Bijdr. i. 82; Br. in Benn. Pl. Jav. Rar. 231 ; Miq. Fl. Ind. Bat. i. pt. 2, 175. *S. angustifolia,* Jack Mal. Misc. ex Hook. Bot. Misc. i. 287. *S. Jackiana,* Wall. Cat. 1134.

In all the Provinces except the Andaman and Nicobar Islands : at low elevations. Common. Distrib. Java and Sumatra, Cochin-China, Burmah.

Var. *glabrescens,* King : leaves 8 to 12 in. long, by 3 to 4·5 in. broad, softly pubescent beneath when young, much less narrowed to the (always sub-cordate or cordate) base than in the type; panicles much branched and sometimes longer than the leaves. *S. angustifolia,* Kurz (not Roxb.), in part, For. Fl.. Burm. i. 138; *S. parviflora,* Kurz (not of Roxb.) Journ. As. Soc. Beng. xliii. pt. 2, p. 116. *S. mollis,* Kurz (? of Wall.) l. c. xlv. pt. 2, p. 120. *S. Balanghas,* Linn. var. *glabrescens,* Mast. in Hook. fil. Fl. Br. Ind. i. 358, in part.

Andaman Islands; Helfer (Kew Distrib. No. 595); Kurz, Prain, Bot. Gard, Collectors. Nicobars, Kurz. Great Cocos, Prain. There are no Mergui or Eastern Peninsula specimens of this at Calcutta, and I believe the variety to be confined to the Islands above named.

In this species the petioles lengthen with the age of the leaf, many young leaves having petioles less than ·25 in. long, while in old leaves the length varies from 1 to 1·5 in. And there is considerable variability in the size of the blade. Moreover, while in some the upper surface of the leaves is perfectly glabrous (except the midrib which is almost invariably rusty-tomentose), in others it is rough and scaberulous from the presence of scattered stellate hairs. The next species (*S. ensifolia,* Mast.) has, in my opinion, a very poor claim to specific rank; and I think it would be better to treat it a shrubby variety of this with narrower leaves and longer flowers. *S. parviflora,* Roxb. also differs very little from this, and might be reasonably enough regarded as a form of it with broader more glabrous cordate leaves with fewer nerves.

13. S. ENSIFOLIA, Mast. in Hook. fil. Fl. Br. Ind. i. 359. A shrub or small tree : young branches and petioles densely ferruginous-tomen-

tose. *Leaves* membranous, oblong-lanceolate or oblanceolate, shortly caudate-acuminate, entire, the base rounded, sometimes minutely cordate, rarely acute; upper surface glabrous, the midrib alone tomentose; under surface sparsely rusty-tomentose, the midrib and 8 to 10 pairs of spreading lateral nerves prominent: length 6 to 12 in., breadth 1·25 to 3·5 in., petiole ·3 to 1·5 in.; stipules erect, linear, half as long as the petiole, deciduous. *Panicles* or *racemes* axillary, solitary, lax, few-flowered, rusty-tomentose, hardly so long as the leaves; bracteoles linear, shorter than the pedicels. *Calyx* ·5 or ·6 in. long, broadly campanulate, pubescent on both surfaces, the tube much shorter than the linear-lanceolate spreading lobes the tips of which curve inwards and cohere. *Male flower;* staminal column longer than the calyx-tube but much shorter than its lobes, glabrous, curved, bearing at its apex 10 2-celled, oblong, nearly sessile, anthers. *Female flower;* gynophore very short: ovaries 5, ovoid, rusty-villous, surrounded at the base by 10 sub-sessile stamens. *Styles* short, united, densely covered with white hairs; stigmas united into a boldly 5-lobed disc. *Follicles* 1 to 5, shortly stalked, narrowly oblong, tapering to each end, the apex with a hooked beak, coriaceous, brownish-velvetty, red when ripe, 2 to 2·5 in. long and ·75 in. broad. *Seeds* oval, black, smooth. Pierre Fl. Forest. Coch-Chine t. 194 C. *S. angustifolia,* Jack (not of Roxb.) Mal. Misc. ex Hook. Bot. Mis. i. 287.

Penang, Perak, at low elevations, common. Distrib. Burmah.

I have no doubt whatever that this is the plant described by Jack as the *S. angustifolia* of Roxb.

14. S. PUBESCENS, Mast. in Hook. fil. Fl. Br. Ind. i. 357. A tree, the younger parts rusty-pubescent. *Leaves* oblong, obtuse, or abruptly acuminate, entire, the base cordate; upper surface glabrous; lower densely and minutely pubescent, the nerves stellate-pilose: length 4 to 6 in., breadth 2 to 2·5 in.; petiole 1·25 in., sulcate: stipules subulate, ·25 in. long. *Panicle* erect, as long as or longer than the leaves, much branched: ultimate pedicels jointed, pubescent, spreading. *Calyx* ·25 in. long, campanulate; the lobes triangular acute, as long as the tube, hairy within. *Ovary* globose, downy.

Malacca, Maingay.

Except by Maingay's two specimens in the Kew Herbarium, this species is unknown. Specimens of it in good flower and in fruit are much wanted, so that a completer description than the foregoing may be prepared.

15. S. COLORATA, Roxb. Hort. Beng. 50. A tree 30 to 60 feet

high; young branches thick, rough, rather pale, glabrous. *Leaves* thinly coriaceous, roundish or reniform, usually palmately 3 to 5-lobed, the lobes triangular, acuminate; base deeply cordate, 5 to 7-nerved; both surfaces pulverulent-pubescent when young, glabrous when adult; length 4·5 to 9 in., breadth 5 to 12 in.; petiole 3·5 to 8 in., puberulous; stipules lanceolate, caducous. *Flowers* in axillary panicles or racemes from the axils of last year's fallen leaves, 2·5 to 4 in. long, densely covered, as is the exterior of the flowers, with coral-red, scaly tomentum. *Calyx* ·75 in. long, funnel-shaped, curved, the mouth with 5 acute short triangular teeth, puberulous internally, villous at the base. *Staminal column* as long as, or longer than the calyx, slightly flattened, minutely furfuraceous-pubescent: anthers 20 to 25, sessile at the apex of the column, oblong, closely surrounding the 5 flask-shaped ovaries; styles 5, short, recurved: stigmas acute. *Follicles* 2 to 3 in. long, membranous, glabrous, veined, stipitate, open from an early age and bearing on their edges usually 2 smooth oval seeds. Roxb. Cor. Pl. i. 26, t. 25; Fl. Ind. iii. 146; Mast. in Hook. fil. Fl. Br. Ind. i. 359; Pierre Fl. Forest. Coch-Chine, t. 199; Kurz For. Fl. Burm. i. 138; Brand. For. Flora N. W. Ind. 34; Wall. Cat. 1119; Hook. Ic. Pl. 143; Dalz. & Gibs. Bomb. Fl. 23; W. & A. Prodr i. 63. *Firmiana colorata.* Br. in Benn. Pl. Jav. Rar. 235; Thwaites Enum. 29. *Erythropsis Roxburghiana,* Scott. & Endl. Melet. Bot. 33.

Andamans; Kurz, Prain. Distrib. India, Ceylon.

16. S. FULGENS, Wall. Cat. 1135. A tree 30 to 70 feet high; young branches rather thick, with smooth dark bark, at first pubescent, ultimately quite glabrous. *Leaves* large and with long petioles, thinly coriaceous, rotund with 5 shallow acuminate lobes, the base cordate: upper surface glabrous, harsh to the touch: lower densely and minutely stellate-pubescent, palmately 7-nerved, the nerves prominent beneath: length and breadth 15 to 18 in.; petiole 15 to 20 in., sulcate, minutely puberulous. *Racemes* or *panicles* 3 to 4 in. long, from the axils of last year's leaves, densely covered with orange or golden-yellowish scurfy tomentum as are the outer surfaces of the flowers. *Calyx* 1·25 in. long, funnel-shaped, slightly curved, the mouth with 5 short triangular teeth; internally minutely velvetty-puberulous with a ring of long matted hair near the base. *Staminal tube* ·5 in., longer than the calyx, 5-angled, sulcate, minutely tomentose; anthers 20, sessile, oblong, 1-celled, embracing the 5 flask-shaped ovaries; styles short, reflexed: stigmas acute. *Follicles* unknown. Mast. in Hook. fil. Fl. Br. Ind. i. 360; Kurz For. Fl. Burm. i. 139: Journ. As. Soc. Beng. pt. 2, 1874, p. 117; Wall. Cat, 1135; *Firmiana colorata* var. β, Br. in Benn. Pl. Jav. Rar. 235; Miq. Fl. Ind. Bat., i. pt. 2, 178.

73

Perak; King's Collector, No. 8673, Scorteclini. Distrib. W. Sumatra,
Forbes, No. 2105 : Java, Burmah ; Wallich.
There is no doubt this comes very close to *S. colorata*, Roxb. of
which it might possibly be better to treat it as a variety characterised
by larger flowers, with much more exserted staminal column, larger
leaves, thinner and dark-coloured branchlets. Wallich, however, who
saw the tree growing, regarded it as a species; and Robert Brown (Pl.
Jav. Rar. p. 235), while treating it as a variety of *colorata*, remarks
that it is probably worthy of specific rank. This plant (whether species
or variety) is never found in British India proper. Its most northerly
limit is Tenasserim, and from thence it extends southward into the
Malayan Archipelago. In the Flora of British India, Dr. Masters gives
the distribution of this as "Tropical Western Himalayan." The plant,
however, which occurs in tropical valleys in that region is just as
different from *S. fulgens*, Wall., as that is from *S. colorata*, Roxb. It is
the tree to which Wallich gave the name *S. pallens;* and which he
published (without describing) in Voigt's Hort. Suburb, Calcutta, p.
105. The leaves of *S. pallens* resemble those of *colorata* in shape; but
their under surface is covered with dense pale yellow stellate tomentum.
The calyx has a much wider mouth than that of *colorata*, and (like the
axis and pedicels of the panicle) is densely covered with a very pale
yellow tomentum, while the tomentum of *colorata* is of a vivid coral red.
*S. pallens* is confined to the Western Himalaya, just as *S. fulgens* is
limited to Burmah and Malaya.

17. S. ALATA, Roxb. Hort. Beng. 50. A tree 80 to 150 feet high ;
young branches rather stout, striate, glabrous. *Leaves* membranous,
broadly ovate or ovate-oblong, acute or shortly acuminate, entire ; the
base deeply cordate, 5 to 7-nerved, some of the basal nerves pinnate on
one side ; both surfaces glabrous; lateral nerves 4 pairs, prominent on
both surfaces as are the midrib and basal nerves ; length 4 to 12 in.,
breadth 3 to 8 in., petiole 1·5 to 7 in.: stipules minute, subulate, caducous. *Racemes* from the axils of previous year's fallen leaves, usually
in pairs, sometimes solitary, rarely terminal, about as long as the
petioles, flocculent, rusty-tomentose, as are the flowers externally ; bracts
3 to each flower, ensiform, caducous. *Calyx* ·75 in. long, campanulate,
deeply divided into 5 or 6 thick, fleshy, lanceolate segments. *Male
flower;* staminal column thin, cylindric, much shorter than the calyx,
glabrous, bearing at its apex 25 elongate anthers in five groups of 5
each; ovaries imperfect. *Female flower ;* staminodes in 5 phalanges,
sessile, embracing the bases of the 5 sub-ovate, multi-ovulate-ovaries ;
stigmas broad, emarginate. *Follicles* pedunculate, woody, pulverulent-
182

pubescent, 5 in. in diam., sub-globular, slightly compressed. *Seeds* oblong, compressed, the testa spongy, 1 in. long, with a large obovate thick spongy terminal wing 2·5 in. long and 1·25 broad. Roxb. Corom. Pl. iii. 84, t. 287; Fl. Ind. iii. 152; Kurz Fl. Br. Burm. i. 134; Pierre Fl. Forest. Coch-Chine, t. 196; Wall. Cat. 1125. *Pterygota Roxburghii*, Schott & Endl. Melet. *P. alata*, Br. in Benn. Pl. Jav. Rar. 234. *S. coccinea*, Wall. Cat. 1122, partly. *S. Heynii*, Beddome Flor. Sylvat. t. 230.

Perak, Scortechini: Andamans, Kurz. Distrib. Brit. India, Cochin-China.

18. S. LINEARICARPA, Mast. in Hook. fil. Br. Ind. i. 360. A tree 60 to 80 feet high : young branches thick, striate, deciduously pulverulent-tomentose, leaf-cicatrices large. *Leaves* coriaceous, ovate-orbicular, blunt or very slightly narrowed at the apex, edges entire, base deeply cordate, 7-nerved; upper surface glabrous, shining; lower deciduously pulverulent, hairy, almost glabrous when old, minutely reticulate, the midrib and 4 or 5 pairs of lateral nerves prominent : length and breadth 6 to 12 in.; petiole 2·5 to 6 in., sulcate, pulverulent-tomentose. *Panicles* axillary, solitary, stout, erect, as long as or longer than the leaves, rusty pulverulent-tomentose as are the outer surfaces of the calyces, the lateral branchlets short ; bracteoles numerous, rotund, concave, caducous ; flower-buds globose, sessile. *Calyx* rotate, the tube ·1 in. long, with 5 slightly longer ovate acute lobes, tomentose externally, glabrous within. *Staminal column* not so long as the calyx-tube, glabrous ; anthers 10, each with a short filament, cuneate, 2-celled ; ovaries (rudimentary in some flowers) about 3, free, each 1 or 2-ovulate ; style short ; stigma entire, small. *Follicles* (? ripe) linear-lanceolate, 3 to 4 in. long and ·6 in. broad, stipitate, longitudinally ridged and covered outside and inside with yellowish tomentum as is also the single oblong seed.

Malacca, Maingay. Perak ; Scortechini, King's Collector.

The flowers and follicles of this are, in my opinion, those of *Scaphium* rather than of *Firmania*, to which section Dr. Masters has referred it.

19. S. SCAPHIGERA, Wall. Cat. 1130. A tree 90 to 120 feet high : young branches rather thick ; the bark pale, minutely warted and striate, glabrous. *Leaves* coriaceous, glabrous, ovate to oblong-ovate, sub-acute or bluntish-acuminate, entire ; the base rounded or sub-truncate, often faintly cordate or emarginate, 3 to 5-nerved ; main nerves 2 to 4 pairs, sub-erect, prominent on both surfaces ; length 5 to 10 in., breadth 2·75 to 4·5 in. ; petiole 2 to 5 in., thickened at both ends. *Panicles* only at

the ends of the branchlets, puberulous, shorter than the petioles, robust, with many short spreading branches, many-flowered; pedicels short, pubescent; bracteoles subulate, deciduous. *Calyx* from ·3 to ·4 in. long, deeply 5-lobed and almost rotate when expanded, stellate-puberulous externally, glabrous internally, the lobes lanceolate. *Male flower* with 15 to 30 anthers almost sessile round the apex of the column and surrounding the rudimentary villous ovary. *Female flower;* ovaries 5, bi-ovulate; styles united; stigma 5-lobed. *Follicles* 1 to 5, on rather stout pubescent stalks, when ripe 6 to 8 in. long and 1·25 to 2·5 in. broad, membranous, boat-shaped, gibbous about the middle, conspicuously veined and more or less puberulous externally especially on the nerves. *Seeds* 1 (rarely 2), ovoid, glabrous, shining, ·5 to 1 in. long, attached to the very base of the follicle. Mast. in Hook. fil. Fl. Br. Ind. i. 361; Kurz For. Fl. Burm. i. 140; Pierre Fl. Forest. Coch-Chine, t. 201. *Scaphium Wallichii,* R. Br. in Benn. Pl. Jav. Rar. 226.

Malacca, Griffith. Distrib. Sumatra, Burmah.

M. Pierre is in doubt whether his fine figure (l. c. t. 201), represents really the true plant of Wallich. In my opinion it does so most decidedly : R. Brown was right in describing the ovaries as five, and there is a specimen in the Calcutta Herbarium with 5 follicles.

20. S. AFFINIS, Mast. in Hook. fil. Fl. Br. Ind. i. 361. A tree : young branches rather stout, rough, dark in colour, the leaf cicatrices large, the very youngest minutely rusty-tomentose. *Leaves* thinly coriaceous, elliptic-oblong, with rather straight edges ; the apex broad, suddenly acute ; the base truncate (sometimes obliquely so), ·3-nerved ; both surfaces glabrous, the upper shining, the lower pale and rather dull : main nerves 6 or 7 pairs, conspicuous beneath as is the midrib; length 5 to 9 in., breadth 4·75 to 5·5 in. ; petiole 4·5 in., thickened at each end. "*Panicle* erect, as long as the leaves, its branches downy, flattened or angular; peduncles thickly striated, angular, sub-pilose, spreading ; ultimate pedicels downy, densely crowded. *Flowers* very small, the buds ovoid. *Flowers* ·25 in. *Calyx-lobes* ovate, longer than the funnel-shaped tube. *Follicle* a span long, falcate, leafy, glabrescent, shining within. *Seeds* ·65 in. long, solitary, oblong, black." *Scaphium affine,* Pierre Fl. Forest, Coch-Chine, t. 195 E.

Malacca ; Maingay, No. 225 (Kew. Distrib.)

The only Maingayan specimen of this in the Calcutta Herbarium consists of leaves only, with a single detached fruit ; and I have seen no specimen from any other collector. The foregoing description (as regards inflorescence, flower and fruit) is therefore copied verbatim from Masters (in F. B. I. l. c.).

21. S. CAMPANULATA, Wall. A tree 50 to 60 feet high : young branches rather slender, rusty-tomentose, soon becoming glabrous. *Leaves* membranous, broadly ovate, shortly acuminate, entire ; the base usually deeply cordate, 3 to 7-nerved ; sometimes 3 to 5-lobed ; lateral nerves 3 or 4 pairs ; upper surface glabrous, the midrib and nerves pubescent or puberulous ; lower surface pubescent ; length 4 to 6 in., breadth 3·75 to 5·5 in. ; petiole 2·25 to 5 in. puberulous : stipules lateral, subulate, caducous. *Panicles* 3 or 4 in. long, in clusters of 2 or 3 at the apices of the branches, few-flowered, glabrous, erect, sub-corymbose ; pedicels jointed, about ·3 in. long, bracteoles caducous. *Calyx* widely campanulate, more than ·75 in. across, green, pruinose, glabrous, veined, its mouth cut half-way down into 5 triangular velvetty-edged lobes : *Staminal column* pubescent below. *Ovaries* gibbous at the apex : styles short, cohering ; stigmas filiform, recurved : ovules 2, erect. *Follicles* 3 to 6, on slender puberulous stalks, membranous, veined, 2 to 3 in. long, boat-shaped, saccate with a sub-terminal lanceolate wing. *Seeds* sub-globose, with a shining crustaceous testa, ·5 in. long or less. Mast. in Hook. fil. Fl. Br. Ind. i. 362 ; Kurz For. Fl. Br. Burm. i. 139. *Pterocymbium Javanicum*, Br. in Benn. Pl. Jav. Rar. 219, t. 45 ; Miq. Fl. Ind. Bat. i. pt. 2, 179. *Pt. campanulatum* and *Javanicum*, Pierre, Fl. Forest. Coch-Chine, t. 195.

Perak ; Fr. Scortechini, King's Collector. Nicobars, Kurz. Distrib. Malayan Archipelago, Burmah.

M. Pierre (l. c.) remarks that, in his opinion, the two species *campanulatum* and *Javanicum*, although closely related, are distinct species ; but he does not mention the characters on which he relies for separating them. After dissecting many flowers of the tree (until recently growing in the Botanic Garden, Calcutta), on which Wallich founded his species *campanulatum*, I cannot see any respect in which they differ from Robert Brown's minute and excellent description and figures of *Pt. Javanicum*. I therefore agree with Dr. Masters in considering the two as one and the same species.

22. S. TUBULATA, Mast. in Hook. fil. Fl. Br. Ind. i. 362. A tree ; young branches about as thick as a goose-quill, tomentose at the very points, the bark dark and rather rough. *Leaves* thinly coriaceous, elliptic-oblong, with a short abrupt rather blunt apiculus ; edges entire ; the base broadly rounded or sub-truncate, very slightly cordate ; when adult both surfaces glabrous except the midrib and main nerves which are minutely rusty-tomentose ; main nerves 5 to 7 pairs, spreading, slightly prominent below : length 4 in., breadth 1·75 in. ; petiole ·75 in. slender, deciduously rusty-tomentose. *Cymes* terminal, as long as the

77

leaves, many-flowered. *Calyx* ·5 in. long, glabrous, narrowly tubular below, the mouth slightly expanded and with 5 ovate-lanceolate lobes shorter than the tube. *Staminal column* pilose; anthers in a ring. *Ovaries* 5; styles inflexed, cohering by their tips. *Follicles* 5, from 2 to 3 in. long and 1 in. broad, on tomentose stalks, oblong, acute, dilated at the base. *Seed* ovoid.

Malacca, Maingay.

At once distinguished by the singular calyx, tubular in its lower, lobed and spreading in its upper, half.

## 2. TARRIETIA, Blume.

Tall trees. *Leaves* digitate or simple, glabrous or scaly. *Flowers* unisexual, panicled. *Calyx* tubular, small, 5-toothed. *Petals* 0. *Staminal-column* short, bearing a ring of 10-15 very densely clustered anthers, cells parallel. *Ovary* of 3-5 nearly free carpels opposite the sepals; styles as many, short, filiform, stigmatose within; ovules 1 in each cell. *Ripe carpels* of stellately spreading samaras with long falcate wings. *Seeds* oblong; albumen bipartible; cotyledons flat; radicle next the hilum.—Distrib. Known species 5 or 6, Australian and Malayan.

Leaves digitately compound.
    Under-surface of leaflets persistently
        stellate-tomentose ... ... 1. *T. Perakensis.*
    Under-surface deciduously tomentose,
        the hairs simple ... ... 2. *T. Penangiana.*
Leaves simple.
    Fruit glabrous ... ... 3. *T. simplicifolia.*
    ,, tomentose. ... ... 4. *T. Kunstleri.*

1. T. PERAKENSIS, King, n. sp. A tree 40 to 60 feet high: young branches, petioles, petiolules, under surface of leaves (when young) and inflorescence with minute deciduous rusty tomentum. *Leaves* digitately compound; leaflets 5 or 6, the lower smaller, obovate-elliptic to obovate-rotund, shortly and rather abruptly acuminate, the edges entire, slightly wavy; the base narrowed; upper surface minutely areolate, glabrous except the very minutely tomentose midrib and nerves; lower glabrous except the midrib: main nerves 10 to 14 pairs, stout and prominent beneath: length of the middle leaflet 4·5 to 5·5 in., of the lower 2 to 3·5 in.: breadth of the middle 2 to 3 in., of the lower 1·25 to 1·5 in.; petiolules ·5 to 1 in.; petioles ·3 to 4·5 in. *Inflorescence* in solitary, axillary, cymose racemes or panicles more than half as long as the

186

leaves, much crowded at the points of the branches. *Flowers* ·15 in long : pedicels slender, three times as long. *Calyx-tube* tomentose externally, sparsely pubescent within ; *staminal tube* less than half its length. *Female calyx* rather longer than the male, otherwise the same : stamens 0 : ovaries 5, obliquely ovoid, glabrous, each with a pubescent conic style crowned by a small hooked stigma. *Ripe fruit* compressed-ovoid, 1·25 to 1·5 in. long, and 1 to 1·15 in. in diam., glabrous, the wing falcate, 2 in. long and ·5 in. broad, striate.

Perak, at low elevations ; King's Collector, Penang, Curtis, No. 2229.

In its leaves this much resembles *T. Javanica*, Bl. (Rumphia iii. t. 127, fig. 1) ; but the leaves of Blume's plant are smaller and have more wavy edges. The flowers, however, of the two differ much in size, those of this being twice as large as the flowers of *T. Javanica*.

2. T. CURTISII, King, n. sp. A tree 20 to 40 feet high ; young branches, petioles, petiolules and under surfaces of leaves densely covered with rusty stellate, non-deciduous tomentum. *Leaves* digitately 5 or 6-foliolate, the lower smaller, obovate, entire, wavy, apex retuse, base acute; upper surface minutely areolate, glabrous except the stellate-tomentose midrib and main nerves ; under surface, and especially the midrib, stellate-tomentose : main nerves 9 or 10 pairs, spreading, prominent beneath : length of the middle leaflet 3·5 to 4·5 in., of the lower 1·5 to 2·5 in. ; breadth of the middle 2·25 to 2·5 in., of the lower ·8 to 1·5 in., petiolules ·5 to ·75 in., petioles 2 to 2·5 in. *Inflorescence* iu solitary, axillary, cymose racemes or few-flowered panicles, more than half as long as the leaves. *Ripe fruit* glabrous, compressed-ovoid, 1 in. long and ·8 in. broad ; wing narrowly falcate, 1·25 iu. long and ·25 in. broad, striate.

Penang at 2000 feet : Curtis No. 1427.

This is known only by Curtis's scanty specimens which are in fruit only. Its flowers are unknown. In leaves it closely approaches *T. Perakensis*, but the tomentum is stellate and persistent ; whereas in *T. Perakensis*, the hairs are simple and deciduous. The leaflets of this are also smaller, fewer-nerved, more decidedly obovate, less elliptic than in *T. Perakensis*, and they are mucronate rather than acuminate.

3. T. SIMPLICIFOLIA, Mast. in Hook. fil. Fl. Br. Ind. i. 362. A tree, young branches pale, sub-glabrous, striate. *Leaves* simple, coriaceous, elliptic or obovate-elliptic, apex truncate or emarginate, shortly mucronate, entire, rather suddenly narrowed at the base or rounded ; upper surface glabrous, shining ; lower dull, rusty, minutely puberulous,

187

and slightly scaly; maiu nerves 16 to 20 pairs, prominent below, spreading; length 4·5 to 7 in., breadth 3 to 4·5 in.; petiole 2 to 3 in, thickened towards the apex. *Cymes* axillary, solitary, many-flowered, 1·5 to 2·5 in. long, minutely rusty-tomentose. *Flowers* ·1 in. long; the pedicels shorter, stout. *Calyx-tube* campanulate, minutely tomentose externally, puberulous within: staminal tube short. *Fruit* (including wing) 3 in. long, obliquely spathulate, glabrous.

Malacca; Griffith, Maingay (Kew Distrib.) No. 231.

4. T. KUNSTLERI, King, n. sp. A tree 50 to 70 feet high : young branches petioles and peduncles minutely stellate-pubescent and lenticellate. *Leaves* elliptic to obovate-oblong, blunt, mucronate, entire, the base rounded or slightly narrowed : upper surface smooth, shining; the lower pale, sparsely stellate-puberulous on the midrib and nerves, otherwise (under a lens) minutely puberulous: main nerves 7 to 10 pairs, ascending, prominent beneath. *Fruit* at the apex of a solitary stellate-hairy peduncle, ovoid with an oblique sub-spathulate wing, minutely but densely velvetty fulvous-tomentose ; length of body 1 iu. or more ; wing about the same length and ·6 in. broad.

Perak, near Laroot ; King's Collector No. 7581.

Flowers of this are at present unknown. The leaves are at once distinguished from those of *T. simplicifolia* by their pale under surface, and the fruits by their tomentum.

### 3. HERITIERA, Aiton.

Trees. *Leaves* coriaceous, simple, scaly beneath. *Flowers* small, unisexual, in axillary panicles. *Calyx* 5, rarely 4-6 toothed or cleft. *Petals* 0. *Anthers* in a ring at the top of the column, cells 2, parallel. *Ovaries* 5-6, almost free : style short, stigmas 5, thick ; ovules solitary in each cell. *Ripe carpels* woody, indehiscent, keeled or winged. *Albumen* 0 ; cotyledons thick ; radicle next the hilum.—A genus of 6 or 7 species, natives of the Tropics of the old world, and of Australia.

H. LITTORALIS, Dryand. in DC. Prod. i. 484. A tree : young branches stout, rough. *Leaves* oblong or elliptic, the apex rounded or acute ; the edges entire ; base rounded or slightly cordate ; lower surface pale ; main nerves 7 to 9 pairs, slightly prominent beneath : length 5 to 10 in., breadth 2·25 to 4 in., petiole ·5 to ·75 in. : stipules lanceolate, caducous. *Flowers* ·2 in. long, in many-flowered axillary cymose panicles shorter than the leaves. *Calyx* 5-toothed, puberulous, half as long as the pedicel. *Ripe fruit* 1·5 to 3·5 in. long, woody, compressed ovoid, boldly keeled at apex and on dorsum, glabrous, shining. Mast. in Hook. fil. Fl.

Br. Ind. i. 363; Kurz For. Fl. Burm. i. 140; Pierre Fl. Forest. Coch-
Chine, t. 203; Miq. Fl. Ind. Bat. i. pt. 2, p. 179; Blumo Bijdr. 84;
Roxb. Fl. Ind. iii. 142; W. & A. Prodr. i. 63 : Thwaites Enum. 28; Br.
in Benn. Pl. Jav. Rar. 237; Miq. Fl. Ind. Bat. i. pt. 2, p. 179. *H.
Fomes*, Wall. Cat. 1139, partly. *Balanopteris Tothila*, Gærtn. Fruct. ii.
94, t. 99.

All the Provinces, on the coasts. Distrib. Malayan Archipelago
and coasts of the tropics of the old world generally, and of Australia.

The plant originally issued by Wallich as *Trochetia contracta* (Cat.
No. 1162) and afterwards named by him *Heritiera macrophylla*, (Pierre
l. c. t. 204) has by some writers been reduced to *H. littoralis*. But
Wallich's species was originally found in the interior of Burmah, and it
has since been found in Cachar, far from the sea coast to which *H.
littoralis* is strictly confined. *H. macrophylla* has moreover leaf-petioles
more than twice as long as those of *H. littoralis*, and its fruit is warted
and not smooth. I believe *H. macrophylla* to be a perfectly distinct
species ; as is also, in my opinion, the other Sylhet and Khasia small-
leaved plant which Wallich issued as *H. acuminata*. (Cat. No. 7836.)

4. KLEINHOVIA, Linn.

A tree. *Leaves* 5 to 7-nerved and often cordate at the base. *Inflorescence* a terminal, lax, cymose panicle. *Bracteoles* small. *Sepals* 5,
much longer than the petals, linear-lanceolate, deciduous. *Petals* 5, unequal, the upper short, ovate-round, saccate, the middle pair concave and
obliquely oblanceolate, the lower pair flat with convolute edges. *Stamens*
20, in 5 phalanges of 3 each with five solitary, free, often non-antheri-
ferous, filaments between the phalanges ; the filaments of all conjoined
below into a long, externally hairy, narrowly cylindric tube which surrounds the gynophore : anthers 4-celled, divergent. *Ovary* at the apex
of the long gynophore and surrounded by the staminal tube, 5-lobed,
5-celled. *Capsule* turbinate-pyriform, membranous, inflated, 5-celled,
loculicidal. *Seeds* 1 or 2 in each cell, tubercled : cotyledons convolute,
radicle inferior. Distrib. One species. Tropics of the old world.

K. HOSPITA, L. Spec. 1365. *Leaves* ovate-rotund, acuminate, entire,
palmately 3-5-nerved at the base, glabrous : length 3 to 6 in., breadth
2·5 to 5 in., petiole 1·5 to 2·5 in. DC. Prodr. i. 488 ; W. & A. Prodr.
i. 64; Roxb. Fl. Ind. iii. 141 ; Miq. Fl. Ind. Bat. i. pt. 2, 186; Blume
Bijdr. 86; Hassk. Pl. Jav. Rar. 313 ; Mast. in Hook. fil. Fl. Br. Ind. i.
364. Pierre Fl. Forest. Coch-Chine, t. 177.

In all the Provinces, but usually planted. Distrib. Malaya, Australasia, Br. India.

81

Apparently a variable plant. Dr. Masters (in Oliver's Flora of Trop. Africa, i. 226), describes the African specimens as having no stamens or staminodes alternating with the 5 phalanges of stamens. A specimen in the Calcutta Herbarium from Java has the under surface of the leaves softly hairy.

## 5. HELICTERES, Linn.

Trees or shrubs, more or less stellate-pubescent. *Leaves* simple. *Flowers* axillary, solitary or fascicled. *Calyx* tubular, 5-fid, often irregular. *Petals* 5, clawed, equal or unequal, the claws often with ear-shaped appendages. *Staminal column* surrounding the gynophore, 5-toothed or lobed at the apex; anthers at the top of the column, 2-celled. Five staminodes below the apex of the column. *Ovary* at the top of the column, 5-lobed, 5-celled; styles awl-shaped, more or less united, slightly thickened and stigmatose at the tips; ovules many in each cell. *Follicles* spirally twisted, or straight. *Seeds* tubercled; albumen scanty; cotyledons leafy, folded round the radicle which is next the hilum.— Distrib. About 30 species, natives of the tropics of both hemispheres.

Fruit spirally twisted ... ... 1. *H. Isora.*
Fruit not twisted.
    Leaves ovate to oblong-lanceolate,
      oblique; fruit more than 1 in. long ... 2. *H. hirsuta.*
    Leaves lanceolate or oblanceolate, not
      oblique: fruit less than 1 in. long ... 3. *H. angustifolia.*

1. H. ISORA, Linn. Spec. 1366. A shrub or small tree; young branches minutely tomentose. *Leaves* ovate-rotund, oblique; the apex rounded, abruptly acuminate; the edges irregularly serrate-dentate, sometimes lobed; the base cordate or rounded, rarely acute, palmately 5- to 7-nerved; upper surface scabrous, minutely hispid; lower pubescent or tomentose; length 2 to 4 in., breadth 1·25 to 3 in.; petiole ·3 in. long, tomentose; stipules linear, about as long as the petioles. *Flowers* axillary, solitary, or in few-flowered minutely bracteolate cymes, 1·5 in. long. *Calyx* narrowly campanulate, laterally compressed, 2-lipped, 5-toothed, tomentose outside. *Petals* reflexed, the lower two much shorter and broader than the three upper. *Staminal column* longer than the petals, curved, very narrowly cylindric, bearing at its apex 10 to 12 elongate-ovate stamens, and more internally 5 flat bifid staminodes. *Ovary* ovoid, sulcate, tomentose: styles slender, glabrous, united. *Fruit* cylindric, twisted, crowned by the persistent styles, pubescent; 1·5 in. long, ·4 in. in diam. Mast. in Hook. fil. Fl. Ind. i. 365; Bl

190

Bijdr. 79; Pierre Fl. Forest. Coch-Chine, t. 208, figs. 12 to 25; DC. Prodr. i. 475; Roxb. Fl. Ind. iii. 143; W. & A. Prodr. i. 60; Wight Ic. t. 180; Miq. Fl. Ind. Bat. i. pt. 2, 169; Kurz For. Fl. Burm. i. 142; Brand. For. Flor. 34. *H. chrysocalyx*, Miq. in Pl. Hohen. *Isora cory-lifolia*, Wight, Hassk. in Tijds. Nat. Gesch. xii. 107.

Perak; and probably in all the provinces. Distrib. Brit. India.

2. H. HIRSUTA, Lour. Fl. Coch-Chine, 648. A shrub 6 or 8 feet high; the young branches velvetty-tomentose. *Leaves* ovate, or ovate-rhomboid, sub-oblique (oblong to oblong-lanceolate in vars.) acuminate, irregularly crose-serrate; the base sub-truncate or rounded, rarely sub-emarginate; upper surface scabrid-pubescent, the midrib and nerves tomentose; lower velvetty-tomentose; nerves 4 or 5 pairs, prominent beneath; length 3·5 to 6 in., breadth 1·75 to 2·5 in.; petiole ·4 in., tomentose. *Cymes* scorpioid, few-flowered, axillary, solitary, twice as long as the petiole. *Flowers* ·75 in. long *Calyx* narrowly cylindric-campanulate, coarsely stellate-tomentose externally, the mouth with 5 acute unequal teeth. *Petals* linear, suo-spathulate, two rather broader than the others with slight horn-like appendages about the middle and all longer than the calyx and about as long as the stamens. *Staminal column* and pistils as in *H. Isora. Fruit* cylindric, acuminate, not, twisted, the carpels firmly coherent; externally densely covered by long villous and stellately pilose soft prickles, 1·2 in. long and ·35 in. in diam. Pierre Fl. Forest. Coch-Chine, t. 208, figs. 1 to 11; Kurz For. Fl. Burm. i. 143. *H. hirsuta*, Bl. Bijdr. 80. *H. spicata*, Colebr. in Wall. Cat. 1182; Mast. in Hook. fil. Fl. Br. Ind. i. 366; *Oudemansia hirsuta*, Miq. Fl. Ind. Bat. i. pt. 2, p. 171; Hassk. Retzia, i. p. 184; *Orthothecium hirsutum*, Hassk. Pl. Jav. Rar. 308.

Selangore, King's Collector. Penang, Curtis; and probably in the other provinces at low elevations. Distrib. Malayan Archipelago China, Brit. India.

Var. *oblonga*, (species Wall. Cat. 1183). *Leaves* oblong, 5 or 6 in. long and 1·35 to 1·75 in. broad, sparsely stellate-tomentose beneath.

Penang, Andamans.

Var. *vestita*, (species Wall. Cat. 1844). *Leaves* oblong-lanceolate, oblique at the base; 3·5 to 5·5 in. long and 1 to 1·5 in. broad.

Burmah: ? Andamans.

There seems to be little doubt that Loureiro and Blume indepen-dently of each other gave this species the same specific name. Wallich's distribution of it under Colebroke's MSS. name *spicata* took place many years subsequently, and that name must (although adopted by Dr. Masters) I think fall to the ground.

3. H. ANGUSTIFOLIA, L. sp. 1366. A shrub 4 to 6 feet high : young branches, petioles, under surfaces of leaves and peduncles minutely and more or less densely pubescent. *Leaves* lanceolate or oblanceolate, acute (or obtuse and mucronate in var. obtusa) ; entire ; the base narrowed 3-nerved ; upper surface glabrescent or glabrous ; lateral nerves 5 or 6 pairs, not prominent; length 1·5 to 2 in., breadth ·4 to ·8 in., petiole ·2 to ·3 in. *Cymes* axillary, solitary, not much longer than the petioles, few-flowered. *Flowers* ·4 or ·5 in. long. *Calyx* densely stellate-tomentose externally, cylindric, the mouth slightly expanded, with 5 acute triangular teeth, 2-lipped. *Petals* longer than the calyx, linear-subspathulate, with 2 or 3 horned appendages below the middle. *Staminal column* shorter than the petals, narrowly cylindric and otherwise as in *H. Isora*, the stamens smaller. *Ovary* inserted near the apex of the staminal tube, sub-globular, ridged, tomentose. *Fruit* ovoid-cylindric, apiculate, not twisted, the carpels closely coherent, ·75 in. long and ·4 in. in diam., densely covered with stellate, villous soft prickles as in *H. hirsuta.* DC. Prodr. i. 476 ; Mast. in Hook. fil. Fl. Br. Ind. i. 365 ; Bl. Bijdr. 80 ; Pierre Fl. Forest. Coch-Chine, t. 210 and 211 ; Wall. Cat. 1180. *H. lanceolata,* DC. Prodr. i. 476 ; Pierre, l. c. 210 B. *H. virgata,* Wall. Cat. 1181. *Oudemansia integerrima,* Miq. Pl. Jungh. i. 296 ; Fl. Ind. Bat. i. pt. 2, 170. *Oud. Javensis,* Hassk. Retzia, i. 134. *Orthothecium Javense,* Hassk. Pl. Jav. Rar. 307.

Malayan Archipelago, China.

Var. *obtusa,* (species Wall. Cat. 1184) ; Pierre, l. c. 211 B, 14 to 25. Kurz in Journ. As. Soc. Beng. 1873, pt. ii. 62. *Leaves* obtuse, mucronate.

Perak ; Nicobar Islands.

6. PTEROSPERMUM, Schreb.

Trees or shrubs, scaly or stellate-tomentose. *Leaves* usually bifarious, leathery, oblique, simple or lobed, penninerved. *Peduncles* 1-3, axillary and terminal. *Bracteoles* entire, laciniate, persistent or caducous. *Calyx* of 5 valvate, coriaceous, more or less connate, sepals. *Petals* 5, imbricate, membranous, deciduous with the calyx. *Staminal column* short, bearing opposite to the sepals 3 linear 2-celled anthers, and opposite to the petals 5 ligulate staminodes ; cells parallel ; connective apiculate. *Ovary* inserted within the top of the staminal column, 3-5 celled ; style entire, stigma 5-furrowed ; ovules many in each cell. *Capsule* woody or coriaceous, terete or angled, loculicidally 5-valved. *Seeds* winged above, attached in two rows to the inner angle of the cells of the capsule; albumen thin or 0 ; cotyledons plaited or corrugated, radicle inferior. Distrib. A genus of about 18 species, confined to tropical Asia.

Flowers 6 in. long ... ... ... 1. *P. diversifolium.*
  ,, 2 in. long.
    Sepals shortly pubescent inside,
    capsule 3 to 4 in. long. 2. *P. Blumeanum.*
    Sepals with silky hairs inside;
    capsule 1·5 in. long, with scaly
    hairs ... ... ... 3. *P. Jackianum.*
  ,, less than 2 in. long; capsule 2 to
    2 5 in., glabrous ... ... 4. *P. aceroides.*

1. P. DIVERSIFOLIUM, Blume, Bijdr. 88. A tree 60 to 100 feet high : young branches, petioles, under surfaces of leaves and outer surface of sepals and fruit covered with a layer of minute, tawny tomentum with many, more or less deciduous, rufous, stellate hairs on its surface. *Leaves* coriaceous, varying from obovate-oblong to elliptic-rotund ; the apex broad, blunt, or sub-truncate, suddenly contracted into a triangular point ; the edges entire or sinuous, rarely lobed ; the base always cordate or emarginate, 3 to 7-nerved and often oblique : upper surface shining, glabrous, except the tomentose midrib : main nerves 8 to 10 pairs, straight, sub-erect, prominent on both surfaces ; length 6 to 9 in., breadth 3·5 to 6 in., petiole 1 to 1·25 or even 2 in., stipules small linear, caducous. *Flowers* 6 to 7 in. long, buds narrowly cylindric, solitary, or in 3 to 4-flowered sub-sessile axillary cymes ; pedicels ·2 in. long, each with a minute recurved lanceolate bracteole. *Sepals* coriaceous, slightly shorter than the petals, linear, blunt, adpressed-sericeous internally. *Petals* membranous, linear, glabrescent. *Staminal tube* and gynophore 2 in. long ; the free part of the filaments slightly longer ; fertile anthers about 10, linear ; staminodes 5, pubescent. *Ovary* fusiform, tomentose, 5-celled. *Style* less than 2 in. long, angled, pubescent ; stigma fusiform. *Capsule* woody, oblong, pointed, acutely 5-angled, suddenly constricted at the base, about 4 to 5 in. long and 1·5 to 2 in. in diam. *Seeds* flattened, 1·5 to 2 in. long. Mast. in Hook. fil. Fl. Br. Ind. i. 367 ; Pierre Fl. For. Coch-Chine, t. 179 ; Miq. Fl. Ind. Bat. i. pt. 2, p. 192 ; Hassk. Pl. Jav. Rar. 316 ; Korth. Ned. Kruik. Arch. i. 312. *P. acerifolium,* Zoll. et Mor. Syst. Verz. p. 27 (excl. syn. Willd.)

Perak, Malacca ; common ; at low elevations. Distrib. Java, Philippines, Cochin-China.

The leaves on young shoots of this are often peltate and deeply lobed.

2. P. BLUMEANUM, Korth. Ned. Kruik. Arch. ii. p. 311. A tree

193

40 to 50 feet high : young branches slender, almost black when dry
when very young covered by deciduous furfuraceous rufous stellate
hairs. *Leaves* thinly coriaceous, very inequilateral, oblong to ovate or
lanceolate-oblong, entire, acuminate ; the base broad, unequally cordate,
one side auriculate or sub-auriculate; upper surface very dark when
dry, glabrous, shining ; the lower densely but minutely tawny or rufous-
tomentose with many deciduous cinnamoneous stellate hairs on the
surface ; main nerves 5 to 7 pairs, prominent beneath ; length 3 to
5·5 in., breadth 1·35 to 2 in. ; petiole ·15 in. ; stipules subulate-lanceo-
late. *Flowers* 2 in. long, solitary, or in 2-3-flowered cymes, axillary,
or (by the suppression of the leaves) in terminal racemes : pedicels
·5 in. long, bracteate, cylindric in bud. *Sepals* coriaceous, narrowly
linear, acute, scurfy, stellate-pubescent externally as are the pedicels
and bracteoles, pubescent internally. *Petals* membranous, obliquely
oblong-oblanceolate or sub-spathulate, shorter than the sepals, glabrescent. *Staminal tube* and gynophore about ·5 in. long, the free part of
the filaments rather longer; fertile anthers about 10 ; staminodes 5,
scaly-pubescent above. *Ovary* ovoid, villous, 5-celled. *Style* shorter
than the staminal tube, glabrous : stigma narrowly ovoid. *Capsule*
woody, oblong, 5-angled, sub-acute, gradually and slightly narrowed at
the base, glabrous when ripe ; 3 to 4 in. long and 1·5 in in diam. *Seeds*
flat, 1·5 in. long. Miq. Fl. Ind. Bat. i. pt. 2, p. 191. *Pterospermum
lanceaefolium*, Bl. (not of Roxb.) Bijdr. 87. *P. cinnamoneum*, Kurz,
For. Fl. Burm. i. 147. *P. Javanicum*, Jungh. Kurz, l. c. i. 147.

Perak, Penang; common at low elevations. Distrib. Sumatra,
Java, Borneo, Burmah, Assam.

A very common tree in Perak. Korthal's Bornean species *P. fuscum*
appears to me to be nothing more than a very cinnamoneous-tomentose
form of this. And the Peninsular-Indian *P. rubiginosum*, Heyne,
(Mast. in Hook. fil. Fl. Br. Ind. i. 368) cannot be very different. I
should be induced to reduce both to the oldest described species which
is this. Of the absolute identity of Kurz's *P. cinnamoneum* with this
I have no doubt whatever.

3. P. JACKIANUM, Wall. Cat. 1164. A tree : the small branches
slender, rather dark, when young covered by a layer of white
minute tomentum with many rufous stellate hairs on its surface.
*Leaves* sub-coriaceous oblong or elliptic-oblong, slightly inequilateral,
entire, or sinuate towards the rather abruptly acuminate apex ; the base
sub-acute, or truncate and minutely cordate or emarginate, never
auricled ; upper surface pale brown when dry, glabrous except the
pubernlous midrib and nerves; under surface pale brown or buff, with

a layer of minute tomentum and on the surface (and especially on the midrib and nerves) many minute deciduous rusty stellate hairs; nerves 10 to 12 pairs, prominent beneath, spreading; length 4 to 5·5 in., breadth 1·5 to 2 in., petiole ·25 in. ; stipules caducous. *Flowers* 2 in. long : the buds cylindric, acute, solitary, axillary ; pedicels ·1 in. long, tomentose like the exterior of the sepals, minute, linear-subulate. *Sepals* linear-lanceolate, adpressed-sericeous within. *Petals* shorter than the sepals, oblanceolate, scaly, puberulous externally. *Staminal tube* and gynophore ·25 in. long, the free part of the filaments more than twice as long ; fertile anthers about 12 ; staminodes 5. *Ovary* fusiform. *Style* longer than the stamens, pubescent below ; stigma cylindric. *Capsule* (*fide* Masters) shortly stalked, ovoid, terete, acute, 1·5 in. long and 1 in. in diam., covered with flat scaly hairs. Mast. in Hook. fil. Fl. Br. Ind i. 367 ; *P. oblongum*, Wall. Cat. 1165.

Penang ; Jack, Wallich, Curtis. Malacca ; Stolickza, at low elevations.

This species does not appear to be a common one. *P. Blumeanum* has probably been mistaken for it.

4. P. ACEROIDES, Wall. Cat. 1171. A tree 35 to 50 feet high : young branches rather slender, covered (as are the petioles and under surfaces of the leaves) by a thin felted layer of minute white tomentum, above which is a superficial deciduous layer of loose stellate rufous hairs. *Leaves* coriaceous, more or less elliptic, sometimes obovate-elliptic, the apex abruptly and shortly acuminate, the edge often straight at the sides, sometimes waved, never lobed : the base sub-truncate, often cordate, 5 to 7-nerved ; upper surface (when adult) glabrous : main lateral nerves 12 to 15 pairs, straight, oblique ; length 5 to 10 in., breadth 3·25 to 5·5 in., petiole ·4 to ·5 in. *Flowers* 1·5 to 1·75 in. long ; solitary, or in 3 to 4-flowered sub-sessile axillary cymes ; pedicels ·2 in. long, each with a deeply lobed tomentose bract ; the buds narrowly cylindric, ribbed. *Sepals* very coriaceous, recurved, longer than the petals, linear, acute, scurfy-tomentose outside, adpressed-pubescent within. *Petals* membranous, obovate, glabrous in the inner, scurfy on the outer, surface. *Stamens* as long as the petals or shorter, the tube only ·25 in. long : fertile anthers about 15, linear. *Style* shorter than the stamens, glabrous ; stigma clavate ; ovary densely sericeous, 5-angled. *Capsule* woody, oblong, pointed at both ends, angled, glabrous, 2 to 2·5 in. long. Kurz in Journ. As. Soc. Beng. 1873, pt. 62 ; For. Flora Burm. i. 145. *P. acerifolium*, Mast. (not of Willd.) in Hook. fl. Fl. Br. Ind. i. 368, in part. Miq. Ill. Arch. Ind. 84, in part.

Andaman Islands ; Helfer, No. 568 (Kew Distrib.), Kurz, King's Collectors. Distrib. Burmah ; Wallich.

The nearest ally of this is no doubt *P. acerifolium*, Willd., to which it has been reduced by Dr. Masters. But (having had living trees of both under observation in the Botanic Garden, Calcutta, for many years) I have no hesitation in saying that the two species are quite distinct. *P. aceroides* has entire, not lobed, leaves; much smaller flowers (less than 2 in. long) which expand during December and January: while those of *P. acerifolium* measure ·6 in. length and open in March or April. The capsule of *P. aceroides* is moreover only 2 to 2·5 in. long and quite glabrous; while that of *P. acerifolium* is 4 to 6 in. long, with a rough densely stellate tomentose exterior.

## 7. MELOCHIA, Linn.

Herbs or undershrubs, more or less downy. *Leaves* simple. *Flowers* small, clustered or loosely panicled. *Sepals* 5, connate below. *Petals* 5, spathulate, marcescent. *Stamens* 5, opposite to the petals, connate below into a tube; anthers extrorse, 2-lobed, lobes parallel. *Ovary* sessile, 5-celled; cells opposite the petals, 2-ovuled; styles 5, free or connate at the base. *Capsule* loculicidally 5-valved. *Seeds* ascending, albuminous; embryo straight, cotyledons flat, radicle next the hilum.— Distrib. Species about 50, natives of the warmer regions of both hemispheres.

1. M. CORCHORIFOLIA, Linn. sp. 944. A pubescent, branching herb or undershrub. *Leaves* membranous, variable, broadly ovate, to ovate-oblong or lanceolate, acute, serrate or obscurely lobed; the base rounded, truncate or sub-hastate, 5-nerved, often plaited; petiole from ·4 to 1 in.; stipules linear, minute. *Flowers* ·2 in. in diam., in crowded terminal or axillary heads with many villous bracteoles intermixed. *Sepals* lanceolate, acuminate, ascending. *Petals* obovate. *Ovary* villous; styles glabrous. *Capsule* pisiform, pubescent, exceeding the calyx. Willd. Sp. Pl. iii. 604; Roxb. Fl. Ind. iii. 139; Wall. Cat. 1196, in part; Mast. in Hook. Fl. Br. Ind. i. 374. *M. truncata*, Willd. Sp. Pl. iii. 601. *M. supina*, L. Sp. Pl. 944. *M. affinis*, Wall. Cat. 1198. *M. pauciflora*, Wall. Cat. 1199. *Riedleia corchorifolia*, DC. Prodr. i. 491; W. & A. Prodr. i. 66; Miq. Fl. Ind. Bat. i. pt. 2, 188. *R. truncata*, W. & A. l. c. 66. *R. supina*, DC. Prodr. i. 491. *R. concatenata*, DC. Prodr. i. 492. *Visenia corchorifolia*, Spreng. Syst. iii. 30. *V. concatenana*, Spreng. Syst. iii. 30. *V. supina*, Spreng. Syst. iii. 31. *Melochia concatenata*, Wall. Cat. 1197. *Sida cuneifolia*, Roxb. Hort. Beng. 50,

In all the provinces, a common weed. Distrib. The Tropics generally.

196

2. M. VELUTINA, Bedd. Fl. Sylvat. t. 5. A large shrub or small tree, all parts pubescent and with many of the hairs stellate. *Leaves* membranous, long-petioled, broadly ovate, acuminate, coarsely and irregularly serrate; the base 5 to 7-nerved, rounded or cordate; 4 to 9 in. long, by 3·5 to 8 in. broad: petioles 2·5 to 4·5 in.; stipules rounded, ·25 in. long. *Cymes* on peduncles longer than the petioles, much branched, spreading, many-flowered, terminal and axillary. *Flowers* ·25 in. in diam., pink. *Calyx* campanulate, with 5 deep broad abruptly acuminate teeth. *Petals* narrowly oblong, longer than the calyx, membranous. *Stamens* inserted on a hypogynous disk as are the petals; filaments flat. *Ovary* villous, as are the lower parts of the styles. *Capsules* ·3 to ·5 in. long, ovoid-cylindric, apiculate, deeply 5-grooved, bristly-tomentose. *Seed* solitary in each cell, its wing ascending. Mast. in Hook. fil. Fl. Br. Ind. i. 374; Kurz For. Fl. Burm. i. 148. *Visenia indica*, Houtt. Linn. Syst. vi. p. 287, t. 46; Miq. Fl. Ind. Bat. i. pt. 2, p. 189. *V. umbellata*, (Houtt.) Bl. Bijdr. 88; Wight Ic. 509. *V. Javanica*, Jungh. in Tijdsc. Nat. Gesch. viii. 302. *Glossospermum velutinum*, Wall. Cat. 1153. *G. ? coraatum*, Wall. Cat. 1155.

In all the Provinces at low elevations—a tree-weed appearing in abandoned fields. Distrib. Malayan Archipelago, British India, Mauritius.

8. WALTHERIA, Linn.

Herbs or undershrubs. *Leaves* simple. *Stipules* linear. *Flowers* small, in dense axillary or terminal clusters. *Sepals* 5, connate below into a bell-shaped tube. *Petals* 5, oblong-spathulate. *Stamens* 5, tubular below; anthers 2-lobed, lobes parallel. *Staminodes* 0. *Ovary* sessile, 1-celled; 2-ovulate. *Styles* 2, distinct, clavate. *Capsule* 2-valved, 1-seeded. *Seeds* ascending, albuminous; embryo straight, cotyledons flat. Distrib. About 15 species, one or two of which are weeds in the Tropics generally; the others are Tropical S. American.

W. INDICA, Linn. sp. 941. A pubescent undershrub. *Leaves* ovate-oblong, obtuse, serrate or crenate, the base rounded or cordate; nerves 5 to 7 pairs, prominent beneath. *Flowers* ·25 in. in diam., sessile; bracts linear. *Calyx* campanulate, villous, 10-nerved, the mouth with 5 acuminate teeth. *Petals* oblanceolate, clawed, longer than the calyx. *Capsule* membranous, pubescent. DC. Prod. i. 493; W. & A. Prod. i. 67; Mast. in Hook. fil. Fl. Br. Ind. i. 374; Miq. Fl. Ind. Bat. i. pt. 2, p. 187; Wall. Cat. 1194. *W. Americana*, L. DC. Prod. i. 492. *W. elliptica*, Cav. Diss. vi. 171; Wall. Cat. 1195.

In all the Provinces: a weed. Distrib. The Tropics generally.

## 9. AEROMA, Jacq.

Trees or shrubs. *Leaves* cordate, ovate-oblong, serrulate, sometimes angled. *Peduncles* opposite the leaves, few-flowered. *Sepals* 5, connate near the base. *Petals* 5, purplish, concave below, prolonged above into a large spoon-shaped lamina. *Staminal-cup* of 5 fertile and as many sterile divisions; fertile filaments opposite the petals, 3-antheriferous; anthers 2-lobed, lobes divergent. *Staminodes* longer than the fertile filaments, obtnse. *Ovary* sessile, pyramidal, 5-lobed; cells many-ovuled, styles 5. *Capsule* membranous, 5-angled, 5-winged, truncate at the apex, septicidally 5-valved, valves villous at the edges. *Seeds* numerous, albuminous; embryo straight, cotyledons flat, cordate, radicle next the hilum.—Distrib. 2 or 3 species, natives of Tropical Asia.

1. A. AUGUSTA, Linn. fil. Suppl. 341. A pubescent large shrub or small tree: young branches pale. *Leaves* 5 to 7-nerved at the base, 3·5 to 6 in. long and 3 to 5 in. broad; petiole 1·75 to 2·5 in., the upper much smaller and narrower. *Stipules* linear, deciduous. *Flowers* 2 in. in diam., peduncles 1·5 in., extra-axillary. *Sepals* 1 in. long, lanceolate, free to nearly the base. *Petals* longer than the sepals, imbricate, deciduous. *Capsule* 1·5 to 2 in. in diam., glabrous or nearly so when ripe. DC. Prod. i. 485; Mast. in Hook. Fl. Br. Ind. i. 375; Bl. Bijdr. 85; Roxb. Hort. Beng. 50; Fl. Ind. iii. 156; Miq. Fl. Ind. Bat. i. pt. 2, 183; Beddome Flor. Sylvat. Anal. Gen. t. 5; W. & A. Prodr. i. 65; Wall. Cat. 1142. *A. angulata*, Lam. Ill. 636. *A. Wheeleri*, Retz. Obs. v. 27; Willd. Sp. Pl. iii. 1425. *A. fastuosum*, Gærtn. Fruct. i. 307, t. 64.

In all the Provinces at low elevations: usually near cultivation. Distrib. Malayan Archipelago, Philippines, China, Brit. India.

The bark yields a stout fibre.

## 10. BUETTNERIA, Linn.

Erect climbing or tomentose shrubs, herbs, or trees; sometimes prickly. *Leaves* various. *Flowers* minute, in axillary or terminal much-branched, umbellate cymes. *Sepals* 5, slightly connate near the base. *Petals* 5, unguiculate, concave, inflexed, with 2 small lateral lobes, and a long sub-terminal simple linear or narrowly lanceolate appendage. *Staminal tube* with 5 broad truncate or emarginate teeth and, between them, five 2-celled extrorse anthers (mouth entire in *B. Curtisii*). *Ovary* sessile, 5-celled, the cells 2-ovulate. *Style* entire, 5-fid. *Capsule* globose, echinate, septicidaly 5-valved, the cells 1-seeded. *Seed* ascending, exalbuminous: cotyledons folded round the radicle. Distrib. About 48 species, mostly tropical American: a few tropical Asiatic and one African.

Leaves longer than broad, their bases not
cordate or only minutely so.
    Staminal tube with entire mouth   ...   1. *B. Curtisii.*
    Staminal tube with its mouth 5-lobed.
        Leaves quite glabrous.
            Capsule less than 1 in. in diam.,
                covered with glandular bar-
                bed spines   ...   ...   2. *B. uncinata.*
            Capsule more than 1 in. in
                diam., covered with short
                subulate spines ...   ...   3. *B. Maingayi.*
        Leaves more or less minutely hispid
            on both surfaces   ...   4. *B. elliptica.*
        ,,   hispid on the upper, hispid-
            tomentose on the lower,
            surface   ...   ...   5. *B. Jackiana.*
Leaves about as broad as long, deeply cordate
    at the base.
        Leaves glabrous, or glabrescent, not
            lobed   ...   ...   6. *B. aspera.*
        ,,   sparsely pubescent, often
            lobed   ...   ...   7. *B. Andamanensis.*

1. B. CURTISII, Oliver in Hook. Ic. Pl. t. 1761. A slender woody
creeper, 10 to 15 feet long: young branches minutely puberulous.
*Leaves* linear-lanceolate or oblanceolate-oblong, rarely ovate-oblong,
acuminate, entire, narrowed to the sub-obtuse, minutely cordate, 5-
nerved base: lateral nerves numerous, unequal and spreading at various
angles, reticulations distinct: both surfaces glabrous, the lower with
tufts of stellate hairs in the axils of the leaves; length 2.5 to 7 in.,
breadth ·5 to 2 in. ; petiole ·25 in., pubescent. *Cymes* in axillary fascicles
of 2 to 4, slender, 3 to 7-flowered, puberulous; peduncles about 1 in.
long : flower pedicels ·25 in. *Flowers* ·45 in. in diam., buds conical. *Calyx*
deeply 5-partite, the segments ovate-lanceolate, acuminate. *Petals* strap-
shaped with 2 rather broad inflexed lateral lobes, and a long cylindric
curved sub-terminal appendage, about as long as the sepals. *Capsule*
globular, about 1 in. in diam., veined, pubescent, and armed with
numerous straight smooth bristles.

    Penang, Curtis, Nos. 817 and 1166 ; Perak, common at low eleva-
tions.

    This is closely allied to the Bornean *B. lancifolia,* Hook. fil. The
leaves vary a good deal in shape, the most prevalent form in the Perak

specimens being linear-lanceolate. The Penang specimens are, on the other hand, as figured by Professor Oliver, oval-oblong.

2. B. UNCINATA, Mast. in Hook. fil. Fl. Br. Ind. i. 377. A woody climber : young branches at first scurfy and hispid, but very soon glabrous. *Leaves* sub-coriaceous, elliptic-oblong; gradually tapering in the upper third to the acuminate apex, entire, the base slightly cuneate, 3-nerved ; both surfaces glabrous and shining, nerves 9 or 10 pairs, spreading, thin but prominent beneath : length 9 to 11 in., breadth 3 in., petiole nearly 3 in., thickened at the apex, glabrous. *Sepals* lanceolate, spreading, hispid. *Fruiting peduncles* (fide Masters) " half the length of the leaves. *Capsule* depressed-spheroidal, the size of a hazelnut, covered with hooked gland-tipped barbed hispid spines, 3-celled."

Malacca, Maingay, No. 242 (Kew Distrib.).

I have seen only Maingay's Malacca specimens.

3. B. MAINGAYI, Mast. in Hook. fil. Fl. Br. Ind. i. 377. A woody climber : young branches glabrous. *Leaves* sub-coriaceous, elliptic to elliptic-oblong, shortly bluntly and rather abruptly acuminate, entire ; the base rounded with 3 bold and 2 minute nerves : both surfaces quite glabrous ; lateral nerves about 2 or 3 pairs, prominent beneath as are the reticulations ; length 7 or 8 in., breadth 3 to 3·5 in. ; petioles 1·2 in., thickened towards the apex, glabrous. *Umbels* in axillary fascicles of 6 or 8, their peduncles about 1 in. long, slender, glabrescent ; pedicels ·25 in. *Sepals* ·25 in. long, ovate-lanceolate. " *Petals* shorter than the sepals, with a long linear appendage. *Staminodes* erect, oblong, obtuse, bifid. *Style* as long as the ovary. *Fruiting peduncle* as long as the petiole. *Capsule* globose, 1·25 in. in diam., obscurely 5-lobed, studded with short subulate prickles."

Malacca ; Griffith, Maingay.

Of this species I have seen no good specimens in flower or fruit, and the above account of these parts is taken from Masters' description.

4. B. ELLIPTICA, Mast. in Hook. fil. Fl. Br. Ind. i. 377. A woody climber ; young branches minutely rusty-tomentose. *Leaves* broadly elliptic, abruptly and shortly acuminate, entire ; the base 5-nerved, rounded or minutely cordate ; upper surface minutely scabrid-hispid, the midrib and nerves hispid-tomentose ; lower minutely pubescent on the veins, the midrib and longer nerves tomentose : lateral nerves 3 pairs, oblique, curving, prominent beneath as are the secondary nerves and reticulations : length 5·5 to 7·7 in., breadth 3·5 to 4·75 in., petiole

2 to 3 in., tomentose. *Umbels* pedunculate, solitary or in fascicles of 5 or 6, axillary, few-flowered, stellate-tomentose : peduncles ·5 to 1 in.; pedicels ·1 to ·25 in., both slender. Buds ·1 in. in diam. *Sepals* ovate, acute. *Petals* rounded, with long cylindric inflexed apices longer than the sepals. *Fruit* unknown.

Malacca, Maingay : No. 241 (Kew Distrib.). Perak ; Scortechini.

Evidently a rare species; for I have seen, besides Maingay's, only Scortechini's solitary specimen.

5. B. JACKIANA, Wall. in Roxb. Fl. Ind. (ed. Carey) ii. 386. A stout woody creeper, the young branches with densely minute ferruginous tomentum some of which is stellate. *Leaves* narrowly or broadly elliptic, acuminate, entire ; the base boldly 3-nerved, rounded, sometimes slightly cordate, rarely acute ; upper surface sparsely and shortly hispid ; under surface rufous, hispid-tomentose especially on midrib and nerves, many of the hairs on both surfaces stellate ; lateral nerves 3 or 4 pairs, curved, spreading; length 3·5 to 6 in., breadth 2 to 3·75 in. ; petiole ·4 to 1 in. tomentose. *Umbels* pedunculate, solitary or in groups of 3 or 4 from the leaf-axils, few-flowered, tomentose : peduncles ·25 to 1 in. long ; pedicels about ·35, slender. *Sepals* linear-subulate, spreading, hispid, about ·4 in. long. *Petals* sub-rotund, lobed, each with a single long cylindric terminal appendage as long as the sepals. *Capsule* globose, slightly 5-furrowed, 1 in. in diam., black, glabrescent, armed with many straight spines. *Seeds* oblong, black. Mast. in Hook. fil. Fl. Br. Ind. i. 376 ; Wall. Cat. 1147.

Penang, Perak and Singapore ; at low elevations.

6. B. ASPERA, Colebr. in Roxb. Fl. Ind. (ed. Carey), ii. 383. A powerful woody climber often with a tree-like stem ; young branches glaucous, minutely and deciduously pubescent. *Leaves* sub-orbicular or ovate-orbicular, shortly acuminate, entire, the base cordate, 5 to 7-nerved ; upper surface glabrous, shining; the lower glabrescent, the midrib and nerves puberulous ; lateral nerves 4 to 6 pairs, prominent beneath as are the stout transverse veins ; length 4·5 to 7·5 in., breadth about the same : petiole 2 to 5 in., glaucous-pubescent at first, afterwards glabrous. *Cymes* axillary, solitary or fasciculate, pedunculate, much branched, many-flowered, pubescent to tomentose, 3 or 4 in. long : ultimate pedicels ·5 in. long, slender. *Sepals* lanceolate, acute, spreading, ·15 in. long, puberulous externally. *Petals* cuneate, shorter than the sepals, 3-lobed, the middle lobe linear-lanceolate, reflexed. *Staminodes* truncate. *Ovary* globular, scabrid. *Capsules* globular, 1·5 to 2 in. in diam., slightly depressed, pubescent when young, glabrous when ripe,

armed with many long, nearly straight, sharp spines. *Seeds* oblong, ·5 in. or more long. Wall Cat. 1144; Mast. in Hook. fil. Fl. Br. Ind. i. 377; Kurz For. Fl. Burm. i. 151; Pierre Fl. Forest. Coch-Chine, t. 206, figs. 1 to 8. *B. grandifolia*, DC. Prodr. i. 486. *B. nepalensis*, Turcz. in Bull. Mosc, 1858, 207.

Andaman Islands. Distrib. Brit. India, China, Cochin-China.

7. B. ANDAMANENSIS, Kurz in Journ. As. Soc. Bengal, 1871, ii. 47. A woody climber: young branches scaberulous. *Leaves* sub-orbicular, crenate and palmately 3 to 5-lobed, the lobes acuminate; or ovate-rotund, acuminate and irregularly serrate-crenate and not lobed; the base always deeply cordate, 5 to 7-nerved; lateral nerves about 5 pairs, opposite, prominent beneath as are the midrib and straight transverse veins; both surfaces sparsely pubescent at first, but afterwards glabrous. *Cymes* umbellate, 2 or 3 times branched, spreading, many-flowered, solitary, or 2 or 3 in a fascicle, axillary: the common peduncle stout, ·6 to ·75 in. long; secondary peduncles about the same length, tertiary half as long: flower-pedicels ·15 in., all slender and slightly pubescent. *Sepals* ovate acuminate or deltoid, puberulous externally. *Petals* with 2 obscure lateral lobes, and a long lanceolate inflexed middle lobe. *Staminodes* truncate. *Capsule* globose, less than 1 in. in diam, glaucous when young, armed with a few unequal, rather short, smooth, stiff spines. Kurz in Flora, 1871, p. 277; For. Fl. Br. Burm. i. 152; Mast. in Hook. fil. Fl. Br. Ind. i. 377; Pierre Fl. Forest. Coch-Chine, t. 207, figs. 1 to 9..

Andaman Islands. Distrib. Burmah, Siam, Cochin-China.

### 11. COMMERSONIA, Forsk.

Trees or shrubs. *Leaves* simple, oblique. *Inflorescence* cymose, terminal or axillary or leaf opposed. *Calyx* 5-cleft. *Petals* 5, concave at the base, prolonged into a long strap-shaped appendage at the apex. *Fertile stamens* 5, opposite the petals; anthers subglobose, 2-celled, cells diverging; staminodes 5, opposite to the sepals, lanceolate. *Carpels* 5, opposite to the sepals, connate; *styles* connate; ovules 2-6. *Capsule* loculicidally 5-valved, covered with bristly hairs. *Seeds* ascending, albuminous, strophiolate; cotyledons flat, radicle next the hilum. Distrib. A genus of about 8 species, some of which are natives of the Malay peninsula and Archipelago, others of Australia.

C. PLATYPHYLLA, Andr. Bot. Rep. t. 603 (note). A low tree; young branches softly rusty-tomentose. *Leaves* membranous, inequilateral, ovate-acuminate, irregularly dentate-serrate; the base more or less cordate, one side sub-auriculate, upper surface sparsely and minutely

stellate-hairy, lower softly hoary tomentose : length 5 to 8 in., breadth
3 to 4·5 in., petioles ·2 to ·3 in.; stipules shorter than the petioles,
scarious, lobed. *Cymes* corymbose, much shorter than the leaves,
spreading, much branched, tomentose. *Flowers* ·2 or ·25 in. in diam
*Calyx* pubescent, cut nearly to the base into 5 ovate-lanceolate lobes.
*Petals* as long as the sepals but much narrower, concave at the base ; the
terminal appendage elongate, narrowly oblong, its edges inflexed.
*Stamens* 5, the anthers broad, extrorse. *Staminodes* 5, lanceolate, spreading, reflexed, shorter than the petals. *Ovary* 5-celled. *Capsule* globose,
·4 or ·5 in. in diam., densely covered with long soft, flexuose, pubescent
bristles. Mast. in Hook. fil. Fl. Br. Ind. i. 378. *C. Javensis*, G. Don.
Gen. Syst. i. 523; Hassk. Pl. Jav. Rar. 312. *C. echinata*, Blume Bijdr.
86 ; Wall. Cat. 1143 ; Andr. Bot. Rep. t. 519, not of Forst. *C. echinata*,
var. *β.* Miq. Fl. Ind. Bat. i. pt. 2, 182. *Buettneria hypoleuca*, Turcz. in
Mosc. Bull. 1858, 207.

In all the provinces except the Andamans and Nicobars. Distrib.
Malayan Archipelago, Philippines.

## 12. Leptonychia, Turcz.

Shrubs or trees. *Leaves* simple, entire. *Flowers* in small axillary
cymes. *Sepals* 5, valvate, united near the base. *Petals* 5, valvate, short,
orbicular, concave. *Andrœcium* tubular below, filamentiferous above,
filaments 3-seriate, outer series of 5 to 10 ligulate staminodes opposite the
petals, middle of 10 fertile stamens also opposite the petals, innermost
of 5 very short fleshy subulate staminodes opposite the sepals ; anthers
linear-oblong, introrse, dehiscing at the sides longitudinally. *Ovary*
sessile, 3-4-celled ; placentas axile ; styles connate, stigmas capitellate ;
ovules many in each cell, anatropous. *Capsule* 2-3-celled, or by abortion
1-celled, dehiscing septicidally or loculicidally, or both simultaneously
or irregularly. *Seeds* black, with a fleshy yellowish arillus; albumen
fleshy, cotyledons foliaceous, radicle superior. Distrib. three or four
species—Indo Malayan and Tropical African.

L. glabra, Turcz. in Mosc. Bull. for 1858, p. 222. A tree : the
young branches glabrous. *Leaves* ovate-oblong or oblong-lanceolate,
rather abruptly acuminate, the base slightly narrowed or rounded, 3-
nerved ; both surfaces glabrous, shining ; main nerves 4 to 7 pairs, thin :
length 4 to 8 in., breadth 1·6 to 3 in. ; petiole ·4 to ·5 in., glabrous.
*Flowers* ·25 in. in diam. ; the buds oblong, obtuse, 5-ridged. *Sepals*
ovate-lanceolate or oblong, rather obtuse, spreading, pubescent on both
surfaces, not veined. *Petals* about one-fourth the size of the sepals,
broad, truncate, villous. *Stamens* 10, in five phalanges of two each,

95

nearly as long as the sepals. *Staminodes* 10 to 20, glabrous, the outer 5 to 15 shorter than, or as long as, the stamens, filiform; the inner invariably 5, short, subulate, internal to, and alternating with, the phalanges of stamens. *Ovary* broadly obovate, obtuse, obscurely 4-grooved, with a few scattered hairs near the apex, 4-celled. *Style* cylindric, tapering, with sparse spreading hairs. *Capsule* coriaceous, depressed-obovoid, pale greyish, ·5 in. long, rugose; within shining pale and wrinkled. *Seed* solitary, oblong, black, less than half covered by a thin arillus proceeding from its side. Mast. in Hook. fil. Fl. Br. Ind. i. 379; Kurz For. Fl. Burm. i. 150; Oudem. in Compt. Rend. Ac. Roy. Sc. Amsterd. 2 Ser., 11, 8, cum ic; Walp. Ann. vii. 449. *Grewia ? caudata,* Wall. Cat. 1099. *L. heteroclita,* Kurz For. Fl. Burm. i. 150. *G. heteroclita,* Roxb. Fl. Ind. ii. 590. *Binniudykia trichostylis,* Kurz in Nat. Tijdsc. Ned. Ind., Ser. 3, iii. 164. *Turrœa trichostylis,* Miq. Fl. Ind. Bat. Suppl. 502.

Malacca, Penang, Perak, Andamans; at low elevations. Distrib. Malayan Archipelago, Burma.

Var. *Mastersiana,* young branches, midribs and petioles of leaves puberulous; flowers ·5 in. in diam., the buds pointed; sepals 3-veined: outer staminodes varying from 5 to 15, often pubescent in the upper half: ovary oblong-ovoid, villous, 3-celled: style glabrous: capsule black. *L. acuminata,* Mast. in Hook. fil. Fl. Br. Ind. i. 379.

Malacca and Perak. Distrib. Sumatra, Borneo, Burmah.

This shrub or small tree is common, and I have thus had the advantage of being able to examine a large number of flowers. The result of my examination of these is that, whereas the inner staminodes are invariably 5 in number, the outer series varies in number in the most perplexing way from 5 to 15. Where there are 10, they are always arranged in pairs united at the base: and where there are 15, they are arranged in threes united at the base. The proper view to take of these staminodes is I believe therefore that they are single organs, but sometimes deeply cleft into 2 or 3 linear and equal segments. On this account, and also on account of the similarity of the other organs, I am induced to think that there is but *one* species of *Leptonychia* and that Masters' species *acuminata* and Beddome's *L. moacurroides* are merely forms of the species on which Turczaninow originally founded the genus.

ORDER XIX. TILIACEÆ.

Trees, shrubs or herbs. *Leaves* alternate, rarely opposite, simple or lobed. *Stipules* free, usually caducous. *Flowers* usually cymose, or in cymose panicles, or racemose. *Flowers* regular, hermaphrodite, rarely unisexual. *Sepals* 3-5, free or connate, valvate. *Petals* as many as the sepals, rarely absent, imbricate or valvate. *Stamens* numerous, rarely
204

definite, usually springing from a prolonged or dilated torus, free or sometimes 5-adelphous, filaments filiform ; anthers 2-celled. *Ovary* free, 2-10-celled ; styles columnar, or divided into as many divisions as there are cells to the ovary, stigmas usually distinct, rarely confluent or sessile. *Ovules* attached to the inner angle of the cells of the ovary ; if few in number, often pendulous from the apex or ascending from the base ; if more numerous, disposed in 2 or more ranks, anatropous ; raphe ventral or lateral. *Fruit* fleshy or dry, dehiscent or indehiscent, 2-10 or by abortion 1-celled (cells sometimes divided by false partitions) ; carpels separable or always united. *Seeds* 1 or many, ascending, pendulous or transverse, with no arillus ; testa leathery or crustaceous or pilose ; albumen fleshy, abundant or scanty, rarely wanting ; embryo straight or slightly curved, cotyledons leafy, rarely fleshy, radicle next the hilum.— Distrib. about 370 species ; most abundant in the tropics of either hemisphere.

Series A. *Holopetalæ. Petals* glabrous or rarely downy, coloured, thin, unguiculate, entire or nearly so, imbricate or twisted in the bud. *Anthers* globose or oblong, opening by slits.

Tribe I. *Brownlowieæ. Sepals* combined below into a cup. *Anthers* globose, cells ultimately confluent at the top.

\* *Staminodes* 5.

| | | |
|---|---|---|
| Carpels distinct, 2-valved ... | ... | 1. *Brownlowia.* |
| Carpels combined, indehiscent, winged | ... | 2. *Pentace.* |

\*\* *Staminodes* 0.

| | | |
|---|---|---|
| Stamens on a raised torus ... | ... | 3. *Schoutenia.* |
| Stamens on a contracted torus ... | ... | 4. *Berrya.* |

Tribe II. *Grewieæ. Sepals* distinct. *Petals* glandular at the base. *Stamens* springing from the apex of a raised torus.

| | | |
|---|---|---|
| Fruit drupaceous, not prickly ... | ... | 5. *Grewia.* |
| Fruit dry indehiscent or 3-5 coccous, prickly | | 6. *Triumfetta.* |

Tribe III. *Tilieæ. Sepals* distinct : petals not glandular. *Stamens* springing from a contracted torus.

| | | | |
|---|---|---|---|
| Herbs or undershrubs with 3 or 5-celled capsules : seeds without hairs | | ... | 7. *Corchorus.* |
| Trees with 2-celled capsules ; seeds with marginal hairs ... | ... | ... | 8. *Trichospermum.* |

Series B. *Heteropetalæ. Petals* usually incised, rarely entire or absent, induplicate or imbricate not twisted : anthers linear, opening by a terminal pore often with an apical awn or tuft of hairs.

Stamens on a raised torus ; fruit drupaceous    9. *Elæocarpus.*

97

## 1. BROWNLOWIA, Roxb.

Trees. *Pubescence* stellate or scaly. *Leaves* entire, 3-5-nerved, feather-veined. *Flowers* numerous, small, in large terminal or axillary panicles. *Calyx* bell-shaped, irregularly 3-5-fid. *Petals* 5, without glands. *Stamens* many, free, springing from a raised torus. *Staminodes* 5, within the stamens, opposite the petals and petaloid. *Anthers* subglobose. *Ovaries* 5, each 2-ovulate; styles awl-shaped, slightly coherent; ovules ascending. *Carpels* ultimately free, 2-valved, 1-seeded. *Albumen* 0; cotyledons thick, fleshy.—Distrib. Nine species confined to Tropical Asia.

*Leaves* not peltate.
Leaves lanceolate ... ... 1. *B. lanceolata.*
  „ broadly elliptic to elliptic-rotund 2. *B. Kleinhovioidea.*
*Leaves* peltate.
  Leaves minutely hairy beneath ... 3. *B. Scortechinii.*
  „ glabrous on both surfaces ... 4. *B. macrophylla.*

1. BROWNLOWIA LANCEOLATA, Benth. in Journ. Linn. Soc. V. Suppl. ii. 57. A tree 25 to 30 feet high; young branches pale when dry, sublepidote. *Leaves* thinly coriaceous, lanceolate or oblong-lanceolate, acuminate, the base obtuse; upper surface when adult glabrous, shining, the lower covered by a dense layer of minute whitish yellow shining scales: main nerves 6 to 8 pairs (1 pair of them basal), not prominent: length 4·5 to 6 in., breadth 1·5 to 1·75 in., petiole ·25 to ·4 in. *Panicles* axillary or terminal, 1 to 3 in. long, and less than 1 in. across, few-flowered. *Flowers* ·25 in. long, their pedicels about as long. *Calyx* ·2 in. long, scaly like the pedicel, its lobes lanceolate. *Petals* longer than the calyx, oblong, blunt, slightly narrowed to the shortly unguiculate base, glabrous. *Anther-cells* sub-divaricate, sub-confluent when adult. *Ovary* deeply 3 to 5-lobed, pubescent, the cells 2-ovuled. *Ripe carpels* distinct, sub-globose, truncate, compressed on their inner surfaces, minutely lepidote and pubescent, ·5 in. in diam. *Seed* solitary, with thin testa and large sub-hemispheric cotyledons. Hook. fil. Fl. Br. Ind. i. 381: Kurz For. Flora Burm. 154.

Malacca, Griffith. Distrib. Burmah and Bengal; in tidal forests and mangrove swamps.

The young parts are covered with rusty or pale brown scales, but the adult branchlets leaves and flowers are as above described.

2. BROWNLOWIA KLEINHOVIOIDEA, King, n. sp. A tree 40 to 50 feet high: young branches rather slender, covered with a dense thin layer of
206

minute pale brown hair. *Leaves* thinly coriaceous, broadly elliptic to elliptic-rotund, slightly narrowed to the obtuse apex, very little narrowed to the more or less cordate base : upper surface glabrous, very sparsely lepidote, the lower covered with a thin layer of very minute pale hair ; basal nerves 4 or 6 (two of them small) : main lateral nerves 3 pairs ; transverse secondary nerves distinct : length 5 to 7 in., breadth 3·5 to 4·5 in. ; petiole 2·5 to 3 in., thickened towards the apex, pubescent like the under surfaces of the leaves. *Panicles* mostly terminal, rarely axillary, 9 to 15 in. long (the axillary ones much smaller) the branches rather few, spreading little, the flowers rather closely clustered on the branchlets. *Flowers* ·25 in. long, on pedicels about half as long. *Calyx* widely campanulate, cut for a third of its length into 5 acute triangular teeth, minutely tomentose externally, glabrous inside. *Petals* longer than the calyx, oblong, very obtuse, slightly narrowed but thickened towards the rather long basal claw. *Staminodes* linear, flat, about as long as the filaments. *Ovaries* 3 to 5, sub-globose, laterally compressed, pubescent. *Styles* subulate, a little longer than the stamens, slightly coherent. *Fruit* unknown.

On Gunong Bubu in Perak, at elevations of 600 to 1000 feet ; King's Collector.

A species with leaves not unlike those of *Kleinhovia hospita :* in many respects closely allied to *B. elata,* but with much smaller flowers.

3.  BROWNLOWIA SCORTECHINII, n. sp., King.  A small slender tree : young branches stout, pale, sparsely lenticellate, pubescent at first but soon glabrous. *Leaves* coriaceous, ovate-elliptic, peltate, slightly narrowed to the acute or sub-acute apex ; the edges sub-undulate ; very little narrowed to the rounded, or sometimes sub-emarginate, base ; upper surface glabrous ; the lower pale from a thin continuous layer of very minute hairs ; petiole attached 2·5 to 3 in. above the base, nerves radiating from it about 9, lateral nerves from the midrib about 4 pairs ; all rather prominent beneath, as are the transverse secondary nerves : length 10 to 15 in., breadth 5·5 to 7 in., petiole 7 to 9 in. long, thickened at both ends. *Panicle* terminal, 6 to 12 in. long and about 6 in. broad, or sometimes small narrow panicles in terminal clusters of 6 to 10 : branches spreading, compressed, puberulous ; bracteoles ovate, fugaceous ; pedicels, stout, ·15 in. long in the bud but lengthening as the flower expands, puberulous. *Flowers* ·6 in. long, crowded. *Calyx* narrowly campanulate ; its teeth half as long as the tube, lanceolate, sub-acute, tomentose-lepidote externally. *Petals* longer than the calyx, narrowly obovate, much narrowed to the clawed base. *Staminodes* linear, about as long as the filaments. *Ovaries* 5, ovoid, compressed, stellate-pubescent.

*Styles* slightly longer than the stamens, subulate, bent at the apex. *Fruit* unknown.

Perak; Scortechini, No. 1918.

Collected only once by the late Fr. Scortechini and referred by him to *B. elata,* Roxb. The species is, however, quite distinct from *B. elata ;* and also from *B. peltata,* which it more resembles in its leaves.

4. BROWNLOWIA MACROPHYLLA, King n. sp. A tree 30 to 40 feet high: young branches very stout, deciduously rufous-puberulous. *Leaves* very coriaceous, rotund, those on the older branches elliptic, the apex rounded or very slightly and shortly apiculate, the edges sub-undulate, the base broad, emarginate or slightly cordate, both surfaces glabrous ; main nerves 7 to 9 basal and about 2 pairs lateral, prominent on both surfaces, secondary nerves transverse and very distinct : length of the rotund leaves 11 to 17 in., breadth 10 to 14 : of the elliptic, length 6 to 10 in., breadth 3·5 to 5·5 in. : petiole 2·25 to 4·5 in., thickened at both ends. *Panicle* terminal, almost as long as the leaves, its branches numerous, compressed, grooved, spreading, scurfy and rusty-pubescent : bracts few, linear-lanceolate, nearly 1 in. long, persistent. *Flowers* ·65 in. long : their peduncles shorter than the calyx, stout, deeply grooved. *Calyx* rather widely cylindric-campanulate, its teeth about half as long as the tube, acute, triangular, rusty-tomentose and scurfy externally. *Petals* oblong, obtuse, very little narrowed to the base and without any very distinct claw. *Ovaries* 3 to 5, narrowly ovoid, compressed, vertically ridged, lepidote as are the conjoined styles. *Fruit* sub-globose, much compressed, covered with a layer of very minute pale hairs, ·75 in. in diam.

Perak, at low elevations and in moist ground ; Scortechini, Wray, King's Collector.

## 2 PENTACE, Hassk.

Trees. Herbaceous portions sometimes pubescent or scaly, ultimately glabrous. *Leaves* entire, leathery, the lower surface (except in one species) pale from a thin layer of minute adpressed scaly hair. *Flowers* numerous, small, in terminal panicles. *Calyx* bell-shaped, usually 5-fid. *Petals* 5, membranous, glabrous, longer than the calyx, glandless. *Stamens* numerous, on a slightly raised torus, usually penta-delphous. *Staminodes* 5, opposite the sepals. *Anthers* subglobose ; pollen globose, 3-pored. *Ovary* 5-celled, cells 2-ovuled ; ovule pendulous, raphe next the placenta. *Styles* united, rarely free. *Fruit* dry, indehiscent, 3-10-winged, 1-celled, 1-seeded by abortion. *Seed* solitary albuminous.—Distrib. About 15 species, all Malayan.

Leaves with pinnate nervation.
Ovary 3-ridged...          ...        ...     1. *P. triptera.*
Ovary 5-ridged.
    Leaves with 6 or 7 pairs of nerves...    2. *P. Hookeriana.*
       „      „    3 or 4      „     ...   3. *P. Kunstleri.*
    Ovary 10-ridged         ...       ...   4. *P. perakensis.*
Leaves boldly 5-nerved at the base, lateral
nerves from the central nerve (midrib)
3 pairs; 7 to 14 in. long; ovary 5-ridged...   5. *P. macrophylla.*
Leaves boldly 8-nerved at the base; the
central nerve (midrib) with 1 or 2 pairs of
lateral nerves: rarely more than 7 in. long.
    Ovary not visibly ridged     ...      ...   6. *P. floribunda.*
    Ovary 8 or 9-ridged.
       Styles quite confluent ...       ...   7. *P. Curtisii.*
       „   free...        ...      ...   8. *P. eximia.*
    Ovary 10-ridged.
       Leaves glabrous on both surfaces ...   9. *P. Scortechinii.*
       „    with a dense layer of minute
          adpressed hair on the under
          surface       ...      ... 10. *P. Griffithii.*
Leaves boldly 3-nerved at the base, the
central nerve (midrib) without lateral
nerves, only 3 or 4 in. long; ovary 5-ridged 11. *P. strychnoidea.*

1. PENTACE TRIPTERA, Mast. in Hook. fil. Fl. Br. Ind. i. 382. A
large tree: young branches pubescent, speedily becoming glabrous,
their bark dark-coloured. *Leaves* ovate to ovate-rotund, sometimes
ovate-oblong, sub-acute or shortly and bluntly acuminate, the margins
undulate, the base rounded; upper surface glabrous, the lower pale,
minutely scaly; basal nerves one or two pairs; lateral 5 to 7 pairs, ascending, straight; length 4 to 5 in., breadth 2 to 2·75 in.; petiole ·6 to
1·2 in. thickened towards the apex, pubescent. *Panicles* terminal and
axillary, 6 to 8 in. long, with short many-flowered branches minutely
and softly stellate-tomentose. *Flowers* nearly ·2 in. long, on pedicels
shorter than the calyx. *Calyx* with 5 lanceolate teeth, tomentose outside.
*Petals* spathulate-oblong, obtuse. *Stamens* 5-delphous, longer than the
style. *Staminodes* subulate, shorter than the stamens. *Ovary* densely
tomentose, shortly 3-winged. *Style* filiform, glabrous, bent at the
apex. *Fruit* oblong, narrow, ·6 in. long, with 3 spreading membranous
rounded wings ·5 in. broad.

101

Malacca ; Griffith, Maingay. Perak : Scortechini.
This approaches the Javan *P. polyantha*, Hassk., which has, however,
larger flowers with a shallower calyx with longer teeth, a shorter style,
and a 5-lobed ovary.

2. PENTACE HOOKERIANA, n. sp., King. A tree 30 to 40 feet high :
young branches cinereous, glabrous. *Leaves* elliptic-oblong, slightly
obovate, acute, the base narrowed and slightly unequal ; upper surface
glabrous, the lower dull ; lateral main nerves about 6 pairs (one of the
pairs basal), prominent on both surfaces; the intermediate nerves,
transverse veins and reticulations prominent only on the lower : length
5 to 7·5 in. ; breadth 2 to 2·75 in. ; petiole ·25 in., stout. *Panicles*
terminal and in the axils of the upper leaves, 2·5 to 5 in. long, the
branches spreading, everywhere scurfy-tomentose. *Flowers* rather
crowded, ·2 in. long, on pedicels shorter than the calyx. *Calyx* campanulate, cut half-way down into 3 or 4 broadly triangular sub-acute
spreading teeth, scaly and minutely tomentose outside. *Petals* narrowly obovate. *Stamens* 15 in 5 bundles of 3 each, very much shorter
than the petals. *Staminodes* thick, orbicular, embracing the ovary.
*Ovary* depressed-globose, densely pubescent, obscurely 5-lobed, 5-celled.
*Styles* 5, free, shorter than the ovary. *Fruit* unknown.
Perak, on the banks of the Kiuta river : King's Collector, No. 815.

3. PENTACE KUNSTLERI, n. sp., King. A tree 30 to 40 feet high :
young branches slender, dark-coloured, glabrous. *Leaves* broadly ovate,
with an abrupt short broad blunt acumen, the base rounded : upper
surface shining, glabrous, the lower dull ; lateral nerves 3 or 4 pairs,
curved, prominent beneath ; sometimes a pair of short slender sub-
marginal nerves at the base : length 4·5 to 6 in., breadth 2·5 to 3·5 in. ;
petiole ·75 to 1 in., stout, thickened at the apex. *Panicles* terminal, 3·5
to 6 in. long, puberulous, much-branched. *Flowers* numerous, ·15 in.
long, the pedicels slightly shorter. *Calyx* tubular-campanulate, minutely
stellate-hairy and lepidote outside ; the teeth triangular, acute, erect.
*Petals* spathulate with a very long claw. *Stamens* in 5 bundles.
*Staminodes* linear-lanceolate, as long as the filaments. *Ovary* depressed-
globose, with 5 blunt angles, lepidote and pubescent, 5-celled. *Style*
straight, glabrous. *Fruit* unknown.
Perak, at a very low elevation ; King's Collector, No. 6871.

4. PENTACE PERAKENSIS, n. sp., King. A tree 30 to 40 feet high :
young branches cinereous, glabrous. *Leaves* ovate-elliptic, slightly
oblique, bluntly acuminate, the base rounded or sub-cuneate : upper
surface shining, glabrous : the lower dull ; lateral nerves about 5 pairs,

210

ascending, curved ; length 5 to 6 in., breadth 2·5 to 3 in. ; petiole ·75 to 1 in., stout, and thickened at the apex. *Panicles* terminal, 4 to 5 in. long and less than 2 in. wide, little branched and few-flowered. *Flowers* about ·1 in. long, their pedicels about as long. *Calyx* rotate, minutely lepidote outside ; the teeth triangular, spreading. *Petals* ovate, narrowed to a short claw. *Stamens* about 30, 5-delphous. *Staminodes* lanceolate, as long as the filaments. *Ovary* globose, slightly pointed, 10-ridged, slightly hairy, 5-celled. *Style* about as long as the ovary, cylindric. *Stigma* terminal, small. *Fruit* unknown.

Perak, King's Collector, No. 3428.

5. PENTACE MACROPHYLLA, n. sp., King. A tree usually from 20 to 30 feet high, but occasionally as much as 50 feet. Young branches rather slender, pale brown, glabrous. *Leaves* large, ovate-elliptic to almost rotund, the apex very shortly and abruptly blunt-acuminate, the base rounded : upper surface glabrous ; the lower dull ; basal nerves 2 pairs, the upper branched on one side ; lateral nerves from the midrib 2 to 3 pairs, all ascending and little curved, prominent beneath ; length 7 to 14 in., breadth 5 to 12 in. ; petiole 2·5 to 3 in., stout. *Panicles* terminal and axillary, 6 to 15 in. long, lax, spreading, minutely yellowish-pubescent and scurfy. *Flowers* ·15 in. long and ·2 in. in diam., on pedicels about as long as the calyx. *Calyx* almost rotate, cut for two-thirds of its length into 5 lanceolate acute teeth, minutely yellowish-tomentose outside. *Petals* oblanceolate, obtuse. *Staminodes* linear, as long as the stamens. *Stamens* in 5 bundles of about 15 each. *Ovary* ovoid, scaly and pubescent, obtusely 5-angled. *Style* rather shorter than the stamens, cylindric, pointed. *Fruit* ·75 in. long with 10 radiating semi-elliptic striate sparsely scaly wings each ·4 in. wide.

Perak at elevations up to 500 feet ; King's Collector, Scortechini, Wray : common.

Distinguished from all the other known species by the large size of its leaves.

6. PENTACE FLORIBUNDA, n. sp., King. A tree 40 to 70 feet high : young branches slender, sparsely stellate-puberulous, the bark dark-coloured. *Leaves* elliptic-oblong to elliptic-rotund, the apex shortly and rather abruptly apiculate ; the base rounded or slightly narrowed ; upper surface glabrous, lower cinereous and with some scattered pubescence ; basal nerves 2 pairs, one of them branching on one side : lateral nerves 2 or 3 pairs, all ascending and all rather prominent : transverse veins not prominent ; length 5 to 6·5 in., breadth 2·5 to 3·75 in., petiole 1 to 2 in., thickened towards the apex. *Panicles* towards the apices of the

branches, axillary and terminal, stellate-pubescent, slender, spreading, many-branched. *Flowers* very numerous, ·1 in. long, the pedicels slender and rather longer. *Calyx* when expanded rotate, cut half way down into triangular very acute or acuminate spreading teeth, densely stellate-tomentose outside. *Petals* broadly oblanceolate, obtuse, narrowed to the base. *Stamens* 5-delphous. *Staminodes* apparently absent. *Ovary* globose, densely tomentose, not visibly ridged, 5-celled. *Style* filiform, tapering, straight, glabrous. *Fruit* unknown.

Perak, at elevations from 600 to 1000 feet : King's Collector, Nos. 7616 and 7730.

A species distinguished by its slender hoary panicles, with flowers by far more numerous than in any of the other species described here.

7. PENTACE CURTISII, n. sp. King. A large tree : the young branches slender, with dark-coloured bark, very minutely adpressed-lepidote, not hairy. *Leaves* ovate elliptic, with a short abrupt blunt acumen, the base rounded : upper surface glabrous, the lower cinereous ; basal nerves 2 pairs, one pair slender and close to the margin, the other branching on one side : lateral nerves 2 or 3 pairs ; all ascending and rather prominent beneath : length 3·25 to 5·5 in., breadth 2·25 to 2·75 in. ; petiole ·75 to 1 in. slender, slightly thickened at apex. *Panicles* mostly terminal (a few smaller axillary) 4·5 to 6 in. long with sparse cinereous stellate tomentum and scales, few-branched, and few-flowered. *Flowers* ·15 in. long, on pedicels shorter than themselves. *Calyx* widely campanulate, stellate-tomentose outside ; its teeth as long as the tube, broadly triangular, rather blunt. *Petals* oblanceolate or obovate-obtuse, much narrowed to the base. *Staminodes* lanceolate, acuminate, half as long as the filaments. *Stamens* in 5 bundles. *Ovary* turbinate, with 8 or 9 blunt ridges, lepidote-pubescent : style rather stout, cylindric, shorter than the stamens. *Fruit* ·5 in. long, with 8 semi-elliptic membranous wings ·2 in. broad.

Penang ; Curtis, No. 1573.

8. PENTACE EXIMIA, n. sp., King. A tree 50 to 70 feet high : young branches slender, dark-coloured, glabrous. *Leaves* ovate-elliptic to ovate-rotund, shortly and abruptly acuminate, the base slightly narrowed or rounded : upper surface glabrous, shining : lower paler and dull ; basal nerves 1 pair, bold and reaching to the apex, often with a slender small sub-marginal pair : lateral nerves usually only one pair, short and curving ; all rather bold beneath : length 4 to 5·5 in., breadth 2 to 4 in. ; petiole ·75 to 1·1 in., thickened at the apex. *Panicles* terminal, 3·5 to 5 in. long, (longer in fruit), minutely scurfy-tomentose, with

212

rather numerous spreading branches. *Flowers* numerous, ·1 in. long, the pedicels about the same length. *Calyx* densely scaly outside, the teeth triangular. *Petals* cuneate, obtuse, narrowed to a broad claw. *Stamens* about 25, in groups of 5. *Staminodes* lanceolate. *Ovary* sub-globular, 10-ridged (the ridges in pairs), scaly and pubescent, 5-celled, the cells with imperfect septa and thus falsely 10-celled. *Styles* 10, much shorter than the stamens, free, or united when young at the base only. *Fruit* about ·5 in. long, with 8 radiating semi-elliptic wings ·1 to ·15 in. broad, minutely adpressed-scaly.

Perak; at elevations under 1000 feet, King's Collector, Nos. 3482 and 3649.

This agrees with *P. Curtisii* in having 8-winged fruit, but the flowers are much smaller. The styles moreover are shorter than the ovary and quite distinct, which is the case in no other species of this genus which I have yet met with.

9. PENTACE SCORTECHINII, n. sp. King. A tree ? young branches slender, glabrous, dark-coloured. *Leaves* elliptic-oblong, shortly cau-date-acuminate, the base more or less cuneate : both surfaces quite glabrous, concolorous ; basal nerves 1 pair very bold, as is the midrib ; lateral nerves (from the midrib) 1 or 2 pairs, not conspicuous : length 7 to 9 in., breadth 2·75 to 3·5 in. ; petiole less than ·5 in., stout. *Panicles* terminal and axillary, slender, only about half the length of the leaves, few-branched, minutely tomentose. *Flowers* rather crowded, ·25 in. long, on pedicels shorter than the calyx. *Calyx* widely tubular-campanulate, minutely scurfy-tomentose outside, cut a third of its depth into 5 small triangular reflexed teeth. *Petals* obovate, obtuse, much narrowed to the base. *Stamens* in 5 bundles of 15 each. *Sta-minodes* lanceolate, half as long as the stamens. *Ovary* ovoid, obscure-ly 5-ridged, scaly, 5-celled. *Style* cylindric, tapering, longer than the stamens. *Fruit* unknown.

Perak, Father Scortechini, No. 119b.

Only once collected and without fruit. A very distinct species.

10. PENTACE GRIFFITHII, n. sp., King. A tree : young branches slender, dark-coloured, glabrous. *Leaves* ovate-elliptic, tapering about equally to the acute apex and base ; upper surface shining, lower dull ; basal nerves 2 pairs, the lower pair slender and sub-marginal, the upper branched on one side and bold (as is the midrib), ascending, curved ; lateral nerves (from the midrib) 2 pairs ; length 4 to 7 in., breadth 2·25 to 3 in. ; petiole nearly 1·5 in. long, thickened at both ends, but especially at the apex. *Panicles* terminal, slender, few-branched, lax, minutely

cinereous-tomentose. *Flowers* not very numerous, large for the genus ('25 in. long and '25 in. in diam.), on pedicels about as long as the calyx. *Calyx* widely campanulate, almost rotate, minutely stellato-tomentose outside; the teeth as long as the tube, spreading. *Petals* ovate, obtuse, rather suddenly contracted into a linear claw. *Stamens* in 5 groups of 12 or 13 each. *Staminodes* lanceolate, as long as the filaments. *Ovary* sub-globose, slightly compressed, minutely stellato-tomentose and scaly, obtusely 5-angled, 5-celled. *Style* cylindric, rather shorter than the filaments. *Fruit* nearly 1 in. long, with 10 radiating membranous, horizontal striate, minutely scaly, semi-elliptic, membranous wings, each '35 in. broad.

Tavoy in Tenasserim; Griffith, Aplin.

A very distinct species only once collected within recent years, by Mr. Aplin. There is, however, in the Kew Herbarium a twig of it collected by Griffith many years ago bearing this note in Griffith's handwriting " *Tiliacearum gen. nov. capsulis pluri-alatis.*" Although this plant has hitherto been found only in territory which is politically Burmese, yet Tavoy (being at the southern extremity of Tenasserim) is practically Malayan in its Flora and Fauna. I therefore include it here.

11. PENTACE STRYCHNOIDEA, n. sp., King. A tree 60 to 80 feet high : young branches slender, cinereous, glabrous. *Leaves* ovate-elliptic rarely ovate-oblong, shortly and abruptly acuminate, the base rounded or slightly narrowed; upper surface shining, glabrous; lower pale and dull; boldly 3-nerved and often with a slender sub-marginal pair of nerves ; length 3 to 4 in., breadth 1·75 to 2·25. in. ; petiole ·75 in. slightly thickened at the apex. *Panicles* terminal, 3 to 6 in. long, few-branched, lax, minutely lepidote-puberulous. *Flowers* rather large for the genus ('2 in. long). *Calyx* cup-shaped, tomentose outside, cut more than half way down into 5 triangular acute teeth. *Petals* oblanceolate, slightly oblique, much narrowed to the base. *Stamens* in 5 bundles of about 20 each. *Staminodes* linear-lanceolate. *Ovary* ovoid-globose, obtusely 5-ridged, minutely tomentose and lepidote, 5-celled. *Style* filiform, as long as the stamens. *Fruit* unknown.

Perak ; at elevations of from 500 to 1000 feet, King's Collector, No. 3478.

### 3. SCHOUTENIA, Korth.

Trees with alternate simple pinnately-nerved leaves. *Flowers* axillary, solitary or in clusters ; or in terminal few-flowered panicles. *Calyx* campanulate, 5-lobed ; lobes valvate, accrescent, coloured. *Petals* small, linear without claw, or absent. *Stamens* numerous, free, sometimes inserted on the apex of a short gynophore; anthers oblong, 2-celled: cells parallel, with longitudinal sutural dehiscence. *Staminodes* 0. *Ovary* sessile or shortly stalked, imperfectly 3 to 5-celled ; cells with 2 ovules from the base of the axile placentas, style filiform ; stigmas 3 to 5, linear fleshy, reflexed. *Capsule* with crustaceous fragile pericarp, dehiscing irregularly, 1-celled (by abortion), 1- to 3-seeded. *Seeds* sub-globose, with leathery smooth testa, exalbuminous : the cotyledons large, leafy, thin, crumpled : embryo straight. Distrib. 5 species, of which 4 are Malayan and 1 Cambodian.

Flowers in panicles or solitary, axillary.

    Calyx very accrescent very deeply lobed     1. *S. Mastersii.*
      ,,    slightly accrescent not deeply lobed   2. *S. Kunstleri.*
Flowers in dense axillary glomeruli      ...   3. *S. glomerata.*

1. SCHOUTENIA MASTERSII, King. A tree 60 to 80 feet high : young branches slender, dark-coloured, at first scaly but soon glabrous. *Leaves* thinly coriaceous, ovate-lanceolate, slightly obovate, shortly and bluntly acuminate, the base rounded ; upper surface glabrous, the lower minutely and softly tawny-tomentose ; nerves slightly prominent beneath, about 3 pairs lateral and 1 pair basal : length ·75 to 3·25 in. ; breadth ·4 to 1·1 in.; petiole less than ·1 in. *Flowers* solitary and axillary, or in terminal leafy panicles ; the pedicels from ·35 to ·75 in. according to age, tawny-tomentose, jointed below the middle. *Calyx* membranous, pink, conspicuously veined, at first widely campanulate, ·35 in. long, with 5 shallow teeth becoming with the ripening of the fruit, rotate, flat 1·5 to 2 in. in diam., and 5-angled ; pubescent outside, glabrous within. *Filaments* very slender, longer than the style. *Ovary* obovoid-globose, tawny-tomentose. *Style* stout, three times as long as the ovary, tomentose : stigmas scaly. *Fruit* depressed-globose, ·3 in. in diam., minutely tomentose. *Chartacalyx accrescens*, Mast. in Hook. fil. Fl. Br. Ind. i. 382.

Malacca, Penang, Perak. Distrib. Borneo.

On this plant the late Dr. Maingay founded his genus *Chartacalyx*. The only points, however, in which it differs from *Schoutenia* (as defined by Bentham and Hooker) are the absence of petals and the presence of a stalk to the ovary on the upper part of which the stamens are inserted ; and these appear to me to be, in this order, differences of quite minor importance. Maingay never saw the fruit of this ; but copious fruiting specimens have recently been collected and the fruit is found to be exactly that of *Schoutenia*. As regards the structure of the seeds of

*Schoutenia,* Korthals (the author of the genus) says nothing ; nor does Bennet who (Pl. Jav. Rar. p. 239, t. 46) describes at greater length than Korthals the species *S. ovata,* the only one then known. Bennett neither describes nor figures albumen in the seed. Hasskarl (Retzia 1, 136) describes the seeds as exalbuminous, and I find none in the seeds of these species of which I have been able to examine ripe fruit. The only other known species are *S. ovata,* Korth. from Java ; and *S. hypoleuca,* Pierre (Fl. Cochin-Chine t. 134) from Cambodia.

2. SCHOUTENIA KUNSTLERI, n. sp., King. A tree 60 to 70 feet high : young branches cinereous, rather rough-glabrous. *Leaves* thinly coriaceous, narrowly obovate-oblong or oblanceolate, acute, the margin slightly waved, slightly narrowed to the rounded 3- to 5-nerved base ; upper surface glabrous, shining : lower sub-silvery ; the lateral nerves 4 or 5 pairs, spreading, curving, inter-arching near the margin, prominent on the lower surface as are the basal nerves and the numerous slightly curved transverse veins. *Flowers* crowded towards the ends of the branches, in numerous short few-flowered scurfy-tomentose racemes or cymes : pedicels from ·5 to ·75 in. long, jointed and bracteolate above the base, the bracteole oblanceolate. *Calyx* campanulate, membranous, coloured and veined, stellate-hairy on both surfaces, ·5 to ·75 in. long, according to age, cut to the base into 5 ovate spreading lobes. *Petals* 0. *Stamens* on a slightly elevated torus. *Ovary* sessile, sub-globose, densely tomentose, 5-celled. *Style* longer than the stamens. *Stigmas* 5, short, fleshy. *Fruit* 1-celled, 1-seeded, surrounded by the slightly accrescent persistent calyx.

Perak at elevations of from 300 to 800 feet : King's Collector, No. 3409 : on Ulu Tupa, Wray, No. 2692.

According to the field notes of Messrs. Kunstler and Wray, the calyx is yellow when young, but becomes brown when the fruit ripens.

3. SCHOUTENIA GLOMERATA, n. sp., King. A tree from 40 to 60 feet high : young branches slender, cinereous, minutely pubescent. *Leaves* membranous, glabrous, elliptic-oblong, acute or shortly and bluntly acuminate, the margins slightly waved ; the base broad, rounded or emarginate, 3-nerved, the upper pair of nerves very strong, running to the apex of the leaf and joined to the midrib by numerous prominent curving transverse secondary nerves, all very prominent on the pale silvery shining under surface : length 10 to 15 in., breadth 3·5 to 5·5 in. ; petiole only ·25 in. long, stout, wrinkled. *Cymes* condensed, very crowded, axillary, 1 to 1·5 in. in diam. *Flowers* ·25 in. long and ·3 in. wide, on tomentose rufous pedicels about ·2 in. long. *Calyx* widely

campanulate, densely rufous-tomentose; teeth 5, broadly triangular, sub-erect. *Petals* 0. *Stamens* numerous; the filaments slender, longer than the calyx. *Ovary* ovoid-globose, densely tawny-tomentose, 5-celled : style longer than the stamens : stigmas short, sub-globose. *Fruit* depressed globose, ·75 in. in diam., sparsely stellate-tomentose, becoming glabrous, covered only at the base by the slightly accrescent calyx.

Johore ; on Gunong Panti, King's Collector, No. 159.

### 4. BERRYA, Roxb.

A tree. *Leaves* alternate, ovate, acuminate, glabrous ; base cordate, 5-7-nerved. *Panicles* large, many-flowered, terminal and axillary. *Calyx* campanulate, irregularly 3-5-lobed. *Petals* 5, spathulate. *Stamens* many, inserted on a short torus; anthers didymous, lobes divergent, opening lengthwise. *Staminodes* 0. *Ovary* 3-4-lobed, cells 4-ovuled ; style consolidated, stigma lobed ; ovules horizontal. *Fruit* loculicidally 3-4-valved, each valve 2-winged. *Seeds* pilose ; albumen fleshy ; cotyledons flat leafy, radicle superior next the hilum.—Distrib. The following is the only species.

BERRYA AMMONILLA, Roxb. Hort. Beng., 42. A large tree, glabrous except the inflorescence. *Leaves* membranous, broadly ovate, acuminate, the base slightly narrowed and cordate : both surfaces shining, minutely reticulate : basal nerves 2 or 3 pairs, lateral 5 or 6 pairs : length 4 to 8 in., breadth 3 to 5 in. ; petiole ·75 to 2·75 in. *Panicles* terminal, or in the upper axils, branching, 6 to 10 in. long, scurfy-pubescent : flowers ·35 in. in diam. ; their pedicels slender, ·3 to ·5 in. long. *Petals* longer than the calyx, narrowly oblong, obtuse, glabrous. *Anthers* half as long as the petals. *Ovary* ovoid, truncate, depressed at the origin of the styles, 6 to 8-ridged, pubescent. *Fruit* with 6 radiating, falcate, membranous, striate, deciduously stellate-tomentose wings ·8 in. long. *Seeds* small : 1 to 4 in. each cell, covered with prurient pale brown hairs. Roxb. Fl. Ind. ii. 639 ; Corom. Plants, ii. t. 264; Wall. Cat. 1068; W. & A. Prodr. i. 81; Wight Ill. t. 34; Thwaites Enum. 32; Beddome Flor. Sylvat. t. 58; Kurz Fl. Burm. i. 155; Hook. fil. Fl. Br. Ind. i. 383.

South Andamans. Distrib. Burmah, Southern Peninsula, India, Ceylon.

### 5. GREWIA, Linn.

Trees or shrubs more or less stellate-pubescent. *Leaves* entire, 1-9-nerved. *Flowers* axillary and few, or more numerous and panicled. *Sepals* distinct. *Petals* 5, glandular at the base, sometimes 0. *Stamens* many on a raised torus. *Staminodes* 0. *Ovary* 2-4-celled, cells opposite the petals, 2-many-ovuled; style subulate, stigma shortly lobed. *Drupe* fleshy or fibrous, entire, or 2-4-lobed; stones 1-4, 1-2-seeded, with false partitions between the seeds. *Seeds* ascending; albumen fleshy or rarely 0; cotyledons flat. Distrib. About 60 species, chiefly tropical.

Sect. I. *Grewia* proper. *Flowers* axillary or terminal. *Fruit* fleshy or crustaceous, usually lobed ... ... ... 1. *G. umbellata.*

Sect. II. *Microcos. Inflorescence* terminal, in
panicled cymes. *Flowers* involucrate.
*Drupe* fleshy, entire ... ... 2. *G. paniculata.*

Sect. III. *Omphacarpus. Inflorescence* terminal,
or terminal and axillary. *Flowers* involucrate. *Drupe* with a corky or fibrous rind.

Fruit minutely tomentose: mesocarp thick,
soft, pulpy, and with many fibres; pyrene single, small.
    Pyrene membranous: leaves softly to-
    mentose beneath ... ... 3. *G. fibrocarpa.*
    Pyrene cartilaginous: leaves sparsely
    stellate-hispid beneath ... ... 4. *G. globulifera.*
Fruit glabrous: mesocarp with thin pulp
and a few fibres: pyrenes 2 or 3, bony.
    Leaves sparsely-stellate pubescent be-
    neath: drupe not narrowed into a
    pseudo-stalk... ... ... 5. *G. latifolia.*
    Leaves glabrescent or pubescent be-
    neath: drupe narrowed into a long
    pseudo-stalk... ... ... 6. *G. antidesmæfolia.*
    Leaves quite glabrous.
        Basal nerves bold and reaching be-
        yond the middle.
            Fruit ·5 in. long, furrowed, not
            compressed ... .. 7. *G. laurifolia.*
            Fruit 1·4 in. long, not furrowed,
            compressed ... ... 8. *G. calophylla.*
        Basal nerves slender, not reaching
        to the middle: drupe ·75 in. long 9. *G. Miqueliana.*

1. GREWIA UMBELLATA, Roxb. Hort. Beng. 42 : Fl. Ind. ii. 591.
A shrubby climber 10 to 20 feet long; whole plant except the upper surfaces of the leaves sparsely stellate-puberulous, the bark of the young

branches dark-coloured. *Leaves* oblong-ovate or elliptic, shortly and bluntly acuminate, minutely serrate; base rounded, 3-nerved; upper surface glabrous; the lower pale with the transverse veins prominent and straight: lateral nerves about 3 pairs: length 3 to 4·5 in., breadth 1·5 to 2 in., petiole ·25 in. *Umbels* pedunculate, axillary or terminal, 6 to 8-flowered; the peduncle from ·6 to 1 in. long, with a whorl of small lanceolate glabrous bracteoles at its apex. *Flowers* ·75 in. long when expanded; their pedicels hirsute, unequal, from ·2 to ·5 in. long. *Sepals* ribbed and tomentose outside, glabrous inside, linear-oblong, reflexed. *Petals* much shorter than the sepals, oblong, each springing from the back of a large orbicular claw with hirsute edges. *Torus* long, ridged, tomentose. *Fruit* depressed-globular, obtusely 2- to 4-angled and with 2 to 4 shallow lobes, pericarp sparsely stellate-puberulous; endocarp pulpy; pyrene 2 to 4-celled; its loculi 1-seeded, the endocarp bony. Wight Ic. 83; Wall. Cat. 1084; Mast. in Hook. fil. Fl. Br. Ind. i. 385.

Malacca, Penang, Griffith, Maingay. Perak, King's Collector, Wray.

Roxburgh has left an excellent coloured drawing of this in the library of the Calcutta Herbarium, and there is no doubt about his plant. I cannot agree in identifying with this *G. pedicellata*, Roxb , which that author received from Amboyna: nor do I think that any *Grewia* from the Peninsula of Hindustan is referable to this species.

2. GREWIA PANICULATA, Roxb. Fl. Ind. ii. 591. A bushy tree 15 to 30 feet high: young branches scurfy stellate-tomentose, ultimately glabrous, their bark brown. *Leaves* coriaceous, cuneate-obovate to elliptic; the apex blunt, shortly and abruptly acuminate, sometimes 3-lobed and unequal, obscurely serrate-dentate; the base rounded, 3-nerved; upper surface powdered with minute sparse stellate pubescence, the midrib and nerves tomentose: lower surface uniformly stellate-tomentose; the veins transverse, little curved, bold; lateral nerves 4 or 5 pairs, ascending, rather straight, prominent beneath : length 3 to 6 in., breadth 1·5 to 2·75 in. ; petiole ·25 in., tomentose : stipules glabrescent, lanceolate, often united in pairs, rather shorter than the petioles. *Panicles* 2·25 to 3·5 in. long, terminal or axillary, rusty-tomentose ; bracteoles numerous, linear, sometimes bifid : branches spreading. *Flowers* ·25 in. long, the pedicels rather shorter. *Sepals* spreading, concave, obovate narrowed to the base, the edges thin ; tomentose on the outer, pilose on the inner, surface. *Petals* shorter than the sepals, oblong, blunt, expanded at the base into a concave claw, hirsute especially outside. *Torus* cup-shaped, short, the lip tomentose. *Ovary* ovoid, stellate-tomentose, 4-celled, each cell with several ovules. *Fruit* ob-

ovoid, recurved, with many curved striae, pericarp membranous, minutely and sparsely stellate-pubescent, the mesocarp fibrous with an outer
layer of pulp : pyrene 1-celled, 1-seeded ; endocarp stony. Wall. Cat.
1097, partly ; Miq. Fl. Ind. Bat. i. pt. 2, 203 ; Mast. in Hook. fil. Fl.
Br. Ind. i. 393. *G. Blumei,* Hassk. Tijdschr. Nat. Gesch. xii. 130 ;
Miq. Fl. Ind. Bat. i. pt. 2, 203. *Microcos tomentosa,* Smith in Rees,
Cycl. *G. affinis,* Hassk. Cat. Hort. Bog. 207, not of Lindl.

Singapore ; Malacca, Maingay, No. 250. Griffith, No. 634 (Kew
Distrib.). Perak. Penang ; common.

I retain for this plant the name adopted for it by Masters in
Hooker's Flora of British India. But Blume's *G. paniculata* (Bijdr.
115) was published seven years before Roxburgh's. I have not seen
any specimen of Blume's plant : but if it be the same as this, then
Blume's name must be substituted for that of Roxburgh as the author
of the specific name. If Blume's plant, however, be different from
Roxburgh's, then some other name must be found for the latter. That
the plant above described is what Roxburgh meant to call *G. paniculata,*
his coloured drawing in the Calcutta Herbarium leaves no room for
doubt.

3. GREWIA FIBROCARPA, Mast. in Hook. fil. Fl. Br. Ind. i. 391. A
tree 15 to 40 feet high ; young branches, under surfaces of leaves,
petioles, inflorescence and fruit densely clothed with yellowish-brown
stellate tomentum. *Leaves* membranous, ovate-oblong or elliptic, shortly and abruptly acuminate, minutely and obscurely serrulate, the base
rounded and boldly 3-nerved ; upper surface scaberulous, the midrib
and nerves tomentose, under surface softly tomentose ; the 5 to 7 pairs
of lateral nerves and the transverse veins rather prominent beneath :
length 4·5 to 9 in., breadth 1·75 to 4 in., petiole ·25 to ·5 in., stout :
stipules deeply and narrowly lobed. *Panicles* terminal and in the upper
axils, crowded, ·5 to 2 in. long : involucres lanceolate, curved, tomentose.
*Flowers* ·25 in. long, their pedicels much shorter. *Sepals* obovate-
elliptic, very tomentose externally, the edges inflexed, sparsely piloso
internally. *Petals* minute, sub-orbicular, sometimes absent. *Torus*
short, hirsute. *Ovary* ovoid-globose, tomentose ; the style short, conical,
glabrous. *Fruit* soft, ovoid or obovoid, compressed, 1·25 in. long and
·75 in. in diam., the pericarp membranous and densely tomentose outside, mesocarp fibrous and pulpy ; pyrene small, solitary, leathery, 1-
celled, 1-seeded. *G. paniculata,* Wall. (Cat. No. 1097 partly) not of
Roxb.

Penang ; Wallich, Curtis. Malacca ; Griffith ; Maingay, No. 248,
(Kew Distrib.). Perak ; Scortechini, King's Collector, Wray. Common.

220

In the fruit both of this and of *G. globulifera*, the mesocarp forms a thick pulp with many fibres intermixed, and the solitary pyrene is small with a soft coat.

4. GREWIA GLOBULIFERA, Mast. in Hook. fil. Fl. Br. Ind. i. 391. A small shrubby tree; young branches densely covered with short yellowish-brown tomentum. *Leaves* thinly coriaceous, broadly elliptic, sometimes slightly obovate and unequal-sided, shortly and abruptly acuminate, entire, the base rounded, boldly 3-nerved: upper surface scaberulous, glabrous except the minutely tomentose midrib and nerves : under surface shortly and sparsely stellate-hispid : main nerves 7 to 8 pairs, spreading, prominent beneath, the transverse nerves rather thin, the reticulations minute but distinct: length 4·5 to 10 in., breadth 3 to 6 in.; petiole ·4 to ·75 in., tomentose. *Panicles* often on long peduncles, axillary and terminal, narrow, few-flowered, covered with soft yellowish stellate tomentum : length 2·5 to 4·5 in. (of which the peduncle may be more than half). *Flowers* ·35 in. long, their pedicels much shorter. *Sepals* oblong, spreading, curved inwards, tomentose on both surfaces, the edges much incurved. *Petals* much shorter than the sepals, glabrous, linear-lanceolate, without any distinct claw but sometimes more or less thickened and hairy at the base. *Torus* a very shallow cup with hirsute edge. *Ovary* ovoid, pointed, tomentose; style as long as the ovary, cylindric, glabrous. *Fruit* usually solitary at the apex of a branch of the panicle, sub-obovoid, compressed, 1·25 in. long and ·65 in. in diam.; pericarp membranous minutely tomentose, the mesocarp pulpy and very fibrous; the single pyrene much smaller, endocarp cartilaginous, 1-celled, 1-seeded.

Malacca; Griffith, No. 635; Maingay, No. 245, (Kew Distrib.) ; Harvey. Perak ; Scortechini, King's Collector, Wray : at low elevations.

In its fruit this much resembles *G. fibrocarpa*. The drupe, however, of this is obovoid not ovoid, and the stone is larger with cartilaginous not membranous endocarp. The leaves also differ in being sparsely shortly hispid-pubescent instead of softly tomentose. A near ally of this species is also *G. latifolia*, Mast. from which this differs in its petals having no distinct claw, whereas in those of *G. latifolia* the claw is larger than the limb. This also differs in the shape of its ovary and style, and in the very different appearance of its drupe.

5. GREWIA LATIFOLIA, Mast. in Hook. fil. Fl. Br. Ind. i. 392. A shrubby tree 20 to 40 feet high : young branches rather stout, minutely but harshly tawny-or cinereous-tomentose. *Leaves* coriaceous, drying a dark brown, broadly elliptic, shortly and abruptly sub-acuminate,

entire, slightly narrowed to the rounded 3-nerved base: upper surface glabrescent, the midrib sub-tomentose, lower surface rather sparsely rusty stellate-pubescent: main lateral nerves 5 to 8 pairs, prominent beneath as are the rather straight transverse veins : length 6 to 9 in., breadth 3·5 to 4·5 in. ; petiole ·5 to ·75 in. stout, tomentose. *Panicles* short, axillary or terminal, rusty-tomentose 1·5 to 2·5 in. long and 1 in. or more broad, few-flowered : involucres ovate-lanceolate. *Flowers* 2·5 in. long, their pedicels shorter. *Sepals* oblong, tomentose on both surfaces. *Petals* shorter than the sepals, oblong, acute, the hirsute claw larger than the glabrescent limb. *Torus* cup-shaped, with hirsute margin. *Ovary* depressed-globose : style cylindric, puberulous. *Drupe* obovoid, ·75 in. long and ·5 in. in diam., pericarp at first sparsely pubescent, afterwards glabrous, mesocarp fibrous and pulpy : pyrene single, 1-celled, 1-seeded : endocarp bony.

Malacca ; Griffith, (Kew Distrib.) 638/1 ; Maingay. Perak ; King's Collector, Scortechini, Wray.

6. GREWIA ANTIDESMÆFOLIA, n. sp., King. A tree usually 30 to 40, but sometimes 50 to 60 feet, high : young branches glabrous, their bark cinereous. *Leaves* membranous, glabrescent when young, when old quite glabrous, elliptic-oblong, acute or shortly acuminate, entire, the base usually cuneate but sometimes rounded, boldly 3-nerved ; lateral main nerves 5 or 6 pairs, little curved, ascending, prominent beneath ; length 4·5 to 8 in., breadth 1·5 to 2·75 in. ; petiole ·4 to ·6 in. slender. *Panicles* pedunculate, axillary and terminal, slender, the branches short, spreading, few-flowered, densely but minutely cinereous, velvetty, 2 to 3 in. long. *Flowers* ·25 in. long, their pedicels shorter. *Sepals* elliptic, slightly obovate, their edges in the upper half much incurved, tomentose outside, pubescent inside. *Petals* much shorter than the sepals, oblong, blunt, the glabrescent limb about as long as the broad thickened claw ; claw pilose behind, with hirsute edges in front. *Torus* cylindric, glabrous, with wide wavy hirsute mouth. *Ovary* ovoid-globose, pilose when young, glabrescent when adult, shorter than the cylindric glabrous style. *Fruit* pyriform, obtusely 3-angled, narrowed to a long pseudo-stalk, ·75 in. long (including the narrowed portion) about ·35 in. in diam. ; pericarp glabrous, mesocarp slightly fleshy with a thin fibrous inner layer. *Pyrenes* 3, with bony endocarp, two of them abortive and the third 1-celled, 1-seeded.

Perak : at low elevations ; common, Scortechini, King's Collector, Wray.

Var. *hirsuta ;* young branches, lower surfaces of leaves, and ovary pubescent to tomentose.

Perak ; King's Collector.

**7.** Grewia laurifolia, Hook. in Hook. fil. Fl. Br. Ind. i. 392. A tree 20 to 30 feet high ; all parts except the inflorescence glabrous : young branches with dark-coloured bark. *Leaves* thinly coriaceous, oblong-lanceolate or lanceolate, acuminate or acute, entire ; the base rounded boldly 3-nerved : both surfaces shining ; lateral nerves 1 or 2 pairs, alternate ; length 4 to 6 in., breadth 1·5 to 2·5 in., petiole ·5 to ·7 in. *Panicles* terminal and axillary, 1·5 to 4 in. long, lax, few-flowered, puberulous : bracteoles few, linear, fugaceous. *Flowers* ·2 in. long, their pedicels about ·15 in. *Sepals* ovate, concave, the edges much inflexed, minutely tomentose on both surfaces. *Petals* much shorter than the sepals, oblong, often absent. *Torus* cup-shaped, its rim hirsute. *Ovary* globose, sub-glabrous, 4-celled. *Style* thick, cylindric, tapering, glabrous. *Drupe* ovoid, ·5 in. long, the pedicel about as long, with 1 or 2 vertical furrows, pericarp glabrous, endocarp fleshy and fibrous : pyrene 1 to 3-celled, but usually only one cell containing a single seed : endocarp bony.

Malacca ; Griffith, Maingay. Penang ; Curtis. Perak ; Scortechini, King's Collector. Distrib. Sumatra.

**8.** Grewia calophylla, Kurz Andam. Rep. App. B. iii ; Flor. Burm. i. 157. A tree 20 to 30 feet high : all parts glabrous except the minutely velvetty tawny inflorescence : young branches slender, dark-coloured. *Leaves* thinly coriaceous, shining, ovate-lanceolate to ovate-elliptic, acuminate, entire ; the base rounded or slightly cuneate, 3-nerved ; lateral nerves 3 or 4 pairs, ascending ; transverse nerves slender : length 4 to 7 in., breadth 1·75 to 3 in., petiole ·3 to ·75 in. *Panicles* pedunculate, axillary or terminal, few-flowered, 1·5 to 3 in. long. *Flowers* ·5 in. long, their pedicels very short. *Sepals* narrowly oblong, the edges much incurved, minutely velvetty, much reflexed. *Petals* about half the length of the sepals and much narrower, lanceolate ; the limb subulate ; the claw ovoid, expanded, thick and densely tomentose at the margin. *Torus* cylindric, puberulous outside. *Ovary* ovoid, pointed, style long filiform, both puberulous. *Fruit* obovoid, compressed, 1·4 in. long and ·75 in. in diam. ; pericarp membranous, glabrous, shining ; mesocarp thick, pulpy and fibrous : pyrenes 3, of which one is 2-celled but contains only a single seed, the others abortive ; the endocarp bony. Mast. in Hook. fil. Fl. Br. Ind. i. 392.

Nicobar Islands, Kurz : S. Andaman, Kurz, King.

This is very near *G. laurifolia*, Hook. but has very much larger fruit. A Malacca plant (Griffith, No. 630/2 Kew Distrib.) resembles this in leaves but not in flower. The only specimens which I have seen are too imperfect for determination.

9. Grewia Miqueliana, Kurz, in Flora for 1872, p. 398. A tree 20 to 40 feet high : young branches at first very sparsely and minutely lepidote, afterwards glabrous, the bark dark brown. *Leaves* thinly coriaceous, glabrous, shining, ovate-lanceolate to lanceolate, shortly acuminate, entire, the base cuneate, faintly 3-nerved; both surfaces glabrescent soon becoming glabrous : main lateral nerves 5 or 6 pairs, not prominent ; length 3 to 5 in., breadth 1 to 1·75 in. ; petiole ·2 to ·3 in., scaly-tomentose ; stipules oblong, blunt, oblique. *Panicles* axillary and terminal, lax, few-flowered, sparsely lepidote and puberulous, 1 to 2 in. long. *Flowers* ·3 in. long, their pedicels very short. *Sepals* oblanceolate, acute, the edges inflexed, minutely tomentose. *Petals* much shorter than the sepals, the glabrescent linear acute limb shorter and narrower than the thickened rounded tomentose claw. *Torus* short, cylindric, puberulous with villous edges. *Ovary* globose-ovoid, tomentose, shorter than the cylindric glabrous style, 2-celled. *Drupe* pyriform, ·75 in. long and ·5 in. in diam., glabrous : pericarp smooth, glabrous, shining ; mesocarp fibrous with a little pulp : pyrenes 2, each 1-celled, one 1-seeded, the other barren : the endocarp bony. *Inodaphnis lanceolata*, Miq. Fl. Ind. Bat. Suppl. 357 ; Ann. Mus. Lugd. Bat. iii. 89 ; Meisn. in DC. Prod. xv. 1, 265.

Malacca ; Maingay, (Kew Distrib.) No. 244. Perak ; Scortechini, King's Collector, at low elevations. Dindings ; Curtis, No. 1613. Distrib. Sumatra.

There is an authentic fruiting specimen in the Calcutta Herbarium of Miquel's *Inodaphnis lanceolata* collected in Sumatra. And there is no doubt whatever that Kurz was right in referring the plant to *Grewia*. Miquel founded his genus on specimens without flowers ; and, apparently from the structure of the fruit, he suggested its affinity to *Inocarpus*. Later on he suggested (Ann. Mus. Lugd. Bat. iii. 89) its affinity with the Rosaceous genera *Chrysobalanus*, *Parastemon* and *Diemenia* ( = *Trichocarya*). Meissner in DC. Prod. (l. c.) briefly described the genus at the end of *Hernandiaceae*, but without indicating his opinion as to its proper place. Had these distinguished botanists had an opportunity of examining flowers, they would doubtless have referred it without hesitation to *Grewia*. The practice (fortunately confined to a few authors) of founding genera on specimens without flowers cannot be too strongly condemned.

6. Triumfetta, Linn.

Herbs or undershrubs, generally more or less covered with stellate hairs. *Leaves* serrate or dentate, simple or lobed. *Flowers* yellowish, in dense cymes. *Sepals* 5, oblong, concave. *Petals* 5. *Stamens* 5-35,

224

springing from a fleshy, lobed, glandular torus. *Ovary* 2-5-celled, cells 2-ovuled ; style filiform, stigma 5-toothed. *Fruit* globose or oblong, spiny or bristly, indehiscent or 3-6-valved. *Seeds* 1-2 in each cell, pendulous, albuminous embryo straight, cotyledons flat. Distrib. A genus of about 40 very variable species, mostly tropical weeds.

Fruit tomentose, bristles shorter than itself ... 1. *T. rhomboidea.*

„ villous „ longer „ ... 2. *T. pilosa.*

„ glabrous „ „ „ ... 3. *T. annua.*

1. TRIUMFETTA RHOMBOIDEA, Jacq. DC. Prod. i. 507 Erect, herbaceous or shrubby, annual, glabrous or pubescent. *Leaves* polymorphous, but usually rhomboid, 3-lobed, coarsely and unequally serrate, the upper more or less lanceolate ; length 1·75 to 3 in., breadth nearly as much in the rhomboid, much less in the lanceolate forms ; petioles ·25 to 1·25 in. *Peduncles* short, 4 to 6-flowered. *Flowers* about ·15 in. long, the buds clavate. *Sepals* apiculate : petals oblong, ciliate at the base. *Stamens* 8 to 15. *Fruit* about ·2 in. in diam., globose, tomentose, covered with short glabrous or pubescent hooked spines. Masters in Hook. fil. Fl. Br. Ind. i. 395. *T. angulata*, Lam. Dict. iii. 41 ; Wight Ic. t. 320 ; W. & A. Prodr. i. 74 ; Thwaites Enum. 31 ; Dalz. & Gibs. Bomb. Fl. 25 ; Wall. Cat. 1075, 2, C ; Miq. Fl. Ind. Bat. pt. i. 197. *T. angulata, β. acuminata*, Wall. Cat. 1075 β. *T. Bartramia*, Roxb. Fl. Ind. ii. 463 ; Wall. Cat. 1075, D, E. *T. trilocularis*, Roxb. Fl. Ind. ii. 462 ; Wall. Cat. 1083. *T. vestita*, Wall. Cat. 1078, in part.

In all the provinces : a weed. Distrib. British India, Ceylon, Malacca, Archipelago, China, Africa.

2. TRIUMFETTA PILOSA, Roth Nov. Sp., 233. Erect, herbaceous or shrubby, annual; the whole plant, but especially the young branches and the under surface of the leaves, villous, stellate-tomentose. *Leaves ;* the lower broadly ovate, sometimes 3-lobed ; the upper ovate to ovate-lanceolate, acute or acuminate, unequally and rather coarsely serrate or dentate ; length 2 to 4·5 in., breadth 1 to 1·75 in ; petiole ·5 to 1 in. *Stipules* linear-subulate. *Peduncles* many-flowered, usually shorter than the petiole. *Calyx* ·25 in. long, sparsely hairy. *Petals* spathulate-oblong, nearly as long as the calyx. *Fruit* globular, about ·25 in. in diam., villous, densely covered with spines longer than itself which are hispid below, glabrous above, and usually hooked at the apex. W. & A. Prodr. i. 74 ; Hook. fil. Fl. Br. Ind. i. 394. *T. pilosa*, var. β, Thwaites Enum. 31 ; Dalz. & Gibs. Bomb. Fl. 25 *T. tomentosa*, Wall. Cat. 1078 C. *T. glandulosa*, Heyne Herb. ; Wall. Cat. 1077, 5. *T. polycarpa*, Wall. Cat. 1079. *T. oblongata*, Link Enum. Pl. Hort. Ber. ii. 5 ; Wall.

117

Cat. 1077, 1, 2, 3. *T. orata*, DC. Prodr. i. 507 ? *T. pilosa*, Wall. Cat. 1080. *T. pilosa*, var. α, Thwaites Enum. 31. *T. vestita*, Wall. Cat. 1078, 1, 2. *T. indica*, Ham. in Wall. Cat. 237, 1078 D ; W. & A Prodr. i. 74. *T. oblonga*, Wall. in Don. Prodr. 227.

Malacca, Singapore : Perak, King's Collector, No. 989 ; and probably in all the provinces. Distrib. British India, Ceylon, Africa.

A common and rather variable weed. The bristles of the fruit are usually hooked at the apex ; but in some specimens they are quite straight. The species *T. tomentosa*, was founded by Bojer on specimens collected in Mombassa, having straight fruit-bristles and the lower leaves broadly oval or oblong and often 3-lobed. Many of the Indian forms have been referred to that, but I think they might very well be included in *T. pilosa*, and in the synonymy above quoted I have adopted this view.

3. TRIUMFETTA ANNUA, Linn. Mant. p. 73. Annual, shrubby, erect, 1 to 2 feet high ; the whole plant with sparse pale straight hairs, the older parts glabrescent. *Leaves* thin, ovate-acuminate, coarsely dentate, 3-nerved, 3 to 5 in. long, by 1·5 to 2 broad : petioles nearly 1·5 in. *Stipules* subulate, minute. *Peduncles* axillary, 3-flowered. *Calyx* ·25 in. long, nearly glabrous. *Petals* shorter than calyx. *Stamens* 10. *Fruit* globose, pitted, glabrous, ·2 in. across, bearing numerous smooth glabrous thin hooked spines longer than the capsule. DC. Prod. i. 507 ; Miq. Fl. Ind. Bat. i. pt. 2, 196 ; Hook. fil. Fl. Br. Ind. i. 396. *T. polycarpa*, Wall. Cat. 1079, partly. *T. trichoclada*, Link. ex DC. Prodr. i. 507 ; Wall. Cat. 1082. *T. indica*, Lam. Dict. iii. 420 ?

Perak : a weed. Distrib. British India, Malay Archipelago, Africa.


7. CORCHORUS, Linn.

Herbs or undershrubs, more or less covered with stellate pubescence, or glabrescent. *Leaves* simple. *Peduncles* axillary or opposite to the leaves, 1-2-flowered. *Flowers* small, yellow. *Sepals* 4-5. *Petals* 4-5, glandless. *Stamens* free, indefinite or rarely twice the number of the petals, springing from a short torus. *Ovary* 2-6-celled, style short, stigma cup-shaped. *Capsule* elongated, slender or subglobose, smooth or prickly, loculicidally 2-5-valved, sometimes with transverse partitions. *Seeds* numerous, albuminous, pendulous or horizontal ; embryo curved. Distrib. 35 species, throughout the tropics.

Capsules globular ... ... ... 1. *C. capsularis.*
    „    cylindric, 10-ridged ... ... 2. *C. olitorius.*
    „    „    6-winged ... ... 3. *C. acutangulus.*

226

1. CORCHORUS CAPSULARIS, L. sp. 746. Annual, shrubby, glabrescent. *Leaves* lanceolate or oblong-lanceolate, acuminate, coarsely serrate, the base rounded and with 2 subulate appendages : length 2 to 4 in., breadth ·75 to 1·5 in., petiole ·5 in. or less ; stipules linear-subulate ·25 to ·5 in. *Capsules* axillary, truncate-globose, ridged, wrinkled, sub-muricate, 5-celled. *Seeds* few in each cell. DC. Prodr. i. 505; Roxb. Fl. Ind. ii. 581 ; W. & A. Prodr. i. 73; Wall. Cat. 1071 A, B, C ; Wight. Ic. t. 311 ; Thwaites Enum. 31 ; Dalz. & Gibs. Bomb. Fl. 25; Miq. Fl. Ind. Bat. i. pt. 2, 194; Hook. fil. Fl. Br. Ind. i. 397. *C. Marua*, Ham. in Wall. Cat. 6311.—Rumph. Amb. v. t. 78, f. 1.

Cultivated here and there in all the provinces for its fibre which is known in commerce as " Jute." Doubtfully wild.

2. CORCHORUS OLITORIUS, L. sp. 746. Annual, shrubby, glabrescent. *Leaves* ovate-lanceolate, serrate, the base rounded and with 2 subulate appendages : length 2 to 4 in., breadth ·75 to 2 in., petiole ·75 to 1·5 in., ; stipules linear, ·5 to 1 in. *Capsules* cylindric, 10 ribbed, 5-celled, 2 in. long. DC. Prod. i. 504; Roxb. Fl. Ind. ii. 581; W. & A. Prod. i. 73; Wall. Cat. 1072; 1, 2, 3, 4, D, E, F ; Boiss. Fl. Orient. i. 845; Dalz. & Gibs. Bomb. Fl. 25; Miq. Fl. Ind. Bat. i. pt. 2, 195; Thwaites Enum. 31 ; Hook. fil. Fl. Br. Ind. i. 397. *C. decemangularis*, Roxb. Fl. Ind. ii. 582 ; Wall. Cat. p. 237, 1072 G.

Doubtfully wild : but occasionally cultivated in all the provinces under the name of " Jute."

3. CORCHORUS ACUTANGULUS, Lamk. Dict. ii. 104. Erect, herbaceous, the stems with a broad line of pubescence interrupted and varying in position at the nodes, otherwise glabrous. *Leaves* ovate to ovate-lanceolate, acute or acuminate, serrate, the base rounded, with or without subulate appendages, sparsely hairy on both surfaces ; length 1·5 to 2 in., breadth ·75 to 1·75 in. ; petiole ·25 to ·75 in. slender, villous at the apex : stipules lanceolate, acuminate, ·5 in. long. *Capsules* 1 to 1·5 in. long, cylindric, 6-winged, with 3 terminal bifid beaks, 3-celled. DC. Prod. i. 505; W. & A. Prodr. i. 73; Wall. Cat. 1069, 1074 D, E ; Wight Ic. t. 739; Thwaites Enum. 31 ; Dalz. & Gibs. Bomb. Fl. 25; Miq. Fl. Ind. Bat. i. pt. 2, 194; Hook. fil. Fl. Br. Ind. i. 398. *C. æstuans?* Ham. in Wall. Cat. p. 237, 1074 C. *C. fuscus*, Roxb. Hort. Beng. 42 ; Fl. Ind. ii. 582 ; Ham. in Wall. Cat. 1069.

Johore : at the base of Gunong Panti, King's Collector, No. 180. Distrib. India, Ceylon, Australia, Africa, W. Indies.

8. TRICHOSPERMUM, Blume.

Trees with penni-nerved, minutely stellate, puberulous leaves.

*Flowers* in axillary or terminal, umbellate, stalked cymes or panicles. *Sepals* 5 valvate, thick. *Petals* 5, membranous with a scale at the base. *Stamens* numerous, free, inserted on the inner surface of an annular marginally villous sub-crenate disk ; anthers broad, short, versatile, the connective sub-orbicular. *Ovary* sessile, 2-celled, with numerous ovules on axile placentas : style short, stigma expanded, papillose. *Capsule* orbicular-reniform, much compressed at right angles to the dissepiments, loculicidally 2-valved, many-seeded. *Seeds* sub-lenticular, with a thin imperfect marginally villous arillus ; albumen fleshy ; embryo central the cotyledons orbicular, foliaceous ; radicle straight. Distrib. 3 species 2 of which are Malayan and Polynesian.

1. TRICHOSPERMUM KURZII, King. A tree 40 to 60 feet high : bark of young branches very dark-coloured, sparsely and minutely stellate-pubescent when young, speedily glabrous. *Leaves* membranous, ovate-elliptic, shortly acuminate, minutely serrate-crenate especially near the apex ; the base rounded, sub-truncate, sub-cordate, boldly 3-nerved : lateral nerves about 4 pairs : the transverse veins sub-horizontal, curved, bold : length 4 to 6 in., breadth 2 to 3 in., petiole about ·5 in. *Panicles* solitary, axillary or terminal, stalked, cymose, 2-3-chotomous, much shorter than the leaves when in flower, nearly as long when in fruit, stellate-tomentose. *Sepals* oblong, acute, stellate-tomentose outside, glabrous inside except a tuft of hairs at the base. *Petals* about the size and shape of the sepals, glabrescent, with a fleshy scale at the base and a transverse belt of long hairs above it. *Ovary* sessile, densely villous ; style shorter than the ovary, cylindric, expanding upwards into the broad papillose stigma. *Capsule* about ·75 in. long and slightly wider, emarginate at the apex and crowned by the persistent style : pericarp leathery, villous and dark-coloured ; inside white, shining and glabrous : placentas broad, seeds sessile or shortly stalked, sub-lenticular, the long hairs of the arillus forming a marginal ring. *Bixagrewia nicobarica*, Kurz, Trim. Journ. Bot. for 1875, p. 325, t. 169.

Nicobars : Kurz. Perak ; King's Collector, Wray.

The genus *Trichospermum* was founded by Blume for his single species *T. Javanicum*. The generic definition which I have given above differs from that of Blume (Bijdr. 56), in these respects. Blume describes (1) the æstivation of the sepals as imbricate ; (2) the style as absent ; (3) the stigmas as two and emarginate. The definition also differs from that given by Benth. & Hook. (G. P. i. 236) inasmuch as these authors describe (1) the petals as naked at the base ; (2) anthers oblong ; (3) style almost none ; (4) stigma sessile, retuse ; (5) the apex of the capsule produced into a short thick leathery expansion ; (6) leaves entire.

## 9. ELÆOCARPUS, Linn.

Trees. *Leaves* simple. *Flowers* usually hermaphrodite, rarely polygamous, in axillary racemes. *Sepals* 5, distinct. *Petals* 5, usually laciniate at the apex, rarely entire, springing from the outside of a cushion-shaped, often 5-lobed torus. *Stamens* usually indefinite, never less than 10, arising from the inside of the torus, and more or less aggregated into groups opposite the petals and alternating with the glands of the torus; anthers innate, linear, opening by a terminal pore. *Ovary* sessile, 2-5-celled, cells 2-many-ovuled; style columnar. *Drupe* with a single bony stone which is 3-5 or, by abortion, 1-celled. *Seeds* pendulous, 1 in each cell, albumen fleshy; cotyledons flat. · Distrib. About 50 species chiefly in the Indian Archipelago and India; a few in some of the South Sea Islands, New Zealand, and Australia.

Sect. I. *Ganitrus.* *Ovary* and *drupe* 5-celled, the latter globular.

    Leaves glabescent or glabrous, without stipules.

        Leaves lanceolate ... ... 1. *E. Ganitrus.*

        „ ovate-oblong ... ... 2. *E. parvifolius.*

    Leaves softly rusty-pubescent or tomentose beneath, stipulate .. ... 3. *E. stipularis.*

Sect. II. *Eu-elæocarpus.* *Ovary* 3-celled : longer cell of anthers usually with an apical tuft of minute hair; petals cunei form, fimbriate.

    Leaves pubescent beneath, elliptic-oblong ... ... ... 4. *E. Scortechinii.*

    Leaves glabrescent beneath; the midrib pubescent.

        Leaves ovate to elliptic-ovate, with black dots beneath ... 5. *E. Wrayi.*

        „ narrowly lanceolate, not dotted beneath... ... 6. *E. salicifolius.*

    Leaves quite glabrous everywhere.

        Leaves with rounded bases.

            Petals glabrous ... ... 7. *E. robustus.*

            „ glandular-pubescent ... 8. *E. nitidus,* var. *leptostachyus.*

        Leaves with their bases much narrowed.

            Petals glandular-pubescent : fruit ovoid or slightly ob-ovoid, blunt ... ... 8. *E. nitidus.*

Petals glabrous except on the edges: fruit ovoid-elliptic, slightly apiculate ... 9. *E. floribundus.*

Sect. III. *Monocera.* Outer cell of anther produced into an awn. *Ovary* 2-celled. *Drupe* 1-celled, 1-seeded.

Petals ovate-acuminate, entire ... 10. *E. paniculatus.*

Petals about equally wide at base and apex; the apex toothed ... ... 11. *E. petiolatus.*

Petals wider at the base than the apex, the edges much incurved below the middle, the apex irregularly toothed or fimbriate.

Apex of leaves acuminate.

Racemes longer than the leaves: stamens 35 to 40 ... 12. *E. Griffithii.*

Racemes usually shorter than the leaves: stamens 20 ... 13. *E. Hullettii.*

Apex of leaves obtuse: stamens about 15 ... ... 14. *E. pedunculatus.*

Petals oblong, slightly obovate, apex obtuse with 6 to 8 broad teeth ... 15. *E. Kunstleri.*

Petals cuneiform.

Apex of petals with 8 to 10 rather broad teeth, sometimes 2-lobed: stamens 30 to 50 ... ... 16. *E. obtusus.*

Petals oblong-cuneiform to cuneiform, with numerous fimbriae ... 17. *E. apiculatus.*

Petals broadly cuneiform, lobed and fimbriate ... ... 18. *E. aristatus.*

Sect. IV. *Acronodia.* Flowers 4-merous, polygamous; anthers not awned and usually not bearded (sometimes slightly bearded in *E. glabrescens*).

Leaves sparsely and minutely pubescent or puberulous beneath, their edges serrulate; petals elliptic, the apex slightly lobed ... ... 19. *E. polystachyus.*

Leaves rufous-tomentose beneath, subglabrescent only when very old, edges quite entire, recurved; petals oblong, obtuse, 8 to 10-toothed ... 20. *E. Jackianus.*

Leaves rufous-pubescent on lower surface
when young: ultimately glabrescent
or glabrous ...        ...        ... 21. *E. glabrescens.*
Leaves glabrous at all stages.
Leaves acute narrowed at the base
into the petiole: fruit oblong-
ovoid, ·5 in. long    ...        ... 22. *E. punctatus.*
Leaves acuminate (often caudate)
base not passing into petiole:
fruit ovoid-globose, ·35 in. long... 23. *E. Mastersii.*

1. ELÆOCARPUS GANITRUS, Roxb. Hort. Beng. 42 : Fl. Iud. iii. 592.
A tree 30 to 60 feet high: branchlets with dark bark, cinereously puberulous when quite young. *Leaves* membranous, lanceolate, acute at
base and apex, obscurely serrulate, glabrescent or glabrous: main
nerves 10 to 12 pairs, spreading, slender : length 3·5 to 5·5 in., breadth
1·25 to 2·25 in., petiole ·3 to ·5 in. *Racemes* from the branches below
the leaves, drooping, shorter than the leaves, crowded, many-flowered.
*Flowers* ·35 in. long, narrow and pointed in bud ; their pedicels rather
longer, puberulous. *Sepals* lanceolate, shorter than the petals, puberulous outside, glabrescent inside. *Petals* obovoid, the base thickened,
rounded and puberulous at the edge ; the limb glabrous, laciniate for
more than half its length. *Torus* short, fleshy, wrinkled, pubescent.
*Anthers* about 30 to 35, sessile, slightly pubescent or glabrous ; the
cells slightly unequal, the longer with 1 (or sometimes 2) short white
terminal hairs. *Ovary* ovoid-conic, with deep vertical grooves, minutely tomentose, 5-celled, each cell with about 4 ovules. *Style* much
longer than the ovary, thin, fluted, puberulous or glabrescent, thickened
towards the base. *Fruit* spherical, ·75 to ·9 in. in diam., glabrous,
bluish-purple ; the stone vertically 5-grooved, tubercled, 5-celled, often
only one cell containing a ripe seed. Mast. in Hook. fil. Fl. Br. Ind.
i. 400 ; Kurz Fl. Burm. i. 13; Wall. Cat. 2660 A to D ; Dalz. & Gibs.
Bomb. Fl. 27. *Ganitrus sphærica,* Gærtn. Fruct. ii. 271, t. 139, f. 6 ;
Wight Ic. i. 66.—Rumph. Amb. iii. t. 101. *E. cyanocarpa,* Maing. in
Hook. fil. Fl. Br. Ind. i. 406.

Malacca ; Maingay, No. 263. Penang ; Curtis. Perak ; King's Collector, Scortechini. Distrib. Java ; British India, in damp tropical forests
as far west as Nepal.

I have dissected flowers of the type specimen (Maingay No. 263) of
*E. cyanocarpa,* Maingay, and I can find no difference in them from those
of the type sheets of *E. Ganitrus* in Wall. Cat. Roxburgh's original
drawing of *E. Ganitrus* in Herb. Calc. is wrong as regards the petals,

which it represents as too broad and with too many fimbriæ : otherwise it is an equally exact representation of the Indian plant described by him as *E. Ganitrus*, and of *E. cyanocarpa*, Maingay.

2. ELÆOCARPUS PARVIFOLIUS, Wall. Cat, 2662 A & B. A tree 30 to 50 feet high : young branches at first minutely pubescent, ultimately glabrous greyish-brown and minutely lenticellate. *Leaves* membranous, ovate-oblong, rather bluntly acuminate, serrulate, the base cuneate : upper surface shining, glabrous ; the lower dull of chocolate brown colour, glabrous or glabrescent, the midrib and 5 or 6 pairs of curved ascending nerves pubescent on both ; length 2·5 to 4 in., breadth 1·1 to 1·4 in. ; petiole ·6 to ·75 in., slender, puberulous. *Racemes* from the branches below the leaves, rather shorter than the leaves, the rachis, flower-pedicels and outside of calyx softly and shortly pubescent. *Flowers* ·3 in. in diam., their pedicels about ·1 in., recurved, buds conical. *Sepals* slightly shorter than the petals, lanceolate, puberulous within and 3-nerved. *Petals* cuneiform, slightly nerved, cut half-way down into numerous narrow laciniæ, almost glabrous. *Torus* of 5 distinct, broad, shallow, fleshy, grooved, pale, velvety glands. *Stamens* 15, shorter than the petals, with short filaments ; the anthers scaberulous, cells equal, obtuse, the outer sometimes with 2 or 3 minute pale apical hairs. *Ovary* globose, 5-grooved, 5-celled, sparsely pubescent. *Style* as long as the stamens, cylindric, faintly 5-grooved, glabrescent or glabrous. *Fruit* globose, sometimes ovoid-globose, ·75 to 1 in. in diam. : stone 5-celled, with fertile seeds in only 2 or 3 cells, ovoid, ·7 in. long, bluntly rugose, and with 5 very faint grooves from base to apex. C. Mull. Annot. de fam. Elæocarp. 24 ; Hook. fil. Fl. Br. Ind. i. 401.

Singapore ; Ridley, King's Collector. Malacca ; Griflith, (Kew Distrib.) 684, Maingay, 254. Penang and Singapore ; Wallich, Curtis. Perak ; King's Collector, Scortechini.

3. ELÆOCARPUS STIPULARIS, Blume Bijdr. 121. A more or less rusty-pubescent tree 40 to 70 feet high : young branches thin, minutely tomentose. *Leaves* coriaceous, ovate to oblong-ovate, acute or acuminate : the edges usually entire, slightly recurved when dry, sometimes waved ; the base slightly cuneate, or sometimes rounded : upper surface at first puberulous, ultimately glabrous, the midrib always pubescent : lower softly rusty-pubescent : main nerves 9 to 12 pairs, spreading, interarching close to the margin : length 3·6 to 6·5 in., breadth 1·75 to 2·5 in. ; petiole ·5 to ·75 in., minutely tomentose, not conspicuously thickened at the apex ; stipules halbert-shaped, tomentose, fugaceous. *Racemes* axillary and from the axils of fallen leaves, usually shorter than, but sometimes as long as the leaves ; the rachises, pedicels

and outside of sepals minutely tomentose. *Flowers* 35 in. in diam., their
pedicels ·2 to ·3 in. long; buds sub-globose, obtusely pointed. *Sepals*
ovate-lanceolate, pubescent inside especially towards the base, the midrib thickened. *Petals* longer than the sepals, cunciform, lobed and cut
irregularly half-way into about 25 slightly unequal fimbriae, veined,
glabrous, the edges villous. *Torus* of 5 distinct, fleshy, sub-globose,
puberulous, transversely oblong, truncate, 2-grooved glands. *Stamens*
25, about half as long as the petals : filaments about half the length of
the scaberulous anthers; cells unequal, the longer with (but sometimes
without) an apical tuft of 4 or 5 stiff white hairs. *Ovary* ovoid-globose,
vertically 5-furrowed, tomentose, 5-celled. *Style* twice as long as the
ovary, conic-cylindric, pubescent at the thickened base, glabrescent
above. *Fruit* globose, smooth; ·8 to 1 in. in diam.; pulp thin : stone
very hard, thick, 1-seeded. Miq. Fl. Ind. Bat, i. pt. 2, p. 209; Mast.
in Hook. fil. Fl. Br. Ind. i. 404; Kurz Fl. Burm. i. 170.

Malacca; Griffith, No. 683, Maingay, No. 255, (Kew Distrib.).
Singapore, Malacca, Penang, Perak; very common at low elevations.
Distrib. Java, Sumatra, Borneo, Burmah.

Var. *latifolia*, King. *Leaves* broadly elliptic to elliptic-oblong 5 to
7 in. long and 2·75 to 3·75 in. broad : petioles elongate, 1·5 to 2·75 in.;
stipules lanceolate.

Perak; Scortechini No. 1991, King's Collector, Nos. 4412, 8176,
10786.

4. ELÆOCARPUS SCORTECHINII, n. sp. King.   A tree 30 to 50 feet
high : young branches and stipules as in *E. stipularis. Leaves* elliptic-
oblong otherwise as in *E. stipularis* except that the main nerves are only
8 to 10 pairs, and the under surface is only softly pubescent, not tomentose: length 5·5 to 7·5 in., breadth 2·25 to 3·25 in.  *Flower pedicels*
longer than in *E. stipularis*, and the flowers the same, except that the
ovary is 3-furrowed and 3-celled.  *Fruit* oval, 1 to 1·25 in. long and ·5
to ·75 in. in diam., glabrous and smooth when ripe, 1-celled, 1-seeded by
abortion.

Perak; Scortechini, No. 1481; Wray, Nos. 1376, 1836, 2251;
King's Collector, Nos. 3483, 10303.

This is one of the few plants to which the lamented Father Scortechini gave a manuscript name. He dedicated it to Jack: but as
Wallich's species, dedicated to the same botanist, has long priority, I
name this after my deceased friend. In everything but its 3-celled
ovary and smooth oval fruit it agrees with *E. stipularis*, Bl.

5. ELÆOCARPUS WRAYI, n. sp., King.  A small tree : leaf-buds,

young branches and inflorescence pale tawny-pubescent. *Leaves* ovate to elliptic-ovate, shortly and bluntly acuminate, the margin cartilaginous, sometimes crenate-serrate, the base always entire and rounded ; upper surface glabrous, shining : the lower dull, pale but not glaucous, with scattered black dots, glabrescent except the puberulous midrib and 6 or 7 pairs of rather prominent sub-ascending main nerves; the reticulations distinct, wide ; length 2·25 to 3·75 in., breadth 1·25 to 1·75 in.; petiole ·75 to 1·25 in., pubescent. *Racemes* mostly from the wood below the leaves (a few axillary) more than half as long as the leaves. *Flowers* ·2 in. in diam., their pedicels ·1 in. long or less : buds ovoid, blunt. *Sepals* lanceolate, sub-acute, outside tomentose, inside pubescent and the midrib thickened; the edges not incurved. *Petals* broadly cuneate, glabrous, cut for a third or a fourth of their length into about 25 narrow fimbriae ; the base truncate. *Torus* of 5 distinct, fleshy, oblong, truncate, several-grooved, velvety glands. *Stamens* 20 to 25, shorter than the petals ; filaments less than half as long as the minutely scaberulous anthers ; cells sub-equal, the longer sometimes with 2 or 3 short white hairs. *Ovary* globose, pointed, grooved, tomentose, 3-celled. *Style* slightly longer than the ovary, conic-cylindric, pubescent at the base, glabrescent above. *Fruit* ovoid-globular, glabrous, slightly rugose, 1 to 1·25 in. long when ripe, and ·8 to ·9 in. in diam. : pulp rather thin : stone bluntly rugose : putamen very hard, thick : 1 cell with a solitary seed, the other 2 cells abortive.

Perak ; on Gunong Bubu at 5000 feet elevation ; Wray, No. 3857 : Gunong Batu Patch, Wray, No. 1107; Scortechini, No. 400.

This resembles *E. parvifolius*, Wall. in some respects ; but its leaves have more rounded bases, their nerves are rather more numerous and the petioles longer; the flower buds are blunt and not pointed as in that species, and they are tomentose rather than pubescent; also the stamens are more numerous and the ovary is 3-celled. This is found moreover at much higher elevations than *E. parvifolius* which is found at elevations under 1000 feet.

6. ELÆOCARPUS SALICIFOLIUS, n. sp., King. A tree 30 feet high : young branches puberulous. *Leaves* thinly coriaceous, narrowly lanceolate, slightly oblique : acuminate, serrulate-crenulate except at the entire cuneate base; upper surface glabrous, shining, olivaceous when dry, the midrib puberulous ; lower dull brown when dry, glabrescent, the midrib puberulous : main nerves about 8 pairs, rather straight, sub-ascending, slender ; length 3 to 3·5 in., breadth ·7 to ·9 in. : petiole about ·5 in., puberulous. *Racemes* from the lower axils and from the axils of fallen leaves, nearly as long as the leaves ; the slender rachises,

and pedicels pubescent. *Flowers* ·25 in. in diam., the pedicels ·15 to ·2 in. *Sepals* lanceolate, spreading, hoary adpressed-tomentose outside, pubescent inside, the midrib slightly thickened, the edges not incurved. *Petals* a little longer than the sepals, cuneiform, contracted into a rather narrow claw, divided more than half-way down into about eight 3-fimbriate lobes, glabrescent outside, pubescent inside. *Torus* of 5 distinct, subglobose, fleshy, externally grooved glands. *Stamens* 25, shorter than the petals : filaments half as long as the minutely scaberulous shining anthers : cells subequal, pointed, the upper with a minute apical tuft of short hairs. *Ovary* globose, pointed, tomentose, 3-celled. *Style* longer than the stamens, thick and tomentose at the base, cylindric and glabrous above. *Fruit* unknown.

Singapore ; King's Collector, No. 1207.

This approaches *E. augustifolius*, Bl. but has smaller more pubescent petals, fewer stamens, and less glabrous leaves. It is also closely allied to *E. hypadenus*, Miq., but has not the characteristic rounded stipules of that species, and the leaf-venation is different. It is also allied to *E. parvifolius*, Wall. from which it differs in its narrower leaves with much more slender veins, and also by its 3-celled ovary.

7. ELÆOCARPUS ROBUSTUS, Roxb. Hort. Beng. 42 ; Fl. Ind. ii. 598. A tree 40 to 60 feet high ; young branches rather stout, at first puberulous ; afterwards glabrous, cinereous, lenticellate. *Leaves* thinly coriaceous, ovate-lanceolate to ovate, acuminate or acute, serrate almost to the slightly narrowed rounded rarely cuneate base ; both surfaces glabrous, the upper shining ; the lower dull, slightly paler, the minute reticulations rather distinct and the 10 to 12 pairs of spreading curving nerves rather prominent: length 3·5 to 9 in., breadth 1·75 to 3·5 in. ; petiole 1 to 2·25 in., thickened at the apex. *Racemes* from the branches beneath the leaves, and a few axillary, often nearly as long as the leaves : rachis, pedicels and outer surface of the sepals pubescent. *Flowers* ·5 in. in diam., the pedicels slightly recurved and about ·3 in. long. *Sepals* lanceolate, glabrous inside except the incurved pubescent edges, the midrib thick. *Petals* broadly cuneiform, much contracted in the lower half, the base acute, cut half-way down into about 30 narrow fimbriae, glabrous except the puberulous edges. *Torus* of 5 fleshy, truncate, cushion-like velvety glands. *Stamens* 30 to 50, shorter than the petals, scaberulous ; the filaments curved, about one-fifth the length of the anthers ; cells subequal, the longer with a small tuft of white hair at its apex. *Ovary* ovoid-globose, with about 6 shallow vertical grooves, tomentose, 3-celled. *Style* cylindric, longer than the ovary, shorter than the petals, pubescent in its lower, glabrous

in its upper half. *Fruit* ovoid-globose, 1 to 1·25 in. long : stone oblong-ovoid, rugose, slightly 3-grooved at base and apex, 3-celled. Mast. in Hook. fil. Fl. Br. Ind. i. 402; Kurz Fl. Burm. i. 169; Pierre Fl. For. Coch-Chine, t. 147; Wight Ic. t. 64; Wall. Cat. 2664. *E._ ovalifolius,* Wall. Cat. 2665; C. Müll. Annot. de fam. Elæocarp. 21. *E. amygdalinus,* Wall. Cat. 6857. *E. serratus,* Wall. Cat. 2666 C. *E. oblonga,* Wall. Cat. 2677. *E. aristatus,* Wall. Cat. 2665 B. ? Wall. Cat. 9027. *E. Helferi,* Kurz MSS. ; Hook. fil. Fl. Br. Ind. i. 402.

Penang; Curtis. Pahang; Ridley. Andaman Islands. Distrib. British India, from Burmah to the tropical forests of the E. Himalaya.

8. ELÆOCARPUS NITIDUS, Jack Mal. Misc. Vol. i. No. 2, 41; Hook. Bot. Misc. ii. 84. A tree 25 to 35 feet high; young shoots deciduously pulverulent-pubescent, speedily glabrous as are all other parts except the inflorescence; young branches with blackish bark. *Leaves* thinly coriaceous, oblong-lanceolate to elliptic-oblong, acuminate, crenate-serrulate, (sometimes obscurely so) the base cuneate (rounded in var. *leptostachya) ;* upper surface shining, the lower dull brown; main nerves 10 to 13 pairs, spreading, forming slender arches a little short of the margin : length 4·5 to 9 in., breadth 1·75 to 2·75 in.; petiole 1·25 to 2 in. thickened at the apex. *Racemes* crowded on the old wood below the leaves and rather more than half as long; rachis, flower-pedicels, and exterior of sepals sparsely puberulous. *Flowers* ·35 in. in diam., their pedicels recurved and rather shorter. *Sepals* shorter than the petals, ovato-lanceolate, acute, puberulous and sometimes lenticellate outside, puberulous inside and the midrib very thick. *Petals* cuneiform, finely and irregularly laciniate for nearly half their length, the entire triangular part with thickened nerves and truncate base, glandular-pubescent especially at the edges. *Torus* of 5 truncate, sub-globular, fleshy, to-mentose, cushion-like glands. *Stamens* 15 to 35; the filaments nearly as long as the scabrid obtuse anthers : cells sub-equal, awnless, but sometimes the longer with 2 or 3 small white hairs. *Ovary* globose, slightly pointed, tomentose, 3-celled; style longer than the ovary, slightly thickened below and puberulous. *Fruit* ovoid, or slightly obovoid, smooth, 1·5 in. long, and 1 in. in diam. when quite ripe : stone 3-celled, only one cell bearing a perfect seed. Wall. Cat. 2670 ; Miq. Fl. Ind. Bat. i. pt. 2, p. 208 ; Mast. in Hook. fil. Fl. Br. Ind. i. 401; Wall. Cat., No. 2678 (*E. pedunculatus*) in part.

Penang; Jack, Curtis, No. 282, 463. Perak; King's Collector, No. 4926.

The anthers are sometimes without any terminal hairs : sometimes there are a few. I have seen no authentic specimen of Jack's naming,

236

and nothing that I have dissected quite fits his description of *E. nitidus*, of which he describes the stamens as 15 : whereas in the plants which I refer to this species they vary from 15 to 35. Jack describes the putamen as 5-ridged and 5-celled : I do not find more than 3 cells in the ovary. In spite, however, of these discrepancies, I believe that Jack's specimen above cited belongs to the species which he named *E. nitidus* Wallich's specimen No. 2679 has leaves which do not well answer to Jack's description "attenuate at the base." They are only slightly attenuate, and correspond rather with those of his own species *E. leptostachyus* which is sufficiently distinct as regards the shape of its leaves to be maintained as a variety, though not in my opinion entitled to specific rank.

Var. *leptostachya*. *Leaves* elliptic-oblong to elliptic-rotund, acute, the edge obscurely serrate-crenate, often sub-entire, the base rounded : length 6 to 9 in., breadth 2·75 to 4·5 in. ; petiole 1 in. to 1·75 in., slightly thickened at the apex. *E. leptostachyus*, Wall. Cat. 2672 ; C. Müll. Annot. de fam. Elæocarp. 23 ; Mast. in Hook. fil. Fl. Br. Ind. i. 403.

Penang, Wallich ; Perak ; King's Collector, Nos. 409, 4905, 10105, 10240 ; Scortechini, Nos. 195, 1752 ; Wray, No. 2313.

9. ELÆOCARPUS FLORIBUNDUS, Blume Bijdr. 120. A tree 30 to 40 feet high : young shoots shortly silky ; otherwise glabrous, except the inflorescence. *Leaves* thinly coriaceous ovate-elliptic to oblong-lanceolate or oblanceolate, shortly acuminate, coarsely crenate-serrate, the base much narrowed ; both surfaces shining, with a blistered appearance when dry : main nerves 5 to 7 pairs ; length 3 to 5·5 in., breadth 1·75 to 2·75 in., petiole 1 to 1·5 in., thickened at the apex. *Racemes* usually from below the leaves, sometimes axillary, usually shorter than, but sometimes nearly as long as the leaves ; rachises, pedicels and outside of sepals puberulous. *Flowers* ·4 in. in diam., their pedicels about ·35 in. long. *Sepals* lanceolate, outside glabrescent and often pustulate ; inside glabrous except the pubescent involute edge, the midrib prominent. *Petals* cuneiform, lobed irregularly half-way down, the lobes divided into about 25 fimbriae, glabrous except the pubescent edges, the lower half veined and thickened, often pustulate. *Torus* of 5 distinct, fleshy, oblong, subglobular, truncate, tomentose glands. *Stamens* about 30, shorter than the petals, scaberulous, the filaments very short, the cells slightly unequal, the longer with a small apical tuft of white hair. *Ovary* ovoidglobose, tomentose, 3-celled. *Style* longer than the stamens, cylindric, puberulous in the lower, glabrous in the upper third. *Fruit* 1 in. long, ovoid-elliptic and slightly apiculate when ripe, oblong and much apiculate when young : stone narrowly ovoid tapering to each end, with 3 vertical grooves and many rather shallow large rugæ, 3-celled, one

129

or two of the cells sub-abortive, the walls thick. Mast. in Hook. fil.
Fl. Br. Ind. i. 401 ; Miq. Fl. Ind. Bat. i. pt. 2, 210 ; Kurz Fl. Br. Burm.
i. 167 ; Pierre Fl. Forest. Coch.-Chine, t. 143 ; Miq. Fl. Ind. Bat. i. pt.
2, 210. *E. serratus*, Roxb. (not of L.) Fl. Ind. ii. 596. *E. grossa*, Wall.
Cat. 2661. *E. serratus*, Roxb. ex Wall. Cat. 2666 A, B. partly. *E.
oblongus*, Wall. Cat. 2677 ; C. Müll. Annot. de fam. Elæocarp. 19, f. 30.
*E. Lobbianus*, Turcz. in Mosc. Bull. 1858, 235.

The Nicobar Islands. Distrib. British India through Burmah to
the E. Himalaya, in tropical forests.

There is no doubt that this is the plant which Roxburgh described
as *E. serratus*, Willd.

10. ELÆOCARPUS PANICULATUS, Wall. Cat. 2663. A tree 15 to 30
feet high : all parts glabrous except the inflorescence, young branches
with dark polished bark. *Leaves* thinly coriaceous, lanceolate or oblanceolate-oblong to ovate-oblong, shortly acuminate ; the edges entire,
slightly wavy ; base slightly cuneate, sometimes rounded : both surfaces glabrous, the upper shining ; the lower paler and rather dull, the
reticulations distinct ; main nerves 5 to 7 pairs, sub-ascending, inter-
arching freely within the margin : length 4·5 to 6·5 in., breadth 1·65 to
2·75 in. ; petiole ·8 to 2 in., glabrous. *Racemes* numerous, from the
axils near the apices of the branches, longer than the leaves, erect,
rachises puberulous, becoming glabrous : pedicels spreading, slender,
minutely pubescent, ·5 to ·65 in. long. *Flowers* about ·5 in. in diam. ;
buds ovoid with long narrow points. *Sepals* ovate, acuminate, adpressed-sericeous outside ; glabrous inside except the pubescent infolded
edges. *Petals* not longer than the sepals, ovate acuminate, entire, outside adpressed-sericeous, inside glabrous in the upper villous in the
lower half and especially on the thickened midrib and infolded edges.
*Torus* a shallow fleshy waved sericeous disk. *Stamens* 50, almost
sessile, nearly as long as the petals ; anthers sericeous, the cells sub-
equal, the outer with a rather thick terminal awn. *Ovary* narrowly
ovoid, sericeous, 2-celled. *Style* longer than the ovary, cylindric, glabrous. *Fruit* ellipsoid, blunt at each end, smooth, glabrous, bluish
when ripe, ·4 to ·5 in. long and ·25 to ·35 in. in diam. ; pulp rather
thick, slightly fibrous ; stone bony, minutely tuberculate, 1-celled, 1-
seeded. C. Müll. Annot. de fam. Elæocarp. 12 ; Mast. in Hook. fil. Fl.
Br. Ind. i. 407. *Monoceras leucobotryum*, Miq. Fl. Ind. Bat. Suppl. 409.
*Monocera Griffithii*, Müll. l. c.

Singapore ; Wallich, Anderson. Malacca ; Griffith, Maingay (Kew
Distrib.) No. 257. Perak; Scortechini, King's Collector ; common at low
elevations.

11. ELÆOCARPUS PETIOLATUS, Wall. Cat. 2673. A tree 20 to 40 feet high; all parts glabrous except the inflorescence; young branches dark-coloured, about the thickness of a goose-quill. *Leaves* coriaceous, elliptic to elliptic-oblong, acute or shortly and bluntly acuminate; edges entire: base slightly cuneate or rounded; both surfaces shining, the lower slightly paler when dry, the reticulations sharply distinct on both surfaces: main nerves 7 or 8 pairs, sub-ascending, curving and interarching a little within the margin: length 4·5 to 6·5 in., breadth 2 to 2·75 in.; petiole 1·4 to 2·4 in. slender, dark-coloured, slightly thickened at the apex. *Racemes* numerous from the old wood just below the leaves, shorter than the leaves, rachises and pedicels deciduously puberulous. *Flowers* ·5 in. in diam., their pedicels ·35 in.; buds ovoid, rather abruptly pointed. *Sepals* lanceolate, acuminate, almost glabrous externally; quite glabrous internally, the infolded edges alone pubescent, the midrib thickened from base to apex. *Petals* about as long as the sepals, oblong, the apex cut into 10 to 13 narrow glabrous teeth, the lower two-thirds sericeous, cucullate at the base from the infolding of the edges, a large fleshy villous gland in the middle near the base with a quasi-cell at each side of it, the hairs on the inner surface retroversed. *Torus* a 10-lobed fleshy glabrescent disk. *Stamens* 18 to 25, shorter than the petals, with sericeous or glabrescent flat or sub-cylindric filaments much shorter than the shortly puberulous anthers: apex of anther deeply cleft, the outer cell with a sub-recurved thick awn shorter than the filament. *Ovary* ovoid, pointed, glabrous, 2-celled. *Style* as long as the stamens and much longer than the ovary, cylindric, grooved, glabrous. *Fruit* elliptic, blunt at each end, smooth, ·4 to ·6 in. long, and ·3 in. in diam.: the pulp thin, with very few fibres; stone very slightly rugose, 1-celled, 1-seeded. *Monocera petiolata*, Jack Mal. Misc. i. No. v, 43; ex Hook. Bot. Misc. ii. 86; Cum. et Zoll. in Bull. Mosc. xix, 495. *Monoceras petiolatum*, Miq. Fl. Ind. Bat. i. pt. 2, p. 212; Kurz Fl. Burm. i. 164; Pierre, Fl. Forest. Coch.-Chine, t. 140. *Elæocarpus integra*, Mast (not of Wall.) in Hook. fil. Fl. Br. Ind. i. 408.

Malacca; Griffith No. 699; Maingay, No. 256, (Kew Distrib.); . Derry. Singapore; Hullett, King. Penang; Curtis, No. 383. Perak; Scortechini, King's Collector, Wray, very common at low elevations. Distrib. Sumatra, Beccari, N. S. No. 668.

This is undoubtedly the *Monocera petiolata* of Jack; that it is the *Elæocarpus integra* of Wall. (Cat. No. 2668) I very much doubt. Wallich's No. 2668 was collected in Sillet from which no specimen anything like this has been collected since his day. In fact there is no evidence to show that this species is found in any part of British India (as distinguished from British Malaya), although Kurz includes it in his

Flora of Burmah. This species is a smaller tree than *E. pedunculatus*, which, however, it closely resembles, differing chiefly in the shape of the leaves, the nearly glabrous sepals and in the larger number of stamens. *E. ovalis*, Miq. (a species from Sumatra) must be very nearly allied to this. I have seen only a fruiting specimen of *E. ovalis*, but, except in having leaves of thicker texture and slightly larger fruit, I see little to prevent its being referred here.

12. ELÆOCARPUS GRIFFITHII, Mast. in Hook. fil. Fl. Br. Ind. i. 408. A tree 30 to 40 or over 70 feet high, all parts glabrous except the inflorescence; young branches almost as thin as a crow-quill, dark-coloured. *Leaves* thinly coriaceous, ovate-lanceolate to lanceolate, acuminate, the edges cartilaginous with shallow mucronate crenulations, or subentire with remote marginal black points, the base sub-cuneate or rounded: both surfaces shining, the reticulations minute and distinct: main nerves 5 or 6 pairs spreading, forking and interarching at some distance from the margin, not prominent: length of blade 2·5 to 3·75 in., breadth ·9 to 1·5 in., petiole ·5 to 1 in. *Racemes* from the upper axils, longer than the leaves, rachises and pedicels softly and minutely pubescent. *Flowers* ·5 in. in diam.; pedicels thickened at the apex, ·6 to ·8 in. long: buds ovoid-conic. *Sepals* lanceolate, acuminate, finely adpressed sericeous externally, glabrous internally except the pubescent infolded edges and the thickened sometimes sericeous midrib. *Petals* about as long as sepals, ovate, acuminate, the apex irregularly 2 or 3-toothed with 2 or 3 lateral fimbriæ, outside minutely adpressed-sericeous, inside retroversed hirsute especially on the large gland near the base; edges in the lower two-thirds much infolded so as to form with the gland 2 quasi-cells. *Torus* a shallow, acutely 10-lobed, fleshy disk. *Stamens* 35 to 40, shorter than the petals: filaments short, sericeous as are the unequally 2-celled anthers: outer cell with a tapering awn ¼ to ⅓ of its own length, the inner with a few apical hairs. *Ovary* narrowly ellipsoid, tapering, glabrous except a few silky hairs, 2-celled. *Style* cylindric, grooved, glabrous, longer than the ovary. *Fruit* ellipsoid, blunt at both ends, smooth, ·5 in. long and ·3 in. in diam.: pulp thin with a few fibres; stone slightly rugose, 1-celled, 1-seeded. Kurz in Journ. As. Soc. Beng. pt. 2, for 1870, p. 63; for 1874, pt. 2, 123; For. Flora Burm. i. 164. *Monocera tricanthera*, Griff. Not. pt. 4, 518, t. 619, fig. 3. *Monocera Griffithii*, Wight Ill. i. 84, (not of Müll.). *Monocera holopetala*, Zoll. et Cum. Bull. Mosc. xix, 496. *Monoceras odontopetalum*, Miq. Fl. Ind. Bat. Supp. 409.

Malacca; Griffith, Maingay, No. 257/2 (Kew Distrib.). Perak, at low elevations; King's Collector, Wray. Penang; King's Collector. Distrib. Tenasserim, Helfer, No. 714, Kew Distrib.

13. ELÆOCARPUS HULLETTII, n. sp., King. A tree 30 to 40 feet high : young branches very slender, dark-coloured; all parts glabrous except the inflorescence. *Leaves* thinly coriaceous, lanceolate to ovate-lanceolate, acuminate; edges slightly cartilaginous, entire or remotely and obscurely serrate ; the base cuneate or rounded : both surfaces shining, the reticulations minute, elongate and rather distinct on the lower; main nerves 7 or 8 pairs curving, interarching within the edge, rather faint; length of blade 2·5 to 3 in., breadth ·75 to 1·4 in.; petiole ·65 to ·9 in., slender. *Racemes* from the leaf-axils below the apex, crowded, usually shorter than, but sometimes as long as, the leaves, the rachises glabrescent or puberulous, the pedicels silky puberulous. *Flowers* ·3 in. in diam., their pedicels ·35 in. long. *Sepals* linear-lanceolate, acuminate ; externally adpressed-pubescent ; internally glabrous below, puberulous near the apex and on the infolded edges. *Petals* ovate, concave at the base, narrowed to the 10 to 12-fimbriate apex ; outside glabrous, inside villous on the much-thickened base of the midrib, otherwise puberulous. *Torus* very shallow, deeply 10-lobed, sericeous. *Stamens* 20, slightly shorter than the petals : filaments nearly as long as the minutely scaberulous anthers, outer cell with tapering awn nearly as long as itself. *Ovary* ovoid, pubescent, 2-celled. *Style* as long as the petals and much longer than the ovary, subulate, puberulous below, glabrous above. *Fruit* ellipsoid, blunt at each end, smooth, ·6 in. long, ·35 in. in diam. ; pulp thin, very slightly fibrous : stone rugulose, rather thick, bony, 1-celled, 1-seeded.

Singapore; Hullett, No. 132. Penang; Curtis, No. 1091, King's Collector, No. 1475. Perak, on low hills ; Scortechini, King's Collector.

A species not unlike *E. Griffithii*, Wall. but with smaller flowers and much shorter racemes.

14. ELÆOCARPUS PEDUNCULATUS, Wall. Cat. 2678 in part. A tree 40 to 80 feet high : glabrous except the inflorescence : young branches nearly as thin as a crow-quill, polished, dark-coloured ; their apices and the older branchlets rough and thickened. *Leaves* coriaceous, oblanceolate or narrowly elliptic-oblong, obtuse or slightly narrowed at the apex, the base very cuneate ; the edges cartilaginous, remotely-mucronate crenate-waved, slightly recurved when dry : both surfaces shining : the lower very slightly the paler and with the minute reticulations distinct ; main nerves 5 to 7 pairs, interarching at some distance from the edge ; length of blade 3 to 4·5 in., breadth 1·25 to 1·8 in., petiole ·75 to 1·2 in. *Racemes* axillary but mostly from axils of fallen leaves, 3 or 4 in. long, rachises and pedicels hoary-pubescent. *Flowers* ·3 in. in diam., buds narrowly ovoid, sub-acute ; pedicels recurved, slightly longer than the

flowers. *Sepals* lanceolate, sub-acute, minutely adpressed-sericeous externally, almost glabrous internally except the pubescent inverted edges, the midrib equally thickened from base to apex. *Petals* slightly longer than the sepals, oblong, expanded at the base, the apex broad, cut into 10 to 15 cylindric filiform glabrous fimbriæ one-fourth of the length of the petals : lower part sericeous on both surfaces but especially on the inner (where the hairs are reversed), cucullate with the edges much infolded, and with a large basal gland at each side of which is an imperfect cell. *Torus* a fleshy deeply 10-lobed glabrescent disc. *Stamens* about 15; slightly shorter than the petals, filaments pubescent, less than half the length of the puberulous sub-equal anthers : outer cell with a short sub-recurved awn. *Ovary* ovoid, pointed, smooth, glabrous, 2-celled. *Style* as long as the stamens, cylindric, grooved, glabrous. *Fruit* ellipsoid, blunt at each end, ·5 in. long and ·3 in. in diam., 1-celled, 1-seeded; pulp rather thin, slightly fibrous, stone minutely rugulose, 1-celled, 1-seeded. Mast. in Hook. fil. Fl. Br. Ind. i. 408.

Singapore; Wallich, Ridley. Malacca; Griffith, No. 698, Maingay No. 258 (Kew. Distrib.). Penang; Curtis, No. 256. Perak; Scortechini, King's Collector, Nos. 269, 6907, 10831.

Miquel's *Monocera Palembanica*, from Sumatra, judging from the only authentic specimen which I have seen (and which has no flowers), if not identical with this must be a very closed allied species. Under his Catalogue, No. 2678, Wallich issued two species, the above described as *E. pedunculatus*, and another which is clearly *E. nitidus*, Jack.

15. ELÆOCARPUS KUNSTLERI, n. sp, King. A tree 50 to 70 feet high : young branches as thick as a goose-quill, polished, thickened and rough at the apex : all parts glabrous except the inflorescence. *Leaves* coriaceous, rotund-obovate, the apex broadly obtuse, sometimes with a short broad apiculus, rather abruptly narrowed from below the middle to the acuminate base; both surfaces, shining, glabrous; main nerves about 10 pairs, ascending, interarching freely inside the entire or crenate-serrate edge : prominent beneath ; the reticulations rather faint; length 5 to 8 in., breadth 2·75 to 3·75 in., petiole ·5 to ·7 in., pubescent. *Racemes* crowded from the axils of fallen leaves and a few axillary, less than half as long as the leaves, 6 to 9-flowered : rachises and pedicels slender, puberulous, glabrous when old. *Flowers* ·6 in. in diam., their pedicels ·5 in. or more long. *Sepals* lanceolate, sub-acute, pubescent on both surfaces, the midrib thickened and villous at the base inside. *Petals* about as long as the sepals, oblong slightly obovate, obtuse, thickened in the lower half, the apex with 6 to 8 rather broad teeth, adpressed-sericeous outside, densely villous inside. *Torus* a shallow

toothed villous cup. *Stamens* 28 to 30, shorter than the petals; filaments nearly as long as the minutely scaberulous anthers, swollen in the lower half, the apex of the outer anther-cell with a short recurved awn. *Ovary* ovoid, pointed, tomentose, 2-celled. *Style* cylindric, as long as the petals, puberulous below, glabrous at the apex. *Fruit* unknown.

Perak; at elevations under 1000 feet, King's Collector, No. 8328.

A species near *E. apiculatus*, Mast. but with broader, blunter leaves more abruptly attenuated to the base and quite glabrous, also with smaller flowers.

16. ELÆOCARPUS OBTUSUS, Blume Bijdr. 125. A tree 30 or 40 feet high: young shoots minutely pale pubescent, ultimately glabrous. *Leaves* coriaceous, oblong-obovate, the apex rounded or retuse, gradually narrowed from above or below the middle into the acute or acuminate base; the edges sub-entire or with shallow mucronate crenations; upper surface glabrous, shining; the lower minutely puberulous at first, ultimately glabrous, the reticulations very minute and rather distinct; main nerves 6 to 8 pairs, sub-ascending, not much curved, rather prominent below, scrobiculate at their origin from the midrib; length 4·5 to 6·5 in., breadth 2·25 to 2·75 in., petiole ·75 to 1·1 in. *Racemes* axillary, sometimes from the axils of fallen leaves, less than half as long as the leaves, few-flowered; rachises and pedicels puberulous when young, often nearly glabrous when old. *Flowers* ·9 in. in diam ; their pedicels slender, ·75 to 1 in. long. *Sepals* oblong-lanceolate, outside minutely pubescent, inside sparsely adpressed-sericeous, the midrib thickened. *Petals* longer than the sepals, cuneiform, the base rather broad; the lower third thickened and its edges infolded, the apex with 8 to 10 rather broad teeth sometimes 2-lobed, sericeous on both surfaces but especially on the thickened lower third. *Torus* a wavy, sub-10-toothed, fleshy, sericeous cup. *Stamens* 30 to 50, shorter than the petals: the filaments slender, slightly swollen in the lower half, as long as the pubescent anthers; outer anther cell with a thin tapering awn about as long as itself. *Ovary* ovoid, sericeous or pubescent, 2-celled. *Style* tapering, cylindric, nearly as long as the petals, slightly grooved, puberulous. *Fruit* ovoid, oblong, not pointed, 1·5 in. long, and ·9 in. in diam., smooth: stone boldly tuberculate, 1-celled, 1-seeded. *Monoceras obtusum*, Hassk. Tijds. Nat. Gesch. xii. 136; Miq. Fl. Ind. Bat. i. pt. 2, p. 212. *E. Monoceras*, Cav. (fide Mast. in Hook. fil. Fl. Br. Ind. i. 405). *E. littoralis*, Kurz (not of Teysm. and Binn.) in Journ. As. Soc. Beng. 1874, pp. 132, 182; For. Fl. Burm. i. 167.

Malacca; Griffith, (Kew Distrib.) No. 700. Perak, at low elevations; King's Collector, Nos. 1006, 4671; Scortechini 1396. Pahang; Ridley, 1312. Distrib. Java, Borneo, Sumatra, Burmah.

*E. Monoceras,* Cav. to which Dr. Masters reduces this, was founded by its author on specimens from the island of Luzon. The species, how- ever, is not given in the latest Flora of the Philippines (that of Sig. Vidal); and, as the original description of Cavanilles does not quite agree with the flowers of the Perak specimens, I think it safer not to go farther back than Blume's name, leaving it to be settled hereafter whether *E. obtusus,* Bl. is really the same plant as the Philippine *E. Monoceras.* The Perak plant is closely allied to *E. littoralis,* T. B. (for which Kurz mistook it); and also to the smaller-flowered Sumatran *E. cuneifolius,* Miq.

**17.** ELÆOCARPUS APICULATUS, Mast. in Hook. fil. Fl. Br. Ind. i. 407. A tree 50 to 60 feet high : young branches glabrous, their apices much thickened, rough and puberulous. *Leaves* coriaceous, obovate or ob- lanceolate-oblong, slightly narrowed to the obtuse, sub-acute, or shortly apiculate apex, and much narrowed to the base, the edges sub-entire or with coarse shallow crenations; both surfaces glabrous, shining, the midrib on the lower glabrescent when young ; under surface pale, the reticulations minute, rather distinct ; main nerves 12 to 14 pairs, slightly prominent beneath and interarching freely within the margin, not scrobiculate; length 7 to 10 in. ; breadth 2·5 to 3·75 in., petiole ·3 to 1 in., thickened at the apex. *Racemes* few, mostly from the axils of fallen leaves, usually about a fourth but sometimes half the length of the leaves ; the rachises and pedicels softly pubescent. *Flowers* ·9 in. in diam.; buds oblong, sub-obtuse or pointed, their pedicels ·75 to 1·25 in. *Sepals* oblong-lanceolate, rufous-pubescent outside, glabrous or glabrescent inside, the edge infolded and pubescent, the midrib thickened from base to apex. *Petals* slightly longer than the sepals, oblong- cuneiform to cuneiform, cut from one-fourth to one-fifth of their length into numerous rather broad fimbriæ ; externally adpressed-sericeous in the lower half, glabrous in the upper; internally thickened and villous in the lower, glabrous in the upper, half. *Torus* a shallow fleshy puberulous cup. *Stamens* 30 to 40, half as long as the petals ; filaments shorter than the minutely scaberulous anthers, bulbous at the base : outer anther-cell with short or long apical recurved awn. *Ovary* ovoid, rufous-tomentose, pointed, 2-celled. *Style* as long as the petals, conic-cylindric and pubescent in the lower half, filiform and glabrous in the upper. *Fruit* (fide Masters) " 1 in. long, resembling the fruit of a *Diospyros.*" *Terminalia moluccana,* Wall. (not of Lamk.) Cat. 3969.

Penang ; Wallich. Malacca; Griffith, Maingay, No. 262 (Kew. Distrib.). Perak; Scortechini, King's Collector ; common at low ele- vations.

Allied to *E. Kunstleri,* King and to *E. rugosus,* Roxb. In fact I am inclined to believe that it is merely a form of the latter, from which it should not be separated specifically. Dr. Prain has called my attention to Wallich's sheet No. 3969, which is unmistakeably this species, and has nothing to do with *Terminalia moluccana,* Lamk. which is *T. Catappa,* Linn.

18. ELÆOCARPUS ARISTATUS, Roxb. Hort. Beng.: Fl. Ind. ii. 599. A tree 30 to 60 feet high : young branches of about the thickness of a swan's quill, smooth, thickened and rough towards the apex. *Leaves* thinly coriaceous, obovate, shortly and bluntly apiculate, remotely crenate-serrate, narrowed to the base, glabrous on both surfaces; main nerves 7 to 10 pairs, slender, curving, scrobiculate at the origin from the midrib; length 6 to 8·5 in., breadth 2·75 to 3·75 in., petiole ·5 to ·7 in. *Racemes* axillary and from the axils of fallen leaves, often nearly as long as the leaves, 3 to 5-flowered, rachises and pedicels puberulous or glabrous. *Flowers* nearly 1 in. in diam. ; buds cylindric, pointed ; pedicels ·8 to 1·25, or longer in fruit. *Sepals* as in *E. apiculatus. Petals* also as in *E. apiculatus* but broadly cuneiform, and lobed as well as fimbriate. *Stamens* 50, otherwise as in *E. apiculatus. Ovary* less velvety, but otherwise as in *E. apiculatus. Fruit* ovoid, smooth, 1·25 to 1·4 in. long and ·8 to ·9 in. in diam., pulp rather thick; stone oblong, flattened, pointed at each end, rugose, slightly ridged in the middle of each side, 1 in. long, 1-celled, 1-seeded. Mast. in Hook. fil. Fl. Br. Ind. i. 405. *E. rugosus,* Wall. Cat. No. 2659 (not of Roxb.).

Andaman Islands ; King's Collector. Distrib. Brit. India in Burmah, Chittagong, Sylhet, Assam, Khasia Hills and base of Eastern Himalaya.

This is very closely allied to *E. rugosus,* Roxb.—a species originally discovered by Roxburgh in Chittagong, but specimens of which from that province are very rare in collections. The plants distributed under this name by Wallich as No. 2659 of his Catalogue were not collected there but in Sylhet, while some of them were taken from trees cultivated in the Botanic Garden, Calcutta. They are not *E. rugosus* at all, but *E. aristatus,* Roxb.; and they differ from true *E. rugosus* in having their young branches thinner and smoother; and in leaves which are always glabrous, not so gradually narrowed to the base and with much longer petioles. Their racemes are also more numerous, the petals more broadly cuneiform and the stamens more numerous, (50 as against 30 to 40). The pulp of the fruit is thicker in Andamans specimens of this than in those from Sylhet and Assam ; and the stone is proportionately smaller. There is in Assam and Burmah a plant closely allied to this which has smaller leaves with very large scrobiculæ on the lower surface at the junction of the petioles with the midrib. This has been named *E. simplex* by Kurz, (Fl. Burm. i. 165.) A similar form occurs in Travancore and has been named *E. venustus* by Beddome (Flora Sylvatica, t. 574).

19. ELÆOCARPUS POLYSTACHYUS, Wall. Cat. 2671. A small tree: young shoots rather stout, minutely tawny-tomentose. *Leaves* coriaceous, pale when dry, elliptic to elliptic-oblong, abruptly and shortly acuminate; the edges rather remotely serrulate except at the base, sub-entire when old; the base broad, rounded; upper surface glabrous; the lower sparsely and minutely sub-adpressed puberulous, the midrib pubescent main nerves 7 to 10 pairs, ascending, curving, prominent beneath, the reticulations minute, faint : length 5·5 to 7·5 in., breadth 2·25 to 3·5 in. ; petioles 2·5 to 4 in., minutely tawny-tomentose, slightly thickened at the apex. *Racemes* slightly longer than the petioles ; the rachises, pedicels and outside of sepals densely minutely tawny-tomentose. *Flowers* ·35 in. in diam., their pedicels ·4 in., recurved, buds sub-globose. *Sepals* ovate, acute ; inner surface glabrous, except the pubescent edges, the midrib thickened. *Petals* elliptic, little longer than the sepals, the apex obtuse, sometimes slightly lobed, not fimbriate : villous on both surfaces, the hairs on the inner reversed. *Torus* of 5 retuse thin densely villous glands. *Smensta* half as long as the petals : filaments nearly as long as the hispid-pubescent anthers ; cells subequal, awnless, beardless. *Ovary* ovoid, blunt, densely villous, 2-celled. *Style* about as long as the ovary, puberulous. *Fruit* oblong, blunt, ·6 in. long and ·35 in. in diam., smooth, glabrous ; stone minutely but sharply rugose, 1-celled, 1-seeded, pulp thin and slightly fibrous. C. Müll. Annot. de fam. Elæocarp. 20, f. 13 ; Mast. in Hook. fil. Fl. Br. Ind. i. 403.

Singapore ; Wallich, Hullett, Ridley. Malacca ; Maingay, Nos. 264, 266, (Kew Distrib.).

20. ELÆOCARPUS JACKIANUS, Wall. Cat. 2679. A tree 40 to 80 feet high : young branches stout, densely rufous-tomentose. *Leaves* coriaceous, ovate-oblong to elliptic, rarely oblong-ovoid, shortly acuminate or acute, edges entire, recurved ; the base rounded or slightly narrowed : upper surface rather dull and pale when dry ; glabrous, the midrib alone sometimes pubescent, the lower softly rufous-tomentose, becoming sub-glabrescent when very old, the minute reticulations distinct ; main nerves 8 to 10 pairs, sub-ascending, curving, prominent on the lower, impressed on the upper, surface : length 4 to 7·5 in., breadth 2·5 to 4·5 in. ; petiole 1·75 to 3 in., stout, thickened at each end, tomentose. *Racemes* crowded on the branches below the leaves, sometimes

138

axillary, shorter than the petioles, the rachises pedicels and outside of
sepals softly rufous-tomentose. *Flowers* ·25 in. in diam.; their pedicels
about ·25 in. long, recurved. *Sepals* 4, ovate, acute; inside puberulous
with infolded tomentose edges, the midrib thickened. *Petals* 4, very
little longer than the sepals, oblong, slightly obovate, obtuse, shortly
8- to 10-toothed, villous outside, glabrescent inside, the edges villous.
*Torus* a shallow rufous-villous cup. *Stamens* about 12, shorter than the
petals, scaberulous, the filaments about half as long as the anthers;
anther-cells slightly unequal, pointed, the longer sometimes with, but
usually without, a minute tuft of white hair. *Ovary* (absent in most
flowers) ovoid-oblong, glabrous, imperfectly 2-celled, one cell only
perfect. *Fruit* ovoid, tapering at each end, smooth, shining; pulp thin,
slightly fibrous : stone sharply rugulose, crustaceous, 1-celled. 1-seeded.
*Monocera ferruginea*, Jack Mal. Misc. ex Hook. Bot. Misc. ii. 86.

Singapore; Jack, Kurz. Penang; Curtis, No. 465. Malacca; Griffith,
No. 693; Maingay, No. 259, (Kew Distrib.). Perak; King's Collector;
common at low elevations.

This species approaches *E. glabrescens*, Mast. but is larger in all
its parts and much more persistently tomentose. This is unrepresented
by any Wallichian specimen at Kew, and is therefore referred to by
Masters in the *Flora of Brit. India* only in a note (i. 409).

21. ELÆOCARPUS GLABRESCENS, Mast. in Hook. fil. Fl. Br. Ind. i.
403. A tree : young branches and petioles densely rufous-tomentose.
*Leaves* coriaceous, ovate to ovate-lanceolate, acuminate; edges entire,
slightly revolute; base rounded or slightly narrowed; upper surface
glabrous, the midrib pubescent; lower surface at first rufous-pubescent
ultimately glabrescent or glabrous, the reticulations minute but distinct :
main nerves 5 or 6 pairs, prominent beneath, spreading, curved : length
2·5 to 3·5 in., breadth 1·2 to 1·6 in.; petiole 1·25 to 1·75 in., slightly
thickened at the apex, glabrescent when old. *Racemes* rather longer
than the petioles, axillary and from the axils of fallen leaves; rachises
and pedicels sparsely pubescent. *Flowers* ·25 in. in diam., the pedicels
about ·2 in. long, recurved. *Sepals* 4, sub-erect, ovate-lanceolate, thickened at the base, pubescent outside, glabrescent inside with puberulous
edges, the midrib thickened. *Petals* 4, slightly longer than the sepals,
oblong, the apex obtuse and with 6 to 12 short unequal teeth : pubescent
outside, glabrescent inside, the edges shortly villous. *Torus* a very
shallow villous cup. *Stamens* 10 or 12, shorter than the petals : filaments
short : anthers scaberulous, the cells slightly unequal, slightly pointed,
usually without small apical tufts of minute hair. *Ovary* (absent in
many flowers) ovoid, glabrous, 1-celled. *Style* short, conic, glabrous.
*Fruit* (*fide* Masters) the size of a cherry, 1-celled, 1-seeded.

247

Malacca ; Maingay No. 256 (Kew Distrib.). Penang ; Stoliczka ; on Government Hill at 2,500 feet, Curtis, No. 1092. The Malacca and Penang specimens agree with a specimen at Kew which Miquel has named *E. tomentosus*, Bl. The two species are no doubt close together : but Blume describes the leaves of his *E. tomentosus*, as "setaceous-denticulate" which is not the case here. This tree appears to be uncommon, for Herbarium specimens of it are very few.

22. ELÆOCARPUS PUNCTATUS, King, n. sp. A small tree ; all parts glabrous except the puberulous inflorescence ; young branches thicker than a crow-quill, rough. *Leaves* coriaceous, oblong-lanceolate, acute, the edges cartilaginous, crenate or serrate, sometimes with a short seta on each tooth, the base much narrowed into the petiole, entire : both surfaces shining ; main nerves 8 to 10 pairs, slender but distinct beneath as are the reticulations : length 1·75 to 3 in., breadth ·5 to 1·1 in. ; petiole ·25 to ·3 in., channelled in front. *Racemes* axillary and from the axils of fallen leaves, much shorter than the leaves ; rachises and pedicels puberulous, becoming glabrescent. *Flowers* ·25 in. in diam., their pedicels ·2 in., recurved. *Sepals* 4, oblong-lanceolate, sub-acute, puberulous on both surfaces, the edges thickened and pubescent but not recurved : midrib thickened inside. *Petals* 4, obovoid-oblong, apex obtuse with 5 to 7 short broad teeth, glabrous. *Torus* a shallow wavy pubescent cup. *Stamens* 8 to 12 ; filaments less than half as long as the scaberulous obtuse beardless awnless anthers : the cells sub-equal. *Ovary* ovoid, glabrous, slightly grooved, 2-celled. *Style* about as long as the ovary, cylindric, grooved, puberulous. *Fruit* oblong-ovoid, pointed, narrowed to both ends, glabrous, shining, pale, ·5 in. long and ·25 in. in diam. ; pulp thin, and slightly fibrous ; stone crustaceous, sharply rugose, 1-celled by abortion, 1-seeded. *Elaeocarpus Acronodia*, Mast. in Hook. fil. Fl. Br. Ind. i. 408 *in part*. *Acronodia punctata*, Bl. Bijdr. 123 ; Miq. Fl. Ind. i. pt. 2, p. 213.

Perak ; on Ulu Batang Padang, at 5000 feet, Wray. Malacca. Distrib. Java, Sumatra.

The leaves of specimens from Java and Sumatra are larger than those from Perak and have numerous black dots on the lower surface, whereas those from Perak have no such dots. In other respects the specimens agree : but the Perak material which I have as yet seen is scanty. The plant issued by Wallich as *E. punctatus*, (No. 2676 of his Catalogue) is not the *Acronodia punctata* of Blume, but an altogether different plant. Wallich's specimens are very bad, and Dr. Masters (Fl. Br. Ind. i. 406) suggests that perhaps the leaves are those of a *Pterospermum ;* in reality they belong to a species of *Parinarium*.

23. ELÆOCARPUS MASTERSII, King. A tree 30 to 50 feet high: young branches as thin as a crow-quill, smooth, puberulous; otherwise glabrous except the inflorescence. *Leaves* thinly coriaceous, oblong-lanceolate to ovate-lanceolate, acuminate, often caudate; the edge slightly cartilaginous, remotely and faintly serrate, the base cuneate; both surfaces shining and with the rather transverse reticulations distinct; main nerves 6 to 8 pairs, faint, spreading, interarching within the edge; length of blade 2·75 to 4·5 in., breadth ·8 to 1·4 in,; petiole ·5 to ·75 in., slender. *Racemes* few-flowered, less than half as long as the leaves, from the axils under the apex; rachises and pedicels puberulous, becoming glabrous. *Flowers* ·2 in diam.; buds narrowly ovoid, pointed. *Sepals* 4, ovate-lanceolate, subacute, puberulous or glabrescent outside: glabrous inside on the lower, often puberulous in the upper half and slightly on the infolded edges. *Petals* 4, oblanceolate or narrowly cuneate, the rounded apex with about 15 short teeth, thickened towards the base, veined, glabrous. *Torus* a very shallow wavy pubescent disk. *Stamens* 8 or 9, shorter than the petals, filaments nearly as long as the sub-scaberulous anthers; the cells blunt at the apex, awnless. *Ovary* (absent in many flowers), ovoid, blunt, glabrous, 2-celled. *Style* about as long as the ovary, thick, cylindric, grooved, glabrous. *Fruit* ovoid-globose, the apex slightly pointed, smooth, ·35 in. long and ·25 in. in diam.; pulp thin and without fibres; stone smooth, cartilaginous, 1-celled, 1-seeded. *Elæocarpus Acronodia*, Mast. in Hook. fil. Fl. Br. Ind. i. 401, in part (excl. syn. *Acronodia punctata*, Bl.).

Malacca; Griffith, No. 681; Maingay, No. 261, (Kew Distrib.). Singapore; Hullett, Ridley. Perak; common at low elevations, King's Collector, Scortechini, Wray.

This is a true *Acronodia* allied to *A. punctata*, Bl. ( = *Elæocarpus punctatus*, King, not of Wall.) but is distinguished by its less acuminate longer petiolate leaves, slightly different flowers and smaller, more globose fruit. This occurs at low elevations and is a tree whereas the other is a shrub and is found as high as 7000 feet.

EXCLUDED SPECIES.

ELÆOCARPUS PUNCTATUS, Wall. Cat. 2676 is, (as Kurz pointed out) no *Elæocarpus* but a *Parinarium*. Maingay's Nos. 621 and 621/2 (Kew Distribution) seem to be conspecific with it.

*Materials for a Flora of the Malay Peninsula.—By* GEORGE KING, M. B., LL. D., F. R. S., C. I. E., *Superintendent of the Royal Botanic Garden, Calcutta.*

## No. 4.

As explained in No. 1 of these papers, I was unable to take up the Natural Family of Anonaceæ in its natural sequence. Having now been able to work it out, I present my account of it to the Society. Another of the *Thalamifloral* families (*Dipterocarpeæ*) still remains to be worked out before beginning the *Disciflora.* In the present paper I have followed, for the most part, the arrangement of tribes and the limitations of genera adopted by Sir J. D. Hooker in his Flora of British India; and in most of the instances where I have not done so the fact has been noted.

ORDER IV. ANONACEÆ.

Trees or shrubs, often climbing and aromatic. *Leaves* alternate, exstipulate, simple, quite entire. *Flowers* 2- rarely 1-sexual. *Sepals* 3, free or connate, usually valvate, rarely imbricate. *Petals* 6, hypogynous, 2-seriate, or the inner absent. (*Flowers* dimerous in *Disepalum*). *Stamens* many, rarely definite, hypogynous, closely packed on the torus, filaments short or 0 ; anthers adnate cells extrorse or sublateral, connective produced into an oblong dilated or truncate head. *Ovaries* 1 or more, apocarpous, very rarely (*Anona*) syncarpous with distinct or agglutinated stigmas, style short or 0 ; ovules 1 or more. *Fruit* of 1 or more, sessile or stalked, 1- or many-seeded, usually indehiscent carpels. *Seeds* large ; testa crustaceous or coriaceous ; albumen dense, ruminate, often divided almost to the axis into several series of horizontal plates ; embryo small or minute, cotyledons divaricating.—Distrib. Tropics of the Old World chiefly ; genera about 45 with 500 or 600 species.

Tribe I. UVARIÆ. *Petals* 2-seriate, one or both series imbricate in bud. *Stamens* many, close-packed ; their anther-cells concealed by the overlapping connectives. *Ovaries* indefinite.

*Sepals* imbricate ; trees or shrubs.

Flowers small, globular, scarcely opening ; often uni-sexual and from the

250

older branches or trunk ; ovules 6 to
8, or indefinite.
Trees ; flowers 1-sexual ; ovules
many ; torus conical or hemispheric ... ... ... 1. *Stelechocarpus.*
Trees or shrubs ; flowers unisexual
or hermaphrodite ; ovules 6 to 8 ;
torus flat ... ... 2. *Sageraea.*
*Sepals* valvate ; climbers.
Flowers small, mostly hermaphrodite ;
petals incurved, ovules 6 to 8 ; torus
flat ... ... ... 3. *Cyathostemma.*
Flowers usually large and from the leafy
branches, petals spreading ; torus flat.
Flowers 2-sexual ; ovules many ... 4. *Uvaria.*
Flowers 1- or 2-sexual ; ovules solitary, rarely 2 ... ... 5. *Ellipeia.*
Tribe II. UNONEÆ. *Petals* valvate or open in bud,
spreading in flower, flat, or concave at the base
only ; inner subsimilar or 0. *Stamens* many,
close-packed ; their anther-cells concealed by
the overlapping connectives. *Ovaries* indefinite.
Flowers trimerous.
*Petals* conniving at the concave base and covering the stamens
and ovaries.
Ovaries 1–3, many-ovuled ; peduncles not hooked ... ... 6. *Cyathocalyx.*
Ovaries many, 2-ovuled : peduncles
hooked ... ... ... 7. *Artabotrys.*
Ovaries many ; ovules 4 or more ;
peduncles straight ... ... 8. *Drepananthus.*
*Petals* flat, spreading from the base.
Ripe carpels indehiscent.
Ovules many, 2-seriate ; petals
lanceolate, stamens with acute
apical appendage ... 9. *Canangium.*
Ovules 2–6, 1-seriate on the
ventral suture ... ... 10. *Unona.*
Ovules 1–2, basal or subbasal... 11. *Polyalthia.*
Ripe carpels follicular ... 12. *Anaxagorea.*
Flowers dimerous ... ... ... 13. *Disepalum.*

3

Tribe III. MITREPHOREÆ. *Petals* valvate in bud,
outer spreading; inner dissimilar, concave, connivent, arching over the stamens and pistils,
(divergent in some *Mitrephoras*). *Stamens* many,
(few in *Orophea*), closely packed; anther-cells
(except in *Orophea*) concealed by the overlapping
connectives. *Pistils* numerous (few in some *Oropheas*).

Inner petals clawed.
  Inner petals connivent in a cone, but
    not vaulted ... ... 14. *Goniothalamus.*
  Inner petals vaulted,
    Stamens about 6, Miliusoid; inner
      petals longer than the outer ... 15. *Orophea.*
    Stamens numerous, Uvarioid; inner
      petals not longer or very little
      longer than the outer ... 16. *Mitrephora.*
  Inner petals not clawed.
    Flowers globose; petals subequal ... 17. *Popowia.*
    Flowers elongate; inner petals much
      shorter than the outer ... ... 18. *Oxymitra.*

Tribe IV. XYLOPIEÆ. *Petals* valvate in bud, thick
and rigid, connivent; the inner similar but smaller,
rarely 0.
    Outer petals broad; torus convex ... 19. *Melodorum.*
    Outer petals narrow, often triquetrous;
      torus flat or concave ... ... 20. *Xylopia.*

Tribe V. MILIUSEÆ. *Petals* valvate in bud, the
outer sometimes very small like the sepals. *Stamens* often definite, loosely imbricate; anther-
cells (except in *Phœanthus*) not concealed by
the connectives. *Ovaries* solitary or indefinite.
Ovaries indefinite.
  Sepals and outer petals similar and minute;
    inner petals very large, often cohering by
    their edges.
      Ovules 1 or 2: stamens numerous,
        quadrate, with broad truncate
        apical processes concealing the
        anther-cells from above ... 21. *Phœanthus.*

252

Ovules 1 or 2, rarely 3 or 4; stamens
few or numerous, compressed, the
apical process of the connective
compressed, rot broad or truncate,
and not concealing the anther-
cells from above ... ... 22. *Miliusa.*
Petals larger than the sepals, often saccate
at the base, subequal or the inner smaller 23. *Alphonsea.*
*Ovaries* solitary.
Outer petals valvate, inner imbricate ... 24. *Kingstonia.*
All the petals valvate ... ... 25. *Mezzettia.*

## 1. STELECHOCARPUS, Blume.

Trees. *Leaves* coriaceous. *Flowers* diœcious, fascicled, on the old
wood. *Sepals* 3, small, elliptic or orbicular, imbricate. *Torus* conical.
*Stamens* indefinite; connective dilated, truncate. *Ovaries* indefinite,
ovoid; stigma sessile; ovules 6 or more. *Ripe carpels* large, berried,
globose, 4–6-seeded.—Distrib. Species 3 or 4, all Malayan.

Leaves pellucid-punctate ... ... ... 1 *S. punctatus.*
Leaves not pellucid-punctate.
   Flowers of both sexes alike ... ... 2 *S. nitidus.*
   Male flowers smaller than the female ... 3 *S. Burahol.*

1. STELECHOCARPUS PUNCTATUS, King n. sp. A tree 20 to 30 feet
high: young branches slender, cinereous-puberulous, becoming glabrous.
*Leaves* membranous, minutely pellucid-punctate, elliptic-ovate, shortly
acuminate, slightly narrowed in the lower fourth to the rounded sub-
oblique base: upper surface shining, glabrous except the pubescent
impressed midrib; lower surface shining, paler than the upper, sparsely
puberulous or glabrous, the reticulations minute and distinct: main
nerves 12 to 14 pairs, bold and prominent on the lower, slightly impressed on the upper, surface: length of blade 7 to 10 in., breadth 3 to 4 in.;
petiole ·15 to ·2 in., stout, pubescent. *Male flowers* in several-flowered
fascicles from woody tubercles on the trunk, pedunculate: buds turbinate, nearly ·5 in., in diam.; peduncles 1 to 1·5 in. long, stout, thickened
upwards, ebracteolate, puberulous. *Sepals* very coriaceous, rotund,
concave, conjoined at the base, spreading, rugose, pubescent outside,
glabrous inside. *Petals* very coriaceous, rotund, concave, glabrous;
the outer 3 puberulous outside; the inner three smaller than the outer,
quite glabrous, otherwise like them and all of a dark brownish colour.
*Anthers* sessile, flat, the cells elongate on the anterior surface, the back
striate: apex without any appendage from the connective. *Female
flowers* and *fruit* unknown.

Perak ; King's Collector, No. 7183.

Although female flowers and fruit of this have not yet been found, I describe it as a new species of *Stelechocarpus* without any hesitation. Its male flowers have exactly the facies of those of *S. Burahol*, Bl. ; but they are larger. They, however, differ as to shape of petals ; the leaves of this species are distinctly pellucid-punctate (while those of *S. Burahol* are not) and they are broader and have slightly more nerves than those of *S. Burahol*. When boiled, the flowers of the two have exactly the same peculiar sweetish smell.

2. STELECHOCARPUS NITIDUS, King, n. sp. A tree 30 to 60 feet high ; all parts glabrous except the inflorescence : young branches darkly cinereous, slender. *Leaves* coriaceous, oblong-lanceolate, shortly acuminate, the base acute ; both surfaces shining, very minutely scaly, the midrib and nerves deeply impressed on the upper, bold and prominent on the lower ; the reticulations distinct on both : main nerves 10 to 12 pairs, curved, sub-ascending, inter-arching within the edge : length of blade 6 to 9 in., breadth 1·8 to 3·25 in., petiole ·35 in. *Male flowers* in many-flowered fascicles from tubercles on the trunk, pedicellate ; buds turbinate ; flowers when open probably nearly 1 in. in diam. : pedicels stout, thickened upwards, 1 to 1·5 in. long, scurfy-puberulous, each with several sub-rotund glabrous bracteoles mostly near its base. *Sepals* very coriaceous, shortly oblong, obtuse, concave, spreading, conjoined at the base, puberulous or glabrescent, warted externally. Outer 3 petals much larger than the sepals and somewhat larger than the inner 3 petals, rotund, concave, very coriaceous, glabrous, with scurfy warts externally near the middle : inner 3 petals coriaceous, rotund, blunt, cucullate, glabrous. *Female flowers* like the males, stamens none : *Ovaries* very numerous, obscurely 3-angled, adpressed-sericeous. *Torus* hemispheric. *Ripe carpels* broadly ovoid, blunt, 2·5 in. long, 1·75 in. in diam., puberulous, minutely warted ; pericarp thick, fleshy. *Seeds* about 8 in 2 rows, flattened, 1·25 in. long, and ·5 in. thick.

Perak ; in dense forest at low elevations, King's Collector, Nos. 7629 and 8224.

This species has the flowers of both sexes alike. The carpels of this species are much larger than those of *S. Burahol*, Bl. ; and its leaves are more thickly coriaceous and shining, the nerves and midrib being much more depressed on the upper and prominent on the lower surface.

3. STELECHOCARPUS BURAHOL, H. f. and T. Fl. Ind. 94. A tree 20 to 60 feet high : young branches slender, dark-coloured, glabrous. *Leaves* thinly coriaceous, oblong-lanceolate, acute or very shortly acuminate, the base cuneate : both surfaces glabrous, shining, the reticulations minute and distinct, the lower with minute black dots,

the upper with very minute scales; main nerves 10 to 12 pairs, sub-ascending, prominent, inter-arching ·2 in. within the margin; length of blade 5 to 8 in.; breadth 1·75 to 2·75 in.; petiole ·3 to ·9 in. *Male flowers* much smaller than the female (only about ·4 in. in diam.), in fascicles of 8 to 16 from minutely bracteolate woody tubercles from the branches and trunk, pedicellate ; the pedicels slender, ebracteolate, tomentose, from ·5 to ·75 in. long. *Sepals* coriaceous, triangular, spreading. *Petals* much longer than the sepals, oblong, sub-acute, warted, pubescent inside : anthers with obtuse terminal, dilated, 2-lobed apical appendages from the connective ; ovaries 0. *Female flowers* three times as large as the males, and on similar pedicels; calyx not persistent ; corolla as in the male. *Ovaries* numerous, on an ovoid-conic torus, oval or obovate, the outer surface compressed, the inner with a vertical ridge and adpressed, pale hairs ; stigma sessile, minutely lobed. *Fruit* on stout peduncles 2 to 3 in. long, thickened upwards. *Ripe carpels* few, shortly stalked, globose, obovate, about 1·5 in. long, and 1·25 in. in diam. ; when young puberulous, verrucose, afterwards nearly smooth ; pericarp pulpy, coriaceous externally. *Seeds* 4 to 6, large, oval, sub-compressed, sub-rugose. Hook. fil. Fl. Br. Ind. I, 47. *Uvaria Burahol*, Blume Bijdr. 14 ; Floræ Javæ Anon. 48, t. 23, and 25 C.; Scheff. in Nat. Tijdsch. Ned. Ind. **XXXI**, 5.

Singapore ; Lobb. Distrib. Java.

There is sometimes a remarkable difference in the length of the petioles in this species, some of those *on the same specimen* being three times as long as others.

## 2. SAGERAEA, Dalz.

*Trees. Leaves* shining, and branches glabrous. *Flowers* small, axillary or fascicled on woody tubercles, 1–2-sexual. *Sepals* orbicular or ovate, imbricate. *Petals* 6, imbricate in 2 series, nearly equal, usually orbicular, very concave. *Stamens* 6–21, imbricate in 2 or more series, broadly oblong, thick, fleshy ; anther-cells dorsal, oblong ; connective produced. *Ovaries* 3–6 ; style short, stigma obtuse or capitate ; ovules 6 to 8, on the ventral suture. *Ripe carpels* globose or ovoid, stalked.— DISTRIB. Species 6, tropical Asiatic.

A genus closely allied to *Bocagea*, St. Hilaire, but differing from that in having its sepals and petals much imbricate instead of valvate ; in bearing more ovules, and more seeds in its ripe carpels ; in its anther-cells being more lateral and not so entirely dorsal as in *Bocagea*, and in the apical process of the connective being truncate. The flowers of *Sageraea* are small and the sepals and petals are very concave ; and in these respects, as well in the comparative fewness of the seeds in their ripe carpels, they diverge from those of typical *Uvarieæ*. Hooker filius and Thomson (in their Flora Indica), Bentham and Hooker (in their Genera Plantarum), and Baillon (in his Histoire des Plantes, Vol. I, 202, 281) retain *Sageraea* as a genus,—an example which I would have followed without any hesitation had not Sir Joseph Hooker united it with *Bocagea* in his Flora of British India. The extreme imbrication both of the sepals and petals appears to me however, in spite of Sir Joseph Hooker's more recent view, so insurmountable an argument against its reduction to a genus in which both these sets of organs are very distinctly valvate, that I adhere to the earlier view that *Sageraea* should remain distinct and be put in the tribe *Uvarieæ*.

1. SAGERAEA ELLIPTICA, Hook. fil. and Thoms. Fl. Ind. 93. A large tree ; all parts glabrous except the ciliate petals ; young branches rather stout, angled. *Leaves* coriaceous, narrowly oblong, acute (obtuse, when very old) ; the base narrowed, obtuse or minutely cordate, oblique : both surfaces shining ; main nerves 14 to 16 pairs, spreading, faint; length 8 to 12 in., breadth 2·25 to 3·5 in. ; petiole ·15 in , very thick. *Flowers* monoecious, solitary and axillary, or fascicled on tubercles on the larger branches, small, red : pedicels ·25 in. long, with several basal and medial bracts. *Sepals* small, semi-orbicular, glabrous, ciliate. *Petals* thick, ovate-orbicular, concave, tubercular outside, glabrous, the edges ciliate, ·25 in. long; the inner smaller than the outer. *Stamens* 12 to 18, the connective sub-quadrate at the apex ; anthers extrorse. *Ovaries* in female flower about 3, glabrous ; ovules about 8. *Ripe carpels* sub-sessile, globose, glabrous, 1 in. in diam., seeds several. *Sageraea Hookeri,* Pierre Flore Forest. Coch-Chine t. 15. *Bocagea elliptica,* H. f. and Th. Fl. Br. Ind. I, 92 ; Kurz F. Flora Burma, I, 50. *Uvaria elliptica, A. DC.* in Mem. Soc. Genev. v. 27 ; Wall. Cat. 6470, 7421. *Diospyros? frondosa,* Wall. Cat. 4125.

Burmah to Penang.

An imperfectly known species, badly represented in collections.

### 3. CYATHOSTEMMA, Griffith.

Scandent shrubs. *Flowers* subglobose in di- or tri-chotomous pendulous cymes from the old wood (flowers dimorphous in sp. 3.) *Sepals* 3, connate, hirsute. *Petals* 6, 2-seriate, short, sub-equal, their bases fleshy, all valvate at the base, the tips imbricate. *Torus* flat, margin convex. *Stamens* many, linear ; anthers sub-introrse ; process of connective oblique, incurved. *Ovaries* many ; style cylindric, glabrous, notched ; ovules many. *Ripe carpels* oblong-ovoid, many-seeded.

The petals in this genus are so unmistakeably imbricate in æstivation, that I remove it from the tribe *Unoneæ* to *Uvariæ*. The ripe carpels moreover much resemble those of some species of *Uvaria*. Of the five species described below, three are quite new. The first (*C. viridiflorum*) is the plant upon which Griffith founded the genus; while the fourth has been hitherto referred to *Uvaria* under the specific name *U. parviflora*. Flowers uniform and hermaphrodite.

Flowers in more or less elongated pendent
    cymes
    Leaves oblong-lanceolate or oblanceolate;
      inner petals contracted at the base ... 1. *C. viridiflorum.*
    Leaves obovate-elliptic to obovate-oblong;
      petals not contracted at the base ... 2. *C. Scortechinii.*
Flowers in stem-fascicles of 10 to 14, or in
    axillary pairs; leaves with pubescent
    midribs ... ... ... 3. *C. Wrayi.*
Flowers in 2- or 3-flowered extra-axillary
    or leaf-opposed fascicles or cymes:
    leaves quite glabrous ... ... 4. *C. Hookeri.*
Flowers dimorphous, the females with a few abor
    tive anthers ... ... ... 5. *C. acuminatum.*

1. CYATHOSTEMMA VIRIDIFLORUM, Griff. Notulæ IV, 707: Ic. Pl. IV, t. 650. Scandent (?) the young branches thin, glabrous, dark-coloured when dry. *Leaves* coriaceous, oblong-lanceolate or oblanceolate, apiculate; the base slightly narrowed, minutely cordate: both surfaces rather dull; the upper glabrous except the minutely tomentose midrib; the lower darker, puberulous on the midrib and 8 to 10 pairs of rather prominent spreading main nerves; length 4·5 to 7·5 in., breadth 1·5 to 2 in., petiole ·2 in. *Cymes* dichotomous, on peduncles several inches long from warty tubercles on the older roughly striate branches, few-flowered, corymbose, minutely rusty-tomentose, with an oblong bract at each bifurcation and another about the middle of each pedicel. *Flowers* ·5 in. in diam. *Sepals* broadly cordate, spreading or sub-reflexed. *Petals* acute, the base contracted especially in those of the inner row, coriaceous, tomentose. *Ovaries* tomentose. *Ripe carpels* stalked, 1 to 1·5 in. long, oblong-ovoid, blunt, glabrous; stalk ·75 in. Hook. fil. Fl. Br. Ind. I, 57; Kurz For. Fl. Burm. I, 33.

Eastern Peninsula; Griffith. Penang; Maingay (Kew Distrib.) No. 36.

A species known by only a few imperfect specimens. According to Griffith, the wood of this species resembles that of a Menisperm.

9

Kurz gives this as a native of the Andamans ; but I have seen no specimen from those islands.

2. CYATHOSTEMMA SCORTECHINII, n. sp. King. A climber 50 to 70 feet long : branches of all ages, except the very youngest, dark-coloured, glabrous ; the very youngest slender and rufous-pubescent. *Leaves* coriaceous, obovate-elliptic to obovate-oblong, shortly apiculate, slightly narrowed to the sub-cuneate, not cordate, base ; upper surface rather dull, glabrous except the minutely pubescent midrib ; lower glabrous, the midrib slightly muriculate, the reticulations fine, distinct : main nerves 8 to 11 pairs, prominent beneath : length 6 to 10 in., breadth 2·5 to 4 in., petiole ·25 in. *Cymes* di- or tri-chotomous, on pedicels 2 to 12 in. long from the older branches ; minutely rufous-tomentose, bracteate in the upper half ; the bracts numerous, ovate to rotund, concave. *Flowers* ·5 in. in diam. *Sepals* sub-rotund, united into an obscurely 3-angled flattish cup. *Petals* equal, not much longer than the stamens, sub-rotund, puberulous, coriaceous. *Connective of stamens* produced at the apex, obliquely truncate. *Ovaries* numerous, cylindric, pubescent : stigmas truncate : ovules numerous. *Ripe carpels* oblong, slightly oblique, apiculate, transversely furrowed, glabrous, shortly stalked, 1·25 to 1·5 in. long ; pericarp thin. *Seeds* 8 to 10, flattened, ovoid, smooth.

Perak ; Scortechini, King's Collector, No. 5857. Singapore : Ridley.

The specimens collected by the late Father Scortechini were referred by him to *Cyathostemma viridiflorum*, Griff., from which species however, this differs by its larger, more obovate, more glabrous, leaves ; flat calyx-cup formed by the entirely connate sepals ; more rotund petals, not contracted at the base ; and narrower shorter-stalked fruit.

3. CYATHOSTEMMA WRAYI, King n. sp. A creeper 20 to 60 feet long : young branches rufous-puberulous, ultimately glabrous and darkly cinerous. *Leaves* membranous, broadly oblanceolate, shortly and rather obtusely acuminate, narrowed below the middle to the rounded base ; both surfaces finely reticulate, the upper dull when dry, glabrous ; the midrib minutely pubescent ; lower surface shining, glabrous except the sparsely puberulous midrib ; main nerves 8 to 9 pairs, oblique, forming double arches inside the margin, impressed on the upper, prominent on the lower surface ; length 7 to 9 in., breadth 2·5 to 3 in., petiole ·2 in. *Flowers* in fascicles of 10 to 14 from tubercles on the older branches, or in pairs from the axils of the leaves, sub-globular, about ·5 in. in diam. ; pedicels ·25 to ·4 in. long, granular, sparsely pubescent and with a small ovate bracteole near the base. *Sepals* broadly ovate, spreading, rufous-puberulous and granular outside, glabrous inside, ·1 in. long. *Petals* concave, cartilaginous, slightly imbricate,

258

minutely puberulous especially towards the edges; the outer row ovate-orbicular, sub-acute, ·35 in. long; the inner row smaller, thicker, blunter and more imbricate than the outer. *Stamens* numerous; the connective with a rather thick truncate, 4- or 5-sided apical process concealing the apices of the linear dorsal anthers. *Ovaries* numerous, obliquely oblong, curved, glabrous, pubescent at the base, 1 to 2-ovuled, with a conical, narrow, inflexed stigma. *Ripe carpels* reddish, ovoid, ·4 to ·6 in. long, glabrous, with a single ovoid or 2 plano-convex shining pale brown seeds : stalks about as long as the carpels, slender.

Perak; Scortechini, Wray, King's Collector.

4. CYATHOSTEMMA HOOKERI, King n. sp. A climber 40 to 80 feet long; all parts, except the inflorescence, quite glabrous. *Leaves* membranous, broadly oblanceolate to oblong or ovate-elliptic, acute or very shortly and obtusely acuminate, the base rounded or sub-cuneate; both surfaces shining, glabrous, minutely reticulate; main nerves 9 or 10 pairs, spreading or ascending, curving, inter-arching within the edge; length 5·5 to 7 in., breadth 2·25 to 2·75 in., petiole ·3 in. *Flowers* ·25 in. in diam., sub-globose, in extra-axillary or leaf-opposed fascicles or cymes of 2 or 3; pedicels slender, puberulous, ·3 to ·4 in long with 1 or 2 bracteoles. *Sepals* spreading, broadly and obliquely ovate, sub-acute, slightly thickened at the base, ·1 in. long. *Petals* concave; the outer row slightly longer than the sepals but narrower, obovate, contracted into a pseudo-claw at the base, sparsely puberulous outside; the inner row narrower, thicker, and more concave, oblique. *Stamens* numerous, short, with a thick incurved apical process from the connective; anther cells dorsal. *Ovaries* numerous, oblong, thickened upwards, puberulous; the stigma large, sub-quadrate, slightly 2-lobed. *Ripe carpels* numerous, oblong to ovoid, blunt at each end, glabrescent, ·75 to 1·75 in. long and ·6 to ·9 in. in diam.; stalk 1·5 to 2 in. stout. *Seeds* 6 in a single row, compressed, oblong, pale brown, shining. *Uvaria parviflora*, Hook. fil. and Thoms. Fl. Ind. 103 ; Fl. Br. Ind I, 51.

Penang; Phillips, Curtis. Perak; Scortechini, Wray, King's Collector.

For upwards of seventy years this plant had been known only by Phillips' scanty specimens from Penang. In 1887 Mr. Curtis sent flowering specimens of it, together with a single ripe carpel from the same island; while copious flowering and fruiting specimens were, about the same time, received from Perak. In all its parts the plant is essentially a *Cyathostemma.*

5. CYATHOSTEMMA ACUMINATUM, n. sp. King. A climber; branches pale brownish, the youngest slender, dark-coloured, rufous-puberulous.

*Leaves* membranous, oblanceolate-oblong, caudate-acuminate, the base acute; both surfaces glabrous shining and minutely reticulate; the midrib depressed above and puberulous, beneath prominent and minutely muriculate: main nerves 10 to 11 pairs, spreading, curved, sub-ascending, prominent beneath, depressed above: length 8 to 9 in.; breadth 2·2 to 2·5 in.; petiole ·15 in., tomentose. *Cymes of hermaphrodite flowers* rufous-pubescent, 4 to 6 in. long; pedicel about as long as the branches, the latter with numerous distichous, oblong, nervose bracts. *Flowers* ·4 to 5 in. in diam., on short pedicels. *Sepals* triangular, blunt, spreading. *Petals* as in *C. Scortechinii*; connective of stamens forming at the apex a thick incurving point. *Ovaries* as in *C. Scortechinii* but with conical stigma. *Cymes of female flowers* much shorter than those of the hermaphrodite, dichotomous, few-flowered, about 1·5 in. long (of which the peduncle is 1 in.); slightly rufous-pubescent; bracts few, lanceolate. *Flowers* about ·4 in. in diam. when open, buds conical. *Sepals* broadly triangular, cordate, acute, spreading, pubescent. *Petals* coriaceous, granular-pubescent, concave; the outer broadly ovate-triangular, the apex sub-acute, incurved in bud; the inner row smaller, narrower, erect, connivent. *Stamens* absent. *Ovaries* as in the hermaphrodite, but the stigma larger, and not conical.

Upper Perak; Wray No. 3468.

A remarkable species of which I have seen only Wray's incomplete specimens. These specimens are accompanied by some loose young carpels, ovate-globular, oblique, with persistent recurved styles, and a single or at most two seeds. If these carpels really belong to the specimen, the definition of the genus will have to be modified. The structure of both the hermaphrodite and pistillate flowers agrees perfectly with that of the other species above described.

### 4. UVARIA, Linn.

Scandent or sarmentose shrubs, usually stellately pubescent. *Flowers* terminal or leaf-opposed, rarely axillary, cymose, fascicled or solitary, yellow, purple or brown. *Sepals* 3, often connate below, valvate. *Petals* 6, orbicular, oval or oblong, imbricate in 2 rows, sometimes connate at the base. *Stamens* indefinite; top of connective ovoid-oblong, truncate or subfoliaceous. *Torus* depressed, pubescent or tomentose. *Ovaries* indefinite, linear-oblong; style short, thick; ovules many, 2-seriate, rarely few or 1-seriate. *Ripe carpels* many, dry or berried, few- or many-seeded.—DISTRIB. About 110 species—many tropical Asiatic, a few African species, and some Australian.

A genus characterised by the usually large showy flowers with imbricate Rosaceous corolla:—allied to the American genus *Guatteria* Ruiz and Pavon (*Cananga*, Aubl.) and distinguished from it chiefly by its multi-ovulate ovaries.

Flowers more than ·5 in. in diam.
  Connective of anthers slightly produced at the
    apex, compressed, oblique.
      Carpels stalked, oblong, rugulose   ...  1.  *U. Larep.*
      Carpels ovoid to sub-globular.
        Carpels 1·5 to 2·25 in. long, not tuber-
          culate, very pulpy, tomentose   ...  2.  *U. Hamiltoni.*
        Carpels not more than 1 in. long, tuber-
          cular, with little pulp.
          Carpels ovoid, oblique ; leaves woolly-
            tomentose beneath, even when old  3.  *U. dulcis.*
          Carpels globular or globular-ovoid,
            leaves glabrous when adult   ...  4.  *U. Lobbiana.*
  Connective produced beyond the apex to about
    half the length of the anther, flattened, ob-
    liquely truncate ; flower 1·5 in. in diam.  ...  5.  *U. macrophylla.*
  Connectives produced, those of the inner an-
    thers truncate, those of the outer flattened
    and oblique : flower 2 to 3 in. in diam. ;
    leaves conspicuously stellate-tomentose be-
    neath  ...      ...      ...      ...  6.  *U. purpurea.*
  Connectives of anthers slightly, or not at all,
    produced at the truncate apex.
      Whole plant stiffly hairy...      ...  7.  *U. hirsuta.*
      Whole plant softly hairy...      ...  8.  *U. Curtisii.*
  Connectives of anthers produced into a broad
    flattened sub-quadrate process ; the outer
    anthers changed into staminodes   ...  9.  *U. Ridleyi.*
  Anthers oblong-cuneate, the connectives pro-
    duced at the apex and always truncate.
      Leaves pubescent beneath.
        Flowers in terminal umbellate ra-
          cemes  ...      ...      ...  10.  *U. pauci-ovulata.*
        Flowers in terminal umbels or in
          many-flowered lateral narrow
          panicles...      ...      ...  11.  *U. Scortechinii.*

13

Leaves glabrous except the midrib, 2·5 to
5 in. long ; flowers less than ·5 in. in
diam. ... ... ... 12. *U. micrantha.*
Flowers small (less than ·5 in. in diam.)
Leaves glabrous except the midrib ... 12. *U. micrantha.*
Leaves pubescent.
Leaves on under-surface stellate rufous-
pubescent ; young branches and flow-
ers outside with scurfy rufous tomen-
tum ... ... ... 13. *U. andamanica.*
Leaves on under surface and young
branches minutely tawny-tomentose... 14. *U. excelsa.*
Species of doubtful position.
Probably near *U. Lobbiana* ... ... 15. *U. astrosticta.*
With axillary flowers ... ... 16. *U. sub-repanda.*

1. Uvaria Larep, Miq. Fl. Ind..Bat. Suppl. 370. A climber 20
to 40 feet long : youngest branches and petioles sparsely covered with
minute scaly stellate hairs; the older cinereous, lenticellate, glabrescent.
*Leaves* membranous, elliptic or sub-obovate-elliptic, shortly acuminate,
slightly narrowed in the lower fourth to the rounded sub-emarginate,
not cordate, base : upper surface glabrous, shining, the midrib minutely
tomentose ; lower surface with a few short spreading hairs on the midrib
and some of the nerves, otherwise almost glabrous ; main nerves 10 to
12 pairs, spreading, interarching within the edge, bold in the lower, impressed on the upper, surface ; length of blade 5 to 8 in., breadth 2 to
3 in., petiole ·2 to ·3 in. *Peduncles* from half-way between the leaves,
·1 in. long, 1- to 2-flowered (one of the flowers often abortive), warted
and yellowish-pubescent ; pedicels ·75 in. long, with 1 or 2 reniform
bracts : flowers 1·5 to 1·75 in. in diam. *Sepals* small, (·2 in. long) reniform,
united at the base, reflexed, pubescent. *Petals* oblong-oblanceolate, sub-
acute, about ·75 in. long, sub-coriaceous, puberulous. *Anthers* sessile in
very few rows, flattened ; the connective slightly produced, flattened,
oblique. *Ovaries* numerous, angled, puberulous, with a few long projecting hairs near the apex. *Torus* of the fruit small, sub-globular,
pubescent. *Ripe carpels* numerous, stalked, cylindric-oblong, oblique,
curved, slightly apiculate, rugulose, minutely rufous-pubescent, 1·25 to
1·5 in. long, and ·5 in. in diam. *Seeds* about 10, in 2 rows, compressed,
shining. *Stalks* 1·25 to 1·5 in. long, rufous-tomentose.
Perak : King's Collector, No. 4011, Wray No. 1826.
2. Uvaria Hamiltoni, Hook. fil. and Thoms. Fl. Ind. 96. A
262

powerful climber: young branches slender, softly rufous-tomentose, becoming glabrous. *Leaves* membranous, elliptic-oblong to elliptic, sometimes slightly obovate, acuminate ; the base narrowed or rounded, sometimes slightly unequal, never cordate ; upper surface adpressed-pubescent, almost glabrous when old, the midrib minutely rufous-tomentose ; lower surface softly stellate-tomentose ; main nerves 14 to 17 pairs, spreading, rather prominent beneath ; length of blade 4 to 8 in., breadth 2·25 to 3·5 in., petiole ·15 to ·2 in. *Peduncles* solitary or 2 to 3 together, ·75 to 1·75 in. long, extra-axillary, 1-flowered ; flowers 1·5 to 2·5 in. in diam. : bract single, sub-orbicular, rufous-tomentose outside, shortly hispid inside : buds turbinate, tomentose. *Sepals* broadly triangular, ultimately reflexed, membranous. *Petals* much longer than the sepals, coriaceous, obovate, the apices obtuse and incurved, minutely tomentose on both surfaces, brick-red. *Anthers* sub-sessile, equal, obliquely truncate at the apex, ·15 to ·2 in. long. *Ovaries* slightly shorter than the stamens, compressed, pubescent. *Torus* hemispheric, tomentose, pitted when adult. *Ripe carpels* on long slender stalks, ovoid to sub-globular, about 1·5 in. long, and 1 in. in diam. when fresh, tomentose, scarlet ; when dry slightly constricted between the seeds ; stalks slender, tomentose, 1 to 1·5 in. long. *Seeds* about 6, flat, shining. Hook fil. Fl. Br. Ind. I, 48. *U. grandiflora*, Wall. Cat. 6485 E.

In the Forests at the base of the Eastern Himalaya ; Madhopore Forest in E. Bengal : Assam ; Khasia ; Shan Hills (Prazer).

Var. *Kurzii*, King. Leaves with broader bases often minutely cordate ; fewer nerves (12 to 14 pairs) ; smaller flowers (1·3 iu. iu diam.) on shorter pedicels (1 to 1·25 in.) ; petals yellowish, ovate-oblong.

South Andaman : Kurz, Kings' Collector.

This was referred by Kurz who first collected it, to *U. macrophylla*, Roxb., then to *U. purpurea*, Bl. : but was finally considered by him as "altogether doubtful." The fuller materials recently received show it to be, in my opinion, a very distinct variety of *U. Hamiltoni*, allied no doubt to *U. purpurea*, Bl., but a much larger plant with smaller flowers and more globular fruit.

3. UVARIA DULCIS, Dunal Anon. 90, t. 13. A powerful creeper often 80 to 100 feet long ; youngest branches softly cinereous-tomentose ; the older sub-glabrous or glabrous, dark-coloured, rather rough. *Leaves* coriaceous, elliptic or oval, sometimes unequal-sided, acute or sub-acute ; the base broad, rounded, or sub-truncate, minutely cordate ; upper surface sparsely adpressed-stellate-pubescent. The midrib ferruginous-tomentose ; lower surface densely sub-ferruginous or cinereous woolly-tomentose : main nerves 8 to 10 pairs, spreading, slightly curving, prominent beneath : length of blade 4·5 to 7 in., breadth 2·5 to 3·5 in.,

petiole ·2 in., stout. *Peduncles* ·5 in. long, lateral, not axillary, 1-flowered, solitary or 2 to 3 together, each bearing a small ovate deciduous bract; buds ovoid-globose, tomentose; flowers 1·25 to 1·5 in. .in diam. *Sepals* broadly triangular, sub-acute, slightly reflexed, fleshy, tomentose on both surfaces. *Petals* much longer than the sepals, sub-coriaceous, broadly ovate, sub-acute, sub-reflexed, minutely tomentose on the outer surface; pubescent on the inner. *Stamens* and *pistils* forming a compact hemispheric mass; anthers sub-sessile, ·1 in. long, the connective much produced at the apex, compressed, oblique. *Ovaries* numerous, densely crowded, slightly shorter than the stamens, tomentose. *Torus* depressed-hemispheric, stellate-tomentose, pitted when adult. *Ripe carpels* numerous, stalked, ovoid, oblique, blunt, much and unequally tuberculate, densely and loosely ferruginous stellate-tomentose as are the 1 in. long stalks. DC. Prod. I, 88; Hook. fil. and Th. Fl. Ind. 98; Miq. Fl. Ind. Bat. I, Pt. 2, p. 24; Ann. Mus. Lugd. Bat. II, 8. *U. javana*, Dunal Anon. 91, t. 14; Blume Bijdr. 12; Fl. Javæ t. 3 and 13 B.; DC. Prod. I, 88? *U. aurita* Blume Fl. Javæ t. 3.

Malacca, Griffith; Maingay (Kew Destrib.), No. 25. Perak, King's Collector. Penang, Curtis, No. 1414.

As regards the size of its leaves and the colour of its flowers (which appear to vary from green though yellow to purple) this is rather a variable species. One of its forms, barely distinguishable from the type, was named *U. javana* by Dunal who also gave a figure of it. Blume, who again figured *U. javana*, distinguished it from *U. dulcis* by the stellate (not simple) hairs on the upper surface of its leaves. But, as Hook. fil. and Th. point out (Fl. Ind. 98), both kinds of hairs occur on the same leaf. In all the specimens named *U. javana*, received from the Dutch Botanists, the leaves are much smaller and less densely woolly below than those collected in the Malay Peninsula. Miquel suggests that *U. aurita*, Bl. is only a form of this. By neither figuring nor describing the fruit of what he understood as *U. dulcis, aurita* and *javana*, Blume neglected one of the best characters in this rather perplexing genus; and it may be that when fruit of the small-leaved Java species issued from the Herbarium of Buitenzorg shall be forthcoming, the reductions above made will have to be cancelled.

4. UVARIA LOBBIANA, H. f. and T. Fl. Ind. 100. A powerful climber, often reaching 100 to 150 feet in length : young branches pubescent, ultimately glabrous and dark-coloured. *Leaves* sub-coriaceous, oblong or oblong-oblanceolate, acute or very shortly acuminate, rarely obtuse, narrowed to the rounded or sub-cordate base ; both surfaces when very young stellate-furfuraceous, speedily becoming glabrous except the puberulous midrib; the upper (when dry) pale green, the lower brown : main

nerves 13 to 16 pairs, curving slightly, spreading below, suberect abovo, thin but prominent beneath; length of blade 4 to 7 in., breadth 1·5 to 2·25 in., petiole ·25 in. *Peduncles* only ·25 in. long or even less, terminal or leaf-opposed, 2-or 3-flowered, tomentose, each flower with a large rotund amplexicaul bract; buds depressed-globose, tomentose : flower 1 to 1·2 in. in diam. *Sepals* conjoined into a wavy cup, tomentose outside, minutely pubescent inside. *Petals* coriaceous, often 7 or 8, slightly unequal, broadly oval, obovate, blunt; slightly warted on both surfaces, minutely tomentose on the outer, pubescent on the inner. *Anthers* sessile, flattened, ·1 in. long, the connectives produced at the apices, compressed, obliquely truncate, the outer row sterile. *Ovaries* 4-angled, pubescent except the truncate lobulate stigma. *Ripe carpels* numerous, stalked, globular or globular-ovoid, slightly oblique, boldly tubercled, pubescent, ·5 to ·75 in. in diam., and sometimes 1 in. long; pericarp thin; stalks slender, 1·5 to 2 in. long, glabrescent. *Seeds* 4 to 10, large, plano-convex, smooth. Miq. Fl. Ind. Bat. I, Pt. 2, 34 : Hook. fil. Fl. Br. Ind. I, 49.

Malacca ; Griffith, Maingay (Kew Distrib.), Nos. 27 and 30. Singapore and Perak ; King's Collector. Penang ; Curtis. Sumatra ; Forbes, No. 3059.

5. UVARIA MACROPHYLLA, Roxb. Fl. Ind. II, 663. Scandent usually to the extent of 15 to 20 feet, but sometimes reaching 50 or 60 feet ; young branches and petioles rusty-tomentose. *Leaves* coriaceous, elliptic-oblong, rarely elliptic-rotund, sometimes slightly obovate, obtuse or shortly and abruptly acuminate, very slightly narrowed to the rounded or minutely cordate base ; upper surface (when adult) glabrescent or glabrous except the tomentose midrib and nerves ; lower with lax, sometimes stellate, rusty tomentum, especially along the midrib and 11 to 18 pairs of prominent spreading or oblique nerves : length of blade 4·5 to 10 in., breadth 2·5 to 4 or (in some Burmese specimens) even 6 in. ; petiole ·25 in. *Peduncles* extra-axillary or terminal, densely rusty-tomentose, 3-to 5-flowered, each pedicel with an oval or rounded bract ; buds globose : flowers 1·5 in. in diam. *Sepals* connate into a cup with wavy obscurely 3-toothed edge. *Petals* much larger than the calyx, sub-rotund, blunt, coriaceous, purple, tomentose outside, pubescent inside ; anthers sessile, ·3 in. long : the connective produced at the apex to nearly half the length of anther, compressed, obliquely truncate. *Ovaries* narrow, compressed, tomentose, the stigmas truncate, *Torus* of fruit woody, hemispheric, 1 in. in diam. sparsely pubescent, pitted. *Ripe carpels* stalked, oblong, blunt at each end, glabrous, ·75 to 1·25 in. long ; pericarp thin ; stalks ·5 to 1 in. long : seeds numerous, oval, compressed, shining. Wall. Pl. As. Rar. t. 122 ; Cat. 6487 (excl. F. in fruit) Hk. f.

17

and Th. Fl. Ind. 97 ; Hook. fil. Fl. Br. Ind. I, 49 ; Miq. Fl. Ind. Bat. l
Pt. 2, p. 23 ; Thwaites Enum. Pt. Ceyl. 6 ; Kurz Fl. Burm. I, p. 28 ; Beddome Ic. Pl. Ind. Or. t. 81. *U. rufescens*, DC. Mem. Anon. 26. *U. cordata*, Wall. Cat. 6486. *Guatteria cordata*, Dunal Anon. 129 t. 30 ; DC.
Prod. I, 93.

Silhet, Chittagong, Burmah, Malayan Peninsula, Java, Ceylon.

One of the most widely distributed species of the genus and
closely allied to *U. ovalifolia*, Bl. I reduce to this species the *Uvaria
cordata* of Wall. Cat., No. 6486 ; but not without some hesitation, as
both Miquel and Kurz referred it to *U. ovalifolia*, Bl.

6. UVARIA PURPUREA, Blume Bijdr. 11 : Fl. Jav. 13, t. 1 and t. 13 A.
A sarmentose shrub, often climbing to 20 or 30 feet : young parts softly
stellate-rufous-pubescent or tomentose. *Leaves* thickly membranous,
oblong-lanceolate to elliptic-oblong, sometimes slightly obovate, acute or
acuminate, the base rounded or slightly cordate, shortly petiolate;
upper surface, when adult, shining, glabrous or glabrescent, the midrib
and sometimes the nerves tomentose ; under surface rather sparsely
but softly stellate-tomentose ; main nerves 14 to 17 pairs, rather
straight, prominent beneath, the lower spreading, the upper sub-erect ;
length 4·5 to 9 or even 11 in., breadth 2·5 to 3·75 in. ; petiole ·15 to
·25 in. *Peduncles* 1 to 1·5 in. long, extra-axillary or terminal, usually
1- sometimes 2-flowered ; flowers 2 to 3 in. diam. ; bracts 2, large, unequal, leafy ; buds turbinate. *Sepals* broadly triangular, sub-concave,
membranous, fulvous-tomentose on the outer, glabrescent on the inner
surface. *Petals* longer than the sepals, coriaceous, oblong to obovate,
obtuse, coriaceous, dark purple, the inner 3 slightly smaller. *Anthers*
sub-sessile, very numerous, equal, about ·3 in. long ; the connective much
produced at the apex, rhomboid in the inner, compressed and oblique in
the outer anthers. *Ovaries* numerous, densely crowded, slightly shorter
than the stamens, tomentose ; ovules numerous. *Torus* depressed-hemispheric, pubescent, pitted when ripe. *Ripe carpels* numerous, stalked, ob-
long-cylindric, blunt at each end with 2, more or less obscure, ridges and
grooves, minutely rufous-tomentose, sub-tuberculate, 1·5 to 2 in. long
and about ·5 in. in diam. ; stalks ·5 to 1 in. long, rufous-tomentose. *Seeds*
numerous, flat. Hook. fil. and Thoms. Fl. Ind. 95 ; Miq. Fl. Ind. Bat. I,
Pt. 2, 22 ; Ann. Mus. Lugd. Bat. II, 6 ; Hook. fil. Fl. Br. Ind. I, 47 ;
Benth. Fl. Hong Kong, 9 ; Vidal y Soler, Revis. Fl. Filipinas, 39 ; Scheffer
Obs. Phyt. I, 4, 26, 65 ; Ann. Jard. Bot. Buitenz. II, 1. *U. grandiflora*,
Roxb. Fl. Ind. II, 665 ; Wall. Pl. As. Rar. II t. 121 ; Wall. Cat. 6485, A.
to D. and H. ; Wight and Arn. Prod. 9. *U. platypetala*, Champ. in Kew
Journ. Bot. III, 257. *U. rhodantha*, Hance in Walp. Ann. II, 19. *Unona
grandiflora*, DC. Prod. I, 90.

In all the provinces. Distrib : Malayan Archipelago, S. China, Phillipines.

Var. *tuberculata* ; fruits prominently tuberculate.

Perak ; King's Collector, Nos. 960, 4786.

A plant collected in the island of Bangka, closely resembling this in leaves, but with larger flowers with yellow petals, has been described by Messrs. Teysmann and Binnendyk under the name of *U. flava* (Nat. Tijds. Ned. Ind. XXIX, 419). It has also been figured by Miquel (Ann. Mus. Lugd. Bat. II, 6, t. 1). I fear it is merely a form of *U. purpurea ;* but not having seen fruiting specimens, I hesitate to reduce it here.

7. UVARIA HIRSUTA, Jack Mal. Misc. (Hook. Bot. Misc. II, 87.) A sarmentose shrub but often climbing to the length of from 15 to 50 feet : young branches and petioles with numerous rather stiff reddish-brown hairs. *Leaves* thinly coriaceous, narrowly elliptic to elliptic-oblong, rarely obovate-oblong, acute or sub-acute, the base rounded or minutely cordate ; upper surface with scattered sub-adpressed, stiff, mostly simple hairs, the midrib tomentose ; lower surface with more numerous stellate and simple hairs : main nerves 9 to 14 pairs, spreading, depressed on the upper surface (when dry) but prominent on the lower ; length 4 to 7 in., breadth 2·25 to 3·25 in., petiole ·2 in. *Peduncles* 1 to 2 in. long, lateral or terminal, not axillary, 1- rarely 2-flowered ; flowers 1·25 to 1·5 in. in diam. ; bract solitary (rarely 2 or 3), lanceolate, deciduous : buds ovoid-globose, stiffly hairy. *Sepals* membranous, broadly ovate, acute, connate, pilose outside, reflexed. *Petals* red, larger than the sepals, broadly ovate, acute ; outside tomentose with stiff hairs inter-mixed, inside sub-glabrous ; anthers ·15 in. long, sub-sessile, the connec-tive at the apex often slightly produced and obtuse. *Ovaries* 4-angled, truncate, rufous-tomentose, shorter than the anthers. *Ripe carpels* numerous, stalked, cylindric, blunt, 1·5 to 2 in. long, covered (as are the stalks and torus) with dense darkly ferruginous tomentum mixed with stiff hairs : stalks 1 to 1·25 in. long : torus hemispheric : seeds numer-ous, ovoid, plano-convex. Blume Fl. Javae, Anon. 22, t. 5 ; Wall. Cat. 6458 (excl. C.) ; Hook. fil. and Thoms. Fl. Ind. 99 ; Hook fil. Fl. Br. Ind, I, 48 ; Miq. Fl. Ind. Bat. I, Pt. 2, p. 24 ; Ann. Mus. Lugd. Bat. II, 8 ; Scheff. in Nat. Tijdsch. XXXI, 2 ; Zoll. in Linnæa XXIX, 304 ; Kurz Flora Burm. I, 28 ; Scheff. Observ. Phyt. I, 2. *U. trichomalla*, Bl. Fl. Jav. Anon. 42, t. 18. *U. velutina,* Blume (not of Roxb.) Bijdr. 13. *U. pilosa*, Roxb. Fl. Ind. II, 665.

In all the provinces. Distrib. Malayan Archipelago and Burmah.

There is some difference amongst individuals as to the breadth of the leaves, and on one of the forms with comparatively short but broad leaves Blume founded his species *U. trichomalla.*

8. Uvaria Curtisii, King n. sp. A large climber : young branches densely rusty-tomentose, slender. *Leaves* oblong-lanceolate, sometimes slightly oblanceolate, acuminate, slightly narrowed to the rounded base; upper surface glabrous except the strong rusty-tomentose midrib and the nerves; under surface stellate-rufous-tomentose, especially on the midrib, reticulations, and 7 to 12 pairs of ascending, curving, bold main nerves : length 4 to 9 in., breadth 1·7 to 3·25 in.; petiole ·15 to ·2 in., stout. *Flowers* 1 to 1·25 in. in diam., solitary or in pairs, axillary : pedicels 1 to 1·75 in., densely tomentose like the outer surface of the sepals, and with an ovate supra-median bracteole. *Sepals* broadly ovate, concave, spreading, puberulous within, ·35 in. long. *Petals* thinly leathery, white, subequal, ovate-oblong, obtuse; the outer rather broader than the inner, ·5 in. long, puberulous on both surfaces but especially on the outer. *Stamens* numerous, all perfect; connective truncate at the apex, not prolonged into a process; the anthers linear, lateral. *Ovaries* numerous, crowded, elongate, 3-angled, tomentose, with 12 ovules in 2 rows : stigma sessile, large, sub-capitate, corrugated, glabrous. *Ripe carpels* unknown.

Perak; on Ulu Bubong, King's Collector, No. 8543. Penang; elev. 2,000 feet. Curtis No. 1415.

9. Uvaria Ridleyi, King n. sp. A strong climber; young branches slender, stellate-rufous-tomentose, ultimately dark-coloured, striate; sparsely lenticellate. *Leaves* sub-coriaceous, elliptic-oblong, acuminate, slightly narrowed to the rounded base; both surfaces with short, stellate, rather pale hairs, scabrid on the upper, soft on the lower surface; the midrib and 10 to 15 pairs of spreading curving slightly prominent main nerves softly rufous-stellate-tomentose on both surfaces; length 3 to 5 in., breadth 1·3 to 2 in.; petiole ·15 in., stellate-tomentose. *Flowers* ·75 to 1·2 in. in diam., 2 or 3 together in short supra-axillary cymes; pedicels stellate-tomentose like the outer surface of the calyx, ·3 or ·4 in. long, with a large orbicular amplexicaul bracteole. *Sepals* orbicular, connate into an obscurely 3-toothed spreading cup ·4 in. in diam., glabrescent inside. *Petals* spreading, sub-orbicular to broadly oblong, very blunt, subequal, rather thin, minutely pubescent on both surfaces but especially on the outer, dark reddish-brown. *Stamens* numerous (the outer row converted into sub-quadrate staminodes) compressed, broad, without filaments; the apical process of the connective broad and flat : anther-cells on the edges of the connective, linear. *Ovaries* numerous, crowded, elongate, narrow, compressed, ridged, minutely stellate-tomentose, the ovules numerous; stigma sessile, short and broad, fleshy, obliquely truncate. *Ripe carpels* ovoid or obovoid, blunt at both ends, minutely pubescent, 1·2 to 1·5 in. long : stalks nearly 1 in., stellate-tomentose.

*Seeds* numerous in two rows, horizontal, oval, compressed, pale brown, shining.

Pahang : Ridley. Perak : Scortechini.

10. UVARIA PAUCIOVULATA, H. f. and T. in Hook. fil. Fl. Br. Ind. I, 51. A sub-scandent shrub : young branches densely stellate rufous-tomentose. *Leaves* coriaceous, rigid, narrowly elliptic or elliptic-oblong, obtuse or obtusely acuminate, the base rounded or cordate ; upper surface (in adult leaves) shining, quite glabrous ; the lower dull, sparsely pubescent; main nerves 10 to 14 pairs, sub-ascending, curving, prominent beneath and impressed above : length of blade 2·5 to 6 in., breadth 1·25 to 3 in., petiole ·2 in. *Racemes* terminal, umbellate, few-flowered, 1·5 to 2·5 in. long, scurfily rufous-tomentose ; bracts numerous and imbricate towards the apex, rotund to ovate, tomentose : buds ovoid-globose : flowers 1·5 in. in diam. *Sepals* small, (·3 in. long) orbicular, sub-acute, connate to the middle and densely tomentose outside, densely and minutely puberulous inside. *Petals* very much larger than the sepals, sub-connivent, coriaceous, ovate-rotund, obtuse, the inner 3 narrower; all scaly-tomentose externally, densely and minutely pubescent and veined internally ; anthers sub-sessile, cuneate ; connective slightly produced at the apex, truncate ; ovaries longer than the stamens, flattened, stellate-hairy ; stigma truncate, ovules 1 to 3. *Ripe carpels* numerous, stalked, sub-globose, mucronate, densely and minutely fulvous-tomentose, ·35 to ·5 in. in diam., 1- to 2-seeded ; stalk ·5 to ·75 in., rather slender. *Seeds* compressed, shining.

Malacca ; Maingay (Kew Distrib.), No. 104. Penang : Curtis, No. 825 : at elevations of 500 to 600 feet.

11. UVARIA SCORTECHINII, King n. sp. A sarmentose, flexuose shrub ; young branches and petioles densely covered with rusty, floccose, rufous tomentum. *Leaves* coriaceous, elliptic to elliptic-rotund, obtuse, very slightly or not at all narrowed to the rounded or minutely cordate base : upper surface shining, glabrescent or glabrous, the deeply impressed midrib and nerves tomentose, transverse veins depressed when dry ; under surface minutely and softly rufous, pubescent especially on the midrib nerves and reticulations which are all bold and prominent : main nerves 10 to 12 pairs, spreading below, sub-ascending above, forming double arches within the edge : length of blade 4 to 7 in., breadth 2·5 to 4 in , petiole ·2 to ·4 in. *Flowers* 1·5 in. in diam., either terminal in umbels of 2 or 3, or in many-flowered lateral panicles 4 in. in length : peduncles ·5 to ·75 in. long ; bracts numerous, but chiefly towards the apices of the peduncles, ovate-orbicular, covered with short rufous flocculent tomentum as are the branches and axes of the panicles. *Sepals* fleshy, triangular, sub-acute, connate in the lower third, concave,

spreading, minutely pubescent. *Petals* fleshy, about 1 in. long, connivent; the outer 3 ovate-rotund, very obtuse, tomentose-pubescent on both surfaces, the outer surface with some small superficial scales, the inner with a round glabrous spot at the base: inner 3 petals obovate, clawed, pubescent outside, glabrous inside except a broad pubescent band near the apex. *Anthers* sessile, angled, the connective projecting beyond the apex, broadly truncate, almost peltate. *Ovaries* (fide Scortechini) "several, with few stellate hairs, 2–3 ovuled: style cylindric, curved, glabrous." *Fruit* unknown.

Perak: Scortechini, No. 1990.

Scortechini's are the only specimens I have seen, and they have flowers only.

12. UVARIA MICRANTHA, H. f. and Th. Fl. Ind. 103. A large climber; young branches slender, softly rufous-tomentose, afterwards glabrous, striate, and dark-coloured with pale warts. *Leaves* thinly coriaceous, oblong-lanceolate, acuminate, the base rounded or slightly cuneate; both surfaces glabrous except the rufous-pubescent midrib: main nerves scarcely visible (even when dry), 12 to 15 pairs, spreading; length of blade 2·5 to 5·5 in., breadth ·8 to 1·4 in., petiole ·15 in. *Peduncles* terminal or extra-axillary, very short, 2-to 4-flowered, softly rufous-tomentose, bracts more or less orbicular; buds globose, slightly pointed, ·15 in. in diam.; flowers ·4 in. in diam. *Sepals* sub-rotund, densely pubescent outside, sub-glabrous inside. *Petals* broadly ovate, sub-obtuse, granular and minutely tomentose outside, pubescent inside. *Ripe carpels* numerous, stalked, ovoid-globose, rounded at each end, glabrous, 2- to 4-seeded. *Seeds* plano-convex, smooth; Hook. fil. Fl. Br. Ind. 1, 51; Kurz Fl. Burm. I, 22; Miq Fl. Ind. Bat. I, Pt. 2, 26; *Uvaria sumatrana*, Kurz Andam. Report, 29; Hook. fil. Fl. Br. Ind. I. 51. ? *Uvaria elegans*, Wall. Cat. 6474 B. *Guatteria micrantha*, A. DC. Mem. 42; Wall. Cat. 6449. *Polyalthia fruticans*, A. DC. l. c. 42; Wall. Cat. 6430. *Anaxagorea sumatrana*, Miq. Fl. Ind. Bat. Suppl. 382.

Burmah, Malacca, Penang. Distrib. Sumatra.

As regards leaves, this closely resembles *Popowia nitida*, King—a plant of the Andaman and Nicobar Islands; and there is reason to believe that some specimens of that *Popowia* from those islands have been issued from the Calcutta Herbarium as *Uvaria micrantha*. 1 am also of opinion that *Uvaria sumatrana*, Kurz Andaman Report, 29, and of Hook. fil. and Thoms. Fl. B. Ind. 1, 51, is possibly *Popowia nitida*, King.

13. UVARIA ANDAMANICA, King n. sp. Scandent: young branches rather stout, scurfily stellate-tomentose. *Leaves* oblong-oblanceolate, shortly acuminate, much narrowed to the rounded, unequal, or minutely

cordate base; upper surface glabrous, the midrib and sometimes the nerves coarsely puberulous; under-surface reticulate, stellate-rufous-pubescent on the midrib and 18 to 22 pairs of spreading curving nerves; length 5·5 to 9 in., breadth 1·75 to 4 in.; petiole ·3 in., tubercular. *Flowers* small, in short terminal or axillary cymes, rarely solitary : pedicels ·3 in. long, densely covered like the outside of the sepals with sub-deciduous coarse, rusty, stellate tomentum; bracteole solitary, orbicular, ovate, close to the flower. *Sepals* valvate, orbicular, partly connate, glabrous inside. *Petals* imbricate, orbicular, fleshy, more or less puberulous outside, glabrous within ; the inner rather smaller than the outer but both under (in the young state) ·25 in. in diam. *Stamens* numerous, narrowly elongate, the apex truncate more or less obliquely ; anther-cells lateral. *Ovaries* absent in the staminiferous flower. *Ripe carpels* oblong, blunt (almost truncate) at each end, slightly tuberculate and densely covered with loose, sub-deciduous, rusty-stellate tomentum : pericarp rather thick. *Seeds* about 8 in 2 rows, plano-convex.

South Andaman ; King's Collector.

This has been collected only on two occasions, once with undeveloped male flowers and once with immature fruit. The full size attained by the flowers is not known, and the measurements of sepals and petals above given are taken from buds. By its leaves and peculiar deciduous rusty stellate tomentum, the species is however readily recognisable.

14. UVARIA EXCELSA, Wall. Cat. 6477. A creeper 30 to 100 feet long : young parts stellate-pubescent; the branchlets tawny-tomentose, speedily becoming glabrous dark-coloured and furrowed. *Leaves* coriaceous, oblanceolate, obovate-oblong to elliptic, the apex acuminate (sometimes very shortly), acute, rarely obtuse, slightly narrowed to the minutely cordate base : upper surface shining, glabrous except the puberulous depressed midrib; lower surface minutely tawny-tomentose ; main nerves 10 to 12 pairs spreading, slender; length 3·5 to 7·5 in., breadth 1·5 to 4 in. ; petiole ·3 to ·5 in. pubescent. *Flowers* white, ·35 to ·4 in. in diam., in contracted cymes from the branches below the leaves, or axillary; pedicels only about ·2 in. long, rufous-tomentose with a large bract close to the flower. *Sepals* semi-orbicular, sub-acute, valvate, concave, spreading, tomentose outside, glabrous within. *Petals* in bud imbricate only at their apices, sub-equal, thick, concave, densely and minutely pubescent on both surfaces : the outer broadly ovate, acute, a little larger than the sepals : inner petals ovate, about as large as the sepals. *Anthers* numerous, narrow, the cells linear, lateral; the apical process of the connective thick, sub-quadrate, obliquely truncate, minutely pubescent. *Ovaries* narrow, elongate, grooved, pubescent; the

23

stigma thick, sub-capitate, sub-truncate; ovules numerous, in two rows. *Ripe carpels* sub-globular, slightly obovoid, blunt at each end, densely and minutely tomentose, 1·1 in. long and ·9 in. in diam. *Seeds* about 14 in two rows, horizontal, half-oval, flat, smooth, brown. *Mitrephora excelsa*, H. f. and T. Fl. Ind. 114 : Hook. fil. Fl. Br. Ind. I, 77 ; Miq. Fl. Ind. Bat. I, Pt. 2, 31.

Penang: Wallich, Curtis. Perak : King's Collector. Scortechini. Malacca : Maingay (Kew Distrib.), No. 36 *in part.*

This plant was originally issued as a *Uvaria* by Wallich. His specimens of it, however, bore no mature flowers ; and Sir Joseph Hooker and Dr. Thomson referred them doubtfully to *Mitrephora*. The excellent specimens recently collected by Mr. Curtis and by the Calcutta Garden Collector show the petals to be sub-equal and concave, imbricate at the apex only, the sepals being quite valvate. This of course is not the typical flower of a *Uvaria*, in which the petals are *much* imbricate. But the stamens, ovaries and ripe fruit are more those of *Uvaria* than of any other genus.

15. UVARIA ASTROSTICTA, Miq Fl. Ind. Bat. Suppl. 370. A climber ? Young branches deciduously rufous-stellate-tomentose with simple hairs intermixed, ultimately glabrous striate and dark-coloured. *Leaves* coriaceous, oblong-lanceolate, sometimes slightly oblanceolate, acuminate, the base rounded or minutely cordate ; upper surface minutely scaberulous, the midrib and sometimes the nerves softly rufous-pubescent ; lower surface at first densely and softly tomentose, ultimately sparsely stellate-pubescent, sub-scaberulous ; main nerves 12 to 16 pairs, spreading, rather prominent on the lower surface : length of blade 4 to 6 in., breadth 1·5 to 1·8 in., petiole ·2 in. *Peduncles* extra-axillary, very short (only ·3 in.), 2-to 3-flowered, rufous-stellate-tomentose as are the 2 or 3 sub-rotund bracts ; buds sub-globular ; flowers ·6 in. in diam. *Sepals* reniform, sub-acute, united half way. *Petals* nearly three times as long as the sepals, sub-coriaceous, broadly oval, slightly obovate, sub-acuto, minutely pubescent. *Anthers* sub-sessile, the connective produced beyond the apices, flattened and truncate, 3 outer anthers barren : torus hispidulous. *Fruit* unknown ; Miq. Ann. Mus. Lugd. Bat. II. 8.

Perak ; Scortechini, No. 121. Distrib. E. Sumatra.

The Perak specimens of this plant agree perfectly with those from Sumatra on which the species was founded. It is allied to *U. heterocarpa* Bl., to *U. rufa* Bl., and also to *U. timoriensis*. I have never seen the fruit, and Miquel's entire description of it consists of the two words " carpella velutina."

*Doubtful Species.*

16. UVARIA SUB-REPANDA, Wall. Cat. 6483. A climber : young
272

branches very slender, rather sparsely scurfy-pubescent. *Leaves* membranous, oblong or obovate-oblong, acute, the base rounded: upper surface shining, glabrous except the pubescent midrib; under-surface pale, yellowish-brown when dry, dull, at first puberulous, ultimately quite glabrous including the midrib, the reticulations distinct; main nerves 10 to 14 pairs, spreading, thin but rather prominent beneath: length of blade 5 to 7·5 in., breadth 2 to 2·25 in.; petiole ·15 to ·25 in., densely scaly-pubescent. *Peduncles* axillary, rufous-stellate-tomentose, 1-flowered; bracts cucullate, sub-orbicular. *Petals* narrowly oblong. *Ripe carpels* unknown. Hook. fil. and Thoms. Fl. Ind. 101: Hook. fil. Fl. Br. Ind. I. 50.

Singapore, Wallich.

A very imperfectly known species, the only specimens being Wallich's which are not good and which are in flower only. The only other specimen which agrees with Wallich's specimens as to leaves and branches is from Penang (Curtis No. 1408): but this has a short 2-flowered, extra-axillary peduncle, and I hesitate to identify it with *U. sub-repanda.*

## 5. ELLIPEIA, H. f. and T.

Characters of *Uvaria*, but with solitary, ventral or sub-basal ovule and 1-seeded carpels, the style sometimes elongate.

Distrib. Malaya: species 10 or 11.

Flowers all hermaphrodite.

Flowers in groups.

Leaves oblong or narrowly obovate-oblong, acuminate, pubescent, puberulous or glaberulous beneath: flowers in short panicles ... ... ... 1. *E. cuneifolia.*

Leaves obovate-oblong, obtuse, softly tomentose beneath, peduncles 3- or 4-flowered 2. *E. leptopoda.*

Leaves oblong or elliptic-oblong, acute, glabrous, cymes 3-to 5-flowered ... 3. *E. glabra.*

Flowers solitary.

Leaves oblong-lanceolate to ovate-lanceolate, acuminate, minutely granular above when dry ... ... ... 4. *E. costata.*

Flowers unisexual or polygamous, solitary or in pairs.

Leaves shortly acuminate, both surfaces minutely granular when dry, not reticulate: stalks of carpels ·15 in. long ... ... 5. *E. pumila.*

Leaves acute, rarely acuminate, not granular,
reticulations transverse and very distinct;
stalks of carpels ·75 to 1 in. long ... 6. *E. nervosa.*

1. ELLIPEIA CUNEIFOLIA, H. f. and T. Fl. Ind. 104. A climber 20
to 100 feet long : young branches at first shortly and densely rufous-
tomentose, ultimately sub-glabrous. *Leaves* thinly coriaceous, oblong or
narrowly obovate-oblong, the apex broadly abruptly and shortly acuminate, the base rounded or sub-cordate : upper surface glabrous, shining,
the midrib and often the main nerves tomentose ; lower minutely rufous-
tomentose to pubescent, very often glaberulous : main nerves 16 to 19
pairs, spreading to sub-ascending, prominent beneath : length of blade
4 to 7 in., breadth 1·5 to 3 in. ; petiole ·15 to ·2 in., tomentose. *Flowers*
·75 to 1 in. in diam., in short few-flowered pedunculate rufous-tomentose
panicles ; bracts at the bases of the pedicels ovate, that at the base of
the flower rotund : pedicels ·25 to ·4 in. long : buds ovoid-conic. *Sepals*
small, fleshy ; sub-orbicular, slightly united below, spreading, coriaceous,
tomentose. *Petals* fleshy, connivent ; outer 3 much larger than the
sepals, rotund, densely pubescent on both surfaces ; inner 3 not much
larger than the sepals, rotund, pubescent externally, glabrous internally.
*Anthers* sessile, short, the cells on the outer surface ; the apex with a broad,
round, oblique, truncate appendage from the connective ; pistils oblong,
tapering to each end, pubescent. *Torus* small, sub-globose. *Ripe carpels*
numerous, on long stalks, ovoid, oblique, blunt, with a faint partial ridge
and a short lateral, conical process, minutely yellowish-tomentose. *Seed*
smooth, ovoid. Hook. Ic. Plant. t. 1025 ; Hook. fil. Fl. Br. Ind. I, 52.

Malacca : Griffith, Maingay (Kew Distrib.) No. 31. Perak, very
common.

In the Perak specimens the tomentum on the under-surface of the
leaves is usually less dense than in specimens from Malacca : moreover
the flowers are smaller in the Perak specimens, and the floral bract is not
close to the calyx but a little way under it. In other respects, however,
they agree.

2. ELLIPEIA LEPTOPODA, King, n. sp. A climber, 50 to 70 feet long :
young branches and petioles densely covered with scurfy cinereous
tomentum. *Leaves* coriaceous, obovate-oblong, rarely elliptic, obtuse, or
with a very short blunt apiculus, narrowed in the lower half to the
minutely cordate, rarely entire, base : upper surface pale-green when
dry, sparsely and minutely stellate-pubescent when young, afterwards
glabrous except the pubescent midrib : lower surface densely covered
with soft, short, dense, pale brown tomentum ; main nerves 10 to 12 pairs,
spreading, obsolete on the upper, slightly prominent on the lower, surface : length of blade 3·5 to 5 in., breadth 2·25 to 2·5 in., petiole ·2 to

·25 in. *Peduncles* extra-axillary, about ·5 in. long; the flowers 3 or 4 on short pedicels, each subtended by a rotund-obovate, cucullate bract; the whole inflorescence and calyx rather sparsely stellate-tomentose: buds depressed-globose : flower '75 in. in diam. *Sepals* often 4 in number, semi-orbicular, very obtuse, slightly united below, spreading. *Petals* coriaceous, three times as long as the sepals, ovate-rotund, obtuse, recurved, minutely pubescent on both surfaces, dark crimson. *Anthers* sessile, very small, the connective produced beyond the apex, flattened, oblique. *Ovaries* about as long as the anthers; the stigmas truncate, hairy. *Torus* hemispheric. *Carpels* numerous, on long slender stalks, ovate-rotund, ·5 in. long, slightly oblique with a slight lateral beak, minutely cinereous-pubescent. *Stalks* slightly thickened and ridged towards the apex, 1·5 to 2·5 in. long. *Seed* ovoid, flattened on one side, smooth.

Perak; at low elevations, King's Collector. Singapore, Ridley.

A species in its leaves resembling *Uvaria heterocarpa*, Bl. but with different fruit : also like *U. timorensis*, Miq., but with much more obovate leaves.

3. ELLIPEIA GLABRA, H. f. and T. Fl. Br. Ind. I, 52. A tree : young branches and inflorescence brown-pubescent. *Leaves* coriaceous, oblong or elliptic-oblong ; the base rounded or acute; both surfaces glabrous, not shining, the upper rigid, the lower paler and reticulate : main nerves about 9 pairs, curved, sub-ascending, prominent beneath ; length 4 to 5·5 in., breadth 1·5 to 2 in., petiole ·25 in. *Cymes* shortly pedunculate, axillary, 3- to 5-flowered, 1 to 1·5 in. long. *Flowers* 1·5 in. in diam.; bracteole oblong, sub-amplexicaul, recurved. *Sepals* ovate-lanceolate, acute, recurved, ·25 in. long. *Outer petals* obovate-lanceolate, sub-acute, flat, without claws, 1 in. long; the inner shorter, obovate, obtuse. *Ovaries* glabrous below, strigose above ; ovule 1, erect (Maingay). *Ripe carpels* sub-globose, ·65 in. long ; pedicels slender, ·75 to 1·25 in. long : pericarp thin. *Seed* oblong, pale, with a deep longitudinal furrow.

Malacca; Maingay No. 66 (Kew Distribution).

Except Maingay's I have seen no specimens of this.

4. ELLIPEIA COSTATA, King. A shrub about 10 feet high : young branches pale, rusty-tomentose. *Leaves* coriaceous, oblong-lanceolate to ovate-lanceolate, acuminate, the base cuneate : upper surface glabrous but rather rough ; lower pale, softly and laxly pubescent, sub-glabrescent when old ; main nerves 8 to 9 pairs, bold, sub-ascending, rather straight : length 4 to 6·5 in., breadth 2 to 2·5 in. ; petiole ·25 in., tomentose. *Flowers* solitary, extra-axillary, ·75 to 1 in. in diam.: pedicels woody, tomentose, ·15 in. long, with 3 ovate acute bracts at their bases. *Sepals* ovate, obtuse, half as long as the petals and, like them, sericeous exter-

27

nally and glabrous or sub-glabrous internally. *Petals* subequal, oblong, obtuse, ·35 to ·45 in. long. *Ripe carpels* ovoid-cylindric, slightly apiculate and shortly stalked, glabrous, ·8 in. long and ·35 in. in diam. ; pericarp thin.

Burmah ; on Moolyet at 5,000 ft. Gallatly.

I have seen no entire fruit of this species but only some loose carpels. When ripe they are said by Mr. Gallatly to be red.

ELLIPEIA PUMILA, King, n. sp. A shrub 2 to 8 feet high : young branches with minute pale rufous tomentum ; when older dark-coloured, glabrous and furrowed. *Leaves* coriaceous, oblong-lanceolate to elliptic-lanceolate, tapering from the middle to the shortly acuminate apex and acute base ; both surfaces minutely granular when dry, the upper glabrous ; the lower sparsely adpressed-pubescent ; the midrib rufous-pubescent ; main nerves about 9 pairs, oblique, rather straight, faint on the lower surface, obsolete on the upper ; length 4·5 to 7 in., breadth 1·5 to 2·25 in. ; petiole ·25 to ·35 in., pubescent. *Flowers* solitary, or in pairs, extra-axillary, sub-sessile, ·75 in. in diam. when expanded, the buds globose ; pedicels ·1 in. long, coarsely hirsute, bracteate. *Sepals* much shorter than the petals, broadly ovate, sub-acute, strigose-pubescent outside and sub-glabrous inside as are the petals. *Petals* imbricate, spreading, lanceolate or oblanceolate-oblong, the outer at first much shorter than, but ultimately sub-equal to, the inner. *Male-flower :* stamens numerous, with transversely elongate, truncate, heads ; pistils 0. *Female flower* like the male but with fewer stamens ; pistils about 10, pubescent, 1-ovuled ; stigma short, flat, pubescent. *Carpels* 4 to 5, sub-cylindric, tapering to each end, ·75 in. long and ·25 in. diam., minutely granular and strigose ; stalks tomentose, ·15 in. long ; torus very small. *Seed* solitary, oblong, pale.

In leaves and in general facies this is very like *Popowia nervifolia,* Maing., but its petals are distinctly imbricate.

Perak on Ulu Bubong ; King's Collector, Scortechini.

6. ELLIPEIA NERVOSA, Hook. fil. and Thoms. Fl. Br. Ind. I, 52. A tree 40 feet high ; young branches glabrous, dark-coloured, slightly ridged. *Leaves* coriaceous with pellucid dots, elliptic-oblong, or lanceolate-oblong, acute or rarely shortly acuminate, the base acute ; upper surface glabrous ; the lower sparsely strigose, the reticulations transverse and very distinct ; main nerves 10 or 11 pairs, oblique, rather straight ; length 8 to 11 in., breadth 2 to 3·5 ; petiole ·35 to ·5 in. glabrous. *Flowers* polygamous, solitary, extra-axillary, rarely in pairs, ·75 in in diam., globose ; pedicels stout, ·1 to ·2 in. long, rufous pilose, bracteate. *Sepals* broadly ovate, acute, pubescent, much smaller than the petals. *Petals* white, spreading, imbricate ; the outer broadly ovate-oblong, ob-

276

tuse; the inner rather shorter and narrower, oblong; all pubescent especially externally. *Stamens* in the male flowers numerous, with roundish flat heads. *Ovaries* in the female flower many, curved. *Carpels* rather numerous, ovoid, slightly apiculate, narrowed into the stalk, rose-red when ripe (Wray), about 1 in. long and ·5 in. in diam., glabrous; their stalks ·75 to 1 in. long.

Malacca; Maingay, (Kew Distrib.), No. 47. Perak; common at low elevations. Penang; Curtis.

In the texture and nervation of its leaves this species has a strong resemblance to *Popowia nervifolia*, Maing. and other species in its neighbourhood. But the petals are not those of a *Popowia*, both rows being distinctly imbricate. The fruit moreover is larger than that of *Popowia*, and the albumen is much more cellular in structure being, in this respect, like that of *Ellipeia cuneifolia*, H. f. & Th.

## 6. CYATHOCALYX, Champion.

*Trees. Leaves* glabrous. *Flowers* fascicled, terminal or leaf-opposed. *Sepals* free or united into a 3-lobed cup. *Petals* 6, 2-seriate, valvate in bud, subequal, bases concave conniving, blade flat spreading. *Stamens* indefinite, long-cuneate, truncate; anther-cells linear, dorsal. *Ovaries* solitary or 2-6, on a concavo torus; stigma large, grooved; ovules many. *Ripe carpels* berried.—DISTRIB. Tropical India and Malaya; species 8.

Ripe carpels ovoid ... ... ... 1 *C. virgatus.*
Ripe carpels globular ... ... ... 2 *C. Maingayi.*

In its petals this genus resembles *Artabotrys* to some extent, but *Polyalthia* still more. The ovaries in the first two species are usually solitary; in the third they are 3 in number: the ripe carpels of all three being large succulent and many-seeded. Baillon admits the genus as it was established by Champion and accepted by Hooker filius & Thomson. In the above diagnosis I have however modified the definition so as to provide for the species with more than one ovary.

1. CYATHOCALYX VIRGATUS, King. A tree 40 to 60 feet high : young branches slender, pale, glabrous, the tips alone pubescent. *Leaves* membranous, elliptic-oblong to oblong-lanceolate, shortly and obtusely acuminate, the base cuneate or sometimes rounded; both surfaces shining, the lower rather darker when dry; the upper glabrous, the lower pubescent on the 8 or 9 pairs of sub-ascending rather prominent nerves : length 4 to 6·5 in., breadth 1·25 to 2·75 in.; petiole ·25 to ·35 in., pubescent. *Flowers* in axillary, sub-sessile fascicles of 2 or 3, about ·75 in. long. *Sepals* united at the base, ovate to ovate-lanceolate, spreading, tomentose, shorter than the inner petals. *Petals* tomentose-sericeous; the outer row much longer than the inner, lanceolate, much acuminate,

about, ·75 in. long.; inner row with orbicular concave base and much acuminate apex, ·5 in. long. Connective of *stamens* slightly produced at apex and obliquely truncate. *Ovaries* 4 to 6, hirsute; ovules many, 2-seriate; stigma thick, discoid, sessile; torus conic, truncate, pubescent. *Ripe carpels* solitary, or in pairs and divergent, oblong-ovoid, blunt at each end, minutely tomentose, 2 to 3 in. long, and 1 to 1·5 in. in diam.; pericarp thick; seeds 8 to 10, compressed, elongate and narrowly sub-reniform, transversely substriate. *Unona virgata*, Blume Bijdr. 14; Fl. Javæ Anon. 43 t. 19 and 25B.; Miq. Fl. Ind. Bat., I. Pt. 2, p. 42. *Meiogyne virgata*, Miq. Ann. Mus. Lugd. Bat. II., 12. *Cananga virgata*, Hook fil. and Thoms. Fl. Br. Ind. I, 57.

Malacca: Maingay (Kew Distrib.), No. 92. Perak; King's collection. Distrib. Java.

Blume describes the carpels as from 3 to 5; but I have never found more than two, and it is difficult to understand how more can come to perfection on the comparatively small torus. In Java this is said often to be a bush from 6 to 8 feet high: in Perak it is a tall tree.

2. CYATHOCALYX MAINGAYI, Hook. fil. and Thoms. Fl. Br. Ind. I, 53. A tree 50 or 60 feet high: young branches rather stout, puberulous, speedily glabrous and dark-coloured. *Leaves* elliptic to oblong, thinly coriaceous, slightly obovate, shortly caudate-acuminate, the base rounded or slightly cuneate; upper surface shining, quite glabrous; the lower puberulous when young, ultimately glabrous; the main nerves 13 to 15 pairs, bold and prominent, spreading, interarching near the edge: length 5·8 to 8·8 in., breadth 2·75 to 3·75 in., petiole ·3 in. *Flowers* 2 to 3 in. in diam., solitary or in short, 2- to 3-flowered racemes, axillary or extra-axillary: pedicels ·5 to ·75 in. long with a large stem-clasping bracteole near the apex. *Sepals* spreading or sub-reflexed, ovate, sub-acute, slightly connate at the base, puberulous on both surfaces, ·4 in. long. *Petals* thinly coriaceous, subequal, puberulous, obovate or broadly obovate-lanceolate, blunt, the base with a short claw, pale greenish with a blotch of reddish yellow at the base, all (but especially the inner row) more or less convex, the inner row slightly concave and glabrous at the base inside. *Stamens* numerous, cuneate, short; the connective produced into a broad, flat, orbicular, oblique expansion which over-hangs the dorsal linear anthers. *Ovaries* 3, narrowly ovoid, pubescent, ovules about 10 in 2 rows: style short, lateral: stigma large, lobed, villous. *Ripe carpels* 1 or 2, globular, 1·5 to 1·75 in. in diam., slightly tubercular when dry and minutely pubescent. *Seeds* 10 in 2 rows, elongated, compressed.

Malacca: Maingay (Kew Distrib.), No. 94. Singapore: Ridley. Perak: King's Collector.

This species is doubtfully referred to *Cyathocalyx* by its authors, and chiefly on the ground that the petals, although valvate at the base, are slightly imbricate above. An examination of the large number of specimens sent from Perak by the Calcutta Botanic Garden Collector enables me to state that in bud the petals are truly valvate, but that as they develope they undoubtedly overlap. The anthers, ovaries and and ripe fruit appear to me to be those of *Cyathocalyx*; and in habit and general appearance of its leaves this plant agrees with the other species above described. In addition to the species above described, there are, in the Calcutta Herbarium, fruiting specimens from Perak of a small tree which is apparently a fourth species of *Cyathocalyx*. The leaves of this are oblong-lanceolate to oblong-ovate, 8- to 10-nerved, glabrous above and puberulous beneath; and the ripe carpels are in pairs, ovoid, puberulous, about 1·5 in. long. None of the specimens has any trace of flower.

### 7. ARTABOTRYS, R. Brown.

Sarmentose or scandent shrubs. *Leaves* shining. *Flowers* solitary or fascicled, generally on woody, usually hooked, recurved branches (peduncles). *Sepals* 3, valvate. *Petals* 6, 2-seriate, bases concave connivent; limb spreading, flat, sub-terete or clavate. *Stamens* oblong or cuneate; connective truncate or produced; anther-cells dorsal. *Torus* flat or convex. *Ovaries* few or many; style oblong or columnar; ovules 2, erect, collateral. *Ripe carpels* berried.—DISTRIB. Tropical Africa and Eastern Asia; described species about 32.

This genus is at once distinguished by the curious hooked flower-peduncles. The petals are thick and mostly narrow, concave and closely connivent at the base, while the limb is spreading. The habit of all is scandent. Besides those described below, there are in the Calcutta Herbarium imperfect materials of five undescribed species from Perak, and of one from the Andaman Islands.

Petals lanceolate to elliptic.
  Flowers less than 1 in. long.
    Petals very fleshy, broadly elliptic, blunt   1. *A. grandifolius.*
    ,,  coriaceous, broadly lanceolate, acu-
       minate ...   ...   ...   2. *A. Scortechinii.*
    ,,  slightly fleshy, elliptic-oblong, ob-
       tuse ...   ...   ...   3. *A. pleurocarpus*
  Flowers about 1 in. long.
    Outer petals ovate-lanceolate; the inner
    lanceolate or linear   ...   ...   4. *A. venustus.*
  Flowers more than 1 in. long.

Leaves elliptic to oblong, obtuse or shortly
and bluntly mucronate, coriaceous ... 5. *A. crassifolius.*
Leaves oblong, acuminate, coriaceous ... 6. *A. oblongus.*
Leaves oblong-lanceolate.
  Leaves shortly caudate-acuminate,
  flower nearly 2 in. long ... 7. *A. Lowianus.*
  Leaves shortly acuminate ; flower 1·5
  to 1·75 in. long ; ripe carpels nar-
  rowly elliptic, tapering to both ends,
  glabrous ... ... ... 8. *A. oxycarpus.*
Limb of petals linear, sub-triquetrous, cylindric, or
sub-clavate.
  Petals thickly coriaceous, linear, blunt, ad-
  pressed-pubescent ... ... 9. *A. speciosus.*
  Petals linear-oblong, obtuse, (glabrous ?) 10. *A. Maingayi.*
  Petals fleshy, the outer 3 flattened ; the
  inner 3 obtusely triquetrous ... 11. *A. gracilis.*
  Petals fleshy, the limb cylindric to clavate 12. *A suaveolens.*
  Imperfectly known species ... ... 13. *A. costatus.*
    *     *     *    ... ... 14. *A. Wrayi.*

1. ARTABOTRYS GRANDIFOLIUS, n. sp. King. A powerful creeper 60
to 80 feet long ; young branches stout, pale, striate, glabrous. *Leaves*
thinly coriaceous, large, minutely pellucid-punctate, pale yellowish-green
when dry, elliptic-oblong to elliptic-obovate ; the apex broad, obtuse or
abruptly sub-acute ; the base cuneate : both surfaces glabrous, distinctly
reticulate, the upper shining, the lower duller : main nerves 10 to 12
pairs, oblique, inter-arching boldly ·25 in. from the edge ; length of
blade 8 to 14 in., breadth 3 to 5 in. : petiole ·4 in., stout. *Petals* very
fleshy, densely and minutely tomentose, unequal ; the outer 3 broadly
elliptic, sub-acute or blunt, slightly concave, ·75 in. long and ·4 in. broad :
inner 3 obovoid, spreading but with incurved apices, slightly shorter
than the outer. *Peduncles* (in fruit) nearly 3 in. long, stout : torus
hemispheric, 1 in. in diam. *Ripe carpels* numerous, glabrous, lenticellate,
elliptic-obovoid, the apex mammillate, narrowed at the base into a short
stout pseudo-stalk nearly ·5 in. long ; length of ripe carpel about 1·5 in.,
diam. 1 in. : pericarp hard, about ·1 in. thick. *Seed* solitary, narrowly
ellipsoid, blunt, 1·1 in. long, and ·6 in. in. diam. ; the testa pale, ruguloso.
*A. macrophyllus*, King MSS. (not of Hook. fil).

    Perak ; at Goping, elevation 500 to 800 feet, King's Collector, No.
4477 ; Scortechini No. 1068.

    Some specimens of this were unfortunately distributed from the
Calcutta Herbarium under the MSS. name of *A. macrophyllus*,—a name

pre-occupied by an African species described by Sir J. D. Hooker (Niger Flora, 207).

2. ARTABOTRYS SCORTECHINII, n. sp. King. A climber. All parts except the flower and possibly the fruit glabrous : young branches slender, dark-coloured. *Leaves* thinly coriaceous, ovate-lanceolate, shortly acuminate, the base cuneate ; upper surface shining ; the lower dull when young, very minutely scaly, afterwards glabrous ; main nerves 9 to 11 pairs, spreading, inter-arching ·1 in. from the edge, slender but rather prominent beneath : length of blade 2·25 to 3·25 in., breadth ·9 to 1·3 in., petiole 2 in. *Peduncle* rather slender, 3-to 4-flowered ; pedicels ·5 in. long, thickened upwards, puberulous, with a small ovate bracteole at the very base. *Flowers* ·6 to ·8 in. long. *Sepals* very coriaceous, triangular, acuminate, the apices slightly reflexed, conjoined at the base only, rugulose and adpressed-pubescent externally, ·25 in long. *Petals* coriaceous, broadly-lanceolate acuminate, tomentose on both surfaces, the inner three smaller than the outer 3. *Anthers* with broad connectival apical appendages. *Torus* rather flat, sericeous : ovaries glabrous. *Fruit* unknown.

Perak, Scortechini.

A species near *A. polygynus*, Miq., but with glabrous leaves and different flowers from that species.

3. ARTABOTRYS PLEUROCARPUS, Maingay in Hook. fil Fl. Br. Ind. I, 54. A large climber ; all parts except the flowers glabrous ; young branches lenticellate, striate, dark-coloured. *Leaves* coriaceous, oblanceolate-oblong, the apex abruptly and shortly acuminate, the base much narrowed : both surfaces shining and reticulate, the upper paler ; main nerves about 10 pairs, spreading, slender : length of blade 4 to 6·5 in , breadth 1·5 to 2·25 in.; petiole ·15 in., thick. *Peduncles* flat, stout, much hooked, bearing several ebracteolate pedicels, ·5 in. long, densely pubescent. *Flowers* 1·5 in. long. *Sepals* broadly ovate, obtuse. *Petals* subequal, flat, elliptic-oblong, obtuse, pubescent on both surfaces, the outer 1 to 1·35 in. long, the inner smaller. *Anthers* with apiculate connectives. *Ovaries* many, slender. *Ripe carpels* broadly elliptic, mammillate, obscurely grooved, narrowed into the short stout stalk, ·75 in. long. *Seeds* 2, with hard testa.

Malacca ; Maingay. Perak, Scortechini, No. 331.

4. ARTABOTRYS VENUSTUS, n. sp., King. A large climber, 30 to 80 feet long ; young branches at first puberulous, afterwards glabrous, dark coloured, striate. *Leaves* coriaceous, elliptic to elliptic-oblong, abruptly and shortly acuminate, the base rounded or very slightly narrowed : both surfaces glabrous, the upper shining, the lower dull, adult leaves pale brown (when dry) : main nerves 7 to 10 pairs, spreading

or sub-ascending, curved, inter-arching freely ·1 to ·2 in. from the edge, prominent on the lower, less so on the upper, surface; length of blade 3·5 to 6 in., breadth 2 to 3 in., petiole ·2 to ·25 in. *Peduncles* extra-axillary, rather slender in flower, (stout in fruit), minutely tomentose, bearing 3 or 4 flowers, ·75 to 1 in. long.; pedicels slender, pubescent or glabrescent., from ·5 to 1 in. long, ebracteate. *Sepals* coriaceous, broadly triangular, sub-acute, slightly conjoined at the base, sub-reflexed, puberulous externally, glabrous within, ·15 in. long. *Petals* coriaceous, minutely tomentose, subequal; the outer 3 with small claw, glabrous inside, ovate-lanceolate sub-acute; the inner 3 shorter than the outer, lanceolate or linear. *Anthers* short, slightly compressed; the apex orbicular, flat. *Ovaries* about 10, oblong, granular. *Carpels* about 6, sessile, narrowly obovoid, apiculate, slightly narrowed to the base, at first puberulous, ultimately glabrous, 1·5 in. long and ·8 in. in diam.; pericarp thin. *Seeds* 2, oblong, plano-convex, about 1 in. long and ·6 in. broad, smooth.

Perak; at elevations up to 1,000 feet, King's Collector, Nos. 3725, 4392, 6499, 6968, King's Collector.

5. ARTABOTRYS CRASSIFOLIUS, H. f. and T. in Hook. fil. Fl. Br. Ind. I, 54. A large climber; young branches minutely rusty-tomentose. *Leaves* very coriaceous when adult, elliptic to oblong, obtuse or shortly and bluntly mucronate, the base acute or rounded: upper surface glabrous, shining: the lower dull, paler in colour when young, sparsely adpressed-pilose, afterwards glabrous; main nerves 9 or 10 pairs, oblique, when dry faintly impressed on the upper and slightly prominent on the lower surface; length of blade 6 to 6·5 in., breadth 1·75 to 2·75 in.; petiole ·3 to ·4 in., stout. *Peduncles* flat, much hooked, stout: each with several stout rusty-tomentose pedicels ·3 to ·4 in. long; bracts few, ovate. *Flowers* 1·25 in. long. *Sepals* ovate-lanceolate, sub-obtuse, softly rusty-pubescent outside, pubescent within. *Petals* coriaceous, oblong-lanceolate, sub-ovate, densely tomentose on both surfaces; the inner 3 smaller than the outer 3. *Fruiting pedicel* very stout; the torus sub-globose. *Ripe carpels* about 8, sessile, sub-obovoid to ovoid, glabrous, slightly rugose, 1·25 to 1·65 in. long and ·75 to 1·15 in. in diam.; pericarp thick, pulpy. *Seeds* 2, collateral, oblong, compressed, grooved along the edge, ·9 in. long and ·6 in. broad. Kurz For. Flora Burma, I, 30.

Burmah; Martaban, King, Brandis. Perak; King's Collector, No. 8384.

6. ARTABOTRYS OBLONGUS, n. sp., King. A climber 50 to 70 feet long, ultimately all parts except the inflorescence glabrous; young branches slender, rufous-pubescent; the bark dark-coloured when very young, afterwards rather pale, striate. *Leaves* when adult coriaceous, oblong, shortly acuminate, the base acute, when adult both surfaces

34

glabrous, the upper shining, the lower dull and when young sparsely
pubescent along the midrib; main nerves 10 to 12 pairs, inconspicuous
on the upper, slightly prominent in the lower surface, spreading, forming 2 or 3 series of arches within the margin; length of blade 6·5 to
9 in., breadth 2·5 to 3 in., petiole ·4 in. *Peduncles* stout, pubescent
when young, bearing 3 or 4 pedicels; flowers 1·35 in. long; pedicels
about 1 in., pubescent, slightly thickened upwards. *Sepals* coriaceous,
triangular, acute, concave, spreading rufous-pilose on both surfaces,
slightly conjoined at the base, ·25 in. long. *Petals* coriaceous, the portion above the saccate base lanceolate, subacute, strigosely tomentose on
both surfaces, the claw partly glabrous and partly covered with minute
white hair. *Anthers* compressed, with oblong, obliquely truncate, flattened heads. *Ovaries* few, oblong, glabrous; the stigma broad, oblique.
*Fruit* unknown.

Perak; King's Collector, No. 6524.

7. ARTABOTRYS LOWIANUS, n. sp., Scortechini MSS. A stout
climber; all parts except the flowers glabrous; young branches slender,
dark-coloured. *Leaves* thinly coriaceous, oblong-lanceolate, shortly
caudate-acuminate, the base cuneate : both surfaces shining, minutely
reticulate; main nerves 8 to 10 pairs, spreading, inter-arching ·2 in.
from the margin, faint; length of blade 3·5 to 6 in., breadth 1·25 to
1·75 in., petiole ·25 in. *Peduncles* extra-axillary, 2- to 3-flowered, glabrous; pedicels thickened upwards, ·5 to 75 in. long, glabrous. *Sepals*
triangular, acute, glabrous, ·25 in. long, enlarging a little with the fruit.
*Petals* fleshy, adpressed-puberulous, elliptic-lanceolate above the concave
base, obtuse; the outer three 1·75 in. long, the inner three smaller.
*Anthers* with a rounded apical process from the connective. *Ovaries*
many, glabrous. *Carpels* (quite young) sessile, ovoid, apiculate; ripe
carpels unknown.

Perak; Scortechini; No. 2012.

This species is near *A. pleurogynus*, Miq , but is perfectly glabrous, not sub-strigose pubescent; its ripe fruit is unknown.

8. ARTABOTRYS OXYCARPUS, n. sp., King. A stout climber, 60 to 80
feet long; all parts except the flower glabrous; young branches slender,
black when dry. *Leaves* oblong-lanceolate, shortly acuminate, the base
cuneate, both surfaces shining, reticulate; main nerves 6 to 8 pairs,
spreading, slender; length of blade 3 to 5·5 in., breadth 1·25 to 1·5 in.
*Peduncles* short (·75 in. long), glabrous, bearing about 2 minutely bracteolate pedicels ·75 in. long. *Flowers* 1·5 to 1·75 in. long. *Sepals*
coriaceous, small, broadly ovate, acute, ·2 in. long, conjoined at the base,
spreading. *Petals* coriaceous, very much longer than the sepals, lanceolate, obtuse; the inner 3 smaller; all adpressed-pubescent, and the

35

saccate base small in all. *Torus* small, sericeous. *Ovaries* glabrous.
*Ripe carpels* numerous, sessile, glabrous, narrowly elliptic, tapering to
each end, the apex caudate, 1 to 1·2 in. long and ·4 in. in diam.; pericarp
thin. *Seeds* 2, plano-convex, compressed, blunt, ·25 in. long.

Perak ; King's Collector, Nos. 5150 and 5605 ; Wray No. 3286.

This species comes near the Bornean *A. polygynus*, Miq. (Ann. Mus.
Lugd. Bat. II, 4). But this species has more pointed and perfectly smooth
ripe carpels ; while those of *A. polygynus* are more ovoid, with shorter
terminal point and have many vertical ridges. *A. polygynus* moreover
is sub-strigosely pubescent, this is glabrous.

9. ARTABOTRYS SPECIOSUS, Kurz in Hook. fil. Fl. Br. Ind. I, 55.
A large climber : young branches slender, dark-coloured, sparsely adpressed-pilose, afterwards glabrous. *Leaves* coriaceous, oblong or oblong-
lanceolate, rarely oblanceolate, shortly and obtusely acuminate, the base
acute ; both surfaces glabrous, shining : main nerves 7 to 10 pairs,
spreading, inter-arching at some distance from the edge, slender : length
of blade 6 to 8 in., breadth 2 to 2·5 in., petiole ·25 in. *Peduncles* extra-
axillary, flattened, short and not much hooked, puberulous, each bearing
several short puberulous 1-flowered ebracteolate flower-pedicels : flowers
from 1·25 to nearly 2 in. long, yellow. *Sepals* ·2 in. long, broadly ovate,
acute, pubescent outside, glabrous inside. *Petals* thickly coriaceous,
adpressed-pubescent, linear above the concave base, rather blunt ; the
inner smaller than the outer ; torus pilose : fruit unknown. Kurz For.
Flora, Burm. I, 32.

Andaman Islands ; along Middle Straits, Kurz. S. Andaman ; at
Caddellgunge, King's Collector.

10. ARTABOTRYS MAINGAYI, H. f. and T. in Hook. fil. Fl. Br. Ind. I,
55. A powerful creeper, 40 to 80 feet long : all parts glabrous except
the flowers ; the young branches slender, dark-coloured. *Leaves* thin,
elliptic, acuminate at base and apex : both surfaces shining, finely reticulate : main nerves 7 to 9 pairs, spreading, faint : length of blade 3·5
to 6 in., breadth 1·35 to 2 in., petiole ·25 to ·5 in. *Peduncles* flat, much
curved, glabrous. *Flowers* 1 in. in diam., fascicled, peduncle ·5 to
1·5 in., hoary-pubescent. *Sepals* small, obtuse, ·2 in. long. *Petals :* the
outer linear-oblong, obtuse, concave the saccate base small and sub-
orbicular, 1 to 1·25 in. long and ·25 to ·35 broad ; the inner smaller and
narrower and much curved. *Ovaries* 3 or 4 ovoid, glabrous. *Ripe
carpels* sessile, elliptic-globose, mammillate, yellow, glabrous, when ripe
2·5 in., long and 1·5 in. in diam. *Seeds* 2, plano-convex, testa stony.

Malacca ; Maingay.

11. ARTABOTRYS GRACILIS, n. sp. King. A slender woody climber,
60 to 80 feet long : young branches dark-coloured : all parts quite
284

glabrous except the petals. *Leaves* thinly coriaceous, ovate-lanceolate, shortly acuminate, the base cuneate; both surfaces glabrous and
shining, the upper when dry tinged with green : main nerves 7 or 8 pairs,
spreading, inter-arching inside the edge, very faint on both surfaces,
reticulations rather distinct : length of blade 2·5 to 3 in., breadth 1 to
1·75 in., petiole ·15 to ·2 in. *Peduncles* extra-axillary, short, much
hooked, glabrous, usually 4- to 6-flowered ; pedicels ·35 in. long, thickened upwards, ebracteolate, glabrous : flower ·3 to ·4 in. long. *Sepals*
very coriaceous, semi-orbicular, slightly pointed at the apex, very little
conjoined at the base, concave, spreading *Petals* fleshy, sub-equal,
curved, spreading, densely tomentose, the outer 3 flattened ; the inner
obtusely 3-angled, tumid at the base, smaller than the outer 3. *Anthers*
with broad apical connectival processes. *Ovaries* 3 or 4, oblong, with
large discoid lobed stigmas, torus villous. *Ripe carpels* 3 or 4, sessile,
obovoid, with several vertical ridges, the base contracted, glabrous, ·8
in. long and ·7 in. in diam. *Seeds* 2, compressed-ovoid, obtuse at each
end, shining.

Perak : at low elevations, King's Collector, Nos. 3746, 4987 and
7543.

Allied to *A. suaveolens*, Bl. ; but with differently shaped petals, pistils
and carpels.

12. ARTABOTRYS SUAVEOLENS, Blume Fl. Javae Anou. 62, t. 30, 31D.
A climber 20 to 30 feet long ; the petals always tomentose, the other
parts mostly glabrous, but sometimes the young branches, peduncles, and
under surfaces of the midribs of the leaves adpressed-puberulous. *Leaves*
thinly coriaceous, oblong-lanceolate to ovate-lanceolate, acute or shortly
acuminate, the base acute ; both surfaces shining, the reticulations rather
distinct, the upper often deeply tinged with green when dry. *Peduncles*
extra-axillary, thin at first, but becoming stout and flat with age, glabrous
or puberulous, bearing from 5 to 15 flowers ; pedicels ·3 to ·45 in. long,
thickened upwards, sparsely adpressed-pubescent, with a small narrowly
ovate bract at the base ; flowers about ·4 in. long. *Sepals* broadly ovate,
the apex pointed, thinly coriaceous, sparsely adpressed-pubescent externally, very slightly conjoined at the base, spreading, ·1 in. long.
*Petals* fleshy, adpressed-tomentose, dilated and thin at the base, the limb
cylindric to clavate, sub-erect, slightly spreading, sometimes with the
apex incurved. *Anthers* short, with a very broad oblique flattened apical
appendage from the connective ; torus slightly pubescent. *Ovaries*
broadly ovoid, sub-compressed, the stigma small. *Ripe carpels* few, ellipsoid, the apex blunt, the base slightly contracted, smooth, glabrous, ·4 to
·5 in. long and ·25 in. in diam. ; pericarp thin, fleshy. *Seed* single, ellipsoid, blunt at each end, the testa granular. Wall. Cat. 6416 ; H. f. & T.

Fl. Ind., 129 ; Hook. fil. Fl. Br. Ind. I, 55 ; Miq. Fl. Ind. Bat. I. Pt. 2, 39
Ann. Mus. Lugd. Bat. II, 43 ; Kurz For. Fl. Burm. I; *Artabotrys
parviflora*, Miq. Fl. Ind. Bat. Supp., 375. *Unona suaveolens*, Blume
Bijdr. 17.

In all the Malayan Provinces at low elevations : common. Sylhet
to Malacca in British India.

This species varies somewhat as to size of flowers and texture of
leaf. The form named *A. parviflora* by Miq. in his Sumatra Supplement was, by himself, subsequently reduced to a variety of this
species (Ann. Mus. Lugd. Bat. II, 38).

13. ARTABOTRYS COSTATUS, n. sp. King. A climber from 15 to 80
feet long : young branches slender, dark-coloured, scantily tawny-pu-
berulous when young, afterwards glabrous. *Leaves* thinly coriaceous,
elliptic-oblong, slightly oblanceolate, abruptly and shortly acuminate,
the base cuneate ; upper surface shining, glabrous except the lower part
of the midrib which is tomentose ; lower surface paler, dull, sparsely
puberulous towards the base when young, afterwards glabrous ; main
nerves 12 to 14 pairs, spreading, forming one series of very bold arches
·3 in. from the margin, with a series of smaller arches outside it, very stout
and prominent on the lower, slightly so on the upper, surface, reticulations distinct on both : length of blade 7 to 9 in., breadth 2·5 to 3·25 in.,
petiole ·2 in. *Peduncles* rather small, much hooked. *Flowers* unknown.
*Carpels* (*unripe*) 2 to 5, sessile, ellipsoid, blunt at each end, about 1 in. long
and ·6 in. in diam, (unripe), glabrous : pericarp thin ; seeds 2, elliptic.

Perak ; on Ulu Bubong at elevations of from 500 to 800 feet, King's
Collector, Nos. 4291 and 10184.

I have ventured to describe this although its flowers are unknown,
and the only fruit collected is unripe. By its oblong costate leaves it
differs from every other described *Artabotrys* except *A. macrophyllus*,
mihi.

14. ARTABOTRYS WRAYI, King. A climber : young branches rather
stout, softly pale rusty-tomentose ; ultimately glabrous pale and furrowed. *Leaves* thinly coriaceous, large, oblong-elliptic to elliptic, shortly
acuminate, the base rounded ; both surfaces boldly reticulate ; the upper
glabrous and shining, sub-bullate when dry ; the lower shortly and
rather softly cinereous-pubescent ; main nerves 10 to 12 pairs, oblique,
curving, inter-arching freely within the edge, depressed above and bold
and prominent beneath like the midrib ; length 8 to 11 in., breadth 2·75
to 5 in., petiole ·35 in., stout, tomentose when young, glabrescent when
old. *Peduncles* extra-axillary, rather short, very thick in fruit, sometimes straight when young and curving only when in fruit, few-flowered,
glabrous ; pedicels 1 in. long, stout, softly tawny-tomentose with several

bracteoles at the base. *Flowers* 1 in. long. *Sepals* broadly ovate at the base, tapering rapidly upwards, acuminate, about ·5 in. long, densely sericeous-tomentose outside, sub-glabrous inside especially at the base. *Petals* thick, sub-equal, ovate-oblong, sub-acute, slightly contracted above the claw, softly adpressed-sericeous except on the glabrous concavity of the claw inside. *Ovaries* numerous. *Ripe carpels* obovoid, tapering much to the base, the apex mucronate, densely tawny-tomentose, sessile ; nearly 1 in long.

Perak ; Wray, King's Collector.

Next to *A. grandifolius*, this has the largest leaves of any of the Asiatic species of the genus, but from that species it differs in having them pubescent beneath. Only a single flower has hitherto been collected.

## 8. Drepananthus, Maingay MSS.

Trees. *Leaves* large, pubescent beneath. *Racemes* very short, fascicled on woody truncal tubercles. *Sepals* 3, nearly free. *Petals* 6, valvate, 2-seriate, subequal ; bases concave, connivent ; limb erect or spreading, broad or narrow. *Stamens* many, cuneate, truncate ; anthers linear, cells lateral ; connective very slightly produced. *Ovaries* 4–12 ; stigma sub-sessile ; ovules 4 or more, 2-seriate. *Ripe carpels* globose, several-seeded. Two species.

This genus differs from *Artabotrys* in its members being trees, not climbers ; and in having 4 or more ovules in its ovaries. Dr. Scheffer (Ann. Jard. Bot. Buitenzorg II, 6) proposed to make it a section of *Cyathocalyx.*

Petals of both rows with more or less ovate limb 1. *D. pruniferus.*
    ,,       ,, with narrowly cylindric limb 2. *D. ramnliflorus.*

1. Drepananthus pruniferus, Maing. in Hook. fil. Fl. Br. Ind. I, 56. A tree 40 to 50 feet high ; branches stout, rufous-pubescent at first, finally glabrescent. *Leaves* coriaceous, elliptic to elliptic-oblong, acute or obtuse, the base rounded or sub-cordate, often unequal ; upper surface glabrous, except the depressed tomentose midrib and main nerves ; lower surface shortly rufous-pubescent when young, glabrescent when adult ; main nerves 14 to 16 pairs, prominent beneath ; intermediate nerves stout, parallel, oblique ; length 7·5 to 14 in., breadth 3 to 6·5 in. ; petiole ·5 to 1·5 in. stout, channelled. *Racemes* 6- to 8-flowered, crowded ; flowers ·75 in. long, their pedicels rufous-tomentose, ·5 to ·75 in. long, each with a large oblanceolate bract. *Sepals* and *petals* subequal, very coriaceous, densely covered (except the inside of the claws of the petals) with a layer of minute whitish tomentum ; sepals united by their base, ovate-oblong, spreading ; petals of outer row broadly ovate,

sub-acute, slightly constricted above the claw; those of the inner row
closely connivent, much constricted above the claw, their apices broad
and emarginate. *Ovaries* oblong, sericeous-tomentose. *Ripe carpels* 6
to 8, sessile, sub-globose, minutely pubescent to glabrescent, 1 to 1·25
in. in diam. *Seeds* numerous, oblong, flat, shining.

Malacca: Maingay (Kew Distrib.) No. 90. Perak; King's Collector,
Scortechini. Penang, Curtis No. 1417.

2. DREPANANTHUS RAMULIFLORUS, Maing. Hook. fil. Fl. Br. Ind. I,
56. A tall tree, the young branches as in *D. pruniferus*. *Leaves* as in *D.
pruniferus*, but slightly broader at the apex and narrowed at the base.
*Flowers* ·4 to ·5 in long, much crowded in very short fascicles from
tubercles on the branches below the leaves: pedicels about ·3 in. long
stout, rufous-tomentose as is the single sub-orbicular bracteole. *Sepals*
much shorter than the petals, broadly triangular, acuminate, spreading,
rufous-tomentose especially outside. *Petals* with concave, connivent,
tomentose claw and fleshy, sub-cylindric, spreading, much curved, adpressed-pubescent limbs. *Ovaries* about 5, sessile, oblong. *Carpels*
(young) ovoid, slightly oblique, densely rufous-tomentose; walls of pericarp very thick: seeds few: ripe fruit unknown.

Malacca: Maingay (Kew Distrib.), No. 91. Distrib. Sumatra;
Forbes, No. 2913.

9. CANANGIUM, Baill. (*Cananga*, Rumph.)

Tall trees. *Leaves* large. *Flowers* large, yellow, solitary or fascicled
on short axillary peduncles. *Sepals* 3, ovate or triangular, valvate.
*Petals* 6, 2-seriate, subequal or inner smaller, long, flat, valvate. *Stamens*
linear, anther-cells approximate, extrorse; connective produced into a
lanceolate acute process. *Ovaries* many; style oblong (or 0?); stigmas
sub-capitate; ovules numerous, 2-seriate. *Ripe carpels* many, berried,
stalked or sessile. *Seeds* many, testa crustaceous, pitted, sending spinous
processes into the albumen.—Two species.

The tree known as *Cananga odorata* H. f. and T. was by Rumphius
(who wrote an account of it in Herb. Amb. II, 195, published in 1750)
named *Cananga* (Latinice) and *Bonga Cananga* (Malaice). Rumphius' description is of the usual pre-Linnæan sort, there being no differentiation
of generic and specific characters and his name of course is not binomial.
In the chapter of his book following that in which *Cananga* proper is
treated of (*l. c.* p. 197), Rumphius proceeds to describe the wild *Canangas*
as distinguished from the *Cananga* proper, which was in his time, (as it is
still) much cultivated by the Malays on account of the fragrance of its
flowers. These wild *Canangas* Rumphius calls *Canangæ sylvestres* and
of them he distinguishes three sorts.

1. *Cananga sylvestris prima sive trifoliata* (Malaice *Oetan*).
2. *Cananga sylvestris secunda sive angustifolia.*
3. *Cananga sylvestris tertia sive latifolia.*

Of the first two Rumphius gives figures on t. 66 of the same volume ; and judging from these figures, the plants fall into the modern genus *Polyalthia.*

Linnæus' Species Plantarum was published in 1753, therefore Rumphius' names are in point of time, as they are in point of form, pre-Linnæan. Linnæus does not accept *Cananga* as a genus and he refers to the *Cananga* of Rumphius only in a note under *Uvaria Zeylanica.* And the first botanists to adopt the *Cananga* of Rumphius as a genus are Hook. fil. and Thomson (in Fl. Ind. 130). But in 1775 Aublet (in his *Histoire des Plantes de la Guiane Francaise,*) published, in regular Linnæan fashion, the genus *Cananga* for the reception of a single species named *C. ouregow* of which he gave a figure (t. 244). Nineteen years later (1794) Ruiz and Pavon, (in their *Prodromus Floræ Peruvianæ et Chilensis,*) published under the name of *Guatteria* a genus with exactly the same characters as Aublet's *Cananga.* Unless therefore Hook f. and Thomson are right in making a special case in establishing, as a genus in the Linnæan sense, the *Cananga* of Rumphius, Aublet's genus *Cananga* must stand, and to it must be relegated all the American species referred to Ruiz and Pavon's genus *Guatteria.* Authorities vary in their treatment of the *Cananga* of Rumphius. Dunal (in his *Monographie de la famille des Anonacees*) pronounces for the suppression of Aublet's *Cananga* in favour of that of Rumphius who, he incorrectly says, assigned *two* species to it ; the fact being as already shown, that Rumphius divided *Cananga* into (*a*) cultivated (with one sort) and (*b*) wild (*sylvestres*) with three sorts. Dunal (and I think wrongly) refers all the *Cananga* of Rumphius to *Unona.* In their Genera Plantarum, Mr. Bentham and Sir J. D. Hooker retain the *Cananga* of Rumphius and reduce *Cananga* of Aublet to *Guatteria.* Baillon, on the other hand, retains the *Cananga* of Aublet as a genus, and to it refers all the S. American species of *Guatteria.* He reduces *Cananga odorata* H. f. and Th. to *Unona* and, altering the termination of its generic name, he makes it a section of *Unona* under the sectional title of *Canangium.*

The grounds for separating *Cananga* from *Unona* as a genus are thus stated by the authors of the Flora Indica. " In habit and general appearance this genus closely resembles *Unona* ; but the indefinite ovules prevent its being referred to that genus. The peculiar stamen (with a long conical apical point) and the seeds are themselves, we think, sufficient to justify us in distinguishing it as a genus." The simplest solution of the synonymic knot, and one for which there is some justi-

41

fication on the ground of structure, appears to lie in the acceptance of Baillon's suggested name, giving up that of the authors of the Flora Indica.

The synonymy of *Guatteria* is further complicated by the fact that a large number of species with valvate æstivation were referred to it by Wallich and others. These, however, were separated by Hook fil. and Thoms. by whom the genus *Polyalthia* was formed for their reception. Sir Joseph Hooker refers to *Cananga*, not only the species *C. odorata*, but another named *C. virgata*. The latter plant appears to me, in the light of full material recently received, to be a typical *Cyathocalyx*, and to that genus I have ventured to remove it. A third species doubtfully referred to the genus *Cananga* under the specific name *monosperma*, appears to me from the description (I have seen no good specimen) to be so doubtful that I exclude it altogether. The seeds both of this species and of *C. Odoratum* are peculiar; I quote the following excellent description of those of *C. odoratum* from Hooker fil. and Thomson's Flora Indica, page 130. " The seeds are pitted like those of the section *Kentia* of *Melodorum*, and of some *Cucurbitaceæ*; and the inner surface of the brownish-yellow, brittle testa is covered with sharp tubercles, which penetrate into the albumen, taking the place of the flat plates which are found in the rest of the order."

Flowers 2 or 3 in. long ... ... 1 *C. odoratum.*
  „  1 to 1·25 in. long ... ... 2 *C. Scortechinii.*

1. CANANGIUM ODORATUM, Baill. Hist. des Plantes, I, 213 (*in note*). A tree 30 to 60 feet high ; young branches rather slender, sub-striate, at first puberulous, slightly lenticellate, dark ashy-coloured when dry. *Leaves* membranous, ovate-oblong or oblong-lanceolate, sometimes broadly elliptic, acute, shortly acuminate or sub-obtuse ; the base rounded or sub-cuneate, unequal; quite glabrous, the midrib and nerves puberulous ; main nerves about 8 pairs, ascending, rather straight and slender: length 3·5 to 8 in., breadth 1·75 to 3 in., petiole ·5 in. *Flowers* 2 to 3 in. long, drooping, in 2- to 3-flowered shortly pedunculate racemes : pedicels slender, 1·5 to 2 in. long, recurved, puberulous, with one median and several basal, small, often deciduous bracts. *Sepals* free or joined at the base only, about ·35 in. long, triangular, tapering to a blunt point, reflexed. *Petals* linear-lanceolate, 3 to 3·25 in. long and ·3 in. wide, adpressed-sericeous when young. *Ovaries* sessile, narrowly oblong : stigma hemispheric. *Ripe carpels* from 10 to 12, pedicellate, oblong-obovoid, glabrous, blunt, ·65 to ·9 in. long, nearly black when ripe, pulpy : stalks from ·5 to 75 in. long. *Seeds* 6 to 12, flattened, sub-ovate. *Cananga odorata*, H. f. and Th. Fl. Ind. 130 ; Fl. Br. Ind. I, 56 ; Miq. Fl. Ind. Bat. I, Pt. 2, 40. Kurz For. Fl. Burm. I, 3. *Uvaria odorata*,

290

Lamb. Ill t. 49°, f. 1; Roxb. Fl. Ind. ii. 661; Wall. Cat. 6457; W. & A.
Prodr 8; Blume Bijdr. 14, Fl. Jav. Anon. t. 9. Pierre Flore For. Coch.
Chine, Anon. t. 18; Griff. Notul. iv. 712. *U. fracta*, Wall. Cat. 6460.
*U. axillaris*, Roxb. Fl. Ind. ii. 667. *Unona odorata* and *U. leptopetala*,
Dunal Anon. 108 and 114; *DC.* Prodr. i. 90 and 91; Deless. Ic. Sel.
t. 88.

In all the provinces, planted. Indigenous in Tenasserim, Java, and
the Philippines. ·

2. CANANGIUM SCORTECHINII, King n. sp. A tree 30 to 40 feet high :
young branches puberulous but speedily glabrous, dark-coloured and
lenticellate. *Leaves* membranous, broadly ovate, sub-acuminate, the
base broad rounded, slightly oblique; both surfaces pubescent when very
young, ultimately glabrescent, the midrib and 6 or 7 pairs of nerves adpressed-pubescent, glandular-dotted; length 2·5 in., breadth 1·5 in. (fide
Scortechini; length 3 to 7 in., breadth 2 to 3 in.) *Cymes* short, from the
axils of leaves or of fallen leaves, few-flowered, shortly pedunculate.
*Flowers* 1 to 1·25 in. long; pedicels under 1 in., pale-pubescent with a
narrow, ovate, obtuse, mesial bracteole ·25 in. long. *Sepals* ovate, sub-
acute, recurved, minutely yellowish-pubescent, ·35 in. long. *Petals*
subequal, linear-obtuse, 1·25 in. long; the claw short, thickened, pubescent
on both surfaces like the sepals. *Stamens* numerous; the connective
with an apical process, bulbous at the base, suddenly tapering into a
sharp point. *Ovaries* numerous, oblong, glabrous except at the pubescent base, with 6 or 8 ovules in two rows; stigma sessile, truncate.
*Ripe carpels* unknown.

Perak : Scortechini.

Scortechini's specimens are in bud only and none of them has any
fruit. The foregoing description has been prepared partly from his notes
and partly from his specimens. The species differs from *C. odoratum* in
having smaller leaves, a different inflorescence, with smaller, quite in-
odorous, flowers. It is also a smaller tree.

*Doubtful Species.*

*Cananga ? monosperma* H. f. and Th. Fl. Br. Ind. I, 57. Of this I
have seen only leaf-specimens.

10. UNONA, Linn.

Trees or shrubs, erect or climbing. *Flowers* often solitary, axillary
terminal or leaf-opposed. *Sepals* 3, valvate. *Petals* 6, valvate or open
in æstivation, 2-seriate; 3 inner sometimes absent. *Torus* flat or slightly
concave. *Stamens* cuneate; anther-cells linear, extrorse, top of connective sub-globose or truncate. *Ovaries* numerous; style ovoid or oblong,
recurved, grooved; ovules 2–8, 1-seriate (rarely sub-2-seriate). *Ripe*

*carpels* many, elongate and constricted between the seeds or baccate.
*Seeds* few or many.—DISTRIB. Tropical Asia and Africa; species about 50.
Sect. I. DESMOS, H. f. and T. Petals 6, in two rows, ripe carpels jointed.

Flowers solitary and always axillary: leaves
elliptic-oblong to oblong-lanceolate ... 1. *U. Dunalii.*
Flowers solitary, and extra-axillary, terminal
or leaf-opposed.
Flower-peduncles 4 to 6 in. long, slender 2. *U. Desmos.*
Flower-peduncles 1 to 2 in long.
Lower surfaces of leaves glaucous;
petals glabrous or at most sparsely
adpressed-sericeous ... ... 3. *U. discolor.*
Flower-peduncles from ·5 to 1 in. long.
Leaves more or less oblong or ovate or
lanceolate, rufous-pubescent or tomentose beneath ... ... 4. *U. dumosa.*
Sect. II. DASYMASCHALON. Petals 3, or sometimes only 2: the inner
row always absent; ripe carpels jointed.
Flowers 3·5 to 6 in. long; petals linear-lanceolate, caudate-acuminate, not constricted between claw and limb ... ... 5. *U. longiflora.*
Flowers 1·5 to 3·5 in. long; petals from ovate
to lanceolate, more or less constricted above
the claw ... ... ... 6. *U. Dasymaschala*
Sect. III. STENOPETALON. Petals 6 in two rows, usually very narrow:
carpels baccate, not jointed.
Flowers solitary ... ... ... 7. *U. Wrayi.*
Flowers in fascicles from the larger branches
or stem.
Petals linear-oblong, 1 to 1·5 in. long; ripe
carpels globose, glabrous, their stalks 1
to 1·5 in. long ... ... ... 8. *U. desmantha.*
Petals narrowly linear, 3 to 3·5 in. long:
ripe carpels globose, densely rufous-
velvetty, shortly stalked ... ... 9. *U. crinita.*
Petals narrowly linear, 1·25 to 3 in. long:
ripe carpels sub-globular or bluntly ovate,
softly tomentose, ultimately sub-glabrous,
sub-sessile ... ... ... 10. *U. stenopetala.*

1. UNONA DUNALII, Wall. Cat. 6425. A climber 60 to 100 feet
long; young branches slender, rather pale, sub-rugose, lenticellate,
glabrous. *Leaves* thickly membranous, pale when dry, elliptic-oblong

to oblong-lanceolate, acute or shortly acuminate, the base rounded, the upper surface glabrous, shining, the lower slightly glaucous, sometimes with a few scattered hairs on the midrib; main nerves 10 to 12 pairs, spreading, not prominent; length 3 to 4 in., breadth 1·2 to 1·75 in., petiole ·2 in. *Flowers* axillary, solitary, 1·25 to 1·4 in. long; pedicels ·35 to ·5 in long, slender, pubescent, with a minute bracteole about the middle. *Sepals* broadly ovate, acute, puberulous, reflexed, ·25 to ·3 in. long. *Petals* narrowly oblong-lanceolate, sub-acute, puberulous to glabrous, 1 to 1·25 in. long, the inner row smaller. *Ripe carpels* numerous, stalked, glabrous, constricted between the 3 to 5 ovoid joints, 1·25 to 1·75 in. long; the stalks about 1 inch. Hook. fil. and Th. Fl. Ind. 131, (excl. the Concan plant); Miq. Fl. Iud. Bat., I. Ft. 2, 41; Hook. fil. Fl. Br. Ind. 1, 58.

Penang; Wallich. Perak; King's Collector.

2. UNONA DESMOS, Dunal Anon., 112. A spreading shrub, often climbing; young branches slender, striate, adpressed, rufous-pubescent, often lanceolate. *Leaves* thinly coriaceous, oblong, acute or acuminate, the base rounded; upper surface glabrous or nearly so, the midrib sparsely pubescent; under-surface paler in colour, puberulous or pubescent; main nerves 12 to 14 pairs, spreading, rather prominent beneath; length 4·8 to 8·8 in., breadth 1·65 to 3·25 in., petiole ·35 in. *Flowers* solitary, extra-axillary, 1·35 to 1·75 in. long; peduncle slender, 4 to 6 in. long, glabrous; bracts few, lanceolate, minute, deciduous. *Sepals* ovate-acuminate, spreading, adpressed-pubescent, ·3 in. long. *Petals* coriaceous, ovate-lanceolate, adpressed-pubescent, nerved; the outer 2 in. long by about ·85 in. broad; the inner smaller. *Ripe carpels* numerous, stalked, ·5 to ·75 in. long, glabrous, constricted between the 2 to 3 oval joints. H. f. and T. Fl. Ind. 134; Miq. Fl. Ind. Bat. I, Pt. 2, 42: Hook. fil. Fl. Br. Ind. I, 59; Kurz For. Fl. Burm. I 34. *U. cochin-chinensis* A. DC Prod. 1, 91; *U. pedunculosa, A. DC* Mem. Anon 28; *U. pedunculosa* Wall. Cat. 6422. *U. fulva*, Wall. Cat. 6427. *Desmos cochin-chinensis* Lour. Fl. Coch. Ch. 1, 352. *U. discolor*, Wall. (not of Roxb.) Cat. 6420 D and E.

From Assam to Singapore. Distrib. Cochin-China.

3. UNONA DISCOLOR, Vahl Symb. II, 63, t. 36. A spreading shrub, often also climbing; young branches slender, sub-rugose, pubescent towards the tips. *Leaves* membranous, oblong or oblong-lanceolate, acute, the base rounded; upper surface glabrous, shining; the lower glaucous, glabrous or pubescent; main nerves 8 to 10 pairs, sub-ascending, slightly prominent beneath; length 3 to 7·5 in., breadth 1 to 2 in., petiole about ·25 in. *Flowers* solitary, extra-axillary, 2 to 2·5 in. long; peduncles 1 to 2 in. long, rather slender, pubescent, with a minute linear

293

45

bracteole below the middle, thickening when in fruit and lenticellate.
*Sepals* ovate-lanceolate, spreading, nearly glabrous, ·4 to ·6 in. long.
*Petals* coriaceous, narrowly lanceolate, 2 to 2·5 in. long, glabrous or
sparsely adpressed-scriccous. *Ovaries* oblong, hairy. *Stigma* laterally
grooved. *Ripe carpels* numerous, stalked, ·75 to 1·5 in. long, glabrous or
pubescent, the constrictions between the 2 to 5 oval joints pubescent;
stalks ·25 in. long. Dunal Anon. 111; DC. Prodr. i. 91; Wall. Cat.
6420 (*partly*); Roxb. Fl. Ind. ii. 669; W. & A. Prodr. 9; H. f. & T.
Fl. Ind. 133; Miq. Fl. Ind Bat. 1, Pt. 2, 41; Beddome Ic. Pl. Ind Or.
t. 51; Bl. Fl. Javæ Anon. 53; A. DC. Mem. 28; W. and A. Prod. 9;
Thwaites Enum. 9; Kurz For. Fl Ind. Burm. I. 34; Hook. fil. Fl. Ind.
I, 59. Scheff. Obs. Phyt. Anon. 5. Nat. Tidsch. Ned. Ind. XXXI, 5.
*U. cordifolia*, Roxb. Fl. Ind. II, 602 ? *U. Dunalii*, H. f. & T. Fl. Ind.
131 (the Concan plant); Dalz. & Gibs. Fl. Bomb. 3 (not of Wallich).
*U. Amherstiana*, A. DC. Mem. 28. *U. biglandulosa*, Bl. Bijdr. 16. *U.
Roxburghiana*, Wall. Cat 6423 B. *U. Lessertiana*, Dunal Anon. 107.
t. 26; DC. Prod. I, 90. *Desmos chinensis* Lour. Fl. Coch. Ch. 1, 352.

Of this variable and abundant species, Sir Joseph Hooker distinguishes four varieties as follows :—

Var. 1, *pubiflora* ; leaves 5–7 in., oblong acute, base often cordate,
flowers silky.

Var. 2, *lævigata* ; leaves 3–4 in., oblong or lanceolate, acute, base
rounded, flowers almost glabrous.—*U. chinensis*, DC. Prodr. i. 90. *U.
undulata*, Wall. Pl. As. Rar. iii. and 42. *U. discolor*, Dalz and Gibs,
Fl. Bomb. 3. t. 265; Wall. Cat. 6428.—Perhaps cultivated only in India,
common in the Archipelago and China.

Var. 3, *pubescens*; leaves as in 1, but densely pubescent beneath.

Var. 4, *latifolia* ; leaves 3–5 by 2–2½ in , broad-oval, acute, flowers
silky. *U. discolor* and var. b, *bracteata* Bl. Fl. Jav. Anon. 53, t. 26
and 31A.

From the base of the eastern Himalaya through the Assam range
to Burmah and the Malayan Peninsula; in tropical forests. Distrib.
The Malayan Archipelago, Chinese Mountains.

4. UNONA DUMOSA, Roxb. Fl. Ind. II, 670. A large bushy climber :
young branches slender, softly rufous-tomentose. *Leaves* membranous,
broadly ovate to oblong-ovate, obovate to oblanceolate-oblong, obtuse,
sub-acute or broadly mucronate, the base rounded or sub-cordate, or
sub-cuneate ; when young rufous-tomentose on both surfaces ; the upper
except the midrib glabrescent when old : main nerves 10 to ·12 pairs,
sub-ascending, rather straight; length 2 to 5·25 in., breadth 1·25 to
2·5 in.; petiole ·15 in., to 3 in., rufous-tomentose. *Flowers* solitary, leaf-
opposed or extra-axillary, 2 to 2·5 in. long ; pedicels ·5 to ·75 in. long,

rufous-tomentose, with a single ovate bract near the base. *Sepals* coriaceous, cordate or ovate, sub-acute or acute, spreading, rufous-tomentose, ·4 in. long. *Petals* obovate-spathulate to broadly ovate-lanceolate, tapering to each end, vertically nerved, densely pubescent at first, less so when old; the inner row smaller. *Ripe carpels* numerous, stalked, glabrous, ·75 to 1·4 in. long, much constricted between the 2 to 3 ovoid joints. *Seeds* shining, the albumen with transverse fibres. Wall. Cat. 6429. H. f. and Th. Fl. Ind. 131; Hook. fil. Fl. Br. Ind. I, 59.

Malacca : Maingay, Nos. 42 and 43 (Kew Distrib.). Perak ; King's Collector, L. Wray Junior. Sylhet ; Roxburgh, Wallich. Assam ; Simons.

The form which occurs in the Malayan Peninsula has narrower petals than that which is found in Assam and Silhet, and its leaves are more oblong and less ovate.

5. UNONA LONGIFLORA, Roxb. Fl. Ind. II, 668. A glabrous shrub or small tree, the leaf-buds silky ; young branches slender. *Leaves* membranous, narrowly oblong or oblong-lanceolate, more or less acuminate, the base rounded or slightly cuneate ; upper surface shining, the lower glaucous : main nerves 12 to 16 pairs, oblique, rather prominent beneath : length 6·5 to 11 in., breadth 1·75 to 3·25 in., petiole ·4 in. *Flowers* solitary, pedunculate, axillary, pendulous, 3·5 to 6 in. long ; the peduncles minutely bracteolate and jointed near the base, slender, from 1·25 to 8 in. long, still longer in fruit. *Sepals* very small, broadly triangular, spreading, mucronate, rufous-pubescent externally. *Petals* linear-lanceolate, much acuminate, cohering by their margins, the base slightly expanded, no constriction between the limb and claw, adpressed-sericeous when young but afterwards glabrous, yellowish ; the inner row absent. *Stamens* with the connective produced and truncate at the apex. *Ovaries* 10 to 20, sessile, hairy ; *ovules* few : stigmas large, recurved. *Ripe carpels* about 10, stalked, moniliform, 3- to 4-jointed, all the joints except the lowest often falling off : individual joints elongated-ovoid, ·5 in. long, glabrous. *Seeds* with thin smooth testa, the albumen intersected by numerous horizontal fibrous processes. Wall. Cat. 6419 ; Hook. fil. and Th. Fl. Ind. 134; Hook. fil. Fl. Br. Ind. I, 61 ; Kurz Fl. Burm. I, 35.

Perak ; in forests under 3,000 feet. E Himalaya ; Assam ; Khasia Hills, Chittagong.

Most of the specimens which I have seen from Assam, the Khasia Hills, and Chittagong have flower-pedicels under 2 inches long, and petals quite 6 inches long. Specimens from Perak, on the other hand, have shorter flowers (3 to 4 in. long) ; and much longer (5 or 6 in.) and more slender peduncles : otherwise the two sets agree. In many of the flowers from both sets of localities there are only two petals.

6. UNONA DASYMASCHALA, Blume Fl. Jav. Anon. 55, t. 27. An erect or sarmentose shrub : young branches sometimes glabrous from the beginning, but usually at first softly rufous-pubescent and sometimes permanently so. *Leaves* thinly coriaceous, elliptic-oblong, oblong, or oblong-lanceolate or oblanceolate, acute or shortly acuminate, the base rounded or narrowed ; upper surface glabrous ; the lower sub-glaucous, glabrous or sometimes puberulous on the midrib and nerves ; length 4·5 to 8·5 in., breadth 1·5 to 3 in., petiole about ·1 in. *Flowers* pedunculate, solitary, axillary, pendulous, 1·5 to 3 in. long ; peduncles 1·25 to 1·75 in. (longer in fruit), minutely bracteolate at the very base. *Sepals* fleshy, very short, broadly triangular, pubescent, reflexed. *Petals* fleshy, varying from ovate-acute to lanceolate-acuminate, concave and (in the narrower forms) expanded at the base, with a constriction between the claw and limb ; the edges united when young, adpressed-puberulous but ultimately glabrous. *Anthers* with the connective expanded at the apex and oblique. *Ovaries* densely villous; the stigma narrow, glabrous. *Ripe carpels* numerous, shortly stalked, moniliform, pubescent to glabrous, the joints oval, about ·35 long. *Seeds* oval, smooth, the albumen with fibrous processes. A. DC. Mem. Anon. 28; Wall. Cat. 6421 ; Hook. fil. and Thoms. Fl. Ind. 135; Miq. Fl. Ind Bat. I, Pt. 2, 42 ; Kurz Fl. Burm. I, 36 ; Hook. fil. Fl. Br. Ind. I, 61. Scheff. Obs Phyt. Anon. 6 ; Nat. Tidsch. Ned. Ind. XXXI, 6

From Burmah to Singpore ; the Andaman Islands.    Distrib.— Sumatra, Java.

Var. *Blumei*, Hook. fil. ; branches glabrous ; leaves pale-yellowish or grey beneath, glabrous or nearly so. Wall. Cat. 6420 B. (*U. discolor*.)

Var. *Wallichi*, Hook. fil. ; branches brown-tomentose ; lower surfaces of leaves glaucous and tinged with purple.

This species, in the absence of the inner row of petals and in other respects, resembles *M. longiflora*, Roxb. ; but the outer petals are neither so long nor so narrow, and there appear always to be three of them, and not often only two as in *M. longiflora*. The peduncles are moreover shorter. The two species, however, are closely allied. In open, exposed situations this is a non-scandent bush ; but under the shade of trees, it often developes into a climber,—a habit which it shares with many species of this family. Blume's figure of this plant (quoted above) is inaccurate as respects the flowers and fruit.

7. UNONA WRAYI, Hemsl. in Hook. Ic. Plant t. 1553. A tree : young branches slender, tawny-tomentose. *Leaves* thickly membranous, elliptic-oblong, shortly acuminate, often obtuse (from the breaking off of the acumen), slightly narrowed to the rounded base ; upper surface glabrous except the puberulous midrib ; lower much reticulate,

puberulous, the midrib pubescent : main nerves 8 to 10 pairs, rather prominent beneath, spreading, and forming two sets of intra-marginal arches :
length 5·5 to 7·5 in., breadth 2 to 2·65 in. ; petiole ·2 in., tomentose.
*Flowers* 3 to 3·5 in. long, solitary or in fascicles from tubercles on the
larger branches : pedicels ·75 to ·9 in., slender. *Sepals* ovate-lanceolate,
sub-acute, about ·3 in. long, puberulous. *Petals* white changing to
deep claret, subequal, rather coriaceous, linear-lanceolate, acuminate,
about 3 in. long, sparsely puberulous outside : breadth about ·3 in.
*Ovaries* numerous, pubescent, with about 4 ovules. *Ripe carpels* red when
ripe, stalked, slightly pulpy, ovoid or oblong, obtuse, glabrous, 1 to 1·25
in. long : stalks ·5 to ·75 in. long. *Seeds* about 3, oval, compressed, rugulose, aromatic, ·6 in. long.

Singapore ; Maingay (Kew Distrib.,), No. 51. Perak ; Wray, No.
560 ; King's Collector. Distrib.—Java.

8. Unona desmantha, H. f. and T. in Hook. fil. Fl. Br. Ind. I, 61.
A small tree : youngest branches with soft yellowish-brown pubescence,
the older with smooth, shining, yellowish-brown bark. *Leaves* coriaceous,
elliptic-oblong, or elliptic-lanceolate, or oblanceolate, shortly and acutely
or obtusely acuminate, the base acute ; upper surface glabrous except
the pubescent midrib ; under-surface paler, puberulous especially on the
midrib and nerves : main nerves 8 to 11 pairs, rather prominent beneath
when dry, oblique. *Flowers* 2·5 in. diam., pale red, densely crowded on
1 to 2 in. broad flat tubercles on the older branches : peduncles ·75 in.,
puberulous, ebracteolate. *Sepals* ovate, acute, ·3 in. long. *Petals* unequal, linear-oblong, tapering to the apex, the base not dilated, sparsely
pubescent, 1 to 1·5 in. long ; the inner rather narrower. *Torus* and
*ovaries* as in *U. pycnantha*, but ovules 3 to 5, superposed. *Ripe carpels*
stalked, globose, dark-coloured, glabrous, nearly 1 in. in diam. : stalk 1
to 1·5 in.

Malacca : Maingay (Kew Distrib.), No. 48.

9. Unona crinita, Hook. fil. and Thoms. Fl. Br. Ind. I, 61. A
tree ? young branches slender ; their bark pale, rugose ; the youngest
densely rufous-tomentose. *Leaves* membranous, oblong, elliptic-oblong
or oblanceolate-oblong, acute or acuminate ; the base rounded ; upper
surface quite glabrous, the lower pubescent especially on the nerves
and veins : the midrib tomentose on both surfaces ; main nerves 10 to
12 pairs, slender, but slightly prominent beneath : length 3 to 8 in.,
breadth 1·25 to 2·5 in. ; petiole ·15 in., tomentose. *Flowers* 3 to 5 in.
long, pedicellate, in dense crowded fascicles from very broad (1 to 2 in.
in diam.) tubercles on the larger branches ; pedicels ·15 in. to ·25 in. long,
rusty-tomentose ; bracteole linear, or absent. *Sepals* ovate-lanceolate,
much acuminate, spreading, ·5 in. to ·75 in. *Petals* subequal in length,

49

narrowly linear, unequal in breadth, 15 in. broad at the base, and at the
middle, narrower between and from the middle upwards; 1-nerved ;
finely pubescent; the inner slightly shorter and narrower. *Torus*
columnar, truncate. *Ovaries* strigose : ovules 3 to 5, 1-seriate : stigma
punctiform. *Ripe carpels* globose, densely rufous-velvetty, shortly stalked.
Malacca : Maingay (Kew Distrib.), No. 41.

10. UNONA STENOPETALA, Hook. fil. and Thoms. Fl. Ind. 136. A
tree 20 to 35 feet high : young branches softly rufous-tomentose ; the
older dark-coloured, glabrous, striate. *Leaves* thinly coriaceous, oblong-
obovate or oblanceolate, more or less acuminate, narrowed below to the
slightly cordate and oblique base : both surfaces glabrous, the midrib
more or less pubescent on the lower ; under-surface faintly reticulate
when dry ; main nerves 7 to 9 pairs, curving upwards, anastomosing
doubly at some distance from the edge, thin but slightly prominent :
length 4 to 7 in., breadth 1·25 to 3 in. ; petiole ·1 to ·25 in., rufous-
tomentose. *Flowers* 1·5 to 2 in. long, almost sessile or shortly pedicelled,
in fascicles of 2 to 4 on minutely bracteate extra-axillary tubercles from
both branches and stem. *Sepals* united at the base, lanceolate, acuminate, the bases broad, ribbed, spreading, pubescent externally, ·4 to ·5
in. long. *Petals* sub-equal, narrowly linear, concave, slightly wider at
the base, keeled, sparsely pubescent, 1·25 to 3 in. long. *Stamens* numerous, short with broad flat apices hiding the lateral anthers. *Ovaries* 4
to 7, villous, 4- or 5-ovuled. *Ripe carpels* few, sub-globular or bluntly
ovate, softly tomentose at first, ultimately sub-glabrous ; the pericarp
thick, ·5 to ·65 in. long and ·5 in. in diam. *Seeds* 1 to 3, thickly discoid,
bi-concave with grooved edge, rugulose. Hook. fil. and Th. Fl. Br. Ind.
I, 60 : Miquel Fl. Ind. Bat. I, pt. 2, 43 : Kurz F. Flora Burma, I, 35.

Singapore : Lobb, Ridley. Penang : King's Collector, Scortechini ;
common. ? Burmah, (in Tenasserim) : Lobb.

This is a rare plant in Burmah, if indeed it occurs there at all. The
leaves of some of the Perak specimens have petioles ·5 in. long : but
usually they are as above described.

11. POLYALTHIA, Blume.

Trees or shrubs with the habit of *Unona*. *Sepals* 3, valvate or
sub-imbricate. *Petals* 6, 2-seriate, ovate or elongated, flat or the inner
slightly vaulted. *Torus* convex. *Stamens* cuneate ; anther-cells extrorse,
remote. *Ovaries* indefinite ; style usually oblong ; ovules 1–2, basal and
erect, or sub-basal and ascending. *Ripe carpels* 1-seeded, berried.—
DISTRIB. Tropical Asiatic sp. about 45 ; African sp. 3 ; Australasian
species 2.

298

Sect. I. Monoon. Ovule solitary, usually basal, erect.
Flowers from the axils of the leaves or of fallen
leaves, not from the trunk.
Flowers solitary.
Leaves under 5 in. in length (7 in. in *P.
Sumatrana*), more or less lanceolate.
Leaves not glaucous beneath; petals
ovate, acute ... ... 1. *P. dumosa.*
Leaves very glaucous beneath; petals
linear-oblong, obtuse.
Ripe carpels smooth ... 2. *P. hypoleuca.*
Ripe carpels vertically ridged ... 3. *P. sumatrana.*
Leaves over 5 in. in length, not glaucous.
Flowers axillary.
Petals more or less narrowly lan-
ceolate.
Leaves ovate-lanceolate, gla-
brous; ripe carpels oblong,
blunt at each end ... 4. *P. andamanica.*
Leaves oblong to obovate-ob-
long, more or less pubescent;
ripe carpels elliptic, mucronate ... ... 5. *P. magnoliaeflora.*
Petals oblong-elliptic, slightly
obovate, 1·3 to 2·25 in. long ... 6. *P. macrantha.*
Flowers terminal; petals ovate-elliptic,
1 to 1·25 in. long ... ... 7. *P. pulchra.*
Flowers solitary or in pairs; ripe carpels little
more than ·25 in. long.
Flowers ·4 in. in diam.; petals broadly
oblong-ovate, obtuse ... ... 8. *P. Kunstleri.*
Petals 1·5 to 2 in. long, lanceolate-oblong; leaves narrowly lanceolate-
oblong or elliptic-oblong ... 9. *P. Scortechinii.*
Petals ·85 to 1·5 in. long, broadly
lanceolate or oblanceolate; leaves
oblong-lanceolate to ovate-elliptic... 10. *P. Jenkinsii.*
Flowers in pairs; petals obovate-oblong, 1 in.
long: ripe carpels ovoid; ·65 in. long ... 11. *P. Hookeriana.*
Flowers always in fascicles or cymes, axillary
or from the branches below the leaves ... 12. *P. simiarum.*
Flowers in fascicles from the young branches

below the leaves, or from the larger branches ;
never axillary.

Leaves 8 to 15 in. long with 12 to 16
pairs of prominent oblique or spreading nerves .. ... ... 13. *P. lateriflora.*

Leaves 6 to 8 in. long with 10 to 12
pairs of slender, spreading nerves... 14. *P. sclerophylla.*

Flowers in fascicles from tubercles on the main
stem, often near its base ; never axillary, and
probably never from the branches.

Inflorescence ærial.

Leaves under 8 in. in length.

Leaves oblong-lanceolate ;
nerves 8 or 9 pairs ; torus
of ripe fruit 1·25 in. in
diam. : stalks of ripe carpels
·75 in. long ... ... 15. *P. macropoda.*

Leaves oblong ; nerves 7
pairs ; torus of ripe fruit ·5
in. in diam. ; stalks of ripe
carpels 1·5 in. long ... 16. *P. clavigera.*

Leaves elliptic to oblong,
slightly oblique ... 17. *P. glomerata.*

Leaves 9 to 16 in. long ; cblong-
elliptic ... ... 18. *P. congregata.*

Inflorescence sub-hypogæal ... 19. *P. hypogæa.*

Sec. II. EUPOLYALTHIA. Ovules 2 (3 in *P. Korinti*), superposed.
Flowers solitary.

Leaves under 5 in. long, not cordate at the
base.

Leaves oblong-lanceolate.

Petals oblong ... ... 20. *P. obliqua.*

Petals broadly ovate or ovate-
orbicular, leaves glaucous ... 21. *P. aberrans.*

Leaves upwards of 5 in. long, cordate at
the base.

Petals narrowly linear ... ... 22. *P. bullata.*

Petals oblong.

Flowers 1 in. diam.... ... 23. *P. subcordata.*

Flowers 1·25 to 1·75 in. in diam. 24. *P. oblonga.*

Flowers in fascicles from the older branches.

Petals linear-oblong, 1 to 1·5 in. long : ripe

carpels ·35 in. long, their stalks ·6 to ·75
in. long ... ... ... 25. *P. Beccarii.*
Petals linear-oblong, 2 to 3 in. long ; ripe
carpels ·75 to 1 in. long, sub-sessile ... 26. *P. cinnamomea.*
Petals oblong-lanceolate or oblanceolate, ·9
to 1·5 in. long ; ripe carpels 1·75 in. long,
their stalks ·25 in. long ... ... 27. *P. pachyphylla.*
Petals linear, obtuse, ·5 to ·75 in. long ... 28. *P. pycnantha.*

1. POLYALTHIA DUMOSA, King n. sp. A shrub; young branches
slender, glabrous. *Leaves* thinly coriaceous, lanceolate or oblong-lanceolate, acuminate, the base rounded ; both surfaces dull, glabrous, very
minutely lepidote ; main nerves 8 or 9 pairs, spreading, faint, inter-arching far from the margin ; length 2·5 to 3·25 in., breadth ·5 to ·9 in.,
petiole less than ·1 in. *Flowers* solitary, leaf-opposed, ·3 to ·35 in. long ;
pedicels slender, glabrous, ·3 to ·4 in. long with a small lanceolate bracteole about the middle. *Sepals* thick, spreading, broadly ovate, acute
or acuminate, ·1 in. long, glabrescent outside, quite glabrous inside.
*Petals* leathery, subequal, narrowly oblong, acuminate, not widened at
the base, sub-corrugated and glabrous outside, puberulous inside, 3 in.
long. *Stamens* numerous, short ; the apical process very broad, rhomboid,
truncate, projecting much over the apices of the short dorsal anther-cells.
*Ovaries* very few, oblong, pubescent ; stigma broad, sessile, hairy. *Ripe
carpels* one or two, ovoid-globose, glabrous, cherry-red when ripe, ·25 to
·3 in. long.

Perak ; elevat. about 1,200 feet ; Wray, Scortechini.

Near *P. suberosa,* H. f. and Th. but with different venation, fewer
carpels, and without hypertrophied bark.

2. POLYALTHIA HYPOLEUCA, Hook. fil. and Thoms. in Fl. Br. Ind.
I, 63. A tree 50 to 80 feet high ; young branches slender, rather pale,
striate ; all parts glabrous except the flowers. *Leaves* coriaceous, oblong-
lanceolate or elliptic-lanceolate, shortly acuminate, the base acute, the
edges slightly recurved when dry, upper surface shining, the lower dull,
pale : main nerves many pairs, invisible on either surface except in
some occasional leaves when dry : length 2·5 to 5 in., breadth ·75 to
1·75 in., petiole ·2 to ·3 in. *Flowers* sub-erect, small (only ·3 to ·4 in.
long) pedicelled, solitary or sub-fascicled, mostly from the axils of fallen
leaves : pedicel stout, about ·15 in. long, tomentose and with about two
cucullate bracts near the base. *Sepals* very small, triangular, pubescent,
deciduous. *Petals* linear-oblong, obtuse, not dilated at the base, grey-
pubescent on both surfaces. *Ripe carpels* few, often solitary, stalked,
elliptic-oblong, obtuse, glabrous, ·8 in. long : stalks ·1 to ·25 in. *Seed*
ovoid-elliptic, blunt, dark-coloured, transversely striate.

53

Singapore : Maingay, No. 50, (Kew Distrib.) Perak; King's
Collector.

This approaches *Guatteria sumatrana*, Miq. in its leaves : but that
species has much larger flowers. But this is still more allied to *Guatteria
hypoglauca*, Miq., from which it differs by its much larger fruit. The
plant named *P. hypoleuca* by Kurz in his Forest Flora of Burmah is, as
he himself informed Sir Joseph Hooker in a letter, really *P. sumatrana*.
Neither species, however, appears to me to occur either in the Andamans
or Burmah.

3. POLYALTHIA SUMATRANA, King (not of Kurz.) A tree 30 to 60
feet high : young branches pale, the older much furrowed : all parts
glabrous except the flowers. *Leaves* coriaceous, oblong-lanceolate, acuminate, the base acute ; upper surface shining, the lower dull glaucous,
both pale (when dry) ; main nerves 15 to 20 pairs, very slender and
little more prominent than the secondary ; length 4·5 to 6·5 in., breadth
1·25 to 1·75 in., petiole ·25 in. *Flowers* 1·4 to 1·75 in. long, solitary or
in fascicles of 2 or 3 from the younger branches below the leaves, or
axillary ; their pedicels ·6 to 9 in. long, minutely bracteolate near the
base, glabrous. *Sepals* very small, half-orbicular-ovate. *Petals* narrowly linear-oblong, sub-acute or obtuse, puberulous, pale green to yellowish,
the outer slightly longer than the inner, 1·35 to 1·75 in. long and ·15 to
·2 in. broad. *Ovaries* glabrous, sub-cylindric, with a single ovule : stigma
hairy. *Carpels* ovoid, tapering to each end, ridged (when dry), pubescent or glabrous, about 1 in. long and ·6 in. in diam. ; their stalks ·5 to
·6 in. long. *Guatteria sumatrana*, Miq. Fl. Ind. Bat. Suppl. 380. *Monoon
sumatranum*, Miq. Ann. Mus. Lugd. Bat. II, 19.

Perak ; at elevations up to 2,500 feet, common. Distrib.: Sumatra,
Korthals, Beccari P. S., No. 613. Borneo, Korthals.

This is allied to *P. hypoleuca*, H. f. and Th.; but has larger leaves,
much larger flowers, and slightly different carpels.

4. POLYALTHIA ANDAMANICA, Kurz Andam. Report (1870) p. 29.
A shrub : young branches slender, tomentose. *Leaves* membranous,
ovate-lanceolate, acute ; the base broad and rounded, slightly unequal ;
some of the larger nerves underneath and the midrib on both surfaces
pubescent near the base, otherwise glabrous and shining ; main nerves
6 or 7 pairs, distant, spreading and forming bold arches far from the
margin : reticulations minute, distinct : length 4·5 to 6 in., breadth 2 to
2·4 in. ; petiole ·2 in., pubescent. *Flowers* axillary or extra-axillary,
solitary, 2 in. in diam.; the pedicel ·4 to ·75 in. long, sub-pubescent,
minutely bracteolate. *Sepals* minute (·1 in. long), broadly triangular,
pubescent. *Petals* thinly coriaceous, sub-equal, oblong, blunt, 1 in. long.
*Ripe carpels* 6 to 8, oblong, smooth, glabrous, slightly apiculate, ·5 or ·6 in.

302

long and ·15 to ·2 in. in diam., their stalks nearly as long. *P. Jenkinsii*, Benth. and Hook. fil. in Hook. fil. Fl. Br. Ind. I, 64 (*in part*) ; Kurz Flora Burm. I, 38.

S. Andaman : Kurz, Man, King's Collector.

Allied to *P. Jenkinsii*, H. f. and T. ; but with much smaller flowers, and leaves with broader bases.

5. POLYALTHIA MAGNOLIÆFLORA, Maing. MSS. Hook fil. Fl. Br. Ind. I, 64. A tree 30 to 40 feet high ; young branches rusty-tomentose. *Leaves* thinly coriaceous, oblong to obovate-oblong, obtuse or acuminate, the base rounded or minutely cordate; upper surface glabrous, the nerves and midrib minutely tomentose ; under surface at first pubescent, ultimately glabrous or glabrescent : main nerves 15 to 20 pairs, rather straight, oblique, prominent beneath, the transverse veins almost straight, distinct ; length 8 to 12 in., breadth 2·5 to 3·5 in. ; petiole ·25 in. stout, tomentose. *Flowers* large, shortly pedunculate, solitary, axillary, 2·5 to 3 in. long; peduncle ·3 in. long, tomentose, with 2 large ovate bracts. *Sepals* coriaceous, short, broadly ovate, acute, spreading, tomentose. *Petals* coriaceous, white, linear-oblong or oblong-lanceolate, sub-acute, tomentose. *Torus* conical. *Ovaries* hirsute. *Carpels* (unripe) stalked, oblong-ovoid, blunt at either end, the apex mucronate, pubescent. *Seed* with smooth shining testa.

Malacca : Maingay. Perak ; King's Collector, No. 10039.

Evidently a rare species. I have seen only Maingay's imperfect specimens from Malacca, and two collected on Ulu Bubong by the late Mr. H. H. Kunstler, Collector for the Bot. Garden, Calcutta. Sir J. D. Hooker states (F. B. Ind. l. c.) on Maingay's authority that the flowers have the colour and odour of those of a *Magnolia*.

6. POLYALTHIA MACRANTHA, King n. sp. A tree 20 to 70 feet high ; young branches rather slender, glabrous. *Leaves* large, thinly coriaceous, oblong to elliptic-oblong, acute, slightly narrowed below the middle to the rounded or minutely cordate base ; upper surface shining, glabrous except the depressed slightly puberulous midrib ; lower surface paler when dry, glabrous, very minutely lepidote ; main nerves 20 to 24 pairs, spreading, thin but prominent beneath ; length 12 to 18 in., breadth 4·5 to 7·5 in., petiole ·4 in., stout. *Flowers* solitary, axillary or slightly supra-axillary, 2·5 to 4·5 in. in diam. ; pedicels 1·5 to 2 in. long (longer in fruit) glabrescent, with a sub-orbicular bracteole about the middle ; the buds conical when young. *Sepals* thick, sub-orbicular, spreading, connate by their edges and forming a cup ·75 in. in diam., puberulous on both surfaces, corrugated outside. *Petals* much larger than the sepals, white, thick, fleshy, flattish, oblong-elliptic, widest above the middle, blunt, puberulous on both surfaces except at the glabrescent

bases, nerved inside; the outer row 1·3 to 2·5 in. long, the inner smaller. *Stamens* numerous, compressed; apical process of connective truncate. *Ovaries* few, oblong, puberulous; stigmas large, capitate-truncate, pubescent. *Ripe carpels* elliptic-ovoid, sometimes oblique, blunt at each end, the apex mucronate, glabrous, 1 to 1·25 in. long, and ·75 in. in diam. *Seed* ovoid, solitary, the testa corrugated.

Perak; King's Collector, Scortechini.

A remarkable species with handsome white flowers, allied in many ways to *P. congregata*; but at once distinguished from it by its axillary, solitary flowers and glabrous ripe carpels.

7. POLYALTHIA PULCHRA, King. A small tree, glabrous except the inflorescence. *Leaves* thinly coriaceous, elliptic to oblong-lanceolate or oblong-oblanceolate, acute or acuminate, the base acute; both surfaces minutely muriculate, the lower paler and dull; length 4·5 to 6 in., breadth 2·5 in. (only 1·75 in. in var. *angustifolia*), petiole ·25 in. *Flowers* large, solitary, terminal, 2 in. or more in diam. when expanded (often 3·5 in. in diam. in var. *angustifolia*): pedicels 1·4 to 1·75 in. long, puberulous, with a lanceolate foliaceous bracteole at the base. *Sepals* ovate, acute or sub-acute, nerved, glabrous, ·6 to ·75 in. long. *Petals* coriaceous, sub-equal, ovate-elliptic, sub-acute, the base slightly cordate (narrowly oblong-lanceolate in var. *angustifolia*) greenish-yellow with a triangular blotch of dark purple at the base. *Stamens* numerous; apical process of connective broad, truncate, sub-orbicular, projecting over the apex of the linear anther-cells, pubescent. *Ovaries* oblong, adpressed-pubescent, 1-ovuled; style short, cylindric, thick, crowned by the convex, terminal, pubescent stigma. *Ripe carpels* numerous, elliptic-ovoid, blunt, slightly contracted at the base, sparsely pubescent but becoming almost glabrous, purple when ripe; pericarp sub-succulent: stalks thick, crimson when ripe, 1·5 in. long. *Seed* solitary, elliptic.

Perak: at Wold's Rest, Scortechini.

Var. *angustifolia*, King. *Leaves* oblong-lanceolate or oblong-oblanceolate, scarcely muriculate; petals lanceolate or narrowly oblong-lanceolate, often 1·75 in. long; sepals often ·75 in. long.

Perak; on Gunong Bubu; elevat. 5,000 feet, Wray.

8. POLYALTHIA KUNSTLERI, King n. sp. A shrub or small tree; young branches puberulous, speedily glabrous. *Leaves* oblong-lanceolate rarely elliptic-lanceolate, shortly and rather bluntly acuminate, the base narrowed and sub-acute or rounded; upper surface glabrous, shining; the lower paler, dull, puberulous on the midrib and nerves; main nerves 6 to 12 pairs, rather prominent beneath, ascending, inter-arching ·1 to ·2 in. from the margin; length 4·5 to 8 in., breadth 1·5 to 2·35 in.; petiole ·2 in., pubescent. *Flowers* ·4 in. in diam., axillary or extra-axillary,

304

solitary or in pairs ; peduncles ·25 in. long, each with two rather large
unequal, broadly ovate bracts above the base. *Sepals* broadly triangular-
ovate, obtuse, nearly as long as the petals and, like them, minutely tomentose. *Petals* sub-equal, broadly oblong-ovate, obtuse. *Ovule* solitary.
*Fruit* 2 in. in diam. ; individual carpels numerous, ovoid-globular, apiculate, ·3 in. long ; stalks slender, ·5 in. long, adpressed rufous-pubescent
like the carpels. *Ellipeia parviflora,* Scortechini MSS.
Perak : King's Collector, Scortechini, Wray.
This much resembles *P. Jenkinsii* and *P. andamanica* in its leaves
and fruit : but its flowers are totally different.
9. POLYALTHIA SCORTECHINII, n. sp. King. A small tree 15 to 20
feet high ; young branches minutely rufous-tomentose, but speedily
glabrous. *Leaves* thinly coriaceous, oblong or oblong-elliptic, acute or
shortly acuminate, the base rounded or sub-acute ; upper surface glabrous, shining, the midrib pubescent ; the lower dull, very minutely dotted,
the midrib and sometimes nerves puberulous ; main nerves 8 to 11 pairs,
bold and prominent on the lower surface, oblique, inter-arching close to
the edge : length 4 to 8 in., breadth 1·15 to 2·25. ; petiole ·25 in., pubescent. *Flowers* pedicelled, solitary or in pairs, from the axils of leaves
or of fallen leaves : pedicels ·5 to ·75 in. long, rufous-tomentose, with a
rather large bract about the middle. *Sepals* small, triangular, pubescent.
*Petals* fleshy, sub-equal, greenish-yellow changing into dark dull yellow,
oblong-lanceolate or oblong-oblanceolate, acute or rather blunt, the edges
wavy, both surfaces minutely pubescent, 1·5 to 2 in. long. *Ovaries*
narrowly elongate-adpressed, pubescent, each crowned by large fleshy
glabrous stigma. *Ovule* solitary, basal. *Fruit* shortly stalked ; ripe
carpels numerous pedicelled, ovoid, crowned by the remains of the
stigma, sparsely pubescent, ·3 in. long ; pedicel slender, pubescent, ·75
in. long. *Seed* with pale smooth testa. *P. Jenkinsii,* H. f. and T. (*in
part*). *Ellipeia undulata,* Scortechini MSS.
Malacca : Griffith, No. 413. Perak, King's Collector, Scortechini.
Distrib. :—Sumatra, Beccari, Nos. 935, 976.
10. POLYALTHIA JENKINSII, Benth. and Hook. fil. Gen. Pl. J, 25.
A tree : young shoots sparsely rufous-pubescent. *Leaves* membranous,
oblong-lanceolate to elliptic-ovate, acute or shortly acuminate, slightly
narrowed to the acute or rounded sub-oblique base ; both surfaces
glabrous, minutely reticulate, the upper shining and the midrib puberulous ; main nerves about 7 pairs, slender, slightly prominent beneath,
inter-arching at some distance from the edge : length 4 to 7 in., breadth
1·35 to 3 in., petiole ·2 to ·3 in. *Flowers* large (1·75 to 3 in. in diam.),
pedicelled, solitary, rarely in pairs, axillary : pedicels ·6 to ·75 in. long,
pubescent, and with several small rounded bracts near the base. *Sepals*

very small, sub-orbicular, puberulous. *Petals* sub-coriaceous, spreading, greenish changing to yellow, broadly lanceolate or oblanceolate, sub-acute or obtuse, the base much narrowed, puberulous or glabrous. *Ripe carpels* numerous, stalked, oblong, slightly apiculate, glabrous, ·4 in. long : stalk slender, ·6 in. long. *Seed* smooth. Hook. fil. Fl. Br. Ind. Ind. I, 64 (*in part*) ; Kurz For. Fl. Burm. I, 375 (*in part*) ; *Guatteria Jenkinsii*, Hook. fil. and Thoms. Fl. Ind. 141 ; Miq. Fl. Ind. Bat. I, pt. 2, p. 46. *Guatteria Parveana* Miq. Fl. Ind. Bat. Vol. I, Pt. 2, p. 48, and Suppl. 378. *Uvaria canangioides*, Reichb. fil. et Zoll. MSS. *Monoon canangioides*. Miq. Ann. Mus. Lugd. Bat. II, 18.

Malacca ; Griffith ; Maingay, No. 46 (and 45 *in part*) (Kew Distrib.). Perak ; King's Collector, No. 3910. Assam and Silhet.

Specimens from Perak have larger flowers than those from Assam ; but otherwise they agree fairly well, and both appear to be specifically identical with the Sumatra plant named *Guatteria* or *Monoon canangioides* by Miquel. The Andaman plant which Kurz originally (Andam. Report (1870) p. 29) named *Polyalthia andamanica*, but which Sir Joseph Hooker (dealing with imperfect materials) reduced (with Kurz's assent) to this species, I have restored to specific rank. Recently received specimens show its flowers to be different from those of true *P. Jenkinsii* (the petals being shorter and narrower), while the carpels are larger.

11. POLYALTHIA HOOKERIANA, King n. sp. A tree 20 to 70 feet high : young branches softly tawny-pubescent, ultimately glabrous and darkly cinereous. *Leaves* membranous, obovate-elliptic or oblanceolate, shortly acuminate, narrowed from above the middle to the sub-cuneate base ; both surfaces reticulate, the upper glabrous except the pubescent midrib and nerves : lower glabrous, the midrib and nerves adpressed-pubescent : main nerves 10 or 11 pairs, oblique, forming imperfect arches close to the edge, prominent beneath ; length 5 to 7 in., breadth 2·25 to 3·25 in. ; petiole ·15 to ·2 in., tomentose. *Flowers* in pairs from peduncles with several aborted flowers near their bases, extra-axillary : pedicels ·5 to ·75 in. long, lengthening in fruit, stout, pubescent, with 1 or 2 small ovate bracteoles at the middle or below it. *Sepals* broadly ovate, concave, free or connate only at the base, pubescent outside, glabrous within, 2 in. long. *Petals* coriaceous, yellowish, subequal, ovate or obovate-oblong, sub-acute, puberulous except at the base inside, only slightly contracted at the base, nearly 1 in. long. *Stamens* numerous, very short, cuneate ; the apical process of the connective thick with a truncate orbicular top hiding the linear dorsal anthers. *Ovaries* short, oblong, puberulous, with 1 ovule : stigma sessile, large, obovate with sub-truncate lobed apex. *Ripe carpels* numerous, ovoid, slightly apicu-

late at the top and somewhat narrowed at the base, ·65 in. long, stalks
1·2 in. long. *Seed* solitary, ovoid, smooth, with a vertical furrow.
Malacca : Maingay (Kew Distrib.). No. 96. Perak ; King's Collector ; Wray.

This is a common tree in Perak. In Malacca, however, it appears
to be rare ; for it is so very imperfectly represented in Maingay's great
Malayan collection (of which the best set is at Kew), that Sir Joseph
Hooker, while recognising it as a *Polyalthia*, had not sufficient material
to enable him to describe it in his Flora of British India.

12. POLYALTHIA SIMIARUM, Benth. and Hook. fil. Gen. Pl. I, 25 ;
Hook. fil. Fl. Br. Ind. I, 63. A tree 50 to 80 feet high ; all parts glabrous except the puberulous leaf buds, under surface of nerves of leaves
and inflorescence ; young branches pale brown, striate, sparsely lenticellate. *Leaves* sub-coriaceous, ovate-oblong to oblong-lanceolate, acute
or shortly acuminate, the base rounded or sub-acute ; upper surface
shining ; lower dull, sometimes puberulous on the midrib and nerves ;
main nerves 12 to 16 pairs, oblique, prominent beneath ; length 5 to
11 in., breadth 2 to 4·5 in., petiole ·25 in. *Flowers* pedicelled, in
few-flowered sessile fascicles from the axils of fallen leaves or from
tubercles on the larger branches : pedicels minutely pubescent, with a
small bract below the middle, 1 to 1·25 in. long. *Sepals* small, bluntly
triangular, recurved, pubescent outside. *Petals* spreading, linear, sub-
acute or acute, greenish-yellow to purplish, puberulous outside, glabrous
inside, 1 to 1·25 in. long, the inner rather the longer. *Ripe carpels* stalked, ovoid-elliptic, slightly mammillate, contracted towards the base,
glabrous and orange-red to bluish-black when ripe, 1·25 to 1·5 in. long :
stalk from 1 to 1·75 in. *Seed* ovoid, grooved, transversely striate.
Kurz For. Fl. Burm. I, 37 ; Hook. fil. Fl. Br. Ind. I, 63. *Guatteria
simiarum*, Ham., Wall. Cat. 6440 ; Hook. fil. and Thoms. Fl. Ind. 142.
*G. fasciculata*, Wall. MSS. ex Voigt Hort. Sub. Calc. 16. *Polyalthia
lateriflora*, Kurz (not of King), Journ. As. Soc. Beng., Pt. 2, (for 1874)
52. *Unona simiarum*, H. Bn., Pierre Fl. Forest. Coch-Chine, t. 23.

Andamans, Bot. Garden Collectors. Perak, King's Collector. Forests at the base of the Eastern Himalaya, the Assam range, Chittagong,
Burmah.

Var. *parvifolia*, King : leaves smaller than in typical form (3·5 to
6 in. long and 1·25 to 2·25 in. broad) puberulous beneath.

Perak ;.at elevation of 3,000 to 4,000 feet. Distrib. Sumatra : on
Goenong Trang, Lampongs. (Forbes, No. 1536).

13. POLYALTHIA LATERIFLORA, King. A tree 50 to 70 feet high :
young branches lenticellate and striate ; all parts except the inflorescence quite glabrous. *Leaves* coriaceous, oblong to elliptic-oblong

abruptly acute or shortly acuminate, slightly narrowed to the rounded rarely sub-cordate and unequal base : upper surface shining, the lower paler, rather dull : main nerves 12 to 16 pairs, rather prominent, oblique spreading, evanescent at the tips : length 8 to 15 in., breadth 2·5 to 7 in. ; petiole ·3 in. stout. *Flowers* in fascicles from tubercles on the stem and larger branches, pedicelled, 1·25 to 2 in. long ; pedicels slender, thickened upwards, pubescent, with 2 bracteoles about the middle, 1·25 to 1·75 in. long. *Sepals* coriaceous, ovate-orbicular, very short, densely and minutely tomentose outside. *Petals* coriaceous, greenish-yellow, dull crimson at the base, oblong-lanceolate, gradually tapering to the sub-acute apex, the outer rather shorter than the inner, minutely pubescent especially on the outer surface. *Ripe carpels* ovoid-elliptic, blunt, slightly narrowed to the base, glabrous, 1·25 in. long and ·7 in. in diam. ; the pericarp thin, fleshy : the stalks stout, glabrous, sub-asperulous, 1·25 to 2 in. long. *Guatteria lateriflora*, Bl. Bijdr. 20 : Fl. Jav. p. 100, t. 50 and 52 D. : Miq. Fl. Ind. Bat. I, pt. 2 p. 47. *Monoon lateriflorum*, Miq. Ann. Mus. Lugd. Bat. II, 19.

Perak ; at low elevations, Wray, King's Collector. Distrib : Java.

This is closely allied to *P. simiarum*, Benth. and Hook. fil. : but has smaller flowers which are often borne on the smaller branches ; smaller leaves ; and shorter stalked carpels. Moreover the leaves and young branches of this are invariably glabrous. The leaves of old trees are very markedly smaller than those on young specimens. Specimens in young fruit of a plant which may belong to this species have been recently received from the Andamans from the Collectors of the Bot. Garden, Calcutta : but, until the receipt of fuller material, I hesitate to include these islands in the geographical area of the species.

14. POLYALTHIA SCLEROPHYLLA, Hook. fil. and Thoms. Fl. Br. Ind. I, 65. A glabrous tree : young branches pale. *Leaves* coriaceous, oblong, ovate or linear-oblong, acute or obtusely acuminate, the base broadly cuneate, shining on both surfaces and with the reticulations distinct ; main nerves about 10 to 12 pairs, spreading, slender : length 6 to 8 in. : breadth 1·5 to 2·6 in., petiole ·5 in. *Flowers* pedunculate, in fascicles from small tubercles on the trunk, 2 in. in diam , greenish : tubercles ·5 to 1 in. in diam. : peduncles 1 to 1·5 in. long, stout, rusty-pubescent, becoming glabrous ; bracts small, orbicular, from about the middle of the peduncle. *Sepals* ovate, obtuse, short. *Petals* linear-oblong, obtuse, the base slightly concave, puberulous on both surfaces, 1·6 in. long, the inner rather smaller. . *Torus* broad, flat, the edge raised. *Ovaries* pilose, shorter than the cylindric style. *Ripe carpels* elliptic-oblong, slightly narrowed at either end, 1 to 1·5 in. long, glabrous, the pericarp thin : stalks 1 to 1·5 in. long. *Seed* oblong, the testa shining, pale.

Malacca; Maingay (Kew Destrib), No. 101.

I have seen only Maingay's Malacca specimens of this plant.

15. POLYALTHIA MACROPODA, King n. sp. A tree 50 to 60 feet high; young branches rather pale, pubescent but speedily glabrous. *Leaves* membranous, oblong-lanceolate, shortly acuminate, the base acute; the edge slightly revolute; upper surface shining, glabrous except the puberulous sulcate midrib; the lower paler when dry, minutely lepidote, sparsely strigose on the midrib and 8 or 9 pairs of curving rather prominent nerves; length 3·5 to 5·5 in., breadth 1·4 to 2·1 in., petiole ·25 in. *Flowers* nearly 1 in. long, in fascicles on short broad rugose woody tubercles from the stem close to its base: pedicels about 1 in. long, woody in fruit and 2 in. or more in length, glabrous; bracteoles (if any) deciduous. *Sepals* broadly ovate, acute, spreading, corrugated and glabrescent outside, glabrous inside, connate at the base to form a cup ·65 in. in diam. *Petals* elliptic, blunt, slightly constricted about the middle, sub-equal, puberulous, coriaceous. *Stamens* numerous, compressed especially the outer rows; apical process of connective transversely elongated, truncate. *Ovaries* numerous, oblong-ovoid. *Ripe fruit* with large woody sub-globular torus 1·25 in. in diam.; *ripe carpels* numerous, oblong-ovoid, tapering to the apex, the base gradually narrowed into a stalk, 2·5 to 3·5 in. long (including the stalk); pericarp rather fleshy, glabrous. *Seed* solitary, elongated-ovoid, grooved vertically.

Perak: King's Collector, Singapore, Ridley.

A species remarkable for its large ripe carpels borne on the stem near the ground. It is possible that Mr. Ridley's plant, collected in Singapore, may really belong to a distinct species, the only specimen of it which I have seen being very imperfect. This comes very near *P. clavigera* King.

16. POLYALTHIA CLAVIGERA, King n. sp. A tree 30 to 40 feet high; young branches slender, at first puberulous but speedily glabrous and pale. *Leaves* thinly coriaceous, oblong, tapering to each end, acuminate; both surfaces reticulate; the upper shining, glabrous except the puberulous sulcate midrib; lower surface slightly puberulous at first but ultimately quite glabrous: main nerves 7 pairs, ascending, curved, not inter-arching, slightly prominent beneath, obsolete above; length 5·5 to 8·5 in., breadth 1·75 to 2·5 in.; petiole ·4 in. slightly winged above. *Flowers* unknown. Peduncle of ripe fruit stout, woody, 2 in. or more in length; the torus depressed-globular, woody, about ·5 in. in diam.: *ripe carpels* ovoid-elliptic, tapering to each end, the base gradually passing into the stout puberulous slightly scabrid stalk, greenish-yellow when dry, glabrous: the pericarp succulent: length 2·25 in., breadth nearly 1 in.; stalk 1·5 in. puberulous; seed solitary, ovoid.

Penang: Pinara Bukit, elevat. 2000 feet. Curtis (No. 2414).
Perak : Waterfall Hill, Wray. Distrib. E. Sumatra, Forbes (No. 1638).
This species is known only by a few fruiting specimens collected
by Messrs. Curtis and Wray Junior. It is nearly allied to P. macropoda,
King; but its leaves have different venation and texture, the torus of
the ripe fruit is smaller, while the carpels themselves are larger and
have longer stalks.

17. POLYALTHIA GLOMERATA, King n. sp. A tree 40 to 50 feet high :
young branches glabrous, pale, rather slender. Leaves membranous,
elliptic to oblong, slightly oblique, acute or shortly acuminate, the base
slightly cuneate or rounded ; both surfaces reticulate, glabrous ; the
midrib alone puberulous on the upper, adpressed-puberulous on the
lower ; main nerves 7 to 8 pairs, curved, ascending, not inter-arching,
thin but slightly prominent beneath ; length 4 to 6 in., breadth 1·8 to
2·6 in , petiole ·25 to ·35 in. Flowers about 1 in. long, in clusters of 20
to 30 from nodulated puberulous tubercles on the stem ; pedicels long
(1·5 to 2·5 in.), slender, puberulous, with an ovate-lanceolate bracteole
about the middle. Sepals thick, lanceolate-acuminate with broad connate bases, sub-erect, puberulous. Petals coriaceous, sub-erect, linear-
oblong, slightly concave and glabrous at the base inside, otherwise
minutely tomentose, the inner slightly smaller than the outer. Stamens
numerous; the connective with an orbicular sub-convex apical expansion concealing the linear dorsal anther-cells. Ovaries much less numerous than the stamens, oblong, hirsute, apparently 1-ovuled; the
stigma small, oblong, slightly pubescent.

Perak ; King's Collector, Wray. Distrib. Sumatra; Forbes, No.
2804.

In all the flowers I have examined the pistils are very small (as if
undeveloped) and I have not been able to find more than one ovule. In
the Sumatran specimens the flowers are much longer than in those from
Perak.

18. POLYALTHIA CONGREGATA, King n. sp. A tree 40 to 60 feet
high ; young branches at first rusty-puberulous but speedily glabrous
and dark-coloured. Leaves thinly coriaceous, oblong-elliptic, acute,
slightly narrowed to the rounded or minutely cordate base ; upper
surface glabrous except the depressed puberulous midrib ; the lower
pale when dry, glabrous, minutely lepidote ; main nerves 13 to 19
pairs, oblique, curving, thin but prominent beneath ; length 9 to 16 in.,
breadth 3·75 to 7 in. ; petiole ·3 or ·4 in. stout. Flowers large, in short,
much divided, rough, tubercular, woody cymes from the stem near its
base ; the pedicels 1·25 to 1·75 in. long, glabrescent ; bracteole single, sub-
orbicular, clasping, infra-median. Sepals thick, broadly ovate-triangular,

spreading, slightly cuneate at the base, concave, corrugated and puberulous outside, glabrous inside, often reflexed, ·5 in. long. *Petals* thick, white, ovate-elliptic, sub-acute, hoary-puberulous except at the base inside on both surfaces; the outer row 1·5 to 3 in. long and ·65 to 1 in. broad, the inner row narrower. *Stamens* numerous, compressed; the apical process of the connective truncate, oblique, granular; anther-cells linear, dorsal. *Ovaries* 20 to 30, oblong, strigose, with a single basilar ovule; stigma oblong, pubescent. *Ripe carpels* elliptic, beaked, 1 in. or more long, hoary-pubescent, narrowed at the base into the short, thick stalk. *Seed* solitary, pale brown, shining, elliptic.

Perak; Scortechini, King's Collector.

This resembles *P. macrantha*, King; but is distinguished from it by its cymose, cauline inflorescence, smaller flowers and puberulous fruit. H. O. Forbes collected in the Lampongs in Eastern Sumatra a plant (No. 1642 of his Herb.) which greatly resembles this.

19. POLYALTHIA HYPOGAEA, King, n. sp. A tree 25 to 30 feet high; young branches rather stout, densely but minutely rufous-tomentose, ultimately rather pale, striate. *Leaves* large, thinly coriaceous, oblong or elliptic-oblong, sometimes slightly obovate, gradually narrowed to the rounded base; both surfaces glabrous when adult, the lower puberulous when young, the veins transverse and, (like the reticulations), distinct; main nerves 18 to 22 pairs, oblique, inter-arching within the edge, thin, prominent on the lower and depressed on the upper surface when dry; length 10 to 20 in., breadth 3 to 7 in.; petiole ·4 in., stout, tomentose, *Flowering branches* from the stem near its base, 1 to 8 feet long, flexuose, rufous-pubescent like the lanceolate bracteoles. *Flowers* ·75 to 1 in. long, cream-coloured; pedicels ·75 to 1·5 in. long, usually with one lanceolate, tomentose bracteole near the middle and a second, sub-orbicular and acuminate, close to the flower. *Sepals* broadly triangular-ovate, acute, spreading, tomentose outside, glabrous inside, ·25 in. long. *Petals* coriaceous, the inner row rather smaller than the outer, narrowly oblong, sub-acute, pubescent outside except the glabrescent base and edges, inside almost glabrous. *Stamens* numerous, short, compressed; apical process of connective broad, slightly convex, slightly oblique, sub-granular, deeply ridged in front, the anther-cells linear dorsal. *Ovaries* few, oblong, villous, 1-ovuled; stigma large, ovoid, granular, sessile. *Immature carpels* narrowly ovoid, sub-compressed, the apex beaked, the base slightly contracted, minutely tomentose. *Seed* solitary, elongated, ovoid, smooth.

Perak; near Laroot, King's Collector. Gunong Batu Putch; elev. 3,400 feet, Wray.

A species remarkable for its hypogœal inflorescence. The flowering branches, which vary from 1 to 8 feet in length, originate from the stem near its base, pass into the soil underneath the surface of which they run for some distance, and bear on their emerging tips the flowers and fruit

20. POLYALTHIA OBLIQUA, Hook. fil. and Thoms. Fl. Ind. 138. A tree : young branches minutely pubescent, lenticellate. *Leaves* subsessile, oblong-lanceolate, acute or shortly acuminate, the base cuneate, minutely and obliquely cordate ; shining and glabrous on both surfaces, the lower pale ; main nerves 7 or 8 pairs, slender, curving and forming bold arches ·15 in. from the margin ; length 4 to 6·5 in., breadth 1·5 to 2·2 in. ; petiole 1 in., very stout. *Flowers* ·4 to 5·4 in. in diam., solitary, pedicellate, extra-axillary ; each pedicel rising from a short conical woody tubercle, curving, ·25 in. long. *Sepals* coriaceous, broadly triangular, blunt, less than half as long as the petals, pubescent. *Petals* coriaceous, sub-equal, oblong, obtuse, sericeous outside. *Ripe carpels* pisiform, with stalks ·5 in. long, dark brown. Hook. fil. Fl. Br. Ind. I, 67 ; Miq. Fl. Ind. Bat. 1, Pt. 2, p. 44.

Malacca; Griffith, Maingay, No. 44 (Kew distrib.). Chittagong Hill Tracts ; Lister. Distrib. Sumatra.

Lister's plant from the Chittagong Hill Tracts agrees well with Griffith's specimens from Malacca.

21. POLYALTHIA ABERRANS, Maing. ex Hook. fil. Fl. Br. Ind. I, 67. A large climber, glabrous except the flowers and fruit : young branches slender, black. *Leaves* membranous, oblong-lanceolate, acuminate, the base slightly cuneate ; both surfaces reticulate, glabrous, the lower glaucous ; main nerves 14 to 18 pairs, very faint, the secondary nerves quite as well marked : length 3·5 to 5 in., breadth 1·4 to 1·8 in., petiole ·2 to ·25 in. *Flowers* 5 to ·75 in. in diam., solitary, axillary ; pedicels slender, 1·25 in. long (longer in fruit), with one minute bracteole below the middle and another at the base. *Sepals* ovate-orbicular, sub-acute, quite connate into a 3-angled glabrous cup ·25 in. in diam. *Petals* leathery, ovate-orbicular, sub-acute, spreading, concave ; the outer row ·35 in. long and ·3 in. broad, yellowish-pubescent on both surfaces except a glabrous patch near the base on the inner : inner petals half the size of the outer but more concave, hoary-puberulous outside, glabrescent inside. *Stamens* numerous ; apical process of connective broad, discoid, depressed in the centre, quite concealing the long linear lateral anther-cells. *Ovaries* narrowly oblong, glabrous, 1 or 2-ovuled : style as long as the ovary, curved : stigma small. *Ripe carpels* ovoid, slightly apiculate, puberulous or glabrescent, ·35 in. long and ·3 in. in diam. ; stalks ·7 to ·8 in., slender, glabrous. *Seeds* solitary, rarely 2, ovoid, shining, smooth. *Melodorum glaucum*, Scortechini MSS.

Malacca : Maingay. Perak ; Scortechini, Wray.

In some carpels there are two seeds, such carpels being about twice as long as those with a single seed. Although referred by the late lamented Father Scortechini to the genus *Melodorum*, this is an undoubted *Polyalthia* in its stamens, in its 1- rarely 2-ovuled ovaries, and in its carpels with usually solitary, ovoid seeds. In externals, save and except the much smaller size of the flowers, this much resembles the plant figured by Pierre under the name of *Unona Mesnyi* (Flore Forest. Coch-Chine, t. 17) to which indeed Pierre reduces *P. aberrans*.

22. POLYALTHIA BULLATA, King n. sp. A shrub 6 to 8 feet high : young branches densely covered with long soft spreading golden hairs. *Leaves* thinly coriaceous, bullate (at least when dry), narrowly oblong, acuminate, narrowed but slightly to the deeply cordate auricled base : both surfaces boldly reticulate, the upper shining, glabrous except the sulcate puberulous midrib ; the lower glabrescent except the midrib and nerves which have sparse hairs like those on the young branches : main nerves 25 to 40 pairs, spreading towards the base, sub-ascending towards the apex, forming a double series of arches within the margin, bold and prominent on the lower, depressed on the upper, surface : secondary nerves and reticulations prominent ; length 12 to 14 in., breadth 2·75 to 3·35 in. ; petiole ·25 in , pubescent like the young branches. *Flowers* solitary, terminal or axillary, 1 in. long ; pedicels slender, 1 in. long, pubescent, bracteole small, mesial. *Sepals* small, lanceolate, spreading, free. sparsely pubescent outside, glabrescent inside, about ·25 in long. *Petals* narrowly linear, slightly wider at the base, subequal, sub-concave, sparsely pubescent. *Stamens* numerous, the apical process of the connective sub-convex, orbicular, slightly granular. *Ovaries* much fewer than the stamens, oblong, pubescent ; the stigma sub-capitate-truncate, puberulous. *Ripe carpels* globular-ovoid, blunt at each end, puberulous, ·4 in. long ; stalks slender, ·2 in. long. *Seeds* 2, plano-convex, the testa rugose, pale : the albumen horny.

Singapore : Ridley. Perak ; King's Collector.

Evidently a rare shrub ; readily recognisable by its elongate very bullate leaves.

23. POLYALTHIA SUB-CORDATA, Blume Fl. Javae, 71 t. 33 and 36 B. A shrub or small tree : young branches sparsely hispid-pubescent, afterwards glabrous and furrowed, not pale. *Leaves* membranous, sub-sessile, oblanceolate-oblong or elliptic-oblong, shortly and obtusely caudate-acuminate ; the base slightly narrowed, sub-cordate, auriculate at one side ; both surfaces glabrous except the sometimes puberulous midrib : main nerves 9 to 12 pairs, slender, the reticulations lax and faint : length 4·5 to 9 in., breadth 1·6 to 3 in. ; petiole ·05 in., pubescent. *Flowers*

about 1 in. in diam., solitary, axillary or extra-axillary ; peduncles slender, ·5 to ·75 in. long, puberulous and with 1 or 2 lanceolate bracteoles. *Sepals* ovate, sub-acute ; united into a cup. *Petals* coriaceous, yellowish, oblong, sub-acute, the inner rather smaller, slightly pubescent outside. *Carpels* numerous, broadly ovoid, not apiculate, furrowed, glabrous, ·4 in. long ; stalks slender, ·25 in. long ; pericarp thin. Miq. Fl. Ind. Bat. I, Pt. 2, p. 44 ; Ann. Mus. Ludg. Bat. II, 14. *Unona subcordata*, Bl. Bijdr. 15.

Perak ; elev. about 800 feet, King's Collector, No. 2373. Distrib. Java.

24. POLYALTHIA OBLONGA, King, n. sp. A shrub or small tree 10 to 15 feet high : young branches at first rufous-tomentose, afterwards glabrous, pale and furrowed. *Leaves* thinly coriaceous, sub-sessile, oblong or oblong-oblanceolate, abruptly and shortly acuminate, narrowed to the minutely cordate, unequal base ; upper surface glabrous, except the pubescent midrib ; lower puberulous, the midrib prominent as are the 14 to 20 pairs of little curving, sub-ascending, main nerves ; reticulations open and distinct ; length 9 to 14 in , breadth 3·5 to 5 in.; petiole ·15 in., tomentose. *Flowers* 1·25 to 1·75 in. in diam., solitary, axillary or extra-axillary, from small tubercles : pedicels 1·25 to 2·5 in. long, puberulous and with 2 lanceolate bracteoles near the base. *Sepals* semi-orbicular, acute, very short, united into a cup, pubescent outside. *Petals* coriaceous, yellow, subequal, oblong, tapering to the sub-acute apex, minutely adpressed-pubescent on both surfaces but especially on the outer, length ·75 to 1·15 in. *Ripe carpels* 10 to 20, ovoid to orbicular, apiculate, ·3 to ·35 in. long, pubescent or sub-glabrous ; stalks slender, ·6 to ·75 in. long. *Seeds* usually solitary and ovoid, or sometimes two and plano-convex.

Perak : very common at elevations of from 1,000 to 2,500 feet.

This plant closely resembles *Guatteria* (= *Polyalthia*) *elliptica* Blume : but its leaves have more numerous nerves and its carpels are stalked, those of *P elliptica* (according both to Blume's description and figure) being sessile and of larger size.

25. POLYALTHIA BECCARII, King n. sp. A tree 15 to 40 feet high : young branches slender, rufous-tomentose ; the older coarsely striate and lenticellate. *Leaves* thickly membranous, narrowly oblong or oblong-lanceolate, acuminate, slightly narrowed to the rounded base ; both surfaces shining and reticulate, the midrib pubescent on the upper tomentose on the lower ; main nerves 6 or 7 pairs, slender, spreading, forming bold arches far from the edge, the secondary nerves distinct ; length 3 to 4·5 in., breadth ·75 to 1·35 in.; petiole ·1 in., tomentose. *Flowers* 1 in. long, in fascicles from bracteolate tubercles on the older

branches, their pedicels slender, pubescent, minutely bracteolate near the base, about 1 in. long. *Sepals* ovate-obtuse, ·15 in. long, pubescent outside. *Petals* coriaceous, dark-yellow, sub-equal, linear-oblong, sub-acute, 1 in. to 1·5 in. long and from ·1 to ·2 in. broad, minutely pubescent especially outside. *Ovaries* pubescent, 2-ovuled. *Ripe carpels* numerous, broadly ovoid, apiculate, glabrous, sub-granular when ripe, ·35 in. long ; their stalks granular, puberulous, ·6 to ·75 in long.

Perak : at low elevations. Scortechini, King's Collector, Wray. Distrib. Sumatra ; Beccari P. S., No. 401. Borneo ; Motley No. 743.

The leaves of this species, although smaller, have much the same venation as those of *P. Teysmannii*, King. The carpels of this are, however, very much smaller than those of *P. Teysmannii*.

26. POLYALTHIA CINNAMOMEA, Hook. fil. and Thoms. Fl. Ind. 138 ; Hook fil. Fl. Br. Ind. I, 65. A tree 50 to 70 feet high ; young branches rusty-tomentose. *Leaves* thinly coriaceous, narrowly oblong to oblanceolate, tapering to each end, acute or shortly acuminate, the base rounded ; upper surface glabrous, shining ; the lower sparsely lucid-pubescent, (glabrescent when old), the midrib tomentose ; main nerves about 12 or 14 pairs, slender, curved, ascending, inter-arching freely ; length 4·5 to 7·5 in., breadth 1·25 to 2·25 in. ; petiole ·2 in., tomentose. *Flowers* sub-sessile, solitary, or in pairs from short woody tubercles from the young branches below the leaves, dull red, 2 to 2·25 in. long; peduncles very short, rusty-tomentose, bracteolate at the base. *Sepals* spreading, sub-orbicular, ·25 in. long, tomentose. *Petals* sub-equal, thick, linear-oblong, sub-acute, slightly narrowed at the base, adpressed-pubescent externally, glabrous within, 2 to 3 in. long. *Anthers* numerous, short, compressed ; connective with broad, flat, apical, truncate process. *Pistils* oblong, pubescent; stigma large, sub-truncate. *Torus* convex, tomentose. *Fruit* globose, 2·5 in. in diam.; the individual carpels pyriform with very short stalks, ·75 to 1 in. long and ·5 to ·75 in. in diam., densely rusty-tomentose ; pericarp thick. *Seeds* 2, plano-convex, with scaly testa, Miq. Fl. Ind. Bat. I, Pt. 2, p. 44. *Guatteria cinnamomea*, Wall. Cat. 6444. *G. multinervis*, Wall. Cat. 6445. *Unona cauliflora*, H. f. and Th. Fl. Ind., 137; Fl. Br. Ind. 2, 60. Miq. Fl. Ind. Bat. I, Pt. 2, 43.

Singapore ; Wallich, Ridley. Penang ; Wallich, Curtis No. 2470. Malacca, Maingay (Kew Distrib.) No. 37.

Apparently not a common species. Maingay's specimens from Malacca have rather larger and smoother leaves than those from Singapore and Penang.

27. POLYALTHIA PACHYPHYLLA, King, n. sp. A tree 50 to 100 feet high ; young branches softly pubescent, afterwards glabrous and furrowed. *Leaves* rigidly coriaceous, elliptic-oblong, sub-acute ; the edge

slightly recurved, the base broad and rounded, or narrowed and sub-acute;
both surfaces glabrous; the lower slightly paler, the midrib tomentose at
the base beneath; main nerves 11 or 12 pairs, spreading, prominent,
evanescent at the tips; length 4·5 to 7·5 in., breadth 1·75 to 3·5 in.,
petiole ·35 to ·5 in., tomentose when young. *Flowers* about 1·5 in. long,
in few-flowered fascicles from small tubercles on the older branches;
their pedicels 2 in. long, bracteolate about the middle, softly tawny-
tomentose. *Sepals* broadly half-orbicular, very short, reflexed, tomentose. *Petals* coriaceous, nerved, pale green, oblong-lanceolate or oblanceolate, sub-acute or obtuse, pubescent on the outer, tomentose on
the inner, surface; the outer slightly shorter and narrower than the
inner, from ·9 to 1·5 in. long and ·3 to ·5 in. broad. *Stamens* numerous,
compressed, the apical process of connective truncate; anthers linear,
dorsal. *Ovaries* numerous, glabrous, vertically striate; stigma sessile,
truncate, puberulous. *Ripe carpels* numerous, crowded when young,
densely covered with minute pale tomentum; when ripe narrowly
obovoid, blunt, narrowed to a short stalk, sub-tomentose, 1·75 in. long
and about 1 in. in diam.; pericarp thick, fleshy; seeds two, plano-convex.

In its leaves this resembles *Guatteria pondok*, Miq. (Fl. Ind. Bat.
Suppl. 380), but that species has carpels with stalks from 2 to 3 in.
long.

Perak; at elevation under 1,000 feet, King's Collector, Nos. 6655
and 7516.

28. POLYALTHIA PYCNANTHA, King. A tree? Young branches
rather stout, covered with soft yellowish pubescence. *Leaves* coriaceous, elliptic-oblong, or oblong-lanceolate, obtusely acuminate, the base
obtuse or rounded : upper surface glabrous; lower paler and puberulous
on the midrib; main nerves arching, prominent; length 6 to 9 in.,
breadth 2·5 to 3·5 in.; petiole ·2 in., pubescent. *Flowers* ·5 to ·75
in. in diam., in fascicles from tubercles on the larger branches, 1 to ·5 in.
in diam.; flower-peduncles ·25 in. long, pubescent, ebracteate. *Sepals*
ovate, acute, ·2 in. long. *Petals* linear, obtuse, flat, sub-equal, the bases
of the inner three concave, ·5 to ·75 in. long, pale sericeous outside,
glabrescent inside. *Torus* columnar-flat-topped, glabrous : ovules 2,
superposed. *Unona pycnantha*, Hook fil. in Fl. Br. Ind. I, 60.

Malacca; Maingay.

## 12 ANAXAGOREA, St. Hilaire.

Trees or shrubs. *Leaves* with pellucid dots. *Flowers* small, greenish,
leaf-opposed. *Sepals* 3, valvate, connate at the base. *Petals* 6 or 3, sub-
equal, 2-seriate, valvate, the inner row sometimes absent. *Torus* convex.

*Stamens* indefinite; anther-cells extrorse or sublateral; connective with a terminal process. *Ovaries* few, style variable; ovules 2, sub-basal, collateral, ascending. *Ripe carpels* follicular; stalk clavate. *Seeds* 1-2, exarillate, testa shining.—Distrib. Tropical Asia and America; species about 8.

Petals 6 ............... 1 *A. luzonensis*

,, 3 ............... 2 *A. Scortechinii*.

1. ANAXAGOREA LUZONENSIS, A. Gray Bot. U. S. Expl. Exped. 27. A shrub; all parts glabrous. *Leaves* membranous, oblong or elliptic-oblong, shortly acuminate, the base cuneate, the under surface pale; main nerves 7 or 8 pairs, spreading, slightly prominent beneath, the reticulations wide, rather distinct; length 5 to 7 in., breadth 1·75 to 2·5 in., petiole ·25 to 35 in. *Flowers* about ·5 in. long, solitary; pedicels ·25 in. long (twice as long in fruit), with 1 or 2 amplexicaul bracteoles. *Sepals* small, ovate-rotund, obtuse *Petals* subequal, elliptic, obtuse, thin, nerved, white. *Ovaries* few. *Ripe carpels* 1 to 3, cuneate-clavate, somewhat compressed, narrowed into a long stalk, 1 to 2-seeded. *Seeds* plano-convex, obovate, black, shining. Hook. fil. Fl. Br. Ind. I, 68. Kurz F. Flora Burm. I, 39. *A. zeylanica*, H. f. and Th. Fl. Ind. 144: Thwaites Enum. 10; Miq. Fl. Ind. Bat. I, Pt. 2, 49; Beddome Ic. Pl. Ind. Or. t. 46. *Rhopalocarpus fruticosus*, Teysm. and Binn. in Miq. Ann. Mus. Lugd. Bat. II, 22 t. 2 fig. B. *Anaxagorea fruticosa*, Scheff. in Nat. Tijdsch. Ned. Ind. XXXI, 9.

Burmah; The Andaman Islands; Malacca; Ceylon. Distrib. Philippines, Cambodia, Sumatra.

2. ANAXAGOREA SCORTECHINII, King, n. sp. A bush or small tree: all parts, except the flower, glabrous; the young branches sub-rugulose, 2-ridged. *Leaves* thinly coriaceous, elliptic-oblong or elliptic-obovate, shortly and abruptly acuminate, slightly narrowed to the rounded or sub-acute base; main nerves 7 to 9 pairs, rather prominent beneath, the reticulations open and distinct: length 6 to 8 in., breadth 2·5 to 3·5 in.; petiole ·3 to ·4 in. *Flowers* ·75 in. long, solitary; pedicels ·3 in. (much longer in fruit) with 1 or 2 amplexicaul bracteoles. *Sepals* membranous, their edges thin, broadly ovate, acute, pubescent outside. *Petals* in a single row, much larger than the sepals, oblong-lanceolate, sub-acute, scurfy-pubescent outside, glabrous within, very fleshy, slightly concave at the base. *Stamens* numerous, those next the pistils barren, elongate and bent over the pistils. *Ovaries* numerous, obovoid, pubescent: styles curved. *Carpels* as in *A. luzonensis*, but two or three times as numerous. *Seeds* obovoid, concavo-convex, compressed, black, shining.

Perak: at low elevations; Scortechini; King's Collector, Wray.

I have altered the diagnosis of this genus as regards the petals to

317

admit this species in which the inner whorl of petals is absent. In other respects the species agrees perfectly with the original diagnosis. Teysmann and Binnindyk's mono-specific genus *Rhopalocarpus* (Miq. Ann. Mus. Lugd. Bat. II, 22, t. 2 fig. B.) is an unmistakable *Anaxagorea* in which the inner petals are narrow and incurved. It is probably near *A. luzonensis*. A. Gray, and *A. javanica*, Bl. (See Benth. and Hook fil. Gen. Plant. I, 957).

### 13. DISEPALUM, Hook. fil.

Trees or shrubs. *Sepals* 2, large, concave, valvate. *Petals* 4, narrowly linear-spathulate, incurved, inserted remotely from each other on the margin of the very broad, sub-concave torus. *Stamens* numerous; the apical process of the connective broadly orbicular, sub-convex. *Pistils* 10 to 15 or numerous, ovoid; style short, terete; stigma small, terminal; ovule solitary. *Leaves* minutely pellucid-punctate. *Flowers* in long terminal peduncles, solitary or in pairs. Distrib. Three species, all Malayan.

1. DISEPALUM LONGIPES, King, n. sp. A glabrous tree 30 to 40 feet high; young branches slender, pale brown. *Leaves* minutely pellucid-punctate, membranous, oblong, sometimes slightly oblanceolate, rarely oblong-elliptic, abruptly and shortly acuminate, the base cuneate; main nerves 7 to 10 pairs, spreading, (sub-horizontal) very faint; length 4 to 7 in., breadth 1·5 to 2·25 in., petiole ·25 in. *Flowers* on long pedicels, dark red, solitary or in pairs terminal, ·5 in. in diam. ; pedicels slender, ebracteolate, 1·25 to 2 in. long. *Sepals* reflexed, concave, broadly ovate, blunt. *Petals* remote from each other, linear-spathulate, sub-incurved, ·2 in. long. *Stamens* numerous; apical process of the connective orbicular, sub-convex. *Ovaries* numerous, stalked, slightly obovoid, glabrescent or sparsely pubescent, 1-ovuled ; style short, straight ; stigma small, terminal. *Immature carpels* ovoid, sub-glabrous, slightly corrugated ; pericarp fleshy, fragrant. *Seed* solitary, ovoid.

Johore ; on Gunong Panti at 1,500 feet ; King's Collector, No. 231. Distrib. Borneo, Beccari (P. B. 1645).

The genus *Disepalum* was founded by Sir Joseph Hooker on a Bornean shrub collected by Lobb, and the only species known to its founder was that described and figured under the name of *D. anomalum* in the Linnœan Transactions (Vol. XXIII, 156, t. 20 A.) The characters which separate the genus from any other in the family are the dimerous symmetry of the sepals and petals, and the small size of the latter, which originate at some distance from each other from the edge of the broad sub-concave torus. The species here described differs from *D. anomalum* in its arboreous habit, larger leaves, and much more numerous

318

ovaries, which are moreover nearly glabrous and have long stalks. Quite ripe fruit is as yet unknown.

## 14. GONIOTHALAMUS, Blume.

Small trees or shrubs. *Leaves* with small nerves, forming intra-marginal loops. *Flowers* solitary or fascicled, axillary or extra-axillary; peduncles with basal, scaly, distichous bracts. *Sepals* 3, valvate. *Petals* 6, valvate in 2 series; outer thick, flat or nearly so; inner smaller, shortly clawed, cohering in a vaulted cap over the stamens and ovary. *Stamens* many, linear-oblong; anther-cells remote, dorsal; connective produced into an oblong or truncate process. *Ovaries* many; style simple or 2-fid; ovules solitary or 2, superposed, sub-basal (4 in *G. uvarioides.*) *Ripe carpels* 1-seeded.—Distrib. About 47 species, natives of Eastern tropical Asia and its islands.

The plants referred to this genus are, by Baillon, treated as part of *Melodorum.*

Ovules 1 or 2.

Style cylindric, slender; stigma subulate, entire ... ... ... 1. *G. subevenius.*

Style very short; stigma funnel-shaped, slit on one side, its edges toothed ... 2. *G. tenuifolius.*

Style cylindric; stigma truncate, entire.

Flowers in fascicles from the stem only; ripe carpels 1·25 in. long ... 3. *G. Prainianus.*

Flowers solitary from the axils of the leaves or fallen leaves; ripe carpels ·4 in. long ... ... ... 4. *G. Kunstleri.*

Style subulate or cylindric; stigma deeply 2-cleft, petals 3 to 5 in. long... ... 5. *G. giganteus.*

Style cylindric; stigma unequally 2-toothed 6. *G. malayanus.*

Style cylindric; stigma minutely and equally 2-toothed.

Flowers axillary or from the axils of fallen leaves; outer petals more than 1 in. long.

Anthers with slightly convex, orbicular apical appendages ... 7. *G. fulvus.*

Anthers with very pointed, conical apical appendages.

Nerves of leaves 28 to 34 pairs 8. *G. Curtisii.*

Nerves of leaves fewer than 20 pairs.

Leaves shining, reticulate,
glabrous; ripe carpels
oblong, ·5 to ·6 in. long     9.   *G. Griffithii.*
Leaves glabrous, opaque,
dull, not reticulate; ripe
carpels        globular-obo-
void ; ·4 in. long.        ... 10.   *G. macrophyllus.*
Flowers in fascicles from tubercles near
   the base of the stem    ...        ... 11.   *G. Ridleyi.*
Style cylindric; stigma 3-toothed; apices
of anthers acuminate.
Leaves thickly coriaceous; nerves inconspicuous ...        ...        ... 12.   *G. Tapis.*
Leaves strongly and prominently nerved.
Sepals large, orbicular-ovate, ob-
   tuse, ·65 to 1 in. long       ... 13.   *G. Scortechinii.*
Sepals small, ovate acuminate,  ·2
   in. long ...        ...        ... 14.   *G. Wrayi.*
Ovules and seeds 4      ...      ...      ... 15.   *G. uvarioides.*

1. GONIOTHALAMUS SUBEVENIUS, King, n. sp.   A shrub or small tree; young branches slender, puberulous; otherwise glabrous except the flower. *Leaves* membranous, narrowly oblong, tapering at each end ; upper surface shining, pale-greenish when dry ; the lower paler, dull ; main nerves 10 to 12 pairs, sub-horizontal, invisible or very faint on either side ; length 3·5 to 6·5 in., breadth 1·25 to 1·75 in., petiole ·2 in. *Flowers* solitary, axillary, ·75 to ·9 in. long ; pedicels ·4 to ·6 in. long, ebracteate. *Sepals* broadly ovate, bluntly acuminate, 3-nerved, minutely pubescent on both surfaces, ·3 in. long. *Petals* thinly coria-ceous, puberulous except towards the base inside, lanceolate, sub-acute ; the inner petals half as large as the outer, slightly clawed. *Stamens* with broad orbicular sub-convex apical process. *Ovaries* narrowly oblong, style cylindric, curved ; stigma subulate, entire. *Ripe carpels* ovoid to oblong, obtuse, tapering very little at the base, glabrous, ·5 to ·75 in. ; stalks ·35 to ·45 in.

Perak ; at low elevations, King's Collector.

2. GONIOTHALAMUS TENUIFOLIUS, King, n. sp.   A shrub 6 to 8 feet high ; glabrous except the petals ; young branches slender, dark-coloured, striate. *Leaves* thinly membranous, lanceolate, or oblong-lanceolate, shortly acuminate, the base acute ; main nerves 8 to 11 pairs, spreading, inter-arching within the minutely undulate margin, faint on both sur-faces ; length 4·5 to 7 in., breadth 1 to 1·75 in., petiole ·2 in. *Flowers* axillary, solitary, drooping ; pedicels slender, bi-bracteolate at the base,

·35 to ·45 in. long. *Sepals* free, large, membranous, green, many-nerved and reticulate, broadly ovate, acute or acuminate, glabrous, ·75 to 1·1 in. long. *Petals* whitish, thinly coriaceous, faintly nerved, broadly lanceolate, acuminate, much contracted at the base, pubescent, 1 to 1·2 in. long, (smaller in var. *aborescens*) ; inner petals less than half as long, ovate, acuminate, the base contracted, pubescent. *Anthers* numerous, compressed, the apices broad, flat, pubescent. *Ovaries* few, narrow, short, 1 rarely 2-ovuled ; the style long, straight, thickened upwards ; stigma hollowed like a funnel, the edges toothed. *Ripe carpels* partly enveloped by the persistent calyx, ovoid, very slightly apiculate, puberulous or glabrescent, ·4 to ·5 in. long ; stalks ·2 in long. *Seeds* usually 1, rarely 2.

Perak ; at a low elevations, King's Collector, No. 3019 ; Wray, Nos. 3379, 3558.

Var. *aborescens*, King ; a small tree 15 to 25 feet high ; *leaves* 4 to 4·5 in. long ; petals coriaceous, adpressed-pubescent, about half as long as in the typical form ; *sepals* only ·3 in. long.

Perak ; elevations from 2,000 to 3,000 feet, King's Collector.

This possibly ought to be considered a distinct species ; but as its anthers and ovaries are exactly the same as in the typical shrubby *G. tenuifolius*, I prefer to consider it a mountain form of that species. Both the typical form and the variety have remarkable stigmas, shaped like funnels and with toothed edges.

3. GONIOTHALAMUS PRAINIANUS, King, n. sp. A tree 50 to 70 feet high : young branches rather slender, pale ; all parts, except the inflorescence, glabrous. *Leaves* membranous, oblong-oblanceolate to ellipticoblong, abruptly shortly and bluntly acuminate, the base slightly cuneate ; main nerves 14 to 18 pairs, oblique, inter-arching within the margin, prominent beneath ; length 7 to 11 in., breadth 2·25 to 2·8 in., petiole ·35 in. *Flowers* 1·25 to 1·5 in. in diam., on long pedicels from large, woody, puberulous tubercles at the base of the stem : pedicels 2 to 4 in. long with two minute bracteoles at the base. *Sepals* coriaceous, united so as to form a spreading cup with three broad sub-acute triangular teeth, puberulous outside, glabrous inside. *Petals* thickly coriaceous, pale yellow ; the outer row large, obovate-rotund, concave, incurved, (ovate-oblong in var.) pubescent on both surfaces, nearly 1 in. long : inner row much smaller, clawed. *Stamens* numerous, the connective prolonged into a blunt, conical, puberulous, apical process. *Ovaries* narrowly oblong, glabrous ; style cylindric, not lobed, truncate. *Ripe carpels* obovoid, slightly apiculate, tapering to the base, glabrous, 1 to 1·25 in. long ; stalks ·25 in. long. *Seed* solitary, smooth.

Perak ; King's Collector, Wray ; at low elevations.

Var. : *angustipetala*, King ; petals oblong-ovate, sub-acute.

Perak : King's Collector.

A species collected by Forbes in Eastern Sumatra (Herb. Forbes, No. 3172) resembles this closely. The specimens are in fruit only, and the individual carpels being a little smaller and less obovoid, it probably belongs to a distinct species. Forbes' specimens have no flowers.

4. GONIOTHALAMUS KUNSTLERI, King. A shrub 4 to 10 feet high : young branches minutely rufous-tomentose, the older pale, glabrous and much striate. *Leaves* thinly membranous, oblanceolate to elliptic-oblanceolate, abruptly and bluntly acuminate, the base cuneate ; both surfaces pale-brown when dry, minutely pellucid-punctate, glabrous ; the midrib alone puberulous on the upper ; main nerves 11 to 13 pairs, spreading, curved and inter-arching boldly a little within the margin, slightly prominent on the under surface : length 6 to 9 in., breadth 2 to 3·25 in. ; petiole ·35 in. puberulous. *Flowers* solitary, slightly supra-axillary ; pedicels ·15 in. long. *Sepals* green, thinly membranous, puberulous, nerved and reticulate, broadly ovate, acute, spreading, very slightly cuneate at the base, ·3 to ·4 in. long. *Petals* sub-coriaceous, yellow or orange-coloured ; the outer lanceolate, acuminate, slightly narrowed at the base, puberulous outside, ·8 to 1·25 in. long : inner petals about one-third as long, ovate, acute, pubescent. *Anthers* many, short, compressed, the tops broad, flat, pubescent. *Ovaries* about as long as the stamens, narrowly cylindric ; style long, straight, thick : stigma notched. *Ripe carpels* crowded, broadly ovoid, slightly apiculate, ·4 in. long.

Perak ; at Goping, King's Collector, Scortechini, Wray.

Var. *marcantha*, King ; leaves narrowly elliptic or oblong, bluntly acuminate, puberulous beneath ; outer petals 1·25 to 1·5 in. long.

Penang and Province Wellesley : Curtis.

5. GONIOTHALAMUS GIGANTEUS, Hook. fil. and Thoms. Fl. Ind., 109. A tree 30 to 70 feet high ; young branches very pale, glabrous. *Leaves* coriaceous, oblong, shortly acuminate, the base cuneate, the edges slightly recurved (when dry) ; upper surface shining, glabrous : the lower dull, puberulous, the midrib very prominent : main nerves 10 to 14 pairs, very slender, spreading, more conspicuous above than below : length 6 to 10 in., breadth 2·25 to 2·75 in. ; petiole ·25 in., deeply channelled. *Flowers* very large, from the axils of fallen leaves and from the younger branches ; peduncles recurved, 1 in., or more, long (elongated in the fruit), pubescent. *Sepals* ovate, acute, pubescent outside, spreading or recurved, about 5 in. long. *Petals* very coriaceous, yellowish tinged with green ; the outer broadly ovate to ovate-oblong, with a dark thick triangular spot at the base, 3 to 5 in. long, minutely pubescent ; the inner only about ·6 in. long, ovate-acute, densely golden sericeous.

*Anthers* very numerous, their apices convex. *Ovaries* hairy, 2-ovuled : style long, slender, much curved ; stigma 2-lobed. *Ripe carpels* oblong, apiculate, tapering much to the stalk, minutely granular and with obscure vertical ridges when dry, 1·25 to 1·5 in. long and ·6 in. in diam. : stalks ·75 in., stout. *Seeds* 1 or 2, oblong, slightly compressed, the testa brown. Hook. fil. Fl. Br. Ind. I, 75 : Miq. Fl. Ind. Bat. I, pt. 2, 28. *Uvaria gigantea*, Wall. Cat. 6469 A. B. (*in part*). *Anonacea* Griff. Icon. Plant. t. 652 ?

Singapore ; Wallich, Ridley, Hullett. Penang ; Curtis. Perak ; King's Collector.

6. GONIOTHALAMUS MALAYANUS, Hook. fil. and Thoms. Fl. Ind 107. A small glabrous tree, 15 to 20 feet high ; bark of branches very pale. *Leaves* coriaceous, oblong to elliptic-oblong, shortly and abruptly acuminate, the base slightly cuneate, rarely rounded, the edges recurved ; upper surface shining, the lower dull, darker (when dry) ; main nerves 12 to 15 pairs, sub-horizontal, faint ; length 5·5 to 9 in., breadth 1·5 to 2·75 in.; petiole ·25 in., deeply channelled. *Flowers* slightly supra-axillary, solitary, greenish ; pedicels ·35 to ·5 in., pubescent, bracteolate at the base. *Sepals* ovate-triangular, acuminate, pubescent, connate at the base, persistent, ·25 in. long. *Petals* coriaceous, the outer broadly ovate, acuminate to ovate-lanceolate, minutely tomentose on both surfaces, with a triangular glabrous basal spot, keeled outside, 1 to 1·25 in. long ; the inner about a third as long, ovate, acuminate, sericeous or tomentose. *Anthers* numerous. *Pistils* about 15, the ovary hairy, ovules 3 to 4 ; style long, slender, much bent outwards ; stigma sub-capitate, unequally 2-lobed. *Ripe carpels* narrowly oblong apiculate, tapering to each end, glabrous, 1·5 in. long, and ·5 in. in diam ; stalks ·1 in., thick. *Seeds* 2 or 3, flattened-ovoid, nearly black. Hook. fil Fl. Br. Ind. I, 75 ; Miq. Fl. Ind. Bat. I, Pt. 2, 28. *Goniothalamus Slingerlandtii*, Scheff. Tijdsch. Ned. Ind. XXXI, 341. *Uvaria* sp. Griff. Notul. IV, 710.

Malacca ; Griffith, Maingay (Kew Distrib.) No. 63. Perak ; common. Distrib. Bangka.

7. GONIOTHALAMUS FULVUS, Hook. fil. and Thoms. Fl. Br. Ind. I, 75. A shrub : young branches slender, dark-coloured, at first rufous-pubescent, afterwards glabrous. *Leaves* membranous, pellucid-dotted, oblong-oblanceolate, obtuse or with a short broad point ; upper surface glabrous, the lower puberulous ; main nerves 14 to 16 pairs, slightly prominent beneath, spreading ; length 7 to 10 in., breadth 2·5 to 3·25 in.; petiole ·3 in., pubescent. *Flowers* solitary, axillary, pedicels ·25 in., puberulous. *Sepals* broadly ovate, obtuse, pubescent, connate at the base, ·25 in. long. *Petals* coriaceous, densely sericeous, the outer oblong-

lanceolate, attenuate to the apex, slightly keeled outside, 1 to 1·25 in. long; inner about ·3 in. long, ovate, acute. *Stamens* numerous, apices of anthers very convex, puberulous. *Ovaries* oblong, pubescent; style cylindric, glabrous : stigma bifid. *Fruit* unknown.

Malacca ; Griffith.

Known only by Griffith's imperfect specimens.

8. GONIOTHALAMUS CURTISII, King, n. sp. A shrub or small slender tree : young branches densely rusty-tomentose, the larger pale and glabrous. *Leaves* stoutly membranous, narrowly oblong to obovate-oblong, more or less abruptly and shortly acuminate, slightly narrowed to the rounded base ; upper surface shining, glabrous except the puberulous midrib ; the lower sparsely puberulous, the midrib and nerves dark rusty-tomentose ; the latter 28 to 34 pairs, sub-horizontal, inter-arching near the margin, very prominent, as is the midrib, on the lower and depressed on the upper surface : length 9 to 15 in., breadth 3 to 5·5 in. ; petiole ·35, channelled, pubescent. *Flowers* solitary, from the stem ; pedicels stout, decurved, with two deciduous bracteoles at the base, ·6 in. long. *Sepals* large, green, rigidly membranous, conjoined into a cup with 3 broadly-ovate, sub-acute teeth, boldly nerved and reticulate, minutely rufous-pubescent, persistent ; length from ·75 to 1 inch. *Petals* coriaceous, velvety-tomentose, yellowish, tinged with red : the outer broadly lanceolate, acuminate, slightly narrowed and thickened at the base, from 1·25 to 1·75 in. long ; the inner rather more than one-third as long, ovate, acuminate. *Anthers* numerous, compressed, linear, with acute granular conical apices. *Ovaries* numerous, narrowly elongate, densely pubescent, 1-ovuled ; style straight; stigma oblique, minutely lobed. *Ripe carpels* obliquely ovoid with long pointed, slightly hooked apices, rufous-pubescent, ·75 in. long : stalks only ·1 in. long, stout.

Selangor ; Curtis, Nos. 310 and 2316. Perak ; King's Collector, No. 10548 : Scortechini, No. 660.

A very distinct species.

9. GONIOTHALAMUS GRIFFITHII, Hook. fil. and Th. Fl. Ind., 110. A large shrub or small tree; all parts glabrous except the ovaries and carpels : young branches dark-coloured. *Leaves* coriaceous, oblong, sub-acute, or shortly and obtusely acuminate, the base cuneate ; both surfaces shining and reticulate ; main nerves 12 to 20 pairs, faint, spreading, inter-arching within the edge : length 7 to 12 in , breadth 1·8 to 3·5 in. ; petiole ·25 to 5 in., thick. *Flowers* solitary, axillary or extra-axillary ; pedicel ·5 to 1 in. long with a few scale-like bracteoles near the base. *Sepals* thinly coriaceous, orbicular-ovate, blunt, connate below, nerved and reticulate, persistent, ·5 to ·75 in. long. *Petals* thickly coriaceous ;

the outer broadly lanceolate, acuminate, 1·5 to 2·5 in. long: the inner ovate, acute, ·6 to ·8 in. long. *Anthers* with an acute apical process. *Ovaries* strigose: style long, subulate; stigma slightly bifid. *Ripe carpels* sub-sessile, oblong, ·5 or ·6 in. long, glabrescent or glabrous. Hook. fil Fl. Br. Ind. I, 73 ; Kurz F. Flora Burma, I, 42.

Burmah: Mergui, Griffith. Moulmein, Falconer.

10. GONIOTHALAMUS MACROPHYLLUS, H. f. and Th. Fl. Ind. I, 74. A glabrous shrub 5 to 15 feet high ; young branches very stout, dark-coloured. *Leaves* coriaceous, large, oblong-lanceolate to oblong-oblance-olate, acute or shortly acuminate, slightly narrowed to the sub-acute or rounded base ; main nerves 16 to 20 pairs, spreading, impressed above and slightly prominent beneath ; length 10 to 18 in., breadth 2·5 to 4·5 in. ; petiole ·6 to 1 in., very stout. *Flowers* slightly supra-axillary or from the branches below the leaves, solitary or in pairs, green ; pedicels ·35 in. long, sub-clavate. *Sepals* broadly ovate, acute, connate at the base, ·65 in , long, slightly puberulous, tinged with purple. *Petals* cori-aceous, the outer oblong-lanceolate, acute or acuminate, 1 to 1·5 in. long ; the inner half as long, ovate, acuminate, the edges ciliate. *Sta-mens* numerous, linear. *Ovaries* 12 to 18, glabrous, 1-ovuled ; style slender, dilated above, stigma 2-lobed. *Ripe carpels* globular-obovoid, slightly apiculate, glabrous, ·4 in. long, *Seed* pale brown. Miq. Fl. Ind. Bat. I, Pt. 2, 28 ; Ann. Mus. Lugd. Bat. II, 38. *Polyalthia macrophylla*, Blume Fl. Jav. Ann. 79 t. 39. *Unona macrophylla*, Blume Bijdr, I, 17.

It is possible that two species may be included here, there being some difference between the specimens in the nervation of the leaves.

Malacca ; Griffith, Maingay, (Kew Distrib.) No. 62. Perak, King's Collector. Penang ; Curtis. Kedah ; Curtis. Distrib. Sumatra, Forbes, 1370.

11. GONIOTHALAMUS RIDLEYI, King, n. sp. A tree : young branches slender, puberulous. *Leaves* membranous, broadly elliptic, shortly and abruptly acuminate, the base sub-acute, pale when dry ; both surfaces reticulate ; the upper dull, glabrous, except the puberulous midrib and nerves , the lower shining, puberulous on the midrib, nerves and reticu-lations main nerves about 6 pairs, curving, ascending ; length about 8 in. ; breadth 4·5 in. ; petiole ·25 in., puberulous. *Flowers* 1·75 to 2 in. long, in fascicles on long pedicels from warted, puberulous, woody tuber-cles on the stem : pedicels 2·5 to 3·5 in. long, minutely bracteolate at the base. *Sepals* coriaceous, broadly ovate-elliptic, obtuse, nerved, ·6 in. long, free, spreading, puberulous. *Petals* coriaceous, pale brown ; the outer elliptic-oblong to ovate, obtuse or sub-acute, with a broad thicken-ed claw, puberulous, 1·65 to 2 in. long ; inner row a little longer than the sepals, obovate, apiculate, with narrow claw. *Stamens* numerous,

325

long, narrow, much compressed ; the apical process of the connective small, sub-conic. *Ovaries* oblong, narrow ; style cylindric, puberulous ; stigma 2-lobed. *Ripe carpels* obvoid-globular, tapering slightly to the short stalk, glabrous, about 1 in. long.

Singapore ; at Sunga Murai, Ridley.

It is possible that in the above description the size of the leaves may be understated, as the only one which I have seen may not be of average size.

12. GONIOTHALAMUS TAPIS, Miq. Fl. Ind. Bat. Suppl. 371. A tree 15 to 40 feet high ; all parts, except the flowers, glabrous ; young branches pale brown. *Leaves* coriaceous, oblong, abruptly shortly and bluntly acuminate, the base rounded or slightly cuneate, the edges recurved (when dry) ; both surfaces dull, brown when dry, the lower paler ; main nerves 10 to 12 pairs, thin, spreading, very indistinct, the midrib prominent beneath ; length 5·5 to 9 in., breadth 2·5 to 3·25 in., petiole ·3 in. *Flowers* solitary and supra-axillary, or in fascicles from tubercles on the branches ; pedicels curved, ·4 in. long, bracteolate at the base. *Sepals* free, ovate, acute, spreading, pubescent, persistent, ·4 in. long. *Petals* coriaceous, puberulous ; the outer ovate-lanceolate, acuminate, contracted and thickened at the base, 1·75 in. long ; the inner ovate, acute, much contracted and thickened at the base, ·65 in. long. *Anthers* numerous and with conical apices. *Ovaries* narrow, hairy ; style straight ; ovules solitary, *Stigma* sub-discoid-capitate, 2- to 3- lobed. *Ripe carpels* crowded, obovoid, smooth, sub-sessile, ·4 to ·5 in. long. Miq. Ann. Mus. Lugd. Bat. II, 35.

Perak ; at low elevations, very common ; Scortechini, Wray, King's Collector. Penang and Pangkore ; Curtis. Distrib. Sumatra, Borneo.

13. GONIOTHALAMUS SCORTECHINII, King, n. sp. A shrub or small tree, glabrous, except the flowers ; young branches with rather pale striate bark. *Leaves* membranous, oblanceolate or oblong-oblanceolate, very shortly acuminate, narrowed from the above the middle to the acute or sub-acute base ; when dry the upper surface greenish, the lower pale brown ; main nerves 18 to 24 pairs, spreading and inter-arching near the edges, slender, slightly prominent beneath ; length 10 to 15 in., breadth 2·75 to 4 in., petiole ·3 in. *Flowers* solitary, rarely in pairs, from the branches below the leaves ; pedicels clavate, decurved, bi-bracteolate at the base, ·5 in. long. *Sepals* rigidly membranous, large, orbicular-ovate, obtuse or sub-acute, much nerved and reticulate, connate below, persistent, from ·65 to 1 in. long (according to age). *Petals* coriaceous, rusty-puberulous ; the outer oblong-lanceolate, sub-oblique, not much longer than the full grown sepals ; the inner broadly ovate, acute, about ·5 in. long. *Anthers* numerous, narrow, with elongate, conical apical pro-

cesses. *Ovaries* narrow, puberulous, 1-ovuled : style straight ; stigma 2- or 3-lobed. *Ripe carpels* crowded, ovoid-oblong, apiculate, glabrous, narrowed to the short stalks, ·45 in. long ; stalks ·2 to ·25 in. *Seed* smooth, pale.

Perak ; at low elevations ; Scortechini, Wray, King's Collector.

The leaves of this species much resemble those of *Polyalthia oblonga*, King.

14. GONIOTHALAMUS WRAYI, King, n. sp. A shrub 3 to 12 feet high, glabrous, except the flowers : young branches slender, very pale. *Leaves* membranous, oblanceolate to lanceolate or oblong, shortly and bluntly acuminate, the base cuneate : both surfaces pale (when dry), obscurely reticulate : main nerves 14 to 18 pairs, spreading, straight, slender and very slightly prominent even when dry : length 4·5 to 9 in., breadth 1·25 to 2 in., petiole ·2 to ·25 in. *Flowers* solitary, slightly supra-axillary ; pedicels slender, decurved, minutely bracteolate, ·35 in. (elongated to ·75 in. in fruit). *Sepals* membranous, slightly nerved and reticulate, ovate, acuminate, spreading or recurved, puberulous outside, ·2 in. long, persistent. *Petals* sub-coriaceous, greenish-yellow, puberulous : the outer narrowly lanceolate, acuminate, the bases thickened and not narrowed to a claw, ·65 to ·75 in. long : inner petals about half as long, ovate-acuminate. *Anthers* numerous, half as long as the ovaries, compressed, their apices with a long thin point from a broad base. *Ovaries* about 20, narrowly cylindric, hairy like the stout, straight style 1- to 2-ovuled : stigma truncate. *Ripe carpels* narrowly obovoid to oblong, apiculate, gradually tapering to the stalk, glabrous, ·6 in. long. *Seeds* usually 1, rarely 2, oblong.

Perak : at low elevations very common ; Wray, Scortechini, King's Collector.

15. GONIOTHALAMUS UVARIOIDES, King, n. sp. A shrub 6 to 15 feet high : all parts glabrous except the flower and fruit ; young branches pale. *Leaves* thinly coriaceous, oblong, slightly obovate, slightly narrow-ed to the minutely cordate base : both surfaces rather dull when dry, the lower pale brown, the edges slightly recurved ; main nerves 22 to 25 pairs, spreading, rather straight, inter-arching near the margin ; length 10 to 15 in., breadth 3 to 6 in. ; petiole ·4 in., stout, channelled. *Flowers* on the trunk, (solitary ?) ; pedicels curved, stout, ·35 in. long. *Sepals* coriaceous, semi-orbicular, blunt, pubescent, 2 in. long. *Petals* very coriaceous, yellow : the outer broadly lanceolate, thickened and truncate at the base, rufous-pubescent, 1·5 in. long : inner petals like the outer but with contracted bases and only 1 to 1·2 in. long. *Anthers* with conical apices. *Ovaries* hairy ; style cylindric ; stigma small, truncate, minutely bifid. *Ripe carpels* oblong, tapering to each end, puberulous,

1·5 in. long, and ·65 in. in diam.; stalks ·7 in. long. *Seeds* 4, compressed, rugose, ·5 in. long.

Perak: Ulu Slim, King's Collector, No. 10664. Ulu Bubong, King's Collector, No. 10126. Distrib., Borneo; Motley, No. 960.

Motley's Bornean specimen above-quoted is in flower only; but it so entirely resembles in leaves and wood those of my collector in Perak which are in fruit only, that I have ventured not only to consider them as belonging to the same species, but to draw up the above description of the flowers from the Bornean and of the fruit from the Perakian specimens. The species resembles *G. fulvus* in leaves and flower and *G. malayanus* in flower. The fruit is more like that of a *Uvaria* than of a *Goniothalamus*, having 4, sub-horizontal, rugose seeds.

## 15. OROPHEA, Blume.

Trees or shrubs. *Flowers* usually small, axillary, solitary, fascicled or cymose. *Sepals* 3, valvate. *Petals* 6, valvate in 2 series; outer ovate; inner clawed, usually cohering by their margins into a mitriform cap; sometimes oblong and slightly approximate below the middle, the apices divergent not vaulted: rarely without claws and in one species slightly imbricate. *Stamens* definite, 6–12, ovoid, fleshy; anther-cells dorsal, large, contiguous, the connective sometimes prolonged into a conical apical point, not truncate. *Staminodes* 0, or 3 to 6. *Ovaries* 3–15; style short or 0; ovules 4. *Ripe carpels* 1- or more-seeded, globular or oblong (very long in several species.)—DISTRIB. Species about 25; all Eastern Asiatic.

Intermediate between *Mitrephora* and *Bocagea*, having the perianth of the former and stamens of the latter.

Inner petals distinctly vaulted, the limbs coherent by their edges.
    Stamens 12 ... ... ... 1. *O. setosa.*
    Stamens 6.
        Leaves glabrous at all ages (see also No. 5) 2. *O. Katschallica.*
        Leaves more or less pubescent (except No. 5).
            Carpels globose when ripe ... 3. *O. hirsuta.*
            Carpels oblong when ripe.
                Carpels under 2 in. in length ... 4. *O. hexandra.*
                Carpels 3 to 5 in. long.
                    Leaves quite glabrous, main nerves 6 or 7 pairs ... 5. *O. enterocarpa.*
                    Leaves puberulous beneath, main nerves 10 or 12 pairs 6. *O. maculata.*
Inner petals slightly vaulted, trapezoid ... 7. *O. gracilis.*

Inner petals spreading, not vaulted and not trapezoid.
Stamens 10 or 12.
    Inner petals hastate; ripe carpels globular   8. *O. hastata.*
    Inner petals linear-oblong, the apices
    divergent and recurved; ripe carpels
    ovoid or slightly obovoid ...      ...  9. *O. dodecandra.*
Stamens 6.
    Inner petals cuneiform or cuneiform-retuse;
    ripe carpels cylindric   ...     ... 10. *O. cuneiformis.*
    Inner petals irregularly oblong, their
    apices broad and curved outwards, ripe
    carpels globular     ...     ... 11. *O. polycarpa.*

    1. OROPHEA SETOSA, King, n. sp. A shrub : young branches densely covered with a layer of minute pubescence with numerous, long, brownish, straight bristles projecting beyond it ; the older branches dark-coloured and almost glabrous. *Leaves* membranous, oblong or oblong-oblanceolate, shortly acuminate, the base rounded : main nerves 8 to 10 pairs, oblique, inter-arching near the edge ; both surfaces sparsely setose, more densely so on the midrib and nerves, the lower also with sparse, minute pubescence ; length 5·5 to 7·5 in , breadth 2 to 2·75 in , petiole ·05 in., setose. *Flowers* solitary, extra-axillary, about ·2 in. in diam. when expanded : pedicels very slender, ·75 in. long, pubescent, with a single minute bracteole below the middle. *Sepals* sub-orbicular, blunt. *Outer petals* much larger than the sepals, broadly ovate, sub-acute, pubescent outside and glabrous inside like the sepals. *Inner petals* longer than the outer, vaulted, ·22 in. long, the limb trapezoid-sagittate, pubescent on the back and edges, glabrous in front ; the claw narrow, shorter than the limb. *Male flower* stamens numerous, cuneate, the connective broadly truncate at the apex. *Ovaries* unknown. *Ripe carpels* 4 or 5, sessile, globose or oblong-globose, ·3 in. in diam., densely and minutely pubescent and with a few long setæ besides. *Seeds* solitary, rarely 2 ; the testa pale, rather rough ; the albumen very dense.

    Perak : at elevations from 800 to 1,200 feet ; King's Collector, Scortechini.

    2. OROPHEA KATSCHALLICA, Kurz in Trimen's Journ. Bot. 1875, p. 323. A small tree 25 to 30 feet high : young branches slightly puberulous at first, ultimately glabrous, black and furrowed. *Leaves* membranous, oblong-lanceolate to oblong or elliptic, shortly and bluntly acuminate, the base sub-cuneate or rounded ; upper surface glabrous, shining ; the lower much reticulate, slightly adpressed-puberlous ; main nerves 3 to 10 pairs, ascending, slender ; length 4 to 7 in., breadth

1·5 to 2·75 in., petiole ·15 in. *Peduncles* extra-axillary, solitary, ·5 to ·75 in. long, with numerous ovate-acuminate, rusty-pubescent bracts. *Flowers* 1 to 4, rather large ; their pedicels about ·4 in. long, pubescent and with a single adpressed ovate-lanceolate bracteole. *Sepals* ovate-acuminate, adpressed-pubescent outside, sub-glabrescent inside. *Outer petals* much larger than the sepals, ovate-orbicular, acute, veined, pubescent on the outer surface and on the upper half of the inner, ·4 in. long. *Inner petals* ·75 in. long, trapezoid, acute, tomentose on both surfaces except a glabrous patch bearing a transverse callosity on the inner ; the claw long, narrow and glabrous. *Stamens* 6 perfect, with a few imperfect in an outer row : anther-cells large, dorsal ; the connective oblique, slightly produced above their apices. *Ovaries* about 3, narrowly ovoid, densely sericeous, 3-ovuled ; stigmas sessile, truncate. *Fruit* unknown.

Nicobar Islands ; Kurz, King's Collector.

3. OROPHEA HIRSUTA, King, n. sp. A shrub 8 to 12 feet high : young branches at first densely rufous-hirsute, afterwards becoming glabrous and dark-coloured. *Leaves* elliptic or elliptic-oblong, often slightly obovate, shortly and bluntly acuminate, narrowed from below the middle to the rounded minutely cordate base : upper surface glabrous, shining, the lower pale, dull, sparsely hirsute, the midrib setose at the base : main nerves 8 to 9 pairs, spreading, very faint : length 3·5 to 4·5 in., breadth 1·24 to 1·75 in. ; petiole ·05, setose. *Peduncles* extra-axillary, about ·5 in. long, 1- to 3-flowered, rufous-hirsute like the pedicels : pedicels about ·75 in. long and with several minute bracteoles. *Flowers* ·5 in. in diam. *Sepals* broadly ovate, acute, coarsely hirsute outside and on the edges, glabrous inside. *Outer petals* much larger than the sepals, broadly obovate, blunt, sparsely pubescent outside and on the edges, glabrous inside, ·15 in. long. *Inner petals* ·25 in. long, vaulted : the limb trapeziform, rather thick, glabrous outside, pubescent inside ; the claw very narrow, longer than the limb, glabrous. *Stamens* 6, in a single row, curved : anthers broad, dorsal, the connective not produced above their apices. *Ovaries* about 6, ovoid, glabrous, 1- to 2-ovuled : stigma sessile, roundish. *Carpels* 4 to 5, globular, yellow when ripe, sparsely hirsute, ·4 in. in diam. ; stalks ·1 in.

Perak : King's Collector, No. 4283.

Only once collected. In its leaves this resembles *Mitrephora setosa*. King.

4. OROPHEA HEXANDRA, Blume Bijdr. 18. A small tree : young branches slender, minutely tomentose, soon becoming dark-coloured, glabrous and furrowed. *Leaves* thinly coriaceous, oblong-lanceolate to elliptic-oblong, rather abruptly acuminate, the base sub-cuneate or

rounded; upper surface glabrous, shining; the lower reticulate, pub rulous, the midrib pubescent; main nerves 7 to 9 pairs, oblique : length 4·5 to 6 in., breadth 1·5 to 2·25 in., petiole ·2 in. *Peduncles* axillary or supra-axillary, slender, 1- to 3-flowered, pubescent; bracts several, subulate, hairy. *Flowers* about ·35 in. long, greenish-white. *Sepals* minute, ovate to ovate-lanceolate, densely pubescent outside. *Outer petals* thin, ovate-cordate, acuminate, pubescent; the inner larger, trapezoid with long narrow claw, glabrous with pubescent margins. *Stamens* 6, in one row. *Ovaries* about 6, pubescent, 2-ovuled. *Ripe carpels* oblong, subsessile, acuminate, minutely adpressed-pubescent, 1·4 to 1·75 in. long. *Seeds* usually solitary, sometimes ·2 in. long, narrowly cylindric. Kurz For. Flora Burma, I, 49 : Miq. Fl. Ind. Bat. I, pt. 2 p. 29. *O. acuminata*, A. D C. in Mem, Soc. Genev. V, 39; Hook. fil. and Thoms. Fl. Ind. 112; Hook. fil. Fl. Br. Ind. I, 91; Wall. Cat. 6432. *Bocagea hexandra*, Blume Fl. Jav. Anon. 83 t. 40.

Burma prov. Tenasserim, Wallich. Great Coco Island; Kurz. S. Andaman; King's Collectors.

Pierre (Flore Forestiere Cochin-Chine t. 44) figures a species called *O. Thorelii* which, as he remarks, must be closely allied to this.

5. OROPHEA ENTEROCARPA, Maingay ex Hook. fil. Fl. Br. India, I, 92. A small tree 15 to 30 feet high; all parts, except the inflorescence, glabrous : young branches slender, black, striate. *Leaves* membranous, ovate or sometimes obovate-lanceolate to elliptic, acuminate (sometimes abruptly so); the base rounded, sometimes sub-cuneate; both surfaces shining : main nerves 6 or 7 pairs, spreading, slender : length 2·5 to 5 in., breadth 1·2 to 2 in., petiole ·1 in. *Flowers* nodding, solitary, extra-axillary : the pedicels very slender, ·75 to 1·25 in. long, glabrous below, pubescent above and with several ovate-lanceolate bracteoles. *Sepals* small, broadly ovate, acuminate, pubescent. *Outer petals* much larger than the sepals, ovate, acuminate, puberulous, the inner a little longer (·6 to ·75 in. long); the limb elongated-trapezoid, puberulous; the claw narrow and glabrous, yellowish with a reddish band; staminodes 6. *Stamens* 6, with broad connective, not apiculate. *Ovaries* 6, cylindric, glabrous, 2- to 7-ovuled; stigma small, sessile. *Carpels* 4 to 6, elongate-cylindric, glabrous, moniliform when dry, 3 to 5 in. long and ·3 in. in diam. *Seeds* 2 to 7, linear-oblong.

Malacca : Maingay. Perak; Scortechini, King's Collector.

6. OROPHEA MACULATA, Scortechini MSS. A shrub or small tree : young branches slender, rusty-tomentose at first, afterwards glabrous, black and striate. *Leaves* membranous, elliptic-oblanceolate, caudate-acuminate, narrowed from below the middle to the rounded or sub-cuneate slightly unequal base : upper surface glabrous, the lower

glabrescent, the midrib and nerves pubescent; main nerves 10 to 12 pairs, spreading, rather faint; length 3·25 to 7 in., breadth 1·5 to 2·25 in , tomentose. *Peduncles* solitary, 1- to 3-flowered, extra-axillary, very slender, ·5 to 1 in. long, pubescent, with numerous, distichous, sub-deciduous, linear-lanceolate, pubescent bracts. *Flowers* large, sub-pendulous. *Sepals* narrowly lanceolate, acuminate. *Outer* petals larger than the sepals, mottled red and yellow, ovate, very acuminate, veined, pubescent on both sides, ·5 in. long. *Inner petals* 1 in. long, with lanceolate, much acuminate, very pubescent limb; the claw long, narrow, pubescent. *Stamens* 6, broad, not apiculate, hairy at the base. *Staminodes* 3, orbicular. *Ovaries* 3 to 6, cylindric, very hirsute, 6- or 7-ovuled : stigma sessile. *Carpels* 4 to 6, much elongate, cylindric, puberulous, 3 to 5 in. long, and about ·3 in. in diam., moniliform when dry. *Seeds* 4 to 7, linear-oblong.

Perak ; Scortechini, King's Collector.

7. OROPHEA GRACILIS, King, n. sp. A tree 20 to 30 feet high ; young branches slender, at first minutely tomentose, afterwards darkly cinereous and glabrous. *Leaves* thinly coriaceous when adult, lanceolate, much acuminate, the base cuneate or slightly rounded, both surfaces glabrous : main nerves 5 or 6 pairs, spreading, inter-arching far from the edge, very indistinct; length 2·5 to 3·5 in., breadth ·9 to 1·2 in , petiole ·05 in. *Flowers* solitary, ·25 in. in diam., extra-axillary ; pedicels ·75 to 1 in. long, very thin, glabrous, jointed, and with several minute, subulate bracteoles above the middle. *Sepals* broadly ovate, sub-acute, connate at the base, spreading or reflexed. *Outer petals* larger than the sepals, ovate, acute, ·15 in. long; both surfaces glabrous, the edges alone minutely pubescent. *Inner petals* ·25 in. long, slightly vaulted ; the limb thick, trapezoid, with pubescent edges ; the claw narrow, not so long as the limb, glabrous. *Stamens* 6, in a single row, the connective much produced above the rather small dorsal anther-cells. *Ovaries* 4 to 10, ovoid, glabrous, 2-ovuled : stigma large, sessile. *Ripe carpels* 6 to 10, globular, glabrous, ·45 in. in diam., their stalks ·25 in. long. *Seeds* solitary or two together, depressed-globose, with a transverse groove and ridge, shining, pale.

Perak : Scortechini, King's Collector.

This is closely allied to the W. Peninsular *O. uniflora*, but that species has twice as many stamens.

8. OROPHEA HASTATA, King, n. sp. A tree 20 to 40 feet high : all parts glabrous except the inflorescence : young branches rather slender, dark-coloured. *Leaves* thinly coriaceous, elliptic to elliptic-oblong, shortly caudate-acuminate; the base cuneate, rarely rounded ; both surfaces shining, the lower pale : main nerves 6 to 8 pairs, spreading,

inter-arching within the edge ; length 3·5 to 5·5 in., breadth 1·6 to 2·4 in., petiole ·2 in. *Peduncles* axillary or supra-axillary, solitary, about ·25 in. long, bearing towards the apex 3 or 4 1-bracteolate, pubescent pedicels. *Flowers* ·4 in. long. *Sepals* broadly ovate, acute, pubescent, outside, glabrous inside as are the outer petals. *Outer petals* twice as large as the sepals, broadly ovate acute. *Inner petals* ·35 in. long ; the limb hastate, triquetrous, thickened, the edges and the base ciliate ; the claw long, narrowed to the base, glabrous. *Staminodes* 0. *Stamens* 10, in 2 rows, curved, slightly apiculate ; the anther-cells large. *Ovaries* about 10, obliquely oblong, curved, pubescent, 2 ovuled ; stigma small, capitate, sessile. *Ripe carpels* 5 or 6, globular, glabrous, ·4 in. in diam., their stalks about ·25 in. *Seeds* solitary.

Perak : Wray, King's Collector, at low elevations.

This is closely allied to *O. dodecandra*, Miq.

9. OROPHEA DODECANDRA, Miq. in Ann. Mus. Lugd. Bat. II, 25. A tree 20 to 40 feet high ; young branches sparsely adpressed-pubescent, afterward glabrous dark-coloured and striate. *Leaves* membranous, elliptic, rarely elliptic-oblong, slightly unequilateral, shortly caudate-acuminate, the base cuneate ; upper surface glabrous, shining, the lower paler with a few scattered, pale, adpressed hairs ; main nerves 5 or 6 pairs, bold beneath, inter-arching ·25 in. from the margin ; length 3·5 to 5·5 in. ; breadth 1·75 to 2·3 in., petiole ·2 in. stout, channelled. *Peduncles* supra-axillary, longer than the pedicels, 3- to 7-flowered, glabrous ; pedicels ·5 in. long, clustered near the apex, bracteolate above the middle. *Flowers* ·5 in. long. *Sepals* smaller than the outer petals, spreading, dotted, conjoined at the base, slightly tubercular outside, glabrous inside. *Outer petals* broadly ovate, acuminate, narrowed at the base, ·15 in. long. *Inner petals* thick, linear-oblong, blunt, puberulous outside, slightly arched below the middle, the apices divergent and recurved. *Staminodes* 0. *Stamens* 12, in 2 rows ; the connective rather narrow, prolonged beyond the apices of the large, broad, dorsal anthers. *Ovaries* 6 to 8, oblong, curved, oblique, glabrous, 2-ovuled ; stigma oblong, sessile. *Ripe carpels* ovoid or slightly obovoid, blunt, glabrous, ·85 in. long ; their stalks ·8 to ·9 in. *Seed* solitary, sub-rotund or oblong, with rugose, pale, scaly testa.

Perak ; Scortechini, King's Collector ; at low elevations.

10. OROPHEA CUNEIFORMIS, King, n. sp. A tree 20 to 40 feet high ; young parts rusty-pubescent or tomentose ; the branchlets rather stout ; ultimately glabrous, dark-coloured and furrowed. *Leaves* thinly coriaceous, oblong, narrowly elliptic or oblanceolate-oblong, more or less sharply acuminate, very little narrowed to the rounded or minutely cordate base ; upper surface at first with many long, thin, pale,

adpressed hairs, ultimately glabrous; lower softly but rather coarsely pubescent, the midrib and 8 to 12 pairs of oblique, rather prominent main nerves rufous-tomentose; length 3·5 to 6 in., breadth 1·1 to 2·2 in.; petiole ·05, tomentose. *Peduncles* 4- or 5-flowered, solitary, supra-axillary, slender, sub-glabrous below, rufous-sericeous above, longer than the pedicels; bracts numerous, linear-lanceolate; pedicels ·3 in. long, rufous-sericeous like the outer surface of the sepals and outer petals, bracteolate at the base. Flower buds globose. *Sepals* ovate, much acuminate, glabrescent inside like the outer petals. *Outer petals* ovate, acute, veined. *Inner petals* with a cuneiform, sometimes retuse, thick limb and a short, narrow claw. *Staminodes* 3, in an outer row, sub-orbicular, fleshy. *Stamens* 6, with broad flat connective, not produced at the apex, and large dorsal anthers. *Ovaries* about 6, oblong, oblique, densely villous, 2- or 3-ovuled, *Stigma* sessile, broad. *Ripe carpels* 2 to 4, sessile, cylindric, tapering a little at each end, puberulous, 1·5 to 1·75 in. long and about 35 in. in diam. *Seeds* 2, oblong.

Perak; Scortechini, King's Collector.

This is readily distinguished from the closely allied species *O. maculata*, by its scorpioid cymes, globular flower-buds, and by the cuneiform (not lanceolate) limbs of its petals.

11. OROPHEA POLYCARPA, A. DC. in Mem. Soc. Genev. V, 39. A large shrub or small tree: young branches slender, pubescent at first, but speedily glabrous, furrowed and dark-coloured. *Leaves* membranous, ovate to ovate-oblong, obtusely and very shortly acuminate, the margins undulate, the base rounded or narrowed; both surfaces glabrous; main nerves 6 to 8 pairs, spreading, faint; length 2 to 4 in., breadth 1 to 1·75 in., petiole ·05 in. *Peduncles* axillary or supra-axillary, slender, 1- to 3-flowered, pubescent; bracteoles several. *Sepals* ovate, acute, very pubescent. *Outer petals* ovate, acuminate, more than twice as large as the sepals, pubescent on the outer, glabrous on the inner, surface. *Inner petals* twice as long as the outer, irregularly oblong, the apices broad and curved outwards, the base slightly narrowed, puberulous outside, glabrous within, ·4 in. long. *Stamens* 6 or 7 in a single row; the anther-cells quite dorsal, separate, the connective flat and very slightly prolonged above their apices. *Ovaries* about twice as many as the stamens, gla-brous, ovate, oblique: stigma small, sessile, sub-capitate. *Ripe carpels* globular, glabrous, shining, ·35 in. in diam.: their stalks ·25 in. long. *Seeds* 1 or 2. Hook. fil. and Thoms. Fl. Ind. 111; Hook. fil. Fl. Br. Ind. I, 91; Kurz F. Flora Burma, I, 49; *Anonacea* Griff. Ic. Pl. Ind. Or. IV, t. 654. Wall. Cat. 6431. *Bocagea polycarpa*, Stend. Nomen. 212. *Melodorum?* *monospermum* Kurz in Andaman Report, App. B. p. 1. *Bocagea polycarpa*, Steud.

S. Andaman; Kurz, King. Burmah : Martaban, Wallich.
*Orophea undulata*, (Pierre Fl. Forest. Coch.-Chine t. 45) must bo
closely allied to this, as must also the same author's *O. anceps*, (l. c. t
46).

## 16. MITREPHORA, Blume.

Trees. *Leaves* coriaceous, strongly ribbed, plaited in vernation.
*Flowers* usually terminal or leaf-opposed, sometimes 1-sexual. *Sepals* 3,
orbicular or ovate. *Petals* 6, 2-seriate, valvate ; outer ovate, thin, veined ;
inner clawed, vaulted and cohering. *Stamens* oblong-cuneate ; the anther-
cells dorsal, remote, the connective broadly truncate at the apex. *Ovaries*
oblong ; style oblong or clavate, ventrally furrowed ; ovules 4 or more,
2-seriate. *Ripe carpels* globose or ovoid, stalked or sub-sessile.—DISTRIB.
Species about 10 ; tropical Asiatic.

Flowers hermaphrodite ... ... ... 1. *M. Maingayi.*
Flowers unisexual.
 Ripe carpels ovoid, apiculate, rugulose ... 2. *M. reticulata.*
  „  „  globular, not apiculate, not rugulose  3. *M. macrophylla*
  „  „  sub-globular, sub-truncate at each
      end, rugulose  ... ... 4. *M. Prainii.*

1. MITREPHORA MAINGAYI, Hook. fil. and Thoms. Fl. Br. Ind. 1, 77.
A tree 20 to 50 feet high : young branches softly rufous-tomentose
afterwards glabrous dark-coloured and striate. *Leaves* coriaceous,
oblong to ovate, (oblong-lanceolate in var. *Kurzii*), acute or shortly and
bluntly acuminate, the base rounded or sub-cuneate ; upper surface
shining, glabrous except the pubescent midrib ; under surface glabrescent, the midrib and nerves thinly adpressed-pubescent ; (pubescent
in var. *Kurzii*) ; main nerves 6 to 10 pairs, oblique, curving, slightly
prominent beneath : length 3 to 5·5 in., breadth 1·5 to 2 in., petiole
·3 to ·4 in. *Flowers* 1 in. or more in diam., axillary or leaf-opposed,
solitary or 2 or 3 in a multi-bracteolate and tomentose raceme ; pedicels ·5 to 1·5 in. (lengthening with age), bracteolate. *Sepals* connate
into a cup, broadly ovate, acute, (or obtuse in var.) tomentose. *Petals*
rather thinly pale yellow mottled with red, all more or less pubescent outside, the outer orbicular or obovate with undulate erose edges, slightly
narrowed at the base, (oblong in var. *Kurzii*) ; inner shorter, the outer very
pubescent inside, vaulted, ovate or cordate with a long linear claw.
*Anthers* numerous, short, with broad flat smooth tops. *Ovaries* gradually
narrowed into the short style ; ovules 4 ; stigma sub-capitate-discoid.
*Ripe carpels* broadly ovoid, blunt at each end, densely tomentose, 1 in.
long, and ·75 in. in diam. : their stalks stout, ·75 in. *Seeds* 4, compressed.
*M. Teysmannii*, Scheff. in Flora LII (1869), 302. *Uvaria obtusa* (not of

Blume), Hook. fil. and Thoms., Fl. Ind. 113; Hook. fil. Fl. Br. Ind. I, 76; Wall. Cat. 6484.

Penang: Wallich, Curtis. Pangkore; Curtis. Malacca; Maingay, (Kew Distrib.) No. 65. Perak: King's Collector, Scortechini, Wray. Burma, Kurz. Distrib. Java.

Var. *Kurzii*, Leaves oblong-lanceolate, acuminate to elliptic: peduncles of racemes woody, 1 in. or more long, tomentose; outer petals narrowly oblong. *M. vandaeflora*, Kurz F. Flora Burma I, 45.

Burma; Kurz, Brandis.

Allied to the Cambodian species *M. Thorellii*, (Pierre Fl. Forest. Cochin-Chine, t. 37).

2. MITREPHORA RETICULATA, Hook. fil. and Thoms. Fl. Br. Ind. I, 77. A tree 20 to 30 feet high : young branches tawny-tomentose, ultimately glabrous and dark-coloured. *Leaves* narrowly oblong, often slightly obovate, acuminate, the base cuneate or rounded ; both surfaces shining, reticulate, glabrous ; the midrib puberulous on the upper, sparsely setose on the lower, surface ; main nerves 12 to 14 pairs, spreading, prominent, distinct beneath ; length 5 to 14 in., breadth 2 to 4·5 in.; petiole ·25 in., swollen. *Flowers* ·2 in. in diam., axillary, solitary or in pairs, or in few-flowered, puberulous cymes ; pedicels long, slender, with many lanceolate bracteoles. *Flowers* as in *M. macro-phylla*, monœcious. *Ripe carpels* ovoid, apiculate, rugose, hoary, ·8 in. long and ·65 in diam. *Seeds* 2.

Kurz F. Flora Burma, I, 44. *Orophea reticulata*, Miq. Ann. Mus. Lugd. Bat. II, 23. *Uvaria reticulata*, Blume Fl. Jav. Anon. 50, t. 20. *Pseuduvaria reticulata*, Miq. Fl. Ind. Bat. i. pt. 2, 30.

Burma: prov. Tenasserim; Helfer. Malacca; Maingay (Kew Distrib.), No. 64. Perak: Wray, King's Collector, Scortechini ; not so common as *M. macrophylla*, Oliver.

This species has the inner petals rather larger than the outer and much vaulted ; and in this respect it conforms to the characters of *Orophea ;* but its stamens are uvarioid in character and they are numerous ; its flowers, moreover, are unisexual. The characters of *Mitrephora* therefore preponderate, and it is better located in the latter genus. But there is no doubt it forms a connecting link between the two genera.

3. MITREPHORA MACROPHYLLA, Oliver in Hook., Ic. Plant, t. 1562. A small tree ; young branches more or less puberulous, speedily becoming glabrous and cinereous. *Leaves* thinly coriaceous, elliptic-obovate or oblong-oblanceolate, acute or shortly acuminate ; the base rounded, slightly oblique ; both surfaces puberulous at first but speedily glabrous, shining, minutely reticulate ; main nerves 14 to 20 pairs, oblique, inter-arching ·15 in. from the margin, prominent beneath ; length 7 to 13

in., breadth 1·75 to 4 in.; petiole ·25 in., swollen. *Flowers* ·25 to ·3 in. in diam., axillary, usually in pairs, or in cymes, 1 to 2 in. long, the cymes minutely pubescent; bracts few, lanceolate; pedicels long, with several broadly lanceolate, partly deciduous bracteoles, or ebracteolate. *Sepals* free, or connate below, reniform, or broadly ovate, puberulous outside and on the edges, glabrous inside. *Outer petals* larger than the sepals, orbicular-ovoid, sub-acute, slightly narrowed at the base, puberulous on both surfaces. ·15 in long. *Inner petals* ·3 in. long, thick, vaulted reniform-sagittate, puberulous, with a glabrous callosity on the inside near the base, the edges pubescent; the claw shorter than the limb, pubescent. *Male flower:* stamens very numerous, short, cuneate; the connective truncate, small and not concealing the tops of the anthers; pistils 3, or a few rudimentary. *Female flower;* staminodes in two imperfect rows. *Ovaries* about 12, ovoid-cylindric, oblique, pubescent, 4-ovuled; stigmas sessile, large, fleshy, truncate, often oblique. *Ripe carpels* globose, densely and minutely tawny-tomentose, ·4 or ·5 in diam.; stalks ·2 in. long. *Seeds* several, compressed, the testa membranous.

Penang; Maingay, Curtis. Perak; Scortechini, King's Collector, Wray.

This species, although rare in Penang, is very common in Perak. Specimens of it vary considerably in several respects. In some plants the young shoots are densely puberulous, in others they are almost glabrous; the leaves also vary in size and in amount of pubescence. In the specimen figured by Professor Oliver (Hook. Ic. Pl. 1562), the flowers are in axillary pairs; but, in the majority of the Perak specimens, they are in cymes. The species is practically dioecious, the staminate flowers having no ovaries at all or only a few rudiments; while the pistillate flowers have rarely a few perfect stamens, and not always any staminodes. The best marks of distinction between this and *M. reticulata*, of which this must be a very close ally, are the smaller number of the nerves in the leaves of this and the ovoid shape of its rugose fruit. In its leaves this plant somewhat resembles some of the species of *Popowia*. And, inasmuch as its inner petals are larger than the outer and are vaulted, it is related to *Orophea*, from which however its numerous uvarioid stamens and unisexual habit exclude it.

4. MITREPHORA PRAINII, King, n. sp. A tree 30 to 40 feet high; young branches tawny-pubescent, speedily becoming glabrous and dark-coloured. *Leaves* membranous, elliptic-oblong, rather abruptly and shortly acuminate, the base cuneate and often slightly unequal-sided; upper surface glabrous except the depressed, strigulose midrib; lower surface much reticulate, glabrous but with a few scattered hairs on the

midrib and 12 to 14 pairs of rather bold, oblique, curving nerves; length 6 to 9 in., breadth 2·25 to 3 in., petiole ·25 in., pubescent. *Flowers* bisexual, from the axis of the fallen leaves, solitary, ·4 in. in diam.; pedicels about ·5 in. long, softly.tomentose, minutely bracteolate at the base. *Sepals* broadly ovate, acute, concave, tomentose outside, glabrous inside. *Outer petals* much larger than the sepals, ovate-orbicular, sub-acute; tomentose outside, glabrous inside. *Inner petals* longer but narrower than the outer; the limb trapezoid, densely tomentose, glabrous inside at the base; the claw narrow, about as long as the limb, tomentose on both surfaces. *Stamens* in the male flower numerous, short, cuneate; the apical process of the connective truncate, concealing the apices of the dorsal anthers. *Pistils* 0. *Female flowers* unknown. *Ripe carpels* sub-globose, rather truncate at base and apex, rugulose, minutely pubescent, ·65 in. in diam. *Seeds* about 5, plano-convex, the testa membranous, rugulose.

Andaman Islands; Prain, King's Collector.

The inner petals of this species are undoubtedly longer than the outer; but they are much narrower. Technically they are the petals of *Orophea* rather than of *Mitrephora*; but the numerous Uvarioid stamens and the unisexual habit are those of the latter, to which I accordingly refer it I have been able to examine only a few flowers of the species, and these are all tetramerous; but whether this arrangement is normal or only occasional I am unable to say until larger suites of specimens are obtained.

## 17. POPOWIA, Endl.

Trees. *Flowers* small, sub-globular, opening but slightly, usually hermaphrodite, sometimes polygamous, extra-axillary or leaf-opposed. *Sepals* 3, ovate, valvate. *Petals* 6, valvate in 2-series, (the inner series imbricate in *Kurzii*), more or less orbicular; outer like the sepals, spreading; inner thick, concave, connivent, acute, the tip sometimes inflexed. *Stamens* indefinite or sub-definite, short, cuneate; anther-cells dorsal, remote. *Carpels* about 6, ovoid; style large, oblong or sub-clavate, straight or recurved; ovules 1-2 on the ventral suture, rarely 1, basal, erect. *Ripe carpels* berried, globose or ovoid, stalked.—DISTRIB. About 20 Asiatic species, 12 Australian and 1 African. (The Australian and African species may be generically separable).

There has been considerable variety of opinion as to the place of the genus *Popowia* amongst the genera of *Anonaceæ*. The genus was founded by Endlicher (Genus No. 4710) to accommodate the species named *Bocagea pisocarpa* by Blume (Flora Javae (Anonaceæ) 90, t. 45).

Endlicher placed it next to *Orophea* from which it is distinguished by its inner row of petals being free and having their apices inflexed in æstivation, while those of *Orophea* are clawed, vaulted, attached by their edges, and not inflexed in æstivation. In their Flora Indica, Hooker filius and Thomson added the species *P. ramosissima* to the original plant of Endlicher, with a remark to the effect that *Uvaria Vogelii* H. f. should be included in the genus. Farther they associated *Popowia* with the genera *Orophea*, *Mitrephora* and *Goniothalamus* in the tribe *Mitrephoreæ*. In their Genera Plantarum, Mr. Bentham and Sir Joseph Hooker take a different view of the position of *Popowia* and, in the arrangement adopted in that great work, *Popowia* is put amongst the *Unoneae*; *Orophea* is relegated to the tribe *Miliuseae*; while *Goniothalamus* and *Mitrephora* are retained side by side in the tribe *Mitrephoreae*. Now the character of the tribe *Unoneae* is :—" petals flat, slightly unequal, or those of the inner row smaller than those of the outer, or absent," while in several of the *Popowias*, e. g., *P. pisocarpa*, *P. ramosissima* the inner petals are longer than the outer. Baillon, whose arrangement of tribes differs from that of Messrs. Bentham and Hooker, puts *Popowia* into *Unoneae*, leaving *Mitrephora* and *Orophea* side by side in his tribe *Oxymitreæ*.

Dr. Scheffer differs from the opinion of the authors of the Genera Plantarum and of Baillon and rather inclines to that of the authors of the Flora Indica. He points out with much force that the proper place for *Popowia* is in the tribe characterised by its " outer petals being open, the inner connivent over the andro-gynœcium, erecto-connivent or connate "—that is to say in the tribe *Mitrephoreæ* of these authors. The stamens of *Popowia* present considerable diversity, but on the whole they have the character of those of *Uvarieæ* rather than those of *Unoneae*. As Scheffer remarks, there is little difference between the genera *Orophea* and *Mitrephora* except that the outer petals of *Mitrephora* are usually larger than those of *Orophea*. And if M. Baillon's plan of reducing the number of the genera in *Anonaceae* were to be carried out, Dr. Scheffer would suggest the union of these two and of *Popowia* into a single genus, from which would be excluded, however, all the African species. Of this new genus *Orophea* would be the typical form, and the other two would form sub-genera.

There is no doubt than in externals many *Popowias* are like *Oropheas*, and the non-unguiculate character of the inner petals of *Popowia* is really the chief character which separates them.

I venture to follow Dr. Scheffer and the authors of the Flora Indica in putting *Popowia*, *Orophea* and *Mitrephora* together in the tribe *Mitrephoreae*.

91

Flowers hermaphrodite.
Both surfaces of leaves glabrous except the
nerves.
Both surfaces minutely granular; nerves
9 or 10 pairs, sparsely pilose beneath ...   1. *P. pauciflora.*
Lower surface granular, the midrib and
6 to 8 pairs of nerves pubescent   ...   2. *P. ramosissima.*
Both surfaces shining, reticulate, glabrous
except the tomentose midrib on the
upper; nerves about 10 pairs, very faint   3. *P. nitida.*
Upper surface of leaves glabrous, the lower
minutely granular and sub-strigose; nerves
4 or 5 pairs   ...   ...   ...   4. *P. Helferi.*
Upper surface of leaves glabrous except the
puberulous midrib, the lower yellowish-tomentose; nerves 11 to 13 pairs; fruit very
large   ...   ...   ...   ...   5. *P. fœtida.*
Upper surface of leaves glabrous except the
tomentose midrib and 8 to 10 pairs of nerves;
lower surface pubescent and sub-granular ...   6. *P. perakensis.*
Both surfaces minutely granular; upper short-
ly puberulous, lower pubescent; nerves 8 to
11 pairs...   ...   ...   ...   7. *P. fusca.*
Both surfaces minutely granular; upper with
a few scattered hairs; lower fuscous, densely
and softly pubescent; the nerves 6 or 7
pairs, tomentose or pubescent ...   ...   8. *P. velutina.*
Both surfaces, but especially the lower, softly
pubescent; nerves about 10 pairs   ...   9. *P. tomentosa.*
Flowers polygamous.
Upper surface of leaves glabrous except the
puberulous midrib; nerves 10 or 11 pairs;
flowers ·5 to ·75 in. in diam ; petals of inner
row larger than those of outer, valvate, their
apices inflexed in bud   ...   ...   10. *P. nervifolia.*
Upper surface of leaves sub-granular, minutely
and sparsely adpressed-pubescent; nerves 9
to 12 pairs; flowers ·4 in. in diam.; inner
petals slightly smaller than the outer, imbricate ...   ...   ...   ...   11. *P. Kurzii.*
Both surfaces of leaves glabrous, the lower
silvery, shining; nerves 7 pairs   ...   12. *P. Hookeri.*

1. POPOWIA PAUCIFLORA, Maingay MSS. Hook. fil. Fl. Ind. I, 69. A tree? Young branches slender, cinereous, strigose. *Leaves* membranous, elliptic-lanceolate, acuminate, the base acute; both surfaces glabrous, minutely granular; the midrib and 9 or 10 pairs of oblique, little curving main nerves sparsely pilose beneath; length 5 to 6 in., breadth 1·5 to 2 in., petiole ·2 in., pubescent. *Flowers* extra-axillary, solitary or axillary, ·25 in. in diam.; pedicels ·15 to ·25 in. long, with a basal bracteole, rusty-strigose. *Sepals* minute, ovate. *Petals;* the outer small and like the sepals; the inner three times as large, sub-orbicular, concave, their apices inflexed. *Stamens* many. *Ovaries* about 6, strigose; ovule solitary, erect. *Ripe carpels* sub-sessile, globular, glabrous.

Malacca: Maingay (Kew Distrib.) No. 56.

Known only by Maingay's imperfect specimens; an obscure species.

2. POPOWIA RAMOSISSIMA, Hook. fil. and Thoms. Fl. Ind. 105. A small spreading tree; young branches at first rufous-pubescent; the older dark-coloured and furrowed. *Leaves* membranous, sub-sessile, narrowly elliptic to lanceolate, sometimes slightly obovate, shortly, bluntly and abruptly acuminate, the base rounded or slightly narrowed; both surfaces glabrous, the lower granular and pubescent on the midrib and 6 to 8 pairs of ascending rather straight nerves; length 2·75 to 4 in., breadth 1 to 1·75 in., petiole ·05 in. *Flowers* globular in bud, leaf-opposed, solitary or in small fascicles, ·2 in. in diam.; pedicels ·15 to ·25 in. long (longer in fruit), minutely bracteolate, rufous-tomentose. *Sepals* broadly triangular-ovate, acute, nearly as large as the outer petals and like them tomentose outside, and glabrous inside. *Petals* sub-equal, coriaceous, rotund, concave; the inner rather larger and with incurved points. *Stamens* short, with very broad truncate concave heads. *Ovaries* 5 or 6, villous; ovules 1 or 2. *Ripe carpels* globose with short stalks, pubescent, ·25 to ·35 in. in diam. Miq. Fl. Ind. Bat. I, Pt. 2, 27; Hook. fil. Fl. Br. Ind. I, 68. *Guatteria ramosissima*, Wall. Cat. 7294, 8006. *Popowia rufula* and *P. affinis* Miq. Ann. Mus. Lugd. Bat. II, 20.

In all the provinces, common. Distrib. Sumatra, Borneo.

3. POPOWIA NITIDA, King, n. sp. A shrub? Young branches sparsely and softly rufous-pubescent, the bark brown. *Leaves* thinly coriaceous, oblong-lanceolate to oblong-ovate, bluntly acuminate, the base rounded; both surfaces reticulate, glabrous and shining, the midrib tomentose on the upper; main nerves about 12 pairs, very faint, spreading and forming double arches inside the edge; length 2·5 to 4 in., breadth ·6 to 1·25 in., petiole ·1 in. *Flowers* few, in short extra-axillary racemes, sub-globular, ·25 in. in diam.; pedicels about as long as the flowers, each with 2 sub-orbicular, stem-clasping, pubescent bracteoles. *Sepals* orbicular, concave, puberulous on both surfaces, about ·15 in. in

93

diam. *Petals* sub-equal, about twice as large as the sepals, orbicular-ovate, sub-acute, cordate at the base, the edges incurved. *Stamens* about 27, in three rows; anther-cells linear, lateral, the apical process of the connective obliquely truncate, papillose. *Pistils* numerous, forming a large mass with their stigmas agglutinated. *Ovaries* sub-cuneate, pubescent especially near the truncate apex; stigma very large and viscous, sessile; ovules 1 to 3, ascending. *Ripe carpels* ovoid, pointed, glabrous, ·4 to ·5 in. long. *Seeds* 1 to 3, compressed, the testa pale brown, shining.

S. Andaman: King. Nicobars: Kurz.

In its leaves this much resembles *Uvaria micrantha*, H. f. and T. as which I have reason to believe some specimens of this have been distributed from the Calcutta Herbarium.

4. POPOWIA HELFERI, Hook. fil. and Thoms. Fl. Ind. I, 69. A small spreading tree; young branches coarsely hairy. *Leaves* membranous, lanceolate or oblong-lanceolate, acuminate, the base narrowed but rounded; upper surface glabrous; the lower granular, sub-strigose, especially on the midrib; main nerves indistinct, about 4 or 5 pairs, ascending: length 2 to 4 in., breadth ·8 to 1·25 in., petiole ·05 in. *Flowers* minute, globose, extra-axillary: peduncles ·05 to ·2 in., tomentose. *Sepals* ovate, strigose. *Outer petals* like the sepals, the inner orbicular, larger than the outer, concave, very strigose, their apices inflexed. *Stamens* 15. *Ovule* solitary. *Carpels* about 6, globular, strigose. Kurz. F. Flora Burm. I, 39.

Andamans; North of Port Mouat; Kurz. Burmah: Tenasserim, on King's Island; Helfer.

A very little known species closely resembling *P. Beddomiana*, H. f. and Th.

5. POPOWIA FŒTIDA, Maingay MSS., Hook fil. Fl. Br. Ind. I, 69. A large tree; young branches tawny-tomentose. *Leaves* sub-coriaceous, elliptic-lanceolate, shortly caudate-acuminate, the base sub-acute; upper surface glabrous except the puberulous midrib, lower densely covered with yellowish-grey tomentum as are the petioles; main nerves 11 to 18 pairs, rather prominent beneath, curved, spreading, inter-arching close to the margin; length 4·5 to 6·5 in., breadth 1·6 to 2 in., petiole ·2 in. *Flowers* solitary, ·35 in. in diam.; pedicels ·2 in., tomentose. *Sepals* minute, ovate, obtuse. *Petals* unequal, the outer ovate-elliptic, obtuse, yellow; the inner slightly larger, apiculate, concave, the margins thick. *Stamens* about 30, the connective large. *Ovaries* about 6, strigose, 2-ovuled. *Ripe carpels* few, very large, oblong-ovoid, obtuse, sessile, densely and shortly yellowish-tomentose, 2·25 in. long, and 1·5 in. in diam. *Seed* solitary, oblong, the testa bony.

Malacca; Maingay, (Kew Distrib.) No. 55.

342

6. POROWIA PERAKENSIS, King, n. sp. A shrub 6 to 15 feet high ; young branches densely and minutely dull rusty-tomentose, the older dark and furrowed. *Leaves* elliptic to oblong-elliptic, very shortly and rather abruptly acuminate, the base slightly narrowed, sometimes sub-oblique ; upper surface glabrous, the midrib and nerves tomentose ; lower pubescent, sub-granular : main nerves 8 to 10 pairs, spreading, slightly prominent beneath ; length 4 to 5·5 in., breadth 2 to 2·5 in. ; petiole ·1 in., tomentose. *Flowers* extra-axillary, usually in pairs (but not contemporaneous) ·3 in. in diam. ; pedicels ·4 in. long, ferrugineous-tomentose, minutely bracteolate. *Sepals* smaller than the petals, semi-orbicular, acute, coarsely tomentose outside, sub-glabrous inside. *Petals* thick, ovoid-orbicular, sub-acute, sub-concave, densely whitish-sericeous outside, glabrous within ; the inner row slightly larger than the outer, neither their edges nor apices incurved. *Stamens* numerous, flattened, with truncate, corrugated heads. *Ovaries* about 10, thin, glabrous, except a few long hairs near the base, 2-ovuled : stigmas large, rounded. *Ripe carpels* few, ovoid, with sub-truncate apices, slightly narrowed to the stalks, glabrous or sparsely pubescent, with several horizontal constrictions when ripe ·5 in. long and ·25 in. in diam. ; stalks ·25 to ·5 in. long. *Seeds* 2, superposed, plano-convex.

This resembles *P. ramosissima* in its leaves but has much larger flowers of which the inner petals are not inflexed and the carpels have 2 seeds.

Perak : King's Collector, Wray ; from 200 to 2,500 feet.

7. POROWIA FUSCA, King, n. sp. A tree 40 to 50 feet high ; young branches densely covered with purplish-brown tomentum ; the older cinerous, sub-pubescent and much furrowed. *Leaves* coriaceous, oval-oblong, obtuse or sub-acute, the base rounded ; both surfaces minutely granular, the upper shortly puberulous, the lower pubescent, the midrib and 8 to 11 pairs of spreading, rather prominent main nerves tomentose on both ; length 2·5 to 3·5 in., breadth 1·4 to 1·8 in. ; petiole ·2 in. purplish-tomentose like the flower pedicels. *Flowers* in small extra-axillary fascicles from small bracteate tubercles, 25 in. in diam. ; pedicels ·15 to ·25 in. *Sepals* ovate-obtuse, tomentose outside, glabrous inside. *Petals* sub-equal, rotund, very thick and fleshy, tomentose outside, puberulous inside. *Ripe carpels* few, globular, densely tomentose, ·25 in. in diam. ; stalks ·1 to ·2 in. long, tomentose. *Seeds* solitary.

Perak, near Ulu Kerling, at an elevation of 500 feet, King's Collector, No. 8602.

This much resembles *P. velutina*, King, but its leaves are more oval, have more nerves, and are not so pubescent.

8. POROWIA VELUTINA, King, n. sp. A tree 20 to 40 feet high ;

young branches covered with minute soft deep brown tomentum. *Leaves* elliptic-oblong, to ovate-elliptic, acute or shortly and narrowly acuminate, slightly narrowed to the rounded sub-unequal base ; both surfaces minutely granular, the upper with a few scattered hairs ; the lower fuscous and more densely and softly pubescent, both the midrib and nerves tomentose or pubescent ; main nerves 6 or 7 pairs, spreading, indistinct ; length 3 to 5 in., breadth 1·4 to 1·8 in., petiole ·1 in. *Flowers* solitary or in pairs, extra-axillary, about ·25 in. in diam., pedicels densely tomentose, ·35 in. long, bracteolate. *Sepals* broadly ovate, sub-acute, densely tomentose outside, glabrous inside, persistent in the fruit. *Petals* sub-equal, thick, sub-orbicular, very tomentose outside, glabrous inside. *Ripe carpels* few, sometimes solitary, ovoid, blunt, slightly oblique at the base and slightly narrowed to the stalk, minutely velvety-pubescent, ·5 in. long and ·35 in. in diam.; stalks ·2 in., tomentose ; torus small. *Seed* solitary, glabrous, rugose, vertically furrowed.

Perak, at Kinta ; at elevations under 1,000 feet ; King's Collector.

A species very like *P. fusca*, but with shorter, fewer-nerved leaves ; evidently not common. None of the collectors' specimens have fully developed flowers, and the foregoing description of these is taken from a bud.

9. POPOWIA TOMENTOSA, Maingay MSS. Hook. fil. Fl. Br. Ind. I, 70 A tree ; young branches softly rusty-tomentose, when older black and rugose. *Leaves* elliptic-oblong to elliptic, acute or shortly acuminate, the base rounded, slightly unequal-sided ; both surfaces, but especially the lower, softly pubescent ; main nerves about 10 pairs, slightly prominent, spreading ; length 4·5 to 5·5 in., breadth 1·75 to 3 in. ; petiole ·1 in., tomentose. *Flowers* extra-axillary, sub-sessile, 25 in. in diam. *Sepals* broadly ovate, connate, slightly smaller than the petals. *Petals* slightly unequal, villous outside, glabrous inside ; the outer ovate, thick ; the the inner larger, very thick and concave, oblong, connivent. *Stamens* about 25. *Ovaries* 7 to 9, oblong, pubescent ; ovules 2. *Ripe carpels* globose, slightly pubescent, ·5 to ·74 in. in diam., 2-seeded ; their stalks ·35 in., pubescent.

Malacca ; Maingay, (Kew Distrib.) No. 54. Penang : Curtis, No. 648. Perak ; Scortechini.

I am not satisfied that there are not two species involved here, the one with broader leaves and shorter pubescence.

10. POPOWIA NERVIFOLIA, Maingay MSS. ex Hook. fil. Fl. Br. Ind. I, 60. A small tree 12 to 25 feet high : young branches at first densely rusty-tomentose, afterwards dark-coloured and furrowed. *Leaves* coriaceous, from oblong-lanceolate or ob-lanceolate to elliptic-oblong, shortly abruptly and bluntly acuminate, the base acute ; upper surface shining,

glabrous except the puberulous midrib; lower paler, sparsely rusty-pubescent; main nerves 10 or 11 pairs, oblique, rather prominent on the lower surface; length 5·5 to 8·5 in., breadth 1·8 to 3 in.; petiole ·35 to ·5 in., rusty-pubescent. *Flowers* polygamous, extra-axillary, solitary or 2 or 3 together, sub-globose, from ·5 to ·75 in. in diam.; pedicels, stout, tomentose, ·15 to ·25 long, with 2 bracts nearly as large as the sepals. *Sepals* ovate-orbicular, acute, slightly smaller than the outer petals, very thick, villous-tomentose outside and glabrous inside as are all the petals: inner petals larger than the outer, their apices much inflexed in bud. *Stamens* numerous, with flat, rhomboid heads. *Ovaries* numerous, hirsute. *Carpels* numerous, cylindric-ovoid, apiculate, narrowed to the stalk, sparsely strigose, ·5 in. long and ·25 in. in diam.; stalks ·2 to ·3 in. long, strigose-pubescent; torus globular, ·4 in. in diam. *Seed* pale, shining.

Malacca: Maingay (Kew Distrib.,) No. 53. Perak: common at low elevations.

Allied to *P. Kurzii*, but with larger flowers which have their inner petals valvate with much inflexed edges.

11. POPOWIA KURZII, King. A shrub or small tree; young branches at first tawny-pubescent, afterwards dark-coloured, glabrous and furrowed. *Leaves* sub-coriaceous, oblong-lanceolate, or elliptic-oblong sub-acute or shortly and bluntly acuminate, narrowed to the sub-cuneate (sometimes almost rounded) base; upper surface sub-granular, minutely and sparsely adpressed-pubescent; lower sparsely pubescent; main nerves 9 to 12 pairs, oblique, inter-arching close to the edge, rather prominent beneath; length 5 to 9 in., breadth 1·5 to 3 in.; petiole ·2 to ·25 in, tomentose. *Flowers* polygamous, solitary, or in pairs, sub-sessile, extra-axillary, sub-globose, ·4 in. in diam.; pedicels tomentose, ·1 to ·2 in. long, bracteolate. *Sepals* smaller than the petals, valvate, semi-orbicular, and, like the petals, tomentose externally and glabrous internally. *Petals* sub-equal, concave, the outer ovate-orbicular, valvate; the inner slightly smaller than the outer, imbricate. *Stamens* numerous, flattened, elongate, with linear, lateral anther-cells and flat, oblique, rhomboid apices. *Ovaries* (often absent) about 10, elongate, pubescent, the stigmas clavate. *Fruit* unknown. *Polyalthia macrophylla*, Hook. fil. and Thoms. Fl. Br. Ind. I, 68. *P. dubia* Kurz F. Flora Burma, I, 38. *Guatteria macrophylla*, Blume Bijdr. 19; Fl. Javae Anon. 96. t. 97; Miq. Fl. Ind. Bat. I, Pt. 2, 47.

South Andaman; Kurz, King's Collector. Burmah; province Tenas-serim; Falconer, Kurz.

This species appears to be practically dicecious. In its flowers the inner petals are distinctly imbricate; they are not connivent, and their points are not inflexed. And in these respects they do not answer to the diagnosis of *Popowia* as heretofore understood. I have therefore ventured to modify the generic character of *Popowia* in these points, and to institute a section of it to receive this and other two species. This species is closely allied to the plant originally described and figured by Blume as *Guatteria macrophylla*, (Fl. Jav. Anon. 96 t. 47,) and to receive which Miquel founded his genus *Trivalvaria* (Ann. Mus. Lugd. Bat. II, 19). But, in Blume's and Miquel's plant, the inner petals are distinctly valvate, although their apices are not inflexed. And in the non-inflection of its petals it also does not conform to the character of *Popowia* as originally defined by its founder Endlicher.

12. POPOWIA HOOKERI, King. A shrub; young branches dark-coloured, glabrous. *Leaves* thinly coriaceous, broadly lanceolate or oblanceolate, acute or acuminate, the base acute : both surfaces glabrous, the lower silvery, shining: main nerves about 7 pairs, spreading, ascending, curving, rather prominent beneath, evanescent at the tips; length 5·5 to 7 in., breadth 1·6 to 2·4 in. *Flowers* solitary or in fascicles of 2 or 3 from short extra-axillary, woody tubercles, polygamous, minute ; "the males as in *Popowia Kurzii* but smaller ; the females with many, densely pubescent ovaries and a few imperfect stamens ; bracts many, minute, strigose. *Carpels* many, ·75 in. long, oblong, granulate, glabrous ; stalk ·35 in." *Guatteria pallida*, H. f. and Th. Fl. Ind., 143 (not of Blume). *Polyalthia argentea*, Hook. fil. and Thoms. Fl. Br. Ind. I, 67.

Assam and Sylhet; in dense forests, Hook. fil. and Thomson ; Naga Hills, Masters. Khasia : Griffith.

A species of which I have seen only imperfect specimens. The description given above of the flowers is copied from Sir Joseph Hooker. In my opinion the plant is a *Popowia* rather than a *Polyalthia* and to the former genus I have ventured to remove it.

*Doubtful Species.*

*Popowia parvifolia*, Kurz in Journ. of Botany for 1875, p. 324. Of this I have seen only leaf specimens with a few detached fruits. It appears to have also had the MSS. name *P. nitida* given to it by Kurz.

18. OXYMITRA, Blume.

Climbing shrubs. *Leaves* parallel-nerved ; nervules transverse, not forming intra-marginal loops. *Flowers* leaf-opposed or extra-axillary. *Sepals* 3, valvate, connate below. *Petals* 6, valvate, in 2 rows, outer large, long, flat or triquetrous and narrow, leathery, more or less spreading or connivent ; inner much smaller, ovate-lanceolate or oblong (long and narrow in *O. filipes* and *O. glauca*), conniving over the stamens and

346

ovaries. *Stamens* many, linear-oblong or cuneate, truncate; anther-cells dorsal, remote (small and ovoid in *O. glauca*). *Ovaries* oblong, strigose ; style oblong or clavate, recurved; ovules 1-2, sub-basal, ascending. *Ripe carpels* 1-seeded, stalked.—Distrib. About 28 species, Asiatic and African.

A genus of which the flowers have some resemblance to those of *Goniothalamus :* but in this the inner petals are not contracted into a claw as in *Goniothalamus* and the calyx in this is smaller and not persistent.

Outer petals flat ... ... ... 1. *O. affinis.*
Outer petals concave.
  Pedicels slender, much longer than the flowers 2. *O. filipes.*
  Pedicels shorter than the flowers.
    Leaves oblong-elliptic, more or less obovate,
      blunt ... ... ... 3. *O. calycina.*
    Leaves oblong-elliptic to oblong-lanceolate
      or elliptic-lanceolate, not obovate, acute,
      or acuminate.
        Outer petals expanded and concave in
        the lower third ; the inner only one
        fourth as long as the outer, very
        acuminate ... ... ... 4. *O. biglandulosa.*
        Outer petals narrowly linear-lanceo-
        late, slightly expanded and concave
        at the very base ... ... 5. *O. glauca.*

1. OXYMITRA AFFINIS, Hook. fil. and Thoms. Fl. Br. Ind. I, 70. A spreading shrub or climber : young branches at first densely rusty to-mentose, afterwards dark-coloured and glabrous. *Leaves* membranous, elliptic to oblong-elliptic, sometimes slightly obovate, acute or very short-ly acuminate, rarely obtuse, the base rounded or slightly narrowed ; upper surface shining, minutely scaly, glabrous except the pubescent midrib ; under surface slightly glaucous, pubescent especially on the midrib and nerves ; main nerves 8 to 14 pairs, spreading, ascending, rather prominent on the lower surface ; length 3·5 to 10 in., breadth 1·25 to 4·5 in. ; petiole ·3 in., tomentose. *Flowers* solitary, extra-axillary ; pedicels ·25 to ·4 in. *Sepals* slightly connate at the base, spreading, broadly ovate or orbicular-ovate, sub-acute, 3- to 7-nerved, adpressed-pubescent, ·5 in. long and slightly narrower than the base of the petals, persistent in the fruit. *Petals* flat, very unequal ; the outer thinly coriaceous, oblong-lanceolate, sub-acute, the midrib thick and with several strong sub-parallel nerves, adpressed-pubescent on both surfaces, 1·5 to 1·75 in. long and ·4 to ·6 in. broad ; inner petals thickly coria-

ceous, ovate, sub-acute, ·5 in. long, pubescent outside, glabrous inside. *Ripe carpels* cylindric, blunt at each end, pubescent, ·5 to ·8 in. long and ·3 in. in diam.: stalks pubescent, ·2 in. long. *Seed* solitary.

Malacca; Maingay, (Kew Distrib.) No. 39. Perak; King's Collector, Scortechini. Distrib., Siam.

2. OXYMITRA FILIPES, H. f. and Th. Fl. Br. Ind. I, 71. A climber: young branches softly brown-tomentose, dark-coloured and lenticellate when old. *Leaves* membranous, oblong-lanceolate or oblong-elliptic, often slightly obovate, acute or shortly acuminate, slightly narrowed to the sub-cordate sometimes slightly oblique base; upper surface glabrous, minutely scaly, sometimes pubescent, the midrib and nerves always so; under surface paler, sub-glaucous, pubescent, the midrib tomentose; main nerves 12 to 14 pairs, spreading, prominent beneath; secondary nerves obliquely transverse, prominent: length 4·5 to 7·5 in., breadth 1·4 to 2·5 in.; petiole ·2 to ·25 in., tomentose. *Flowers* very long and narrow, often curved, 1·75 to 2·5 in. long, solitary on slender extra-axillary pedicels 3 or 4 in. long, which are pubescent and have a subulate bract near the middle. *Sepals* ·25 in. long, spreading, ovate, acute, pubescent. *Petals* very unequal; the outer fleshy, very narrow, triquetrous, expanded and concave at the base, pubescent; the inner less than one fifth of the outer in length, lanceolate with caudate-acuminate apex, glabrous. *Stamens* numerous: ovaries 1-ovuled. *Ripe carpels* numerous, ovate-cylindric, shortly apiculate, softly pubescent, ·5 in. long and ·25 in. in diam.; stalks ·3 in. long, pubescent. *Seed* solitary, pale.

A species readily distinguished in this genus by the extreme length and narrowness of the outer petals. Evidently closely allied to *O. cuneiformis*, Miq. (*Polyalthia cuneiformis*, Bl. Fl. Javae Anon. 75 t. 35, 36D, 37), which it resembles in that respect as also in its filiform, elongated pedicels.

Malacca; Maingay, (Kew Distrib.) No. 60. Perak: King's Collector.

3. OXYMITRA CALYCINA, King, n. sp. A slender, woody creeper; young branches densely rusty tomentose. *Leaves* coriaceous, oblong and sub-acute or cuneiform-oblong, very blunt or even emarginate, always slightly narrowed to the rounded or minutely cordate base; upper surface glabrous, shining, the midrib sometimes rufous-pubescent; under surface pale, glaucous, pubescent especially on the midrib and nerves: main nerves 7 to 14 pairs, prominent on the under, impressed on the upper, surface, spreading; the secondary nerves obliquely transverse, prominent: length 6 to 12 in., breadth 2·65 to 7·5 in., petiole ·2 to ·4 in., rufous tomentose. *Flowers* solitary, extra-axillary; pedicels ·3 to 1 in.,

rufous-tomentose, bearing two bracts, one small, the other large, obovate, ribbed. *Sepals* free, nearly half as long as the outer petals, elliptic, sub-acute ; the edges undulate, rufous-tomentose on both surfaces. *Petals* thick, lanceolate, caudate-acuminate, the midrib prominent, the base concave, both rows glabrous inside, the outer about 1 to 1·25 in. long, tomentose outside ; the inner about ·5 in. shorter, connate into a narrow, acute cone, puberulous outside. *Ovaries* 1-ovuled. *Ripe carpels* elliptic, apiculate, pubescent, ·35 in. long : stalks ·2 in., pubescent.

This closely resembles *Oxymitra cuneiformis*, Miq. of which Blume (under the name of *Polyalthia cuneiformis*) gives an excellent description and three admirable figures (Fl. Javae Anon. 75 t. 35, 36D. and 37. But in Blume's plant the flowers are much larger, the petals are falcate, while the sepals are much smaller and have caudate apices : the pedicels too are much longer and have smaller bracteoles.

Perak : Ulu Bubong at elevations of 500 to 1,000 feet, King's Collector, No. 10604. Singapore : Ridley. Penang ; Curtis.

4. OXYMITRA BIGLANDULOSA, Scheffer in Nat. Tijdsch. Ned. Ind. XXXI, 341. A creeper 50 to 100 feet long ; young branches minutely rufous-sericeous, afterwards dark-coloured and glabrous. *Leaves* coriaceous, elliptic to elliptic-oblong, acute or shortly acuminate, the edges slightly recurved when dry, the base rounded or slightly cuneate ; upper surface glabrous, the midrib puberulous ; the lower paler, subglaucous, puberulous or glabrescent ; main nerves 7 to 9 pairs, ascending, prominent beneath ; length 3·5 to 7·5 in., breadth 2 to 3·5 in., petiole ·2 to ·4 in. *Flowers* shortly pedicelled, solitary, extra-axillary, 1 to 1·15 in. long : pedicels ·4 in. long (elongating in fruit) angled, slender, with 1 subulate bracteole. *Sepals* fleshy, ovate, much acuminate, spreading or reflexed, adpressed, rusty-puberulous. *Petals* fleshy, yellow, very unequal : the outer lanceolate-oblong, obtuse, expanded and concave in the lower third, rusty adpressed-pubescent ; the midrib prominent, sub-glabrous inside ; the inner only as large as the sepals, with broad bases (cleft in the middle) and long acuminate points. *Ripe carpels* oblong-ovoid, blunt at each end or slightly apiculate at the apex, yellow when ripe, puberulous or glabrous, ·75 in. long : stalks ·5 in. *Polyalthia biglandulosa*, Hook. fil. Fl. Br. Ind. I, 65. *Guatteria biglandulosa*, Blume Fl. Javae Anon. 102, t. 51 ; Miq. Fl. Ind. Bat. I, Pt. 2, p. 48 ; Hook. fil. and Thoms. Fl. Ind. 143.

Malacca ; Griffith, Maingay, (Kew Distrib.) No. 49. Selangor ; Ridley. Perak, King's Collector. Distrib. : Malayan Archipelago.

The structure of the flowers of this species appears to me to be that of an *Oxymitra* rather than of a *Polyalthia* or *Guatteria*, and therefore I have transferred it to this genus.

5. Oxymitra glauca, H. f. and Th. Fl. Ind. 146 ; Hook. fil. Fl. Br. Ind. I, 71. A slender woody climber : young branches slightly tomentose, soon becoming glabrous. *Leaves* thinly coriaceous, elliptic, elliptic-lanceolate to lanceolate, obtuse, acute or shortly acuminate ; the base rounded, sometimes slightly narrowed ; upper surface glabrous, the midrib and sometimes the nerves pubescent ; the lower very pale, glaucous, glabrous or sparsely puberulous, the midrib pubescent ; main nerves 8 to 12 spairs, spreading, prominent beneath : length 4 to 6 in., breadth 1·5 to 2 in. ; petiole ·2 in., pubescent. *Flowers* solitary, extra-axillary, narrow and elongate ; pedicels slender, ·5 in. long, with a median subulate bract, longer in fruit. *Sepals* connate at the base, broadly ovate, much acuminate, adpressed-pubescent, ·25 in., long. *Petals* very unequal : the outer thickly coriaceous, linear-lanceolate, sub-acute, slightly expanded and sub-concave at the base, outside minutely pubescent; inside glabrous, the midrib prominent : inner petals with sub-orbicular bases (cleft in the middle), and long acuminate points, glabrous, only about one-fifth as long as the outer. *Ovaries* hairy ; ovule solitary. *Carpels* many, ovoid, slightly apiculate, ·4 in. long and ·25 in. in diam., minutely tomentose ; stalks slender, ·75 in. long. Miq. Fl. Ind. Bat. I, Pt. 2, 50.

Penang, Malacca : Maingay (Kew Distrib.) No. 58. Perak ; common at low-elevations. Distrib. : Sumatra, Beccari, No. 626.

### 19. Melodorum, Dunal.

Climbing shrubs. *Flowers* terminal, axillary and leaf-opposed, fasci cled or panicled ; buds triquetrous. *Sepals* 3, small, valvate, connate below. *Petals* 6, valvate, in 2 rows ; outer plano-convex or trigonous : inner triquetrous above, hollowed below on the inner face. *Stamens* many ; anther-cells dorsal, contiguous ; top of connective more or less flattened, triangular, quadrate or orbicular. *Pistils* many, free ; style oblong ; ovules 2 or more. *Ripe carpels* berried.—Distrib :—species about 35. Tropical Asia and Africa ; Australia.

Section I. Melodorum proper. *Outer petals* oblong-ovate ; ovaries hairy, ovules usually more than 4. *Seeds* smooth (unknown in *M. litseae-folium*).

Flowers not more than ·4 in. long (often ·5 in. in *M. fulgens*), flower-buds broadly pyramidal.

Flowers ·2 to ·25 in. long, in few-flowered, lax, axillary racemes ; leaves beneath hoary-pubescent with a superficial layer of flexuose hairs : ovules 4   1. *M. litseaefolium*.

Flowers ·4 to ·5 in. long; solitary, or in

few-flowered terminal or leaf-opposed cymes; leaves beneath sparsely and minutely strigose: ovules 4 ... 2. *M. fulgens.*
Flowers ·5 in. or more in length (see also *M. fulgens*).

Flower-buds broadly pyramidal.

Flowers racemose, rarely solitary. Leaves glabrous above except the midrib, beneath densely golden-brown sericeous. Ripe carpels ovoid-globose, 1·25 in. long, their stalks 2 to 3 in. long ... 3. *M. manubriatum.*

Flowers in axillary or terminal panicles. Leaves minutely pubescent above, softly brown-tomentose beneath: ripe carpels globose to ovoid, velvetty-tomentose, 1 to 2·25 in. long; stalks ·75 to 1·75 in. ... ... ... 4. *M. latifolium.*

Flowers always solitary and axillary. Ripe carpels cylindric, sub-tubercular, 1 to 1·75 in. long 5. *M. cylindricum.*

Flower-buds narrowly pyramidal, racemose or paniculate.

Leaves glabrous above except the midrib, beneath glaucous hoary-puberulous. Ripe carpels globose or ovoid-globose, tubercled, 1 in. long, their stalks 1 in. ... 6. *M. hypoglaucum.*

Leaves glabrescent or glabrous above, except the midrib; beneath softly rufous-pubescent. Ripe carpels globular, densely and minutely dark brown-tomentose, ·8 in. in diam.; their stalks slightly longer ... ... 7. *M. parviflorum.*

Leaves harshly pubescent above, uniformly and softly pubescent beneath. Ripe carpels globose, harshly and minutely pubescent, 1·1 in. in diam.; stalks slender, twice as long ... ... 8. *M. sphaerocarpum.*

351

103

Section II. PYRAMIDANTHE. Outer petals very long, linear-lauceo-late, 1·2 to 5 in. long. *Flowers* solitary or in pairs, axillary, rarely leaf-opposed (cymose in *M. lanuginosum* and *M. rubiginosum.*)
Ovules more than 4.

  Flowers 1·25 to 1·5 in. long; outer petals
  rufous-lanate externally ; ripe carpels sub-
  globose, ·79 in. in diam. ... ... 9. *M. lanuginosum.*
  Flowers 1·25 to 1·5 in. long; outer petals
  minutely rufous-tomentose externally ; ripe
  carpels oblong, tapering to both ends, 1·5
  to 2 in. long ... ... ... 10. *M. Maingayi.*
  Flowers 1·5 to 2 in. long; outer petals minutely
  rufous-tomentose outside ; ripe carpels
  ovoid, tuberculate, 1·4 in. long ... 11. *M. prismaticum.*
Ovules 4.

  Flowers 3 to 5 in. long; outer petals adpressed-puberulous externally ... 12. *M. macranthum.*
  Section III. KENTIA. Outer petals not much longer than broad,
broadly ovate or sub-orbicular, with broad thick margins : flowers
axillary ; ovaries glabrous, 2 to 8-ovuled : seeds pitted.

  Ovules about 8 : ripe carpels ovoid or ovoid-
  globose ; leaves oblong-lanceolate ... 13. *M. elegans.*
  Ovules 2 ; ripe carpels globular : leaves
  elliptic or elliptic-oblong, sometimes obovate ... ... ... 14. *M. pisocarpum.*
  1. MELODORUM LITSEÆFOLIUM, King, n. sp. A powerful climber :
young branches densely but minutely rusty-tomentose, afterwards
tuberculate and sub-glabrous. *Leaves* coriaceous, oblong-ovate to oblong, acute, the base rounded or slightly cuneate ; upper surface greenish
when dry, glabrous, shining except the rufous-pubescent midrib ; lower
reticulate ; uniformly hoary-pubescent with a superficial layer of deciduous yellowish or reddish flexuose hairs ; main nerves 8 to 10 pairs,
oblique, curving, prominent beneath ; length 2·75 to 4·25 in., breadth
1·35 to 1·6 in. *Flowers* ·2 to ·25 in. long, in few-flowered lax axillary
rufous-tomentose racemes or in terminal panicles ; pedicels ·25 to ·35
in. long with a single small median bracteole. *Sepals* broadly ovate-
acute, concave, connate at the base, spreading, ·1 in. long. *Petals*
broadly ovate-oblong, acute, leathery ; outer ·3 in. long, slightly concave and glabrous at the base, otherwise puberulous inside, rufous-
tomentose outside ; the inner petals much smaller, hoary-puberulous
except the pitted glabrous concavity at the base inside. *Stamens* numerous, apical process of the connective broadly and bluntly triangular ;

352

filaments short. *Ovaries* few, oblong, oblique, rufous-pubescent, 4-ovuled; stigma lateral, oblong. *Ripe carpels* unknown.
Perak : King's Collector, Nos. 4063 and 4986.
The flowers of this resemble those of *M. fulgens*, H. f. and Th., but they are smaller and more numerous than those of *M. fulgens;* the petals of this species also are thinner and the apical process of the anthers is broader and blunter. The leaves too of this are broader and, in the indumentum on their lower surface, they differ considerably from those of *M. fulgens*. Fruit of this species is as yet unknown. The ovaries have only 4 ovules.

2. MELODORUM FULGENS, Hook. fil. Fl. Br. Ind. 120. A large climber; young branches minutely tawny-pubescent, speedily becoming glabrous and dark-coloured. *Leaves* oblong-lanceolate, acuminate, the base rounded or sub-acute; upper surface pale olivaceous when dry, glabrous, the midrib strigose ; under surface brown when dry, sparsely and minutely strigose, especially on the midrib ; main nerves 11 to 13 pairs, oblique, curving ; length 3 to 4·5 in., breadth 1·2 to 1·5 in. ; petiole ·25 to ·4 in. pubescent. *Flowers* ·4 to ·5 in. long, solitary or in terminal or leaf-opposed, few-flowered cymes : pedicels ·3 to ·4 in. long, adpressed tawny-pubescent with one sub-medial and one basal bracteole. *Sepals* broadly ovate, sub-acute, connate at the base, spreading, ·1 in. long, pubescent outside, glabrous inside. *Petals* thick ; the outer flat, ovate-oblong, sub-acute, tawny-pubescent outside, glabrous at the base inside, ·5 in. long; inner petals like outer but concave at the base, only ·3 in. long and glabrous, except near the apex outside. *Stamens* numerous ; apical process of connective of the outer lanceolate and as long as the anthers, that of the inner shorter. *Ovaries* narrowly oblong, oblique, curved, minutely pubescent, with 4 ovules in two rows : style lateral, half as long as the ovary, stigma small. *Ripe carpels* ovoid-globose densely and minutely silky tawny-tomentose like the stalks, 1 to 1·5 in, long, and ·9 in. in diam.; stalks ·85 to 1·5 in. long, stout. *Seeds* oblong, plano-convex, brown, shining. Hook. fil. Fl. Br. Ind. I, 82. Miq. Fl. Ind. Bat. I, Pt. 2, 35. *Uvaria fulgens* and *Myristica Finlaysoniana*, Wall. Cat. 6482 and 6793.
Malacca, Perak, Singapore. Distrib. Borneo, Philippines.

3. MELODORUM MANUBRIATUM, Hook. fil. and Thoms. Fl. Ind. 118. A large creeper: young branches minutely rufous-pubescent. *Leaves* thinly coriaceous, oblong-lanceolate, acuminate, the base rounded or slightly narrowed ; upper surface olivaceous when dry, glabrous, the midrib rufous-pubescent; lower uniformly covered with rather thin brown or golden sericeous tomentum ; main nerves 12 to 18 pairs, oblique, slightly curved, rather prominent beneath ; length 2 to 4·5 in.,

105

breadth ·75 to 1·5 in. ; petiole ·3 in., tomentose. *Flowers* ·6 to ·75 in. long. leaf-opposed or extra-axillary, in short racemes, rarely solitary ; pedicels ·25 to ·75 in., softly pale rufous-tomentose, with one broad clasping bracteole near the base. *Sepals* broadly ovate, shortly sub-acuminate, spreading, connate at the base, sericeous outside, glabrous inside. *Petals* leathery, ovate-lanceolate, sub-acuminate, concave, tho outer ·6 to ·75 in. long, outside sericeous, inside puberulous in the upper half, glabrous in the lower ; the inner petals smaller, minutely pubescent in the upper half outside and near the apex inside, otherwise glabrous, the base very concave. *Stamens* numerous, the connective bluntly tri-angular at the apex. *Ovaries* numerous, oblong, densely sericeous ; ovules 8 in 2 rows ; stigma sessile, glabrous, bifid. *Ripe carpels* numer-ous, ovoid-globose, with thick pericarp, about 1·25 in. long, densely rufous-tomentose ; stalks 2 to 3 in. long. *Seeds* about 8, in two rows. Hook. fil. Fl. Br. Ind. I, 79 ; Miq. Fl. Ind. Bat. I, Pt. 2, 35. *Melodorum bancanum*, Scheff. Nat. Tijds. XXXI, 343. *Uvaria manubriata*, Wall. Cat. 6456.

Penang, Malacca, Singapore. Perak : very common. Distrib. : Bangka.

4. MELODORUM LATIFOLIUM, Hook. fil. and Thoms. Fl. Ind., 116. A large climber ; young shoots velvety rufous-tomentose. *Leaves* coriaceous, oblong or narrowly elliptic, sub-acute or obtuse, the base rounded ; upper surface minutely pubescent, the midrib tomentose ; lower surface uniformly covered with short, soft, brown tomentum ; main nerves 16 to 24 pairs, spreading, bold, not inter-arching : length 3 to 7·5 in., breadth 1·75 to 2·5 in. ; petiole ·4 to ·7 in., stout, channelled, to-mentose. *Flowers* from ·6 to 1·25 in. in diam. when expanded, brown, in lax axillary or terminal racemes or panicles ; pedicels ·35 to ·5 in. with bracteole at the base. *Sepals* broadly ovate, blunt, connate into a flat triangular cup, ·25 in. wide, tomentose outside, glabrous within like the outer petals. *Petals* thick, fleshy, ovate, acuminate, ·4 to ·7 in. long ; the inner much smaller. *Stamens* very numerous, the apex of the con-nective triangular, acute ; anther-cells linear, lateral, *Ovaries* about 6, obliquely oblong, densely sericeous, 6- to 8-ovuled ; stigma small, sessile. *Ripe carpels* globose to ovoid, slightly apiculate and slightly tapering to the base, densely velvety and minutely tomentose, 1 to 2·25 in. long and 1 to 1·2 in. in diam. : stalks stout, velvety, ·75 to 1·75 in. long; Hook. fil. Fl. Br. Ind. I, 79 ; Miq. Fl. Ind. Bat. 1, pt. 2, 35; Wall. Cat. 9411. *M. mollissimum*, Miquel Fl. Ind. Bat. Suppl. 374. *Uvaria latifolia*, Blume Fl. Jav. Anon. t. 15. *Unona latifolia*, Dunal Anon. 115. *Uvaria longifolia*, Bl. Bijdr. 13.

Malacca ; Griffith. Singapore ; Maingay, Hullett. Perak : very common. Distrib. :—Sumatra, Java, Philippines.

354

*Uvaria latifolia*, Blume, as described and figured by that author has larger flowers than the common Perak plant and its carpels are globular, whereas those of the Perak plant are ovoid and apiculate. The plant figured by Blume does, however, occur there, but it is not common. The forms may be characterised thus :—

Var. *typica :* flowers ·7 in. long : fruit globular, not apiculate, 1 in. in diam. *Uvaria latifolia*, Blume l. c. t. 15. Perak, Java.

Var. *ovoidea :* flowers ·5 in. long : fruit ovoid, slightly apiculate, often oblique, as much as 2·25 in. long, very oblique and warted when young. *M. latifolium*, H. f. and Th. Fl. Br. Ind. 79. Malacca, Perak, Singapore. The common form in the Malay Peninsula.

5. MELODORUM CYLINDRICUM, Maingay in Hook. fil. Fl. Br. Ind. I, 80. A climber : young branches minutely rusty-pubescent, speedily glabrous and dark-coloured. *Leaves* coriaceous, elliptic-oblong, brownish when dry, acute or acuminate, the base rounded or slightly narrowed ; upper surface quite glabrous, the lower paler, minutely pubescent ; main nerves 8 to 10 pairs, spreading, very faint ; length 2·5 to 4·25 in., breadth 1·6 to 1·8 in., petiole ·5 in. *Flowers* ·5 in. long, solitary, axillary, drooping ; buds short, pyramidal, adpressed, brown-pubescent : pedicel short, stout, with minute bracteole. *Sepals* small, triangular, connate, forming a flat spreading cup. *Outer petals* triangular-ovate, triquetrous with an excavated base ; the inner very small, triangular, glabrous. *Stamens* numerous, the apex of the connective orbicular. *Ovaries* 4 to 6, sericeous. *Ripe carpels* cylindric, curved, both ends obtuse, sub-tubercular, minutely brown-pubescent, 1 to 1·75 in. long and ·35 in. to ·75 in. in diam. ; pericarp thin ; stalk ·5 in. long, stout. *Seeds* many, horizontal, in two series, compressed, ·65 in. long, shining, with a small cartilaginous arillus.

Malacca ; Maingay (Kew Distrib.) No. 78. Singapore : Ridley, No. 2115.

6. MELODORUM HYPOGLAUCUM, Miquel in Ann. Mus. Lugd. Bat. II, 37. A strong creeper : young branches minutely rufous-pubescent, ultimately glabrous, rather pale and much tubercled. *Leaves* thinly coriaceous, oblong-lanceolate to oblong-elliptic, acute or shortly acuminate, the base rounded or cuneate ; upper surface glabrous except the rufous-puberulous midrib ; lower minutely hoary-puberulous, the 10 or 12 pairs of bold oblique curving main nerves ultimately glabrous and darker-coloured ; length 3 to 5·5 in., breadth 1·35 to 2·2 in., petiole ·25 in. *Flowers* ·5 to ·8 in. long, in lax, 2-to 3-flowered, axillary racemes or (by abortion of the leaves) in lax, terminal, 10- to 12-flowered panicles ; pedicels as long as the flowers, slender ; bracteoles 1 or 2, minute. *Sepals* ovate, acute, concave, conjoined only at the base, rufous-pubescent outside ; puberulous within. *Petals* leathery, linear-lanceolate,

107

the base expanded and concave : the outer minutely rufous-tomentose on the external surface, paler and pubescent on the internal, ·5 to ·8 in. long, concave for their whole length : the inner one-third shorter with a glabrous concavity at the base only, the rest triquetrous, and puberulous. *Stamens* numerous ; apical process of connective large, broader than the anther-cells, sub-globular. *Ovaries* about 12, oblong, golden-silky : with 4 to 6-ovules in 2 rows : stigma large sub-capitate ; style short. *Ripe carpels* globose or ovoid-globose, tubercled, puberulous or glabrescent, 1 in. long ; stalks about the same length, striate. *Seeds* about 4 or 5, oval, compressed, smooth, brown, shining.

Perak : Scortechini, King's Collector.

This plant agrees fairly well with the only specimens of *Melodorum hypoglaucum*, Miq. which I have been able to consult. It also agrees fairly with Miquel's description of that species. But its petals and stamens,and its ovaries externally are rather those of *Xylopia* than of *Melodorum ;* although its habit, its torus and carpels are emphatically those of the latter genus  In the number of ovules it agrees with the majority of the species of *Melodorum*. It thus forms a connecting link between the two genera.

7. MELODORUM PARVIFLORUM, Scheffer in Nat. Tijdsch. Ned. Ind. XXXI, 344. A powerful climber ; young shoots minutely rusty-tomentose, the bark dark-coloured. *Leaves* coriaceous, more or less broadly elliptic, abruptly acute ; the base broad, rounded : upper surface pale yellowish-green when dry, when young minutely stellate-pubescent, when old glabrescent or quite glabrous, the midrib always tomentose ; under surface softly rufous-pubescent, the nervation and venation very prominent ; main nerves 13 to 15 pairs, oblique, curving, inter-arching close to the edge ; length 3 to 6 in., breadth 2·25 to 3·2 in., petiole ·4 in. *Flowers* ·5 in. long, in lax axillary or terminal rusty racemes often more than half as long as the leaves : pedicels ·4 to ·6 in. long with 1 or 2 small bracteoles. *Sepals* triangular, spreading, connate at the base, rusty-tomentose outside, glabrescent inside like the petals, ·1 in. long. *Petals* thick, leathery, oblong-lanceolate with broad bases; the outer ·5 in. long ; the inner smaller, concave at the base, triquetrous in the upper half. *Stamens* numerous, the connective with compressed sub-quadrate apical appendage. *Ovaries* narrow, elongate, densely sericeous, 6- to 8-ovuled. *Ripe carpels* globular, sometimes very slightly apiculate, densely but minutely dark-brown tomentose, ·8 in. diam. ; stalks rather longer, slender, tomentose.

Perak : King's Collector.—Distrib. : Bangka.

A species closely allied to *M. sphaerocarpum*, Blume. The leaves of this are, however, larger, the upper surface is stellate-tomentose
356

when young and dries a pale yellowish-green; the flower-racemes are much longer and laxer, and the flowers larger.

8. MELODORUM SPHAEROCARPUM, Miq. Fl. Ind. Bat. I, pt. 2, p. 35. A strong climber: young branches and all others parts more or less dark rusty-velvety tomentose. *Leaves* elliptic-oblong, obtuse and very slightly apiculate, slightly narrowed to the rounded base; upper surface with harsh, short pubescence, the midrib tomentose; lower surface uniformly and minutely soft-pubescent: main nerves 8 to 12 pairs, oblique not inter-arching at the tips, prominent beneath; the connecting veins transverse oblique, rather prominent, length 2·5 to 4·5 in., breadth 1·25 to 2 in., petiole ·35 in. *Flowers* ·6 or ·7 in. in diam., in axillary or terminal racemes or panicles; pedicels ·35 to ·5 in. long with a small supra-basal bracteole. *Sepals* ovate-acuminate, connate at the base, spreading, minutely tomentose outside, glabrescent inside. *Petals* thick, leathery, brown outside, pink within, ovate, acuminate, slightly pouched at the base; the outer ·3 to ·35 in. long, tomentose outside, puberulous within: the inner smaller than the outer, more concave at the base, glabrous or glabrescent, the upper part very thick. *Stamens* numerous, the apex of the connective thick, obliquely triangular; anther-cells linear, lateral. *Ovaries* about 6, elongate, oblique, pubescent, with 6 to 8 ovules: style short, glabrous: stigma small. *Ripe carpels* globular, harshly and minutely pubescent, 1·1 in. in diam.: stalks rather slender, about twice as long. *Unona sphaerocarpa*, Blume Bijdr. 12: Fl. Javae Anon. 79 t. 16.

Perak: King's Collector.

This is allied to *M. latifolium*; but has smaller leaves with fewer nerves; its pubescence is very dark rusty, not tawny; and the apices of the anthers are truncate, not bearing a broad triangular, acute point. It is also allied to *M. parviflorum*, Scheff.

9. MELODORUM LANUGINOSUM, Hook. fil. and Thoms. Fl. Ind. 117. A strong creeper; young branches softly rufous-tomentose. *Leaves* coriaceous, oblong, sometimes sub-obovate-oblong, abruptly acute or shortly acuminate, rarely obtuse, the base rounded; upper surface glabrous, the midrib rufous-tomentose, olivaceous when dry; lower surface densely rufous-lanate; main nerves 12 to 20 pairs, oblique, curving, inter-arching close to the edge, prominent beneath; length 3·5 to 9 in., breadth 1·9 to 3·5 in.; petiole ·4 to ·6 in., stout, tomentose. *Flowers* 1·25 to 1·5 in. long, axillary or leaf-opposed, solitary, or in short 2- to 4-flowered cymes; pedicels stout, lanate, ·5 in. long, with a single basal bracteole. *Sepals* ovate, spreading, slightly connate, golden or rufous-lanate outside, glabrous inside like the outer petals. *Petals* thick, leathery, oblong-lanceolate from a broad base, sub-acute, the outer 1·25

357

to 1·5 in. long; the inner smaller, glabrescent or glabrous, concave at the base. *Stamens* numerous, the connective obliquely triangular at the apex; the anther-cells very narrow, lateral. *Ovaries* obovoid, oblique, curved, densely sericeous, 4- to 6-ovuled; style glabrous. *Ripe carpels* sessile, shortly stalked, sub-globose, narrowed to the base; densely and softly rufous-tomentose, about ·75 in. in diam. when ripe; seeds about 4. Miq. Fl. Ind. Bat. I, Pt. 2, 35; Hook. fil. Fl. Br. Ind. I, 79. *Uvaria tomentosa*, Wall. Cat. 6454.

Penang: Wallich, Curtis. Singapore; Wallich. Pangkore: Curtis. Penang; Scortechini, Wray, King's Collector.

At once distinguished by its large flowers, lanate leaves and sessile, or shortly stalked, rufous-tomentose fruit.

10. MELODORUM MAINGAYI, Hook. fil. and Thoms. Fl. Br. Ind. I, 80. A climber: young branches pubescent, dark-coloured. *Leaves* coriaceous, reddish-brown when dry, broadly elliptic or oblong, rounded at both ends, the tip sometimes minutely apiculate; upper surface glabrous except the puberulous midrib; lower glaucous and finely pubescent; main nerves 14 to 16 pairs, spreading, slightly prominent and dark-coloured beneath; length 3 to 6 in., breadth 1·5 to 2·35 in.; petiole ·6 in. *Flowers* 1·25 to 1·5 in. long, solitary, axillary; buds swollen at the base, narrowed and triquetrous above: pedicels ·25 to ·5 in., stout; bracteoles several, small. *Sepals* orbicular, sub-acute, quite connate into a disk, ·35 in. in diam. *Petals* leathery; the outer oblong-lanceolate, with broad base, flat but keeled down the middle inside, outside minutely rufous-tomentose, inside hoary-pubescent; inner very small, triangular-ovate, glabrous. *Stamens* numerous, small, with a broad rounded apical process, convex. *Ovaries* about 6, sericeous on one side; stigma sub-sessile. *Ripe carpels* oblong, tapering to each end, the apex shortly beaked, rusty-puberulous; the pericarp thick, 1·5 to 2 in. long and ·75 in. in diam.; stalks ·5 in. long, stout. *Seeds* many, in horizontal rows, ·5 in. long testa shining, not margined.

Penang; Maingay (Kew Distrib.,) No. 108, Curtis, No. 1046. Perak: Wray, 1112.

11. MELODORUM PRISMATICUM, Hook. fil. and Thoms. Fl. Br. Ind. 121. A large creeper; young branches glabrous, dark-coloured. *Leaves* coriaceous, oblong, elliptic-oblong, rarely obovate-oblong, abruptly and shortly acuminate; the base broad, rounded: upper surface glabrous except the minutely puberulous midrib; lower surface glaucous, reticulate, finely pubescent especially on the midrib; main nerves 12 to 18 pairs, spreading, faint especially near the tip, the secondary nerves prominent; length 4·5 to 8·5 in., breadth 2·3 to 3·3 in., petiole ·5 to ·7 in. *Flowers* 1·5 to 2 in. long, axillary, solitary; pedicels ·3 to ·6 in. long,

rufous-tomentose, with 1 large bracteole above the middle and several smaller near the base. *Sepals* quite connate into a flat, obtusely 3-angled disk, ·3 in. broad, pubescent outside, glabrous and tubercled inside. *Petals* very thick : the outer linear-lanceolate, 1·5 to ·2 in. long, trique-trous, rufous-tomentose outside, puberulous inside : the inner thinner and only about ·3 in. long, triangular, ridged outside, much excavated and glabrous at the base inside, otherwise puberulous. *Stamens* numerous, with very short filaments, anthers linear, apex of connective obliquely triangular. *Ovaries* elongate, oblong, tapering to the apex, shortly pubescent : ovules about 14, in 2 rows ; style short, lateral ; stigma sub-capitate, lobulate. *Ripe carpels* ovoid, blunt, tuberculate, puberulous, becoming sub-glabrous, 1·4 in. long and ·8 in. in diam. : stalks ·8 to 1 in., stout. *Seeds* in 2 rows, horizontal compressed, oval, black, shining. Hook. fil. Fl. Br. Ind. I, 81 ; Miq. Fl. Ind. Bat. I, Pt. 2, 36. *Pyramidanthe rufa*, Miq. Ann. Mus. Lugd. Bat. II, 39. *Uvaria rufa*, Wall. Cat. 6455. *Oxymitra bassiæfolia*, Teysm. and Binnin. in Tijdsch. Ned. Ind. XXV, (1863), 419.

Penang, Malacca, Perak, Singapore : common. Distrib. : Borneo.

Authentic specimens both of *Pyramidanthe rufa* and of *Oxymitra bassiæfolia*, T. and B. shew that they unmistakably belong to this species. Specimens of the former from Bangka and from the Buitenzorg Botanic Garden have, however, their leaves rather more hairy beneath than is usual in Perak specimens and their flowers are also rather longer.

12. MELODORUM MACRANTHUM, Kurz in Journ. As. Soc. Bengal, 1872, Pt. II, 291 ; 1874, Pt. II, 56 ; F. Flora Burma, I, 42. A small tree : all parts except the young leaf-buds and the flower glabrous ; young branches dark-coloured, rather slender. *Leaves* membranous, elliptic-oblong, sometimes slightly obovate, shortly and abruptly acuminate, the base cuneate ; upper surface shining, the lower dull ; main nerves 12 to 16 pairs, faint and much more prominent than the secondary, forming a double set of intra-marginal arches : length 6 to 8 in., breadth 2·5 to 3·5 in., petiole ·3 to ·4 in. *Flowers* solitary, axillary or from the branches below the leaves, 3 to 5 in. long, drooping ; pedicels ·5 to ·75 in. long, obscurely bracteolate at the base only. *Sepals* broadly ovate, sub-acute, coriaceous, pubescent at the edges inside, glabrous outside, connate for half their length, ·45 in. long. *Petals* greenish-white, becoming yellowish, coriaceous ; narrowly linear-lanceolate, acuminate, the outer row flat, adpressed-puberulous with a glabrous patch at the base inside, 3 to 5 in. long; the inner row only 1 to 1·25 in. long, cohering by their edges, vaulted at the base and with a glabrous patch ; the limb keeled inside, puberulous on both surfaces. *Stamens* numerous, the anther-cells linear, elongate ; apical process of connective narrowly tri-

angular, pointed. *Ovaries* numerous, narrowly oblong, adpressed-rufous-pubescent, 4-ovuled : style nearly as long as the ovary, cylindric, bent outwards, glabrous ; stigma small, slightly bifid. *Ripe carpels* oblong, blunt, tapering at the base, slightly rugose, glabrous, 1·25 to 1·5 in. long and about ·5 or ·6 in. in diam.: stalk ·4 to ·5 in. *Seeds* 1 or 2, compressed, ovoid, smooth. *Unona macrantha*, Kurz. in Andam. Report, Ed. I, App. B. I : *Pyramidanthe macrantha*, Kurz. l. c. Ed. 2, p. 29.

S. Andaman ; Kurz, King's Collector.

In some of its characters, (*e. g.*, the erect habit, the fewness of the ovules, and the thin texture and flatness of the much elongated outer petals) this does not quite conform to the characters of typical *Melodorum*. By its thin elongated outer petals, it approaches the *Dasymaschalon* section of *Unona ;* but the fewness of its ovules excludes it therefrom. From *Xylopia*, which it in some respects resembles, it is chiefly excluded by the very convex torus of its flowers, and by the very pointed apical appendage of its stamens. The stamens on the other hand are those of *Melodorum*, and the petals resemble those of *M. prismaticum* (*Pyramidanthe rufa*, Miq.). On the whole therefore, I think, it best to leave this plant in the genus to which Kurz finally referred it.

13. MELODORUM ELEGANS, Hook. fil. and Thoms. Fl. Ind. 122. A large climber : young branches slender, puberulous at first, ultimately glabrous, dark-coloured. *Leaves* thinly coriaceous, oblong-lanceolate, acuminate, slightly narrowed to the rounded base : upper surface olivaceous when dry, glabrous : lower paler, puberulous, minutely reticulate, the 12 or 13 pairs of main nerves spreading, faint : length 2·5 to 3·5 in., breadth 1 to 1·25 in., petiole ·25 to ·35 in. *Flowers* axillary, solitary or 2 or 3 in a fascicle, ·35 to ·65 in. long : pedicels slender, ·35 to ·6 in. long often deflexed, with 2 or 3 minute basal bracteoles. *Sepals* ovate, acute, united at the base only, spreading, outside tubercular and pubescent, inside glabrous and concave, ·1 in. long. *Petals* leathery, the outer broadly ovate, sometimes minutely ovate-oblong, silky, rufous-tomentose outside, hoary-puberulous within, with a perfectly glabrous patch at the concave base, ·35 to ·6 in long : inner petals only ·25 in. long, very thick, triquetrous and puberulous above, concave and glabrous at the base, inside. *Stamens* numerous, with filaments half as long as the anther-cells ; apical process of connective short, thick, obliquely triangular. *Ovaries* narrowly oblong, glabrous, with 8 ovules in 2 rows : style short, lateral. *Ripe carpels* ovoid or ovoid-globose, blunt at each end, glabrous, ·35 to ·5 in. long : stalks slender, ·25 in. long, compressed, black, shining, pitted. Hook. fil. Fl. Br. Ind. I, 82 : Miq. Fl. Ind. Bat. I, pt. 2, p. 36. *Uvaria elegans*, Wall. Cat. 6474A.

This is closely allied to *M. fulgens*, H. f. and T. ; but its flowers have

360

more slender and usually longer pedicels : the ovary of this is moreover glabrous, while that of *M. fulgens* is pubescent and the carpels of this are under half an inch in length, while those of *M. fulgens* are three times as long. This is also allied to *M. Kentii*, H. f. and Th., the ovaries of which have, however, never more than two ovules.

Penang : Wallich. Malacca : Maingay (Kew Distrib.,) No. 75. Perak : King's Collector, Wray, Scortechini.

14. MELODORUM PISOCARPUM, Hook. fil. and Thoms. Fl. Ind. 123. A powerful climber : young branches glabrous, black. *Leaves* coriaceous. elliptic or elliptic-oblong, sometimes obovate-elliptic, shortly and abruptly acuminate ; the base rounded or sub-cuneate : upper surface olivaceous when dry, glabrous, shining ; the lower glaucous, slightly puberulous when young : main nerves 10 to 12 pairs, spreading, very indistinct ; length 2·5 to 4 in., breadth 1·25 to 1·8 in., petiole ·35 in. *Flowers* ·3 to ·65 in. long, axillary, solitary or in pairs ; pedicels rather stout, deflexed, rufous-puberulous, bi-bracteolate at the base, ·25 to ·35 in. long. *Sepals* broadly ovate, acute, concave, connate into a triangular cup, rufous-puberulous outside, glabrous inside, persistent. *Petals* thick : the outer flat, oblong-ovate, acute, minutely silky, rufous-tomentose outside, hoary pubescent inside except on the glabrous basal excavation, ·3 to ·65 in, long : *inner petals* less than half as long, with a large glabrous basal concavity and a short, thick, triquetrous point, hoary-puberulous. *Stamens* numerous, filament very short, apical process of connective orbicular. *Ovaries* narrowly oblong, glabrous, pitted, 2-ovuled : style lateral, nearly as long as the ovary. *Ripe carpels* globular, slightly tubercled, glabrous, ·25 in. in diam.: stalks about as long. *Seeds* 2, plano-convex, dark-brown, shining, pitted. Hook. fil. Fl. Br. Ind. I, 82 ; Miq. Fl. Ind. Bat. I, Pt. 2, 37. *M. pyramidale*, Maingay MSS. *Uvaria mabiformis*, Griff. Notulae, IV, 709.

Malacca ; Griffith, Maingay (Kew Distrib.) No. 77. Singapore ; Ridley. Penang ; Curtis. Perak ; common. Distrib. Sumatra, Forbes, No. 2182.

Only two species of *Melodorum* besides this have glabrous ovaries (*M. Kentii* and *M. elegans*) ; but whereas those of this and *M. Kentii* are 2-ovuled, the ovaries of *M. elegans* have 8, or, according to Sir Joseph Hooker, sometimes 10 ovules. This species has however different leaves from the two above mentioned, and its carpels are much smaller and quite globular. As in other species of *Melodorum*, there is considerable variability in the size of the flowers in this species.

20. XYLOPIA, Linn.

Trees or shrubs. *Leaves* coriaceous. *Flowers* axillary. solitary

113

cymose or fascicled; buds triquetrous, conic, often slender. *Sepals* 3, valvate, connate. *Petals* 6, elongate, valvate, in 2 series; outer flat or concave; inner nearly as long, trigonous, concave at the base only. *Torus* flat, or hollow and enclosing the carpels. *Stamens* oblong, truncate or connective produced; anther-cells remote or contiguous, often septate and with a large pollen-grain in each cellule. *Ovaries* 1 or more; style long, clavate; ovules 2–6 or more, 1- to 2-seriate. *Ripe carpels* long or short, continuous or moniliform, usually several-seeded.—Distrib. Tropics generally; species 60 to 70.—Closely allied to *Melodorum*, but very different in habit.

Leaves quite glabrous.
  Leaves 6 or 7 in. long ... ... 1. *X. oxyantha.*
  Leaves between 3 and 5 in. long.
    Ripe carpels cylindric, boldly tubercled 2. *X. dicarpa.*
      „    „    „ smooth ... 3. *X. malayana.*
  Leaves between 2 and 3 in. long.
    Flowers always solitary; pedicels with 2 or 3 orbicular bracteoles, apical process of stamens rounded, anther-cells septate ... ... ... 4. *X. Maingayi.*
    Flowers solitary or in pairs, ·5 in. long: pedicels with orbicular basal bracteoles; apical process of stamens rounded; anther-cells septate... ... 5. *X. pustulata.*
    Flowers in fascicles or solitary, ·75 in. long: pedicels ebracteolate; apical process of stamens oblong: anther-cells not septate ... ... 6. *X. fusca.*
Both surfaces of leaves glabrous, the midrib alone pubescent in its lower half on the upper surface; length 5·5 to 9·5 in. ... ... 7. *X. Curtisii.*
Leaves glabrous on the upper surface (the midrib pubescent in *X. caudata*), the lower slightly pubescent or puberulous.
  Leaves more or less lanceolate, acute or acuminate, not at all obovate.
    Leaves 2 or 3 in. long.
      Leaves not glaucous beneath.
        Flowers ·5 to ·57 in. long, solitary, axillary, obtuse ... 8. *X. elliptica.*
        Flowers ·2 to ·25 in. long, axillary, solitary, or 2 to 3 together ... ... 9. *X. caudata.*

362

Leaves glaucous beneath ... 10. *X. stenopetala.*
Leaves 3·5 to 5·5 in. long, leaves glaucous beneath ; petals very long and
narrow ... ... ... 10. *X. stenopetala.*
Leaves more or less obovate or oblanceolate,
4 to 7 in. long.
Leaves 1·75 to 4 in. broad ; flower pedicels ·2 to ·25 in. long ; ripe carpels
broadly ovoid, blunt, sub-glabrous ... 11. *X. Scortechinii.*
Leaves 1·75 to 2·5 in. broad ; flower
pedicels ·5 to ·8 in. long ; ripe carpels
globular, densely and minutely yellowish-tomentose ... ... 12. *X. olivacea.*
Upper surfaces of leaves glabrous (the midrib
alone pubescent in some) : under surfaces uniformly pubescent.
Under-surface of leaves adpressed-rufous-
sericeous ; length 2 to 3 in. ... ... 13. *X. obtusifolia.*
Under-surface of leaves deep brown, the
pubescence slightly paler ; length 3 to
4·5 in. ; ripe carpels obovoid-oblong, blunt 14. *X. magna.*
Under-surface of leaves purplish-brown, pubescent ; length 3·5 to 5·5 in. ; main nerves
10 to 12 pairs ; ripe carpels much elon-
gate, cylindric, many-seeded ... ... 15. *X. ferruginea.*
Under-surface of leaves brownish-tomen-
tose ; length 6·5 to 8·5 in. ; nerves 12 to
14 pairs ... ... ... 16. *X. Ridleyi.*

1. XYLOPIA OXYANTHA, Hook. fil. and Thoms. Fl. Br. Ind. I, 85.
A tree : young parts puberulous ; the branchlets rather stout, striate.
*Leaves* coriaceous, ovate or oblong, abruptly and shortly acuminate,
glabrous, glaucous on the lower surface ; main nerves 12 to 15 pairs,
spreading, thin ; length 6 to 7 in., breadth 2·5 to 3 in., petiole ·35 in.
*Peduncles* axillary, in fascicles, ·35 to ·5 in. long, adpressed-pubescent.
*Sepals* broadly ovate. *Outer petals* narrowly linear, tapering at the apex,
yellowish pubescent, slightly keeled at the back, 1·25 to 1·5 in. long.
*Stamens* and *ovaries* as in *X. ferruginea.* *Habzelia oxyantha,* Hook. fil
and Th. Fl. Ind. 124 ; Miq. Fl. Ind. Bat. I, pt. 2, 37. *Uvaria oxyantha,*
Wall. Cat. 6478.

Singapore : Wallich.

2. XYLOPIA DICARPA, Hook. fil. and Thoms. Fl. Br. Ind. I, 85. A
tree 20 to 25 feet high ; branches glabrous, dark-coloured, minutely

dotted. *Leaves* coriaceous, elliptic-lanceolate, acute or acuminate, the base acute ; both surfaces glabrous, minutely reticulate ; main nerves about 10 pairs, spreading, very faint, the secondary nerves almost as distinct ; length 3 to 4·5 in., breadth 1·5 to 1·75 in., petiole ·25 in. *Flowers* solitary or in pairs, pendent, 1·5 in. long : pedicel very short with 1 to 3 orbicular, amplexicaul, glabrous bracteoles. *Sepals* ovate, obtuse, tubercled, connate to the middle. *Petals* linear oblong, slightly expanded and concave at the base, hoary, pubescent ; the inner narrower and shorter than the outer, sub-trigonous. *Stamens* numerous, the inner rudimentary : apical process rounded ; anthers linear, septate. *Ovaries* 2 to 4, pilose, multi-ovular : style short. *Ripe carpels* cylindric, blunt at each end, much tubercled, puberulous, 1·5 in. long and about ·75 in. in diam. *Seeds* 7 or 8, compressed, the testa pale, scaly.

Singapore : Maingay (Kew Distribution *in part*) No. 84, King's Collector No. 7079.

3. XYLOPIA MALAYANA, Hook. fil. and Thoms. Fl. Ind. 125. A slender tree : young branches thin, glabrous, the buds pubescent. *Leaves* thinly coriaceous, shortly and bluntly acuminate, the base cuneate ; both surfaces glabrous ; main nerves about 8 pairs, faint, spreading ; length 3·5 to 5 in., breadth 1·5 to 2 in., petiole ·2 in. *Flowers* ·6 to ·9 in. long, solitary or in pairs, axillary ; pedicels rufous-pubescent, ·1 in. long, with several bracteoles at the base. *Sepals* broadly ovate, sub-acute, puberulous outside and on the edges, glabrous inside, ·15 in. long and as broad. *Petals* linear-oblong, tapering to the apex, concave and glabrous at the slightly expanded base, densely pubescent elsewhere ; the inner slightly narrower and shorter than the outer and more concave at the base. *Stamens* numerous, the apices rhomboid, papillose ; the anthers long, lateral, with transverse divisions. *Pistils* about 6 ; the ovaries oblong, densely pale-hirsute, about as long as the stamens, 2-ovuled ; styles about as long as the ovaries and projecting far above the stamens, glabrous, sub-cylindric, clavate. *Ripe carpels* (fide Maingay) ·35 to 1 in., several-seeded ; stalk short, thick. Hook. fil. and Thoms. Fl. Br. Ind. I, 85 ; Miq. Fl. Ind. Bat. I, Pt. 2, 38. *Parartabotrys sumatrana*, Miq. Fl. Ind. Bat. Suppl. 374 ; Scheffer in Nat. Tijdsch. Ned. Ind. XXXI, 15.

Malacca ; Griffith, Derry, Maingay (Kew Distrib.) No. 81 Singapore, Ridley. Perak ; Scortechini. Distrib., Sumatra.

4. XYLOPIA MAINGAYI, Hook. fil. and Thoms. Fl. Br. Ind. I, 85. A tree ? Young branches rusty-pubescent, afterwards glabrous and with white dots. *Leaves* small, coriaceous, elliptic or elliptic-oblong, subacute or obtusely acuminate, the base sub-cuneate : both surfaces glabrous and reticulate, the upper pale, the lower dark ; main nerves slender ; length 2

to 3 in., breadth 1 to 1·25 in. ; petiole ·25 to ·3 in. *Flowers* solitary, pendent, pale-orange ; pedicels very short, stout, curved ; bracteoles 2 or 3, orbicular, rusty-tomentose. *Sepals* broadly ovate, connate to the middle, rusty-tomentose. *Petals* flat, linear-oblong, sub-acute, softly tomentose except the glabrous concave base ; the inner narrower, almost as long, trigonous. *Stamens* with rounded apiculus : the anthers narrow, septate. *Ovaries* about 9, with 6 ovules ; style glabrate. *Ripe carpels* unknown.

Malacca : Maingay.

5. XYLOPIA PUSTULATA, Hook. fil. and Thoms. Fl. Br. Ind. I, 85. A tree : young branches pale, glabrous, minutely white-dotted. *Leaves* coriaceous, small, elliptic, sub-obtuse, the base acute, both surfaces glabrous, the lower reddish brown and reticulate : main nerves faint, not more prominent than the secondary. *Flowers* solitary or in pairs, axillary, ·5 in. long, pendent ; pedicels very short, with orbicular, ciliate, deciduous basal bracteoles. *Sepals* short, ovate, sub-acute, rusty-pubescent, united to the middle. *Petals* linear, sub-acute, densely adpressed-pubescent ; the outer obtuse with a rather broad concave base, the inner shorter and much narrower with a broader concave base. *Stamens* linear with rounded apiculus : the anthers long, septate. *Ovaries* 5 to 8, hirsute ; the style slender with clavate stigma ; ovules several. *Ripe carpels* unknown.

Malacca : Maingay (Kew Distribution) No. 86.

6. XYLOPIA FUSCA, Maingay ex Hook. fil. Fl. Br. Ind. I, 85. A tree ; young branches rather stout, glabrous, black : buds silky. *Leaves* coriaceous, oblong, obtuse, the base cuneate ; upper surface glabrous shining ; the lower dull, dark, reticulate ; main nerves 8 or 9 pairs, very faint ; length 2 to 3 in., breadth ·75 to 1 in. ; petiole ·2 in., stout. *Flowers* ·75 in. long, supra-axillary, solitary, racemed, or fascicled ; peduncle ·25 to ·75 in. with several bracts ; pedicels ·25 in., puberulous, ebracteolate. *Sepals* ovate, acute, connate into a cup with 3 spreading, acute teeth, puberulous outside. *Petals* linear-oblong, tapering to the sub-acute apex : the outer adpressed golden-sericeous outside ; the inner narrower and shorter, concave at the base. *Stamens* with an oblong apical process ; anthers linear, lateral, not septate. *Ovaries* 4 or 5, cohering into a cone, golden-silky ; ovules 10 to 16, in two rows. *Ripe carpels* unknown.

Malacca : Maingay, (Kew Distribution) No. 86.

7. XYLOPIA CURTISII, King, n. sp. A tree 30 feet high : young branches stout, glabrous, striate, dark-coloured. *Leaves* very coriaceous, oblong, acute or shortly acuminate ; the base cuneate, slightly oblique : upper surface glabrous, shining ; the lower dull, darker (when dry),

117

puberulous on the midrib near the base; main nerves 12 to 20 pairs, very prominent beneath and connected by straight transverse veins; length 5·5 to 9·5 in., breadth 2 to 3 in.; petiole ·35 in., stout. *Flowers* 1 or 2, on stout woody extra-axillary peduncles; pedicels ·2 in. long, rufous-pubescent, with a single large bracteole. *Sepals* thick, spreading, broadly ovate, sub-acute, minutely tomentose on both surfaces but especially on the outer. *Petals* thick, subequal, linear-oblong, obtuse, keeled outside; the claw orbicular, vaulted over the andro-gynœcium and glabrous inside, otherwise minutely tomentose, ·75 in. long. *Stamens* numerous, the heads obliquely truncate and concealing the linear, lateral anthers. *Ovary* solitary, cylindric, fluted, glabrous, multi-ovulate. *Ripe carpel* ovoid, compressed, silvery-grey, many-seeded, 3 in. long, and 2·5 in. in diam.

Penang: Curtis, No. 1569.

8. XYLOPIA ELLIPTICA, Maingay ex Hook. fil. Fl. Br. Ind. I, 86. A tall tree : young branches dark-coloured, glabrous, the youngest pubescent. *Leaves* membranous, small, elliptic, obtusely acuminate, the base rounded or acute : upper surface glabrous, pale ; the lower brown, minutely adpressed-pubescent ; both reticulate : main nerves 6 or 7 pairs, oblique, very faint ; length 1·5 to 2 in., breadth 1 to 1·25 in. ; petiole ·2 in., slender. *Flowers* solitary, erect, axillary, ·5 to ·75 in. long : peduncle about half as long, rusty-pubescent like the calyx, bracteoles minute. *Sepals* ovate, sub-acute, united to the middle. *Petals* pale brownish-tomentose ; the outer linear-subulate with a broader concave base : the inner trigonous, shorter and narrower than the outer. *Stamens* numerous, minute, the apex rounded ; anthers linear. *Ovaries* 1 to 3, densely hairy, 4- to 6-ovuled. *Ripe carpels* unknown.

Malacca : Maingay (Kew Distrib.,) No. 82. Perak : Wray No. 3194. Penang : Curtis, No. 2482:

9. XYLOPIA CAUDATA, Hook. fil. and Thoms. Fl. Ind. 125. A shrub or small tree : young branches very slender, minutely pubescent. *Leaves* thinly coriaceous, lanceolate, long and obtusely acuminate, the base cuneate ; upper surface glabrous except the pubescent midrib ; the lower sparsely adpressed-sericeous : main nerves about 10 pairs, spreading, faint ; length 2 to 2·25 in., breadth ·6 to ·8 in.; petiole ·1 in., slender. *Peduncles* 1 to 3, axillary, very short, minutely bracteolate at base and apex. *Flowers* ·2 to ·3 in. long. *Sepals* ovate, sub-acute, connate at the base, adpressed-pubescent outside, glabrous inside. *Petals* linear-oblong, obtuse, pubescent except a small glabrous concave spot at the base, the inner about as long as, but narrower than, the outer. *Anthers* rather numerous, compressed, the apical process narrow. *Ovaries* 2, elongate, sericeous, 2-ovuled : style long, pointed, glabrous, exserted.

*Ripe carpels* (fide Hooker) 2 or 3, sub-globose or ovoid, pubescent, ·5 in. long, 2-seeded. Hook. fil. Fl. Br. Ind. I, 85; Miq. Fl. Ind. Bat. I, Pt. 2, 38. *Guatteria (?) caudata*, Wall. Cat. 6452.

Singapore: Wallich, Maingay (Kew Distrib.) No. 79. Malacca; Griffith.

10. XYLOPIA STENOPETALA, Oliver in Hook. Ic. Plantar. t. 1563. A tree 50 to 60 feet high: young branches dark-coloured, glabrescent, minutely lenticellate. *Leaves* thinly coriaceous, elliptic-oblong, shortly and obtusely acuminate, the base sub-cuneate; upper surface glabrous, shining; the lower glaucous or glaucescent, sparsely adpressed-pubescent; both reticulate; main nerves 10 or 12 pairs, spreading, inter-arching close to the edge, faint: length 2·5 to 4·5 in., breadth 1·1 to 1·6 in., petiole ·25 in. *Flowers* axillary, solitary or in fascicles of 2 to 5; pedicels slender, often decurved, puberulous, with one minute bracteole, ·5 to ·75 in. long. *Sepals* united to form a small puberulous cup with acute, spreading teeth. *Petals* fleshy, very narrow, slightly expanded and concave at the base, minutely tawny-pubescent, the inner slightly shorter and narrower. *Stamens* linear, the connective prolonged into a cylindro-conic apical appendage; the anthers fusiform, lateral. *Ovaries* numerous, elongate, pubescent, 6-ovuled; style filiform: stigma sub-clavate. *Ripe carpels* oblong, sub-terete, narrowed to the stalk, 2 to 2·5 in. long and ·5 in. diam.: pericarp fleshy. *Seeds* 1 to 4: stalks thick, ·3 in. long.

Penang; on Government Hill at 600 feet: Curtis Nos. 857 and 880.

11. XYLOPIA SCORTECHINII, King n. sp. A tree 50 to 60 feet high: young branches rusty-tomentose, ultimately glabrous, much striate and pale brown. *Leaves* coriaceous, obovate-elliptic to elliptic-oblong, very shortly and abruptly acuminate, slightly narrowed to the sub-cuneate rounded slightly oblique base: upper surface glabrous, the midrib slightly rufous-puberulous near the base: lower surface pale, sparsely rufous-pubescent especially on the midrib and 10 to 14 pairs of oblique, rather straight, prominently raised main nerves; length 4 to 7 in., breadth 1·75 to 4 in.; petiole ·35 in., pubescent. *Flowers* rarely solitary, usually in fascicles of 2 to 5 on tubercles in the axils of leaves or of fallen leaves; pedicels short, (·2 to ·25 in.), stout, rusty-tomentose with a sub-mesial bracteole. *Sepals* quite free, broadly ovate, blunt, pubescent outside, glabrous inside. *Petals* thickened, linear-obtuse with an orbicular concave claw, vaulted over the stamens and pistils, 1·25 to 1·75 in. long, pubescent everywhere except on the glabrous concavity of the claw. *Stamens* numerous, with truncate 4- or 5-angled apices concealing the lateral anthers. *Ovaries* few, short, oblong, pubescent, 4- or 5-ovuled; stigma large, oblong. *Ripe carpels* broadly ovoid, blunt, rufous-pubes-

cent when young, glabrescent when old, ·8 in. long and ·6 in. in diam. *Seeds* about 4, discoid, pale brown, shining. *Drepananthus stenopetala,* Scortechini, MSS.

Perak : Scortechini, No. 1781; King's Collector, No. 8241.

A species allied to *X. olivacea,* King; but with broader leaves, shorter flower pedicels, narrower petals and ovoid sub-glabrous fruit.

12. XYLOPIA OLIVACEA, King n. sp. A shrub or small tree : young branches pubescent, ultimately brown, striate and glabrous. *Leaves* thinly coriaceous, elliptic-oblong, sometimes slightly obovate, shortly and abruptly acuminate, the base cuneate ; both surfaces dull oliva-ceous when dry ; the upper glabrous, the lower paler, slightly scurfy ; main nerves 6 to 8 pairs, oblique, curving, inter-arching boldly ·15 in. from the margin, prominent beneath ; length 3·5 to 7 in., breadth 1·75 to 2·5 in., petiole ·25 in., swollen, puberulous, black when dry. *Flowers* solitary or in pairs, supra-axillary; pedicels rather stout, ·5 to ·8 in. long, cinereous-tomentose with an ovate-lanceolate, mesial bracteole. *Sepals* thick, especially at the base, ovate, acute, connate below the middle, pale cinereous-puberulous on both surfaces. *Petals* sub-equal, fleshy, narrowly linear with a tapering limb and slightly expanded concave vaulted claw, densely and minutely cinereous-tomentose, 1 to 1·5 in. long, the inner shorter. *Stamens* short, cuneate, the broad oblique heads covering the apices of the linear anthers. *Ovaries* few, oblong, densely sericeous, 6- to 8-ovuled ; style short, cylindric : stigma large, fleshy. *Ripe carpels* few, globular, with slightly flattened minutely apiculate apex, and an imperfect lateral ridge, densely and minutely yellowish-tomentose, ·6 in. in diam., stalks very short. *Seeds* 4 or 5, discoid, smooth, pale brown, shining, separated from each other by imperfect dissepiments.

Perak : up to elevations of 3,000 or 4,000 feet, common. Scorte-chini, Wray, King's Collector.

13. XYLOPIA OBTUSIFOLIA, Hook. fil. and Thoms. Fl. Br. Ind. I, 85. A tree : young branches glabrous, dark-coloured, striate : buds silky. *Leaves* coriaceous, oblong, obtuse or retuse, the base cuneate, upper surface glabrous, shining ; the lower adpressed rufous-sericeous : main nerves 8 or 10 pairs, oblique, very faint ; length 2 to 3 in., breadth 1 to 1·5 in., petiole ·25 in. *Flowers* ·5 in. long, axillary, solitary or 2 or 3 in small sub-racemose cymes ; pedicels ·2 to ·25 in., rufous-pubescent with a single bracteole. *Sepals* thick, broadly ovate, acute, united to the middle, pubescent outside, glabrous inside. *Petals* linear-oblong, tapering towards the blunt apex ; the outer petals adpressed-rufous-pubescent outside, puberulous within, slightly concave and glabrous at the base ; the inner smaller, more concave at the glabrous base, puberu-

lous elsewhere. *Stamens* numerous, elongate, narrow, with an acute apiculus; the anther-cells linear, lateral. *Pistils* one or two, conical, adpressed-pubescent; the style short, thin. *Ripe carpels* oblong, cylindric, sub-oblique, blunt, 1·25 in. long ·7 in. in diam. *Seeds* 3 or 4, globular.

Malacca: Griffith. Perak: King's Collector, No. 2816.

14. XYLOPIA MAGNA, Maingay ex Hook. fil. Fl. Br. Ind. I, 84. A tree: young branches tomentose, becoming glabrous and darkly cinereous. *Leaves* coriaceous, ovate-lanceolate to elliptic, sub-acute, the base rounded, the edges slightly revolute when dry; upper surface shining, reticulate, glabrous except the pubescent midrib; under surface deep brown, with rather pale pubescence; main nerves about 10 pairs, spreading, inter-arching some way from the edge, faint: length 3 to 4·5 in., breadth 1·25 to 2 in.; petiole ·25 in., pubescent. *Flowers* 2 to 2·5 in. long, solitary or in pairs, axillary: pedicels stout, tomentose, with a single large, ovate, acute, often bifid bract. *Sepals* thick, ovate acute, connate into a 3-toothed cup, adpressed-pubescent outside, glabrous inside. *Petals* sub-equal, the inner narrower and shorter, narrowly linear, slightly expanded and concave at the base, tapering towards the apex, pubescent except in the basal concavity. *Stamens* numerous, elongate, with an oblong obtuse apical process; the anthers lateral, linear, septate. *Pistils* about 15, narrowly oblique, hirsute on the outer side, 4-ovuled. *Style* filiform, long. *Ripe carpels* obovoid-oblong, compressed, blunt, minutely tomentose, 1·4 in. long and ·65 in. diam.; stalks thick, only ·15 in. long. *Seeds* about 4, in two rows, arillate, the testa bony.

Malacca: Maingay (Kew Distrib.) No. 83. Singapore; Ridley. Perak; Scortechini.

15. XYLOPIA FERRUGINEA, Hook. fil. and Thoms. Fl. Br. Ind. I, 85. A tree 20 to 60 feet high; young branches brownish-pubescent. *Leaves* coriaceous, narrowly oblong, acute; the base slightly narrowed and oblique, rounded or minutely sub-cordate; upper surface glabrous, shining; the lower glaucous and softly purplish-brown pubescent: most densely so on the midrib; main nerves 10 to 12 pairs, oblique, inter-arching near the edge, prominent beneath; length 3·5 to 5·5 in., breadth 1·1 to 2 in.; petiole ·2 in., channelled. *Flowers* solitary or in pairs, axillary or extra-axillary, erect or pendulous, yellow; pedicels ·5 to ·75 in., rusty-pubescent; bracteoles 1 to 3, small, lanceolate. *Sepals* broadly ovate-acuminate, connate at the base, spreading, small, pubescent outside, glabrous within. *Petals* linear, fleshy, tapering at the very apex, very long; the outer rufous-pubescent outside, cinereous-puberulous inside, concave at the very base, 1·25 to 2 in. long; inner petals much

narrower and thinner and a little shorter than the outer, cinereous-puberulous. *Stamens* about 24, narrow: anthers linear, lateral, the connective ending in a broadly oblong apical process. *Ovaries* numerous, narrowly oblong, pointed, densely rusty-hirsute, multi-ovular: style short, filiform, glabrous ; stigma minute. *Ripe carpels* numerous, much elongate, cylindric, glabrescent, with transverse partitions between the seeds, many-seeded, sub-moniliform when dry, 2 to 5 in. long. *Seeds* oblong, rugose, minutely pellucid-dotted, ·3 in long. *Habzelia ferruginea*, H. f. and T. Fl. Ind. 123. Miq. Fl. Ind. Bat. I, Pt. 2, 37. *Artabotrys malayana*, Griff. Notul. IV, 713.

Malacca: Griffith. Maingay (Kew Distrib.) No. 85. Perak: Scortechini, King's Collector, Wray : common. Selangor : Curtis.

16. XYLOPIA RIDLEYI, King n. sp. A tree ? Young branches stout, densely rusty-tomentose. *Leaves* coriaceous, obovate-elliptic, abruptly and very shortly acuminate, narrowed from below the middle to the slightly cuneate base : upper surface glabrous except the rufous-puberulous midrib : lower softly rusty-tomentose with longer, superficial, paler hairs : main nerves 12 to 14 pairs, oblique, inter-arching boldly within the margin, prominent on the lower, depressed on the upper, surface ; length 6·5 to 8·5 in., breadth 2·75 to 3·5 in. ; petiole ·5 to ·6 in. stout, tomentose. *Flowers* in extra-axillary (often leaf-opposed) fascicles of 3 to 5 : pedicels stout, rufous-tomentose, with a single bracteole, ·25 to ·3 in. long. *Sepals* broadly ovate, long-acuminate, rufous-pubescent outside, glabrous within, ·35 in. long. *Petals* filiform, triquetrous, with expanded concave vaulted bases concealing the andro-gynœcium, and glabrous inside, otherwise pubescent, 2·5 to 3·5 in. long. *Stamens* numerous, with truncate 4- or 5-angled heads concealing the elongate, lateral anthers. *Ovaries* obliquely ovoid, densely sericeous, 4- to 6-ovuled : stigmas fleshy, agglutinated. *Ripe carpels* unknown.

Singapore : Ridley.

21. PHÆANTHUS, H. f. and T.

Trees or climbers. *Flowers* solitary, terminal or in extra-axillary fascicles. *Sepals* 3, small, valvate. *Petals* 6, valvate in 2 rows ; outer small like the sepals ; inner large, flat, coriaceous. *Stamens* numerous, oblong or quadrate, truncate ; anther-cells dorsal, distant. *Carpels* numerous ; style cylindric or clavate, sometimes grooved ventrally. *Ovules* 1–2, sub-basal, ascending. *Ripe carpels* staked, 1-seeded.—DISTRIB. Species about 6 ; one in Southern Peninsular India, the rest Malayan.

Leaves softly pubescent ... ... ... 1. *P. nutans*.
Leaves glabrous.
    Ovules and seeds solitary ... ... 2. *P. lucidus*.
    Ovules and seeds in pairs ... ... 3. *P. andamanicus*.

1. PHÆANTHUS NUTANS, H. f. and Th. Fl. Ind. 147. A small tree : young branches rusty tomentose. *Leaves* membranous, oblong-lanceolate or oblanceolate to obovate-elliptic, caudate-acuminate, the base always narrowed and sometimes acute ; upper surface glabrous, the midrib and main nerves tomentose ; lower softly pubescent, the midrib tomentose : main nerves 10 to 14 pairs, spreading, prominent beneath, inter-arching near the edge : length 5 to 9 in., breadth 1·3 to 4·5 in. ; petiole ·3 in., tomentose. *Flowers* fœtid, solitary or 2 or 3 together, drooping, extraaxillary ; pedicels ·5 to 1·5 in. long with 1 or 2 linear bracteoles, pubescent. *Sepals* linear-lanceolate, spreading, tomentose, ·2 in. long. *Petals* very unequal ; the outer small like the sepals ; inner ovate-oblong, acute, yellow, pubescent, 5- to 7-ribbed, ·75 to 1 in. long. *Ripe carpels* ovoid, pubescent, beaked, ·6 in. long and ·35 in. in diam. ; stalk nearly as long. Hook. fil. Fl. Br. Ind. I, 72 ; Miq. Fl. Ind. Bat I, pt. 2, 51. *Uvaria nutans*, Wall. Cat. 6481. *U. tripetala*, Roxb. Fl. Ind. ii, 667. *U. ophthalmica*, Roxb. ex Don Gen. Syst. i, 93.

Singapore ; Wallich and others. Penang ; Curtis. Malacca ; Maingay, (Kew Distrib.) No. 67. Perak ; at low elevations. Sugei Ujong ; Ridley. Distrib. Moluccas, Sumatra.

2. PHÆANTHUS LUCIDUS, Oliver in Hook. Ic. Pl. t. 1561. A tree 40 to 50 feet high : young branches minutely rusty pubescent or almost glabrous, dark-coloured and furrowed. *Leaves* thickly membranous, oblong-elliptic to lanceolate, acuminate, the base cuneate ; both surfaces shining, glabrous except occasionally the puberulous midrib ; main nerves about 8 pairs, oblique, rather prominent beneath : length 4·5 to 6·5 in., breadth 1·25 to 2·25 in. ; petiole ·2 in. *Flowers* solitary, rarely in fascicles of 2 or 3, extra-axillary, erect, ·6 in. to 1 in. in diam., buds triquetrous ; peduncles 1 to 1·25 in. long, slender, puberulous, with 2 minute bracteoles. *Sepals* ovate, acute, less than ·1 in. long. *Outer petals* like the sepals but a little longer : inner petals thick, greenish-yellow, oblong-ovate, acute, about ·5 in. long, glabrescent with puberulous edges. *Anthers* with square truncate heads. *Ovaries* numerous, 1-ovulate. *Ripe carpels* oblong. ·6 in. long and ·3 in. in diam., minutely granular, sub-glabrous as are the ·5 to ·6 in. long stalks.

Penang : Curtis. Perak : at low elevations : King's Collector, Nos. 7275 and 10044.

3. PHÆANTHUS ANDAMANICUS, King n. sp. A small glabrous shrub : young branches pale brown, slender. *Leaves* membranous, elliptic or elliptic-lanceolate, acute, slightly narrowed to the rounded base, both surfaces rather pale when dry ; main nerves 15 to 20 pairs, faint, slender, horizontal, forming double loops near the margin, the reticulations faint ; length 4 to 7·5 in., breadth 1·75 to 2·5 in., petiole ·35 in. *Flowers*

·5 to ·75 in. in diam., campanulate, solitary, rarely in pairs, extra-axillary : pedicels ·2 in. long, bracteolate at the base. *Sepals* very small, semi-orbicular. *Outer petals* slightly larger than the sepals and about ·1 in. long; inner petals united at the base, oblong-ovate, sub-acute, ·5 to ·7 in. long, 4 or 5 nerved. *Anthers* numerous, flattened from front to back, about as broad as long with truncate not apiculate heads. *Ovaries* numerous, elongate, narrow, 2-ovuled : stigmas elongate. *Ripe carpels* sub-globular, ·5 in. in diam. : stalks ·5 to ·7 in. *Seeds* two, plano-convex, pale.

South Andaman, King's Collector.

This is a very distinct species recognisable at once by the un-usual character of having its petals united at the base and by its 2-seeded carpels.

## 22. MILIUSA, Leschenault.

Trees or shrubs. *Flowers* usually bi-sexual (diœcious or polygam-ous in No. 1), green or red, axillary or extra-axillary, solitary, fascicled or cymose. *Sepals* 3, small, valvate. *Petals* 6, valvate in 2 series ; outer smaller, like the sepals ; inner cohering when young by the margins, at length free. *Torus* elongated, cylindric. *Stamens* definite or indefinite ; anthers subdidymous ; cells contiguous, ovoid, extrorse ; connective more or less apiculate. *Ovaries* indefinite, linear-oblong ; style oblong or very short ; ovules 1–2, rarely 3–4. *Ripe carpels* globose or oblong, 1- or 2- or many-seeded.—Distrib. Species 8 ; all Indian.

Flowers diœcious or polygamous ... 1. *M. Roxburghiana.*
Flowers hermaphrodite ... ... 2. *M. longipes.*

1. MILIUSA ROXBURGHIANA, Hook. fil. and Thoms. Fl. Ind. 150. A small tree ; young branches softly pubescent, ultimately glabrous, striate and pale. *Leaves* thinly coriaceous, oblong or oblong-lanceolate, shortly acuminate, the base rounded ; upper surface glabrous, the lower sparsely adpressed, pubescent to tomentose ; main nerves about 10 pairs, spread-ing, inter-arching ·15 in. from the base ; length 2·5 to 4 in., breadth ·85 to 1·4 in. ; petiole ·05 in., pubescent. *Pedicels* 1 to 3 together, axillary, slender, ·5 to 1·5 in. long, sometimes on a short peduncle ; bracteoles several, linear. *Flowers* diœcious or polygamous, about 5 in. long. *Sepals* and outer petals subequal, lanceolate or linear, rusty-tomentose. *Inner petals* ·5 to ·6 in. long, ovate or oblong-lanceolate, sub-acute, nerved, red. *Stamens* in male flower numerous, with obliquely truncate, broad apices. *Ovaries* (in female flower) oblong, glabrous ; style oblong, ovules 1 or 2. *Ripe carpels* ovoid or oblong, blunt, glabrous, granulate, ·25 to ·35 in. in diam. ; stalk ·4 in. long, slender. *Seeds* 1, rarely 2. Hook. fil. Fl. Br. Ind. I, 87 ; Kurz F. Flora Burma, I, 47. *M. Wallich-*

*iana*, H. f. and T. l. c. 149. *M. tristis*, Kurz F. Flora Burma, I, 47; *Uvaria dioica*, Roxb. Fl. Ind. ii. 659. *Phwanthus dioicus*, Kurz in Flora LIII. (1870) 274. *Guatteria globosa*, A. DC. Mem. Soc. Genev. V, 43; Wall. Cat. 6448. *Hyalostemma Roxburghiana*, Wall. Cat. 6434; Griff. Ic. Pl. Ind. Or. iv. t. 653.

Sikkim, Himalaya; Assam Hill ranges; Chittagong Hills: Burma; Singapore up to 4,000 feet.

Kurz's species *M. tristis*, (F. Flora Burma, I, 47) appears to be a form of this with larger leaves and flowers than usual. The only specimens of it extant are very poor and better material may shew it to be, as Kurz thought, a distinct species. According to M. Pierre, his Cambodian species *M. mollis* (Fl. Forest. Coch.-Chine, t. 40) is closely allied to *M. Roxburghiana*. The same author's species *M campanulata* (l. c. t. 41) is also allied to *M. Roxburghiana* and to *M. macrocarpa*.

2, MILIUSA LONGIPES, King, n. sp. A small tree 15 to 30 feet high : young branches dark-coloured; all parts glabrous except the edges of the sepals and outer petals. *Leaves* membranous, shining, oblong-oblanceolate, acuminate, the base sub-cuneate or rounded; main nerves about 12 pairs, spreading, faint : length 5·5 to 7 in., breadth 1·75 to 2·75 in., petiole ·1 to ·15 in. *Flowers* ·5 to ·65 in. long, axillary, solitary; pedicels slender, ·5 to ·75 in. long, (larger in fruit) with 3 or 4 lanceolate bracteoles at the base. *Sepals* and *outer petals* sub-equal, minute, ovate, sub-acute, the edges ciliate. *Inner petals* very much larger than the outer, ovate-oblong, veined, sub-acute, greenish-yellow, ·5 or ·6 in. long. *Stamens* about 18, compressed, short, often bent, the apiculus broad, shallow. *Ovaries* numerous, elongate, glabrous; *stigma* large, capitate, sessile. *Ripe carpels* numerous, globular-ovoid, blunt, glabrous, subgranular, ·25 to ·3 in. long; stalks ·75 to 1 in., slender. *Seeds* ovoid.

Perak : at low elevations, Scortechini, King's Collector.

This species approaches *M. macropoda*, Miq : but its leaves are more narrowed to the base and more acuminate.

### 23. ALPHONSEA, H. f. & T.

Lofty trees. *Leaves* more or less coriaceous, glabrous, shining. *Flowers* small or middle-sized, in leaf-opposed, rarely extra-axillary, peduncled fascicles; buds conical. *Sepals* 3, small, valvate. *Petals* 6, valvate in 2 series, often saccate at the base, larger than the sepals, equal or the inner rather smaller. *Torus* cylindric or hemispheric. *Stamens* indefinite, loosely packed; anther-cells dorsal, contiguous; connective apiculate. *Ovaries* 1 or more; style oblong or depressed; ovules 4–8, in 2 series on the ventral suture. *Carpels* sub-sessile or stalked.—Distrib. Species 9, all Indian or Malayan.—Baillon Hist. 215 unites this genus with *Bocagea*.

125

Leaves rusty-pubescent beneath at all stages ... 1. *A. Maingayi.*
Leaves glabrous on both surfaces (puberulous on
the lower in *A. elliptica*).
  Leaves more than 3 inches long.
    Buds conical ; ripe carpels ovoid or glo-
    bose.
      Leaves glabrous on the upper sur-
      face, puberulous on the lower
      when young, elliptic or ovate-
      elliptic ; main nerves 6 to 8 pairs 2. *A. elliptica.*
      Leaves quite glabrous, broadly ellip-
      tic, shortly acuminate : main ner-
      ves 7 to 8 pairs ... ... 3. *A. lucida.*
    Buds globose ; ripe carpels cylindric ... 4. *A. sub-indehiscens.*
  Leaves 3 inches long or less : ripe carpels
  cylindric ... ... ... 5. *A. cylindrica.*
  Of uncertain position (fruit unknown)... 6. *A. Curtisii.*

1. ALPHONSEA MAINGAYI, Hook. fil. and Thoms. Fl. Br. Ind. I, 90.
A tree : branches rusty-tomentose, ultimately dark-coloured and glabrous. *Leaves* coriaceous, elliptic-oblong or oblong-lanceolate, shortly, and
often obtusely, acuminate, the base rounded ; upper surface shining,
glabrous except the midrib, puberulous near the base ; lower surface
rusty, conspicuously reticulate, pubescent, the midrib tomentose ; main
nerves 8 or 9 pairs, oblique, inter-arching far from the edge ; length
5 to 7 in., breadth 1·5 to 2·7 in., petiole ·25 in. *Flowers* ·75 in. in diam.,
supra-axillary, solitary or in small racemes ; pedicels ·1 in. long, rusty-
tomentose, bracteole small. *Sepals* sub-orbicular, very small. *Petals*
ovate, pubescent outside, glabrous within, the outer recurved, the inner
smaller. *Stamens* with broad short filaments ; the anther-cells small,
diverging below. *Ovules* about 20. *Ripe carpels* ovoid, short-stalked,
2 in. long, by 1 in. in diam. *Seeds* many, smooth.
  Malacca, Maingay (Kew Distrib.) No. 98.
  2. ALPHONSEA ELLIPTICA, Hook. fil. and Thoms. Fl. Br. Ind. I, 90.
A tree ? Young branches rather stout, grey, glabrous. *Leaves* coriaceous, elliptic or ovate-elliptic, shortly and bluntly acuminate or acute,
the base abruptly cuneate; upper surface glabrous, shining ; the lower
reticulate, puberulous when young, glabrous when adult, slightly paler
than the upper ; main nerves 6 to 8 pairs, spreading, slightly prominent beneath ; length 3·5 to 5 in., breadth 1·25 to 1·75 in., petiole ·2 in.
*Flowers* ·8 in. in diam., axillary, solitary or 2 to 3, in short racemes ;
peduncles very short, multi-bracteate, pedicels ·25 to ·35 in. long, with
1 or 2 minute bracteoles. *Sepals* sub-orbicular, obtuse, recurved, con-
374

nate at the base. *Petals* adpressed-pubescent; the outer ovate-lanceolate, reflexed: the inner rather smaller. *Stamens* in several rows, apiculate. *Oraries* linear-oblong, pubescent; stigma sub-sessile, subcapitate. *Ovules* numerous, in two series. *Ripe carpels* unknown.

Malacca; Maingay (Kew Distrib.) No. 99.

3. ALPHONSEA LUCIDA, King, n. sp. A shrub 6 to 8 feet high: all parts glabrous except the flower; young branches slender, rather dark-coloured. *Leaves* thinly coriaceous, broadly elliptic, shortly, abruptly and rather obtusely acuminate, the base cuneate; under surface very minutely scaly; main nerves 7 or 8 pairs, oblique, curving, depressed on the upper, bold and prominent on the lower, surface; length 4·5 to 5·5 in., breadth 1·75 to 2·5 in.; petiole ·3 in., stout. *Flowers* extra-axillary, solitary or 2 or 3 in racemes: peduncle of raceme short, pedicels shorter than the peduncle, puberulous, ebracteolate, ·3 to ·4 in. long. *Sepals*, triangular-ovate, connate at the base, reflexed, puberulous outside, glabrous inside. *Petals* yellowish-white, subequal, oblong, oblique, tapering gradually to the sub-acute apex, the base broad, suddenly narrowed and slightly pouched, puberulous, ·5 in. long, the inner slightly smaller. *Stamens* in 3 rows; filament very short, connective with a short apiculus. *Oraries* 4 or 5, oblong, adpressed-pubescent; ovules many, in two rows: stigma sessile, sub-capitate. *Ripe carpels* unknown.

Perak: elevat. 500 feet. King's Collector, No. 5387.

4. ALPHONSEA SUB-DEHISCENS, King, n. sp. A shrub or small tree: young branches rather slender, puberulous at first but speedily becoming glabrous. *Leaves* thinly coriaceous, oblong-lanceolate to elliptic, shortly and rather bluntly acuminate, the base rounded or sub-cuneate; upper surface glabrous except the puberulous midrib, the lower reticulate, sparsely puberulous or glabrous; main nerves about 10 pairs, spreading, very faint; length 4 to 6 in., breadth 1·75 to 2·3 in.; petiole ·25 in. *Flowers* globular, scarcely opening, ·25 in. in diam., solitary or in pairs, slightly supra-axillary, on short pedicels, with several large sub-orbicular pubescent bracteoles. *Sepals* thick, fleshy, connate into a flat cup, ·3 in. in diam., with three broad obtuse, spreading lobes. *Petals* larger than the sepals, thick, hard and fleshy, valvate, orbicular, acute, concave, outside tawny-pubescent, inside glabrous except near the apex; the outer ·2 in. in diam., the inner row rather smaller than the outer. *Stamens* numerous; the apical process large, fleshy, conical, concealing the apices of the narrow, linear anther cells: torus conical. *Pistil* solitary, clavate, minutely puberulous, many-ovuled: stigma minute. *Ripe carpels* elongate-clavate, puberulous, 1 to 1·25 in. long, tapering into a stalk, ·25 to ·3 in. long. *Seeds* about 10.

Perak: King's Collector.

127

The dried fruits of this species sometimes open longitudinally by a sort of quasi-suture—hence the specific name.

5. ALPHONSEA CYLINDRICA, King, n. sp. A small tree 20 to 30 feet high; young branches with long, soft, pale brown pubescence, ultimately glabrous, cinereous, striate, *Leaves* thinly coriaceous, ovate-lanceolate, sometimes oblanceolate, shortly and bluntly acuminate; the base rounded or sub-cuneate, slightly oblique; upper surface glabrous, shining; the midrib pubescent, the lower dull sparsely pubescent on the midrib and nerves; main nerves 7 to 9 pairs, spreading, faint; length 2·5 to 3·5 in., breadth 1·1 in. to 1·5. in., petiole ·15 in. *Flowers* ·35 in. long, single or 2 or 3 from leaf-opposed or extra-axillary peduncles; peduncles ·15 to ·4 in. long, with deciduous, distichous, sub-orbicular bracts : pedicels 2 to ·35 in. long, pubescent, with 1 bracteole near the base. *Sepals* semi-orbicular, blunt, connate at the base, tomentose outside, glabrous within, reflexed. *Petals* subequal, oblong-ovoid, tapering from the sub-saccate base to the sub-acute apex, tomentose outside, pubescent minutely inside except a glabrous patch at the base, 4 in. long. *Stamens* in 3 rows with short, broad filaments : anthers ovate, the connective very slightly apiculate. *Ovaries* 3, oblong, densely pale yellowish sericeous, with many ovules in two rows : style short, stigma bifid, sub-capitate. *Ripe carpels* 1 or 2, elongate, terete, tapering to the apex, pubescent or puberulous, nearly 1 in. long and only ·2 in. in diam.

Perak : on Ulu Bubong, elevat. 400 to 600 feet. King's Collector, No. 10633.

A species resembling *A. sub-dehiscens* in its narrow cylindric fruit.

6. ALPHONSEA CURTISII, King, n. sp. A scandent shrub : young branches yellowish-pubescent, speedily becoming glabrous and dark-coloured. *Leaves* coriaceous, oblong-lanceolate, acute at base and apex; upper surface glabrous shining, the lower minutely, sparsely adpressed-puberulous or glabrous, darker than the upper when dry, minutely reticulate; main nerves about 12 to 15 pairs, sub-horizontal, very faint, inter-arching far from the edge; length 4 to 5·5 in., breadth 1·2 to 1·75 in., petiole ·2 in. *Peduncles* extra-axillary, 1- or 2-flowered; flowers about ·5 in long, conical in bud : pedicels about ·3 in. long, tawny-to-mentose; bracteoles 1 or 2, sub-orbicular. *Sepals* connate into a spreading cup, ·25 in. broad, tomentose outside and glabrous inside, with 3 broad, sub-acute teeth. *Petals* much larger than the sepals, fleshy, oblong, ovate, sub-acute; the outer tomentose on both surfaces, ·4 in. long; the inner narrower, glabrous inside. *Stamens* numerous, with short thick filaments : apical process of connective small, not concealing the short perfectly dorsal anther-cells. *Pistils* about 3, oblong, tomentose, many-ovuled : stigma large, broad, sessile. *Ripe carpels* unknown.

Penang : Curtis, No. 1410.
376

25. KINGSTONIA, H. f. and T.

Trees. *Flowers* fascicled on cauline tubercles, bisexual. *Sepals* 3, persistent, ovate, acute, the bases connate. *Petals* 6 ; outer valvate ; inner smaller, oblong, imbricate. *Stamens* about 12, the filament half the length of the extrorse anther-cells ; connective obliquely truncate. *Ovary* 1 ; stigma sessile, peltate, crenate : ovules few. *Ripe carpels* globose. *Seeds* several, 2-seriate.

1. KINGSTONIA NERVOSA, Hook. fil. and Thoms. Fl. Br. Ind. I, 93. Young branches rusty-pubescent. *Leaves* thinly coriaceous, oblong, rarely elliptic, shortly acuminate, the base rounded ; both surfaces glabrous, the nerves and midrib puberulous beneath when young ; main nerves 12 to 14 pairs, oblique, rather straight, depressed on the upper, strong and prominent on the lower, surface ; length 4 to 8 in., breadth 1·5 to 3·25 in. ; petiole ·4 in., puberulous. *Flowers* ·25 in. long, in extra-axillary fascicles of 8 or 10 : pedicels ·35 to ·5 in., slender, rusty-pubescent ; bracteoles orbicular, one close to the flower, the others basal and imbricate. *Sepals* ovate, connate at the base, spreading, pubescent outside, glabrous within. *Outer petals* oblong-elliptic, concave, obtuse, cinereous-tomentose outside, pubescent inside ; *inner petals* smaller, thick, concave and very tomentose, in the upper half. *Stamens* about 15, the connective with a broad truncate apex. *Ovary* one, oblong, angled, pubescent ; ovules 4 to 6. *Ripe carpels* broadly ovoid, blunt, minutely velvety pale-rusty tomentose, 1·5 in. long and 1·1 in. in diam.; pericarp woody. *Seeds* about 4, oblong, compressed, separated by dissepiments.

The species above described has only a single pistil. But there are, in the Calcutta Herbarium, specimens from Sumatra (Forbes No. 2713, in fruit but without flower) of what appears to be a second *Kingstonia*, and in these there are two carpels. If this plant proves to be a *Kingstonia*, the diagnosis of the genus will have to be amended.

Malacca : Maingay, (Kew Distrib.) No. 22. Perak : Wray, No. 3376.

26. MEZZETTIA, Beccari.

Trees. *Flowers* small, greenish, axillary or from the axils of fallen leaves, fasciculate or umbellate. *Sepals* 3, ovate, valvate. *Petals* 6, valvate, opening late and accrescent, flat, linear, the inner petals smaller than the outer. *Stamens* 9 to 12, in two rows ; anther-cells lateral, introrse ; connectives produced beyond their apices, truncate. *Torus* small, slightly concave, pubescent. *Ovary* solitary, ovate, glabrous, contracted into a very short style ; stigma sub-capitate ; ovules 2, superposed. *Carpel* coriaceous, elliptic or globose. *Seeds* 2, large, compressed. Five species, all Malayan.

129

1. MEZZETTIA LEPTOPODA, Oliver in Hook. Ic. Pl. t. 1560. A tree : young branches dark-coloured, glabrous, striate, rather stout. *Leaves* coriaceous, oblong or narrowly elliptic, obtusely acuminate or acute ; the base rounded or acute ; upper surface glabrous, shining ; the lower dull, obscurely reticulate ; main nerves 8 or 9 pairs, forming wide arches far from the margin, very faint ; length 2·5 to 4 in., breadth 1 to 1·75 in., petiole ·35 in. *Flowers* ·5 in. long, on long slender pedicels in axillary fascicles of 2 to 6 ; pedicels ·5 to ·75 in., pubescent : bracteols minute. *Sepals* broadly ovate, connate at the base, tomentose, reflexed. *Petals* tomentose, on both surfaces ; the outer linear, obtuse, ·2 in. long ; the inner shorter and broader. *Ovary* ovoid. *Ripe carpels* unknown, *Lonchomera leptopoda*, H. f. and Th. Fl. Br. Ind. I, 94.

Malacca : Maingay (Kew Distrib.) No. 102.

This plant is very imperfectly known. The carpels associated with Maingay's specimens do not agree with his description of them (Fl. Br. Ind. I, 94) and they are evidently those of some species of *Polyalthia*.

2. MEZZETTIA HERVEYANA, Oliver Hook. Ic. Plant. t. 1560. A tree ; young branches rather stout, nodose, glabrous. *Leaves* coriaceous, elliptic-oblong, shortly acuminate, the base cuneate, both surfaces glabrous, the upper shining ; main nerves about 10 pairs, spreading, inter-arching within the margin, faint ; length 2·5 to 3 in., breadth 1 to 1·25 in., petiole ·25 to ·35 in. *Flowers* ·4 in. long, rather crowded, in sessile axillary or extra-axillary fascicles of 3 to 8 : pedicels ·3 in. long, puberulous, ebracteolate. *Sepals* broadly ovate, obtuse, connate at the base, pubescent like the petals. *Outer petals* ovate-lanceolate, obtuse, flat, the inner smaller, broadly elliptic, obtuse, the tips incurved. *Anthers* sessile, obovate-quadrate, about 12. *Ovary* oblong, tapering into the style : ovules 2, superposed. *Ripe carpels* unknown.

Malacca : Hervey.

3. MEZZETTIA CURTISII, King n. sp. A tree, 30 to 40 feet high : young branches cinereous, rugose. *Leaves* thinly coriaceous, oblong-lanceolate or oblong, more or less acuminate, the base acute ; both surfaces glabrous ; the upper shining, the lower dull ; main nerves about 10 pairs, spreading, faint ; length 2·5 to 5 in., breadth ·5 to 1·5 in., petiole ·25 in. *Flowers* ·25 in. long, in crowded, sessile, axillary or extra-axillary fascicles of 5 to 10 ; pedicels slender, ebracteolate, scurfily pubescent, ·35 to ·6 in. long. *Sepals* semi-orbicular, with reflexed tips, connate and forming a spreading, shallow cup, densely and minutely tomentose. *Outer petals* ligulate, acute, tomentose like the sepals but with a glabrous patch at the base inside. *Inner petals* like the outer, but less acute and one-third shorter. *Stamens* about 12, short, about as broad as long, the connective very broad, truncate at the apex. *Ovary* solitary, broadly ovoid,

378

*From the Journal, Asiatic Society of Bengal, Vol. LXII, Part II, No. 2, 1893.*

tapering to the curved, truncate stigma, 2-ovuled. *Ripe carpels unknown.* Penang: on Government Hill at 1,200 feet; Curtis, No. 2266. A species with rather longer, thinner leaves than *M. Herveyana,* and a different calyx.

---

*Materials for a Flora of the Malayan Peninsula.—By* GEORGE KING, M. B., LL. D., F.R.S., C.I.E., *Supdt. of the Royal Botanic Garden, Calcutta.*

[Read June 7th].

No. 5.

ORDER XVI. DIPTEROCARPEÆ.

Resinous trees, rarely climbing shrubs. *Leaves* alternate, simple, quite entire, rarely sinuate-crenate, penni-nerved, the main nerves bold; stipules usually small and inconspicuous, sometimes larger and persistent, or fugitive, leaving an annular scar, (absent in *Ancistrocladus*).` *Flowers* in few- or many-flowered, axillary and terminal racemes or panicles. *Bracts* usually minute or 0, rarely larger and persistent. *Sepals* free, or cohering into a tube surrounding but free from, or more or less adnate to, the base of the ovary and fruit. *Petals* contorted, connate at the base, or free. *Stamens* ∞, 15, 10 or 5, hypogynous or sub-perigynous, free, connate, or adnate to the petals; filaments short, often dilated at the base; anthers 2-celled, the outer valves sometimes larger, connective often aristate or with an obtuse appendage. *Ovary* slightly immersed in the torus, usually 3- rarely 2- or 1-celled; style subulate or fleshy, entire or with 3 minute stigmatic lobes; ovules anatropous, 2 in each cell, pendulous or laterally affixed (solitary and erect in *Ancistrocladus*). *Fruit* usually nut-like, its pericarp leathery or woody, 1- rarely 2-seeded, surrounded by the variously accrescent calyx of which two or more sepals or lobes are usually developed into linear wings. *Seed* exalbuminous (albumen fleshy and ruminate in *Ancistrocladus*); cotyledons fleshy, equal or unequal, straight or more or less plaited and crumpled, sometimes lobed; radicle directed towards the hilum, usually included between the cotyledons.—DISTRIB. Confined (except a few Tropical African species) to Tropical Eastern Asia; genera about 18, species about 250.

Sect. I. EU-DIPTEROCARPEÆ. *Ovaries* 3-celled, each cell 2-ovuled : stigmas united, more or less 3-lobed : seeds usually exalbuminous the outer segments of the fruiting calyx usually enlarged : trees or erect shrubs, mostly stipulate.

Fruiting calyx with 2 or more of its segments

or sepals produced into long membranous,
reticulate, nerved wings much longer than
the fruit; pericarp leathery, (woody in some
sp. of *Shorea*).

Fruiting calyx with a distinct tube.

Calyx-tube quite free from the
fruit ...     ...     ...     1 *Dipterocarpus.*

Calyx-tube adherent to the fruit     2 *Anisoptera.*

Sepals united at the base only, the short
calyx-tube either quite free from the
fruit or slightly adherent to it, the
calyx-segments or sepals valvate or
nearly so.

Stamens with a single, long apical,
appendage from the connective     3 *Vatica.*

Stamens with 4 apical appendages from the anthers and 1
from the connective     ...     4 *Pentacme.*

Sepals free, imbricate.

The three outer sepals always, and
one or both of the inner two occasionally, winged in the fruit;
anthers with a short apical
appendage from the connective     5 *Shorea.*

The two outer sepals winged in
the fruit, the three inner not
longer than the fruit and closely embracing it; stamens with
a terminal appendage from the
connective longer than the
anther     ...     ...     6 *Hopea.*

Sepals of fruiting-calyx all enlarged but not
exceeding, or only slightly exceeding, the
fruit; pericarp leathery or woody.

Fruiting calyx embracing the fruit but
not adherent to it.

Sepals of fruiting calyx slightly
thickened.

Sepals of fruiting-calyx oblong, nearly equal, usually
shorter than the fruit, re-
flexed or erect     ...     7 *Retinodendron.*

Sepals of fruiting-calyx
rotund, unequal (the inner
two smaller), reflexed ... 8 *Isoptera.*
Sepals of fruiting calyx much
thickened and woody at the
base.
  Calyx forming a cup at the
  base of the fruit, but not
  adhering to it : pericarp
  woody ... ... 9 *Balanocarpus.*
  Calyx adherent to the fruit : pericarp
  thickly leathery ... ... 10 *Pachynocarpus.*

Sect. II. ANCISTROCLADEÆ. *Ovary* 1-celled with
a single ovule ; stigmas 3, distinct : *Seeds*
with copious ruminate albumen. Exstipulate
climbers. ... ... ... ... 11 *Ancistrocladus.*

## 1. DIPTEROCARPUS, Gærtn. f.

Lofty trees, stellately pubescent or more or less clothed with
fascicled hairs. *Leaves* coriaceous, entire or sinuate-crenate; lateral nerves
connected by marginal loops and transverse reticulations ; stipules large,
valvate, enclosing the terminal bud, finally caducous and leaving an
annular scar. *Flowers* large, white or reddish. *Calyx-tube* free. *Petals*
usually pubescent externally, especially on the outer margin. *Stamens*
∞ ; anthers linear, equivalved, acuminate. *Ovary* 3-celled ; style filiform ;
ovules 2 in each cell. *Fruit* nut-like, 1-seeded, enclosed in the accrescent calyx-tube, free ; accrescent calyx-lobes 2, erect. *Seed* adnate to
the base of the pericarp ; cotyledons large, thick, unequal ; radicle
inconspicuous.—DISTRIB. Tropical E. Asia ; species about 60.

Ripe fruit sphæroidal or ellipsoidal, neither angled nor winged.
  Young branches, petioles, under surfaces of the midribs, and
  nerves of the leaves covered with coarse stiff fasciculate hairs.
    Fruit glabrous ... ... ... 1. *D. crinitus.*
    „ stellate-pubescent ... ... 2. *D. Scortechinii.*
  Young branches deciduously pubescent.
    Leaves with 12 or more pairs of nerves.
      Leaves oblong-elliptic, their under sur-
      faces sparsely stellate-pubescent ... 3. *D. Skinneri.*
      Leaves elliptic or ovate-elliptic, their
      under surfaces puberulous or quite
      glabrous ... ... ... 4 *D. turbinatus.*
381

Leaves with 8 to 10 pairs of nerves.

All parts quite glabrous ...          ... 5. *D. Kerrii.*

Ripe fruit with 5 angular tuberosities on its
upper portion     ...        ...        ... 6. *D. cornutus.*

Ripe fruit 5-angled :

Calyx-tube glabrous ; leaves 2·5 to 3·25 in.
long        ...        ...        ... 7. *D. fagineus.*

Calyx-tube densely stellate-tomentose ;
leaves 6 to 8 in. long     ...        ... 8. *D. oblongifolius.*

Ripe fruit with its 5 angles produced into wings :

Leaves glabrous :

Young branches at first scurfy-puberulous,
ultimately quite glabrous : buds ovoid,
minutely pale canescent        ... 9. *D. grandiflorus.*

Young branches as in the last, but with
conspicuous tawny-tomentose, oblique
annuli ; buds cylindric, hoary-canes-
cent        ...        ...        ... 10. *D. Kunstleri.*

Young branches minutely tawny-pubescent, not annulated and never gla-
brous ; buds ovoid, densely sericeous 11. *D. Griffithii.*

Leaves minutely stellate-pubescent on the
lower surface :

Flowers about 1 in. long ; leaves with
rounded or sub-cordate bases ; young
branches very stout, with ovoid buds :
the accrescent lobes of the calyx
1·5 in. broad        ...        ... 12. *D. incanus.*

Flowers 1·5 in. long ; leaves with rounded
or cuneate, not sub-cordate, bases :
young branches moderately stout with
cylindric buds : accrescent calyx-lobes
·7 to ·8 in. broad        ...        ... 13. *D. alatus.*

1. DIPTEROCARPUS CRINITUS, Dyer in Hook. fil. Fl. Br. Ind. I. 296.
A tree 90 to 150 feet high : young branches, petioles, under surface of
midrib and nerves, pedicels and outer surface of bracts of inflorescence
clothed with stiff yellowish-brown fascicled hairs. *Leaves* very coriaceous, ovate or more usually obovate, acute, the base rounded or sub-
acute ; the edge entire, fringed with fascicled hairs, recurved (at least
when dry) ; both surfaces sparsely hispid when young, glabrescent when
old ; main nerves 12 to 18 pairs, spreading, rather straight, very
prominent on the lower, depressed on the upper, surface ; length 3 to
5 in., breadth 1·75 to 2·75 in., petiole 1 to 1·25 in. *Racemes* about 6-

flowered. *Flowers* nearly 2 in. long. *Calyx* glaucous, glabrous. *Petals* puberulous, linear, blunt. *Stamens* 15. *Fruit* (immature) ellipsoid, wingless, glaucous, smooth ; the enlarged calyx-lobes linear-oblong, blunt, 3-nerved, inconspicuously reticulate, shining, 3·5 in. long and ·6 to ·8 in. broad. Dyer in Journ. Bot. 1874, p. 103. *D. hirtus*, Vesque, Comptes-Rendus, 1874, 78, p. 627 ; Journ. Bot. 1874, p. 151 ; Dyer l. c. 154.

Malacca ; Maingay (Kew Distrib.) No. 196.

Perak : Scortechini, No. 1955. DISTRIB. Borneo : (fide Dyer), Beccari, 779, 1883.

Burck (Ann. Jard. Bot. Buitenzorg, Vol. 6, p 196) reduces this to *D. Tamparan*, Korth. Korthals however describes the fruit of that species as having accrescent calyx-lobes 13 inches long by 3 broad.

2. DIPTEROCARPUS SCORTECHINII, King, n. sp. A large tree : young branches rather stout, densely clothed, (as are the short cylindric buds, the petioles and racemes) with large tufts of coarse, brownish, shining hairs. *Leaves* coriaceous, elliptic-ovate, or sometimes elliptic-sub-ovate, sub-entire, abruptly and shortly acuminate, slightly narrowed to the rounded base ; upper surface glabrous or glabrescent, the nerves sparsely stellate-pubescent, the midrib tomentose ; under surface sparsely stellate-pubescent, the nerves (and especially the midrib) with long silky hairs intermixed : main nerves 16 to 18 pairs, straight, oblique, very prominent beneath : length 6 to 7·5 in., breadth 3 to 3·5 in , petiole 1 to 1·2 in. *Racemes* few-flowered, short. *Fruit* (? immature) ovoid, contracted under the mouth, glaucous, stellate-pubescent, ·75 in. long and ·5 in. in diam ; accrescent calyx-lobes linear-oblong, reticulate, slightly narrowed in the lower half, the apex obtuse, obscurely 3-nerved (the middle nerve bold, the two lateral faint), 4 to 5 in. long and ·8 to 1 in. broad.

Perak : Scortechini, No. 1813.

This is closely allied to *D. crinitus*, Dyer, to which Scortechini doubtfully referred it. It differs from *D. crinitus* in its larger leaves and stellate-pubescent fruit. It has also a different time of flowering ; for, as Scortechini remarks in his field notes, this is in immature fruit in the beginning of March, while *D. crinitus* does not come into flower until the end of April.

3. DIPTEROCARPUS SKINNERI, King, n. sp. A tall tree ; young branches thin, deciduously tawny-pubescent. *Buds* cylindric, narrow, golden-sericeous. *Leaves* oblong-elliptic, narrowed in the upper half or third to the acute or shortly acuminate apex, slightly narrowed to the rounded base, upper surface glabrous or sparsely adpressed-pubescent, the midrib tomentose, the lower sparsely stellate-pubescent, the midrib and 16 to 19 pairs of straight oblique nerves adpressed-sericeous ; nerves prominent on the lower, faint on the upper, surface when dry :

length 5 to 8 in., breadth 2·25 to 3 in. ; petiole ·7 to ·9 in., tomentose. *Racemes* simple, short, 2- or 3-flowered, pubescent. *Flowers* 2·5 in. long. *Calyx* with narrowly campanulate tube, covered outside with minute, pale, stellate tomentum. *Petals* linear-oblong, blunt, more or less pubescent outside. *Fruit* (? immature) globular-ovoid, glabrous, ·65 in. in diam. : accrescent calyx-lobes glabrous, reticulate, linear, blunt, contracted at the very base, nearly 5 in. long and about ·75 in. broad.

Penang; at the back of West Hill, at an elevation of 1,000 feet. Curtis No. 1403.

A very distinct species known only by Mr. Curtis' scanty specimens. I have named it in honour of Mr. Skinner, Resident Councellor of Penang.

4. DIPTEROCARPUS TURBINATUS, Gaertn. f. Fruct. III. 51, t. 188. A tree 80 to 100 feet high : young shoots rather slender, at first minutely velvety, pale grey, afterwards glabrous : buds cylindric, softly pale pubescent. *Leaves* thinly coriaceous, elliptic or ovate-elliptic, acute or shortly acuminate, the base rounded or sub-cordate, the edges slightly undulate, sometimes sub-crenate ; both surfaces glabrous, or the lower puberulous especially on the midrib and nerves : main nerves 12 to 18 pairs, straight, oblique, prominent on the lower surface ; length 4·5 to 11 in., breadth 2·5 to 5·25 in. ; petiole 1 to 1·5 in., glabrous or pubescent : stipules tawny-velvety in the lower part but pubescent towards the apex. *Racemes* 3- to 5-flowered. *Flowers* 1·25 to 1·5 in. long. *Calyx*-tube obconic, glabrous, smooth, not winged. *Petals* linear-oblong, obtuse, more or less canescent. *Fruit* ellipsoid-ovoid, tapering to each end when young : globular when ripe and ·75 in. in diam., with neither wings nor ridges ; the two accrescent calyx-lobes glabrous, conspicuously reticulate, obscurely 3-nerved, oblong-lanceolate, obtuse, 4 to 4·5 in. long and 1·25 in. broad ; the three small lobes of the calyx deltoid, very short. Roxb. Hort. Beng. 42 ; Fl. Ind. II. 612; Corom. Plants III. 10 t. 213. Ham. in Mem. Wern. Soc. VI. 300 : Wall Cat. 952; A. DC. Prod. XVI. 2, 607 ; W. and Arn. Prod. 85 ; Dyer in Hook. fil. Fl. Br. Ind. I, 295 : Journ. Bot. 1874, p. 102 t. 143, fig. 13: Kurz. For. Fl. Burm. I. 114. *D. laevis*, Ham. l. c. 299. ; A. DC. l. c. 607. W. and A. Prod. 85 : Kurz, l. c. 114. ?*D. indicus*, Bedd. Forest. Rep. 1864-5, 17 cum tab.; Flora Sylvat. t. 94.

Assam, Cachar, Chittagong, Burmah, S. India.

VAR. *andamanica* : enlarged calyx-lobes linear-oblong, not oblanceolate, ·75 in. broad ; leaves broadly ovate, sub-cuneate at the base.

South Andaman : common.

Following Dyer, I have included under this the plant named *D. laevis* by Buchanan Hamilton in the Memoirs of the Wernerian Society,

384

93

Vol. VI. p. 299. Hamilton distinguishes his species *D. laevis* by its flattened branchlets, and perfectly glabrous leaves and petioles, while *D. tuberculatus* Gaertn. has terete branches and pubescent leaves and petioles. The former (called *Dulia Garjan*, by the natives of Chittagong) yields, he says, no wood-oil; while the latter (called *Telia Garjan*) does. The materials before me do not enable me to differentiate the two as species. Moreover, specimens sent to me by Dr. E. Thurston, Reporter on Economic Products to the Government of India, (and which had been collected by the Forest Officer of Chittagong under the vernacular names *Dulia* and *Telia Garjan*) appear exactly alike. Careful investigation in the field may however prove that there is some better basis for Hamilton's view than the trifling differences which he has noted in the outline of the branchlets and the pubescence of the leaves. I am not at all satisfied that the Southern Indian tree named *D. indicus* by Beddome is rightly reduced here. Better Herbarium specimens than any which I have seen, and investigation in the field, are I think required to settle this point also.

5. DIPTEROCARPUS KERRII, King, n. sp. A tall tree; all parts, except the petals, glabrous; young branches thin, slightly flattened at the tips, not annular. *Buds* narrow, cylindric. *Leaves* coriaceous, ovate-elliptic, acute or very shortly and bluntly acuminate, the edges undulate, the base cuneate; main nerves 8 to 11 pairs, oblique, straight, bold and shining on the lower surface; length 3 to 4 in., breadth 2 to 2·5 in., petiole ·9 to 1·1 in. *Panicles* short, spreading, few-flowered. *Flowers* 1·5 in. long. *Calyx-tube* glaucous. *Petals* linear-oblong, obtuse, more or less pubescent or tomentose towards their middle externally. *Fruit* turbinate, smooth, 1 to 1·15 in. in diam.; accrescent calyx-lobes linear-oblong, blunt, reticulate, 3-nerved, 4·5 to 5 in. long, and 1·25 to 1·5 in. broad : minor lobes very short, broad, rounded.

Malacca ; Maingay (Kew Distrib.) No. 199, Griffith 727, Derry 1032. Pangkore ; on Gunong Yunggal, Curtis No. 1561.

Mr. Curtis describes this as a very large tree yielding an oil. It resembles *D. Hasseltii*, Bl., but has much smaller leaves.

I have named this species in honour of Dr. Kerr, an enthusiastic Botanist much interested in the Malayan Flora. Closely allied to this, and perhaps identical with it, is the tree represented by Mr. Curtis' specimen (Waterfall, Penang) No. 1653. The young wood of the latter is however paler than that of *D. Kerrii* from Pangkore and Malacca, and the leaves are puberulous, not glabrous, beneath. I have seen no flowers of it.

6. DIPTEROCARPUS CORNUTUS, Dyer in Hook. fil. Fl. Br. Ind. I, 296. A tree 50 to 70 feet high: young branches stout, compressed, minutely

385

rufous-tomentose with a few scattered longer hairs. *Leaves* large, coriaceous, oblong, blunt at each end, the edges undulate or obscurely sinuate-crenate: upper surface glabrous, the midrib and nerves pale when dry : under surface densely covered with minute, pale, stellate tomentum : main nerves 16 to 20 pairs, prominent, spreading, straight, the transverse veins rather distinct : length 9 to 14 in., breadth 5 to 8 in., petiole 2 to 3 in. ; stipules rufous-sericeous, the hairs fascicled. *Racemes* 7- or 8-flowered. *Flowers* 1·75 in. long. *Calyx-tube* 5-winged, canescent, the short lobes very obtuse. *Petals* oblong or sub-spathulate, stellate-canescent. *Fruit* about 1 in. long, sub-globular, with 5 thick short wings in its upper half; enlarged calyx-lobes linear, obtuse, 5 or 6 in. long and 1·25 to 1·75 in. broad, shining, boldly 3-nerved, reticulate. Dyer in Journ. Bot. 1874, p. 103, t. 143. fig. 15. *Parinarium dilleni-folium*, R. Br. Wall. Cat. No. 7520. *Petrocarya dillenifolia*, Steud. Nomencl. II, 309.

Singapore: Wallich. Malacca: Maingay (Kew Distrib.) No. 197. Penang : Curtis No. 1402. Perak : Wray, No. 4160.

It was Sir Joseph Hooker who first pointed out that the Walli-chian plant No. 7520, issued as *Parinarium*, belongs really to this species.

7. DIPTEROCARPUS FAGINEUS, Vesque in Comptes-Rendus, tome 78, p. 626: Journ. Bot. for 1874, p. 149. A tree 40 to 80 feet high : young branches slender, at first minutely pulverulent tawny-pubescent, ulti-mately glabrescent or glabrous and dark-coloured, the buds cylindric. *Leaves* coriaceous, elliptic-ovate to elliptic-lanceolate, acute, the edges entire or sub-undulate-crenulate, the base cuneate, both surfaces puberu-lous especially on the midrib and nerves ; main nerves 10 to 13 pairs, straight, oblique, prominent on the sub-glaucous lower surface ; length 2·5 to 3·25 in., breadth 1·3 to 1·75. *Racemes* slender, 1- to 4-flowered. *Flowers* about 1·25 in. long. *Calyx-tube* campanulate, not constricted at the mouth, 5-angled. *Ripe fruit* ellipsoid, tapering more at the base than at the apex, 5-angled, glaucous, 1 in. long: accrescent calyx-lobes linear-oblong, obtuse, contracted at the base, 3-nerved, 2·5 to 3 in. long and about ·75 in. broad. *D. prismaticus*, Dyer Journ. Bot. 1874. pp. 104, 152. t. 144 fig. 17. *Dipterocarpus*, sp. Hook. fil. in Linn. Trans. XXIII, 161.

Perak : King's Collector No. 3527, Scortechini. Penang ; Curtis No. 1401.

*D. fagineus*, Vesque, has been collected hitherto only in Borneo (Beccari No. 3008 and Motley No. 143,) and the leaves are described by Dyer as being papyraceous in texture and having about 8 pairs of lateral nerves. The leaves of the Perak tree which I now refer to this species, are coriaceous and have 10 to 13 pairs of nerves. The Perak plant may therefore belong to a distinct, but closely allied, species. Curtis' Penang specimens (No. 1401) are quite glabrous in all parts except the petals.

8. DIPTEROCARPUS OBLONGIFOLIUS, Blume, Mus. Bot. Lugd. Bat. II, 36 A tall tree: young branches glabrous, dark-coloured, sparsely lenticellate; buds cylindric. *Leaves* coriaceous, oblong or elliptic-oblong, shortly and bluntly acuminate, the edges sub-undulate, the base cuneate; both surfaces shining, glabrous, the midrib and 13 to 16 pairs of straight bold nerves with a few stellate hairs along their sides: length 6 to 8 in., breadth 2 to 2·75 in., petiole ·9 to 1·1 in. *Racemes* slightly supra-axillary, densely tawny-tomentose, bifurcating, each branch with 3 to 5 flowers and several linear membranous decid-uous bracts. *Flowers* about 2·5 in. long. *Calyx-tube* fusiform, slightly contracted at the mouth, 1 in. long, boldly 5-angled, densely stellate tawny-tomentose as are the 3 minor calyx lobes; the 2 larger linear-oblanceolate lobes sparsely stellate-pubescent, boldly 1-nerved and with 2 obscure lateral nerves. *Ripe fruit* unknown. Miq., Fl. Ind. Bat. I. pt. 2, p. 498; A.DC. Prod. XXI. 2, 614; Dyer in Journ. Bot. 1874, 105. *D. stenopterus*, Vesque, Comptes-Rendus, tome 78, p. 625; Journ. Bot. 1874, p. 150.

Perak, Scortechini. DISTRIB. Borneo, Sumatra.

Except as regards inflorescence, the Perak specimens of this are practically glabrous. In Bornean specimens, however, the young parts, buds and petioles are fusco-tomentose. (Dyer l. c.)

9. DIPTEROCARPUS GRANDIFLORUS, Blanco, Fl. Filipp. Ed. 2, 314. A tree 80 to 120 feet high: young branches rather stout, sub-compressed, at first hoary-puberulous, but finally quite glabrous, nearly black when dry; leaf-buds shortly ovoid, minutely pale-canescent. *Leaves* coriaceous, ovate-elliptic, shortly acuminate; the base broad, rounded or sub-truncate, sub-cordate; the edges entire or obscurely undulate-crenate, both sur-faces glabrous; main nerves 14 to 16 pairs, spreading, rather straight, prominent on the lower, obsolete on the upper, surface; length 6 to 9 in., breadth 3·5 to 5 in.; petiole 2 to 3 in. long, glabrous. *Racemes* about 4-flowered. *Flowers* articulated to the rachis, 2 in. long. *Calyx-tube* 5-winged from base to apex. *Petals* linear-oblong. *Fruit* oblong, 2·5 in. long, wings stout, ·5 in. or more in width; the 2 accrescent lobes of the calyx oblong, obtuse, glabrous, reticulate, 3-nerved, the mesial nerve the longest and most distinct, 7 to 9 in. long and 1·5 to 2 in. broad, the smaller calyx lobes sub-orbicular. A.DC. Prod. XVI., 2 p. 612; Dyer in Journ. Bot. 1874, p. 106, t. 145, fig 19; Burck in Ann. du Jard. Bot. Buitenzorg, vol 6, 201. *D. Blancoi*, Bl., Mus. Lugd. Bat. II.

35. *D. Motleyanus*, Hook. fil. in Trans. Linn. Soc. XXIII. 159. A.DC. in DC. Prod. XVI., pt. 2, 611. *D. pterygocalyx*, Scheff. Obs. Phyt. II. 35; Dyer in Hook. fil. Fl. Br. Ind. I, 298. *Mocanera grandiflora*, Blanco, Fl. Filipp. Ed. I, 451. *Anisoptera?* Turcz. in Bull. Soc. Nat. Mosc. 1858, I, 233.

Malacca: Maingay (Kew Distrib.) No. 198. Penang: Curtis 424. Perak: Scortechini 152 b. DISTRIB. Bangka, Teysmann. (?) Philippines.

The late Father Scortechini's field notes contain the following account of the flower: "The petals of this are red inside in the middle, but pale towards the margins; the stamens are numerous, 2-seriate, united in a ring by their enlarged bases, falling off together: staminodes many, short, adpressed to the ovary. Ovary pubescent, scaly towards the base. Fruiting-calyx reddish." The species comes near *D. Griffithii*: but is distinguished from it by the characters which I have noted under that species. Flowers of *D. Griffithii* are, however, wanting for comparison.

10. DIPTEROCARPUS KUNSTLERI, King, n. sp. A tree 80 to 120 feet high; young branches flattened, at first sparsely covered with minute scurfy deciduous pubescence, ultimately glabrous, but always with oblique tawny-tomentose annuli. *Buds* narrowly cylindric, hoary-canescent. *Leaves* elliptic or sub-rotund-elliptic, very shortly acuminate, the base rounded or sub-cuneate, the edges undulate or sub-crenate, both surfaces glabrous: main nerves 16 to 18 pairs, oblique, straight, prominent on the lower surface: length 7·5 to 11 in., breadth 4·5 to 7 in., petiole 1·5 to 2 in. *Racemes* 6 to 8 in. long, often bifid, 4- to 6-flowered, glabrous. *Flowers* 2·5 to 3 in. long, glaucous. *Calyx-tube* narrowly obconic, 5-winged, glaucous. *Petals* linear, obtuse, glaucous. *Fruit* sub-globular, an inch or more long, with 5 wings about ·25 in. wide: accrescent calyx-lobes oblong, obtuse, slightly narrowed towards the base, glabrous, reticulate, 3-nerved, 6 or 7 in. long and about 1·25 in. broad.

Perak: King's Collector, Nos. 3638, 3798, 7508 and 7606.

Allied to *D. grandiflorus*; but with larger leaves, smaller fruit and different buds. Allied also to *D. Griffithii* but with smaller fruit and different buds. This species has leaves like *D. trinervis* Bl. and *D. retusus* Bl., but differs from these in having winged fruit: it also resembles *D. Dyeri*, Pierre, which, however, has longer leaves with hairy petioles and more narrowly winged fruit.

11. DIPTEROCARPUS GRIFFITHII, Miq. Ann. Mus. Lugd. Bat. I, 213. A tree 100 to 125 feet high: young branches stout, sub-compressed, minutely tawny-canescent; the leaf buds ovoid, densely covered with

yellowish-brown shining hair. *Leaves* coriaceous, broadly ovate, usually slightly narrowed to the rounded base, but sometimes the base truncately sub-cordate, the apex acute or shortly acuminate, both surfaces glabrous, the upper shining; main nerves 12 to 14 pairs, spreading, straight, slightly prominent on the lower surface : length 5 to 11 in., breadth 3 to 5·5 in., petiole 2·25 to 3·5 in. *Racemes* 3- or 4-flowered. *Flowers* 1·5 in. long. *Calyx* ob-conic, sub-glabrous, 5-winged. *Fruit* oblong, 2·5 in. long, the wings extending from base to apex, stout, ·5 in. or more broad : accrescent lobes of calyx oblong, obtuse, glabrous, reticulate, boldly 3-nerved, 5 to 7 in. long and about 1·75 in. broad. A. DC. in DC. Prod. XVI, Pt. 2, 611 ; Dyer in Hook. fil. Fl. Br. Ind. I, 299 : Journ. Bot. 1874, 107. Kurz For. Flora Burm. I, 116. *D. grandiflorus* Griff. Notul. IV, 515 (not of Blanco).

S. Andaman : Kurz, King's Collector.

This closely resembles *D. grandiflorus*, Blanco, but the two may be readily distinguished by their young branches and leaf-buds. The young branches of this species are pale canescent and its leaf-buds broad and golden sericeous ; while the branchlets of *D. grandiflorus* are quite glabrous and dark-coloured and the buds are narrow and pale canescent.

12. DIPTEROCARPUS INCANUS, Roxb. Hort. Beng. 42 ; Fl. Ind. II. 614. A tall tree : young shoots terete, stout, densely but minutely tawny-tomentose ; the buds short, ovoid, thick, with longer tomentum than the branchlets. *Leaves* coriaceous, broadly ovate, acute or sub-acute, the base rounded or sub-cordate, the edges undulate ; upper surface glabrous, the midrib alone slightly pubescent : under surface uniformly pale, shortly but softly stellate-pubescent, the midrib and nerves tomentose. main nerves 12 to 15 pairs, oblique, straight, prominent on the lower surface ; length 5 to 8 in., breadth 2·5 to 4·75 in.; petiole ·8 to 1·25 in., pubescent. *Flowers* about 1 in. long, usually in racemes but occasionally in short 7- or 8-flowered panicles. *Calyx-tube* ob-conic, 5-winged, minutely tomentose. *Petals* oblong, obtuse. *Fruit* sub-globose, about 1 in. in diam., 5-winged from base to apex ; the wings thin, from ·25 to ·5 in. broad ; the 2 accrescent lobes of the calyx narrowly oblong, obtuse, glabrous, much reticulate, 3-nerved in the lower half, when mature 5·5 in. long and nearly 1·5 in. broad ; the 3 minor lobes sub-orbicular. Wight & Arn. Prod. 84 ; A. DC. Prod. XVI. 2, 611 ; Dyer in Hook. fil. Fl. Br. Ind. I, 298; Journ. Bot. 1874, p. 106.

S. Andaman : common. DISTRIB. Burmah, Kurz, Herb. No. 2109 (in part).

The plant here described under the name *D. incanus* closely resembles *D. alatus*, Roxb.; but its flowers are shorter, the leaves are more broadly ovate, and have rounded or cordate, not cuneate, bases, while the pubescence of the lower surface is paler and more uniform and the young branchlets and leaf-buds are stouter. Moreover the accrescent lobes of the calyx are longer and nearly twice as broad : the 5 wings of the calyx-tube are also broader. Roxburgh's description of his species *D. incanus* is very brief ; he left no drawing of it at Calcutta ; and no authentic specimens of his own naming appear to exist. It is therefore impossible to decide with absolute certainty what Roxburgh's *D. incanus* is. At Kew Mr. Dyer accepts Kurz's Pegu specimen No. 2109 as belonging to it, and the specimens recently brought from the S. Andaman by my collectors agree with that number of Kurz's.

13. DIPTEROCARPUS ALATUS, Roxb. Hort. Beng. 42 ; Fl. Ind. II 614. A tree 80 to 125 feet high : young branches terete, rather stout, softly and minutely pubescent; the buds narrow, rufous-sericeous. *Leaves* coriaceous, ovate-elliptic, the apex acute, the base cuneate, the edges undulate : upper surface glabrous except the minutely tomentose nerves and midrib : lower sparsely and minutely stellate-pubescent, the 10 to 14 pairs of oblique rather straight prominent main nerves densely tomentose ; length 5 to 8 in., breadth 2·75 to 4·5 in. ; petiole 1 to 1·5 in , pubescent : stipules sericeous-pubescent. *Panicles* 6- or 7-flowered. *Flowers* about 1·5 in. long. *Calyx-tube* ob-conic, 5-winged, stellate-pubescent, as are the linear-oblong petals. *Fruit* globose, 1 in. in diam., puberulous, 5-winged from base to apex ; the wings glabrous, thin and about ·5 in broad ; the 2 accrescent lobes of the calyx linear-oblong, obtuse, glabrous, much reticulate, 3-nerved in the lower half, 4·5 in. long and ·7 or ·8 in. broad : the 3 unenlarged lobes obtuse. Wall. Cat. 953 : A. DC. Prod. XVI. 2, 611 *in part :* Dyer in Hook. fil. Fl. Br. Ind. I, 298 ; Journ. Bot, 1874, p. 106 (excl. syn. *D. costatus*, Gaertn.) Kurz For. Flora Burm. I. 116 ; Pierre Flore Forest. Coch-Chine, t. 212. *Oleoxylon balsamiferum* Wall. Cat. p. 157.

Burmah : Wallich, Brandis, Helfer No. 730, Kurz. Andamans ?

Gærtner's figure and description of his *D. costatus* are confined to the fruit only. The former is that of a *Dipterocarpus* with the elongated calyx-lobes of *D. alatus*, Roxb., but with the 5 wings on the tube of the calyx very narrow, whereas those of Roxburgh's *D. alatus* are very broad. Dyer (F. B. I. i, 298) expresses his belief that Gaertner's figure is a bad representation of *D. alatus*, Roxb., and he reduces Gaertner's *D. costatus* to Roxburgh's *D. alatus*. M. De Candolle, on the other hand, retains *D. costatus*, Gaertn. as a good species and in this he is followed by Kurz ; but Messrs. Dyer and De Candolle agree that the *D. costatus* described by Roxburgh is a different plant from Gaertner's. For Mr. Dyer it is still a doubtful species; while M. De Candolle reduces it to *D. angustifolius* W. & A., which for Dyer is in its turn a doubtful species. A careful examination of the material now collected at Calcutta and at Kew leads me to believe that *D. costatus*, Gaertn., is a perfectly good species, and that the best character to distinguish it from Roxburgh's *D. alatus* is the narrowness of the wings of the calyx-tube. Specimens collected in Burmah by Kurz (No. 113 of his Herbm.) and by Brandis, have fruits exactly like that figured by Gaertner. Moreover I see no reason for thinking that the tree described by Roxburgh (Fl. Ind. II; 614) as *D. costatus*, Gaertn., is anything else than Gaertner's plant. Mr. Dyer (Journ. Bot. 1874, p. 153) expresses the opinion that *D. Lemeslei*, Vesque—a species collected on the island of Pulo Condor off the Cambodian coast—is reducible to *D. alatus*, Roxb.

It is very doubtful whether *D. alatus*, Roxb., occurs in the Andamans. I have seen no specimens of it from these islands, and I give it as an Andaman plant on the authority of the "Flora of British India."

Besides the preceding, there are various other species of *Dipterocarpus* in the Calcutta Herbarium from localities within the British Malayan region which, for want of sufficient materials, I am unable to describe. Chief amongst these are :—

(1) Curtis No. 1560 from Penang, a species with winged calyx-tube.

(2) A species from Perak, represented in Scortechini's collection (without number) by fruits resembling those of *D. Lowii* H., f., *D. intricatus*, Dyer, and *D. lamellatus*, Hook. fil.

(3) A species from the Andamans with leaves resembling those of *D. Griffithii*, Miq., but with globular fruit which has neither angles nor wings on the calyx-tube. This possibly may be a form of *D. pilosus*, Roxb.

(4) A Perak species (Herb. Scortechini mixed with No. 1478) represented by fruits something like those of *D. fagineus*, Vesque, but with the calyx-tube winged, not angled.

(5) A Perak species represented by leaf-twigs and loose fruit of a species resembling both *D. fagineus*, Vesque, and *D. gracilis*, Bl., but differing from both.

(6) A species from Perak (Wray No. 4031) having leaves like *D. Griffithii*, Miq., but with shorter petioles, and having also fruit rather like *D. Griffithii*, but the calyx-tube with narrower wings, and the minor calyx-lobes smaller.

2: ANISOPTERA, Korth.

Resinous trees. *Leaves* coriaceous, entire, feather-veined and finely reticulate ; stipules small, fugacious or inconspicuous. *Flowers* in lax terminal panicles *Calyx-tube* very short, adnate to the base of the ovary ; the segments imbricate, then subvalvate. *Stamens* ∞ ; anthers ovoid with a long subulate connective, outer valves larger. *Ovary* 3- (rarely 4- 5-) celled ; style fleshy, ovoid or oblong, with an attenuate 3–5-fid apex ; ovules 2 in each cell. *Fruit* adnate to the calyx-tube, indehiscent, 1-seeded, crowned by the accrescent calyx-segments, of which 2 form linear-oblong lobes. *Cotyledons* fleshy, unequal ; radicle superior. —DISTRIB. Malay Peninsula and Archipelago to New Guinea. Species about 6.

1. ANISOPTERA CURTISII, Dyer MSS. A tree 80 to 120 feet high : young branches slender, minutely scurfy-tomentose. *Leaves* oblong, tapering to both ends, the apex sub-acute or acute, the base narrowed but rounded ; the upper surface glabrous, shining, the lower densely ochraceous-lepidote and sparsely stellate-pubescent ; main nerves 18 to 20 pairs, spreading : length 2 to 3·5 in., breadth ·75 to 1·25 in., petiole ·5 to ·75 in. Accrescent calyx-lobes 3·5 to 4·5 in. long, linear-spathulate, shining, 3-nerved : the transverse veins bold and numerous.

Penang : Curtis. Perak : King's Collectors.

*Var. latifolia :* leaves broadly elliptic, blunt, the bases rounded but narrowed.

Penang : Curtis, No. 1400.

The vernacular name of this in Penang is *Ringkong.*

3. VATICA, Linn.

Large or moderately sized resinous trees. *Leaves* coriaceous, entire, feather-veined and finely reticulate ; stipules small, fugacious or inconspicuous. *Flowers* in axillary and terminal panicles, usually tomentoso before expansion. *Calyx-tube* short, free, or adnate to the base of the ovary ; segments somewhat acute, imbricate, then sub-valvate. *Stamens* 15 ; anthers oblong, external valves larger, connective apiculate. *Ovary* 3-celled ; style short, subulate, or apex clavate or capitate; stigma entire or 3-toothed ; ovules 2 in each cell. *Fruit* leathery, indehiscent, 1-seeded, surrounded by and sometimes partly adnate to the accrescent, membranous, nerved and reticulate calyx-lobes, two of which expand into narrow wings 2 or 3 in. long, the other three being much smaller. *Cotyledons* fleshy.

DISTRIB. Tropical Asia and chiefly Malaya ; species about 10.

*Synaptea* is a genus established by Griffith (Notulæ IV., 516, Tab. 585 A, fig V.) for a tree collected at Mergui, and named by him *Synap-*

*tea odorata.* This plant has been named *Synaptea grandiflora* by Kurz, (Journ. A.S., Beng., 1870, 2, 65), and *Anisoptera odorata* Kurz, (For. Flor. Burm. I, 112), while Dyer has identified it with *Hopea grandiflora*, Wall, Cat. 958, and reduced it to *Vatica grandiflora* (F.B.I., i., 301).

The characters of the genus *Synaptea*, as given by its author, are practically those of *Vatica*, Linnæus (Mantissa II., p. 152-3, No. 1311), except that, whereas in the Linnæan description nothing is said about the fruit or its relation to the calyx, Griffith distinctly explains that he has given the name *Synaptea* because the ovary is adnate to the calyx. He does not say to what extent adnate, but, in fruiting specimens of his *Synaptea odorata*, the adhesion extends to the lower part only. In the "Mantissa" of Linnæus, only one species of *Vatica* is described, *viz., V. chinensis;* and of the specimen thus named in the Linnæan Herbarium, Sir J. G. Smith publishes a figure (Smith Ic., ined., t. 36.). This figure however does not show clearly whether the base of the ovary is, or is not, adherent to the calyx, and the fruit is not figured at all. A reference to Linnæus' speci-men ought to settle what *V. chinensis* really is; but unfortunately it has not settled it. I have not myself examined the actual Lin-næan specimen; but the opinions of botanists who have examined it vary as to its identity. The plant is generally admitted not to be of Chinese origin, for no Dipterocarp is known to inhabit China. Wight and Arnot are of opinion (Prod. 84) that *Vatica chinensis* is the same as *Vatica laccifera*, W. A. (*Shorea Talura*, Roxb.—*fide* Dyer). Alph. De Candolle (Prod. XVI., 2, p. 619) keeps up the species *V. chinensis*, while Dyer (Fl. Br. Ind., I, 302) reduces it to *Vatica Roxburghiana*, Blume (Mus. Bot. Lugd. Bat. II, 31. t. 7.), Blume's *Vatica Roxburghiana*, being, as the citations and figure given by that author show, the *Vateria Roxburghiana* of Wight's Illustrations, p. 87, and Icones t. 26. It cannot be demonstrated, therefore, either from Linnæus' description or specimen, or from Smith's figure of the latter, whether Linnæus intended his genus *Vatica* to include only plants with the ovary and fruit free from the calyx, or whether plants in which there is such partial adhesion might not also be admitted. If the latter were the case there would be no occasion to keep up the genus *Synaptea.* This is the view adopted by Messrs. Hooker and Bentham, who remark of *Synaptea*, "*ex descriptione auctoris verisimiliter ad Vaticam referenda est.*" This view is also adopted by Dyer, in "Hooker's Flora of British India," where he reduces *Synaptea odorata*, Griff., to the genus *Vatica*, Section *Eu-Vatica.* This view is also to a certain extent adopted by Burck who (Ann. Jard. Bot. Buitenzorg) makes *Synaptea* a section of *Vatica*, characterised by having the lobes of the fruiting calyx unequally accrescent, two of them being much elongate, and *the fruit being partly inferior*; while the section *Eu-Vatica*, as proposed by Bentham and Hooker originally, and adopted by Burck, is characterised by having the same fruiting calyx as *Synaptea* ; nothing being said about the adhesion between the calyx and the fruit. Pierre, on the other hand, keeps up *Synaptea* as a genus on account of the presence of albumen and the structure of the embryo (characters not easily worked in herbarium specimens of this family). In my own opinion it appears advisable to admit *Synaptea* as a section of *Vatica*, but to exclude *Isauxis*, *Retinodendron*, and *Pachynocarpus*, retaining these as distinct genera. *Vatica* would, according to this scheme, be divided into two sections :—

    I.   *Eu-Vatica :*—Fruit free from the accrescent calyx, *i.e.*, fruit superior.

    II.   *Synaptea :*—Fruit adnate in its lower part to the accrescent calyx, *i e.*, fruit half inferior.

Sect. I. Eu-VATICA.—Fruit quite free from the calyx.

    Inflorescence and ripe fruit pale tomentose ;
        flowers ·4 in. long ... ... ...   1. *V. perakensis.*
    Inflorescence and ripe fruit rusty-tomentose.
        Flowers ·25 in. long ; nerves of leaves 13
        to 15 pairs; petioles ·3 to ·4 in. long ...   2. *V. Lowii.*
        Flowers ·45 in. long ; nerves of leaves 9
        to 12 pairs ; petioles ·6 to 1·5 in. long...   3. *V. Maingayi.*

Sect. II. SYNAPTEA —Calyx-wings adherent to the ripe fruit for nearly half its length.

    Leaves 9 to 10 in. long and with 18 to 20 pairs
        of nerves ... ... ...   4. *V. nitida.*
    Leaves 2·5 to 7 in. long, with 6 to 13 pairs
        of nerves.
        Larger lobes of calyx of fruit obovate and
        very blunt.
            Leaves with 6 to 8 pairs of faint
            nerves ... ... ...   5. *V. cinerea.*
            Leaves with 11 to 13 pairs of bold
            nerves ... ... ...   6. *V. Curtisii.*
        Larger lobes of calyx narrowly oblong.
            Leaves oblong or elliptic-oblong,
            with 9 to 11 pairs of nerves ;
            petals narrowly oblong ...   7. *V. faginea.*
            Leaves broadly elliptic, with 11 to 13
            pairs of nerves ; petals broadly
            elliptic ... ... ...   8. *V. Dyeri.*

Leaves 2·5 to 3·5 in. long, with about 7 or 8
pairs of faint, main nerves, minutely reticulate. 9. *V. reticulata*.

1. VATICA PERAKENSIS, King, n. sp A tree 60 to 80 feet high ;
young branches slender, deciduously scurfily stellate-pubescent, the
bark rather pale. *Leaves* thinly coriaceous, oblong-lanceolate, rarely
oblanceolate, more or less bluntly acuminate, sometimes caudate, the
base cuneate ; both surfaces glabrous, the midrib on the upper puberulous ; main nerves 10 to 12 pairs, rather prominent beneath ; length 2·5
to 4 in., breadth ·8 to 1·3 in., petiole ·4 to ·5 in. *Panicles* axillary and
extra-axillary, crowded near the ends of the branches, 1 to 2 in. long,
minutely pale tomentose, as are the ovate-lanceolate calyx-lobes.
*Flowers* ·4 in. long. *Petals* narrowly oblong, obtuse, glabrous. *Stamens*
slightly apiculate. *Ovary* minutely tomentose ; stigma conical. *Ripe fruit*
·3 in. in diam., globose, the style persistent, minutely tomentose, quite
free from the calyx ; the two accrescent calyx-lobes oblong-ob-lanceolate,
obtuse, obscurely 5-nerved, 2·5 in. long and ·5 in. broad ; minor lobes
unequal, lanceolate-acuminate, the largest about ·85 in long.

Perak : King's Collector, Wray ; a common tree. Pangkore : Curtis.

The nearest ally of this is *Vatica Bantamensis*, Benth. and Hook. ;
but that has rather larger and more coriaceous leaves, which are perfectly glabrous ; larger flowers with petals scaly externally and a more scurfy
inflorescence ; moreover the whole of the accrescent calyx-lobes of its
fruit are more coriaceous and the minor lobes are blunter.

2. VATICA LOWII, King, n. sp. A tree 60 to 80 feet high : young
branches, petioles, inflorescence and calyx densely rusty, scurfy-tomentose with stellate hair intermixed, the branches ultimately glabrous
and with dark bark. *Leaves* coriaceous, oblong, sub-acute, the base
rounded ; both surfaces glabrous, the midrib puberulous on the upper ;
main nerves 13 to 15 pairs, spreading, slightly prominent beneath ;
length 2·5 to 3·5 in., breadth 1 to 1·5 in., petiole ·3 to ·4 in. *Panicles*
axillary and terminal, much crowded towards the ends of the branches ;
·75 to 1·5 in. long. *Flowers* ·25 in. long. *Calyx-lobes* lanceolate, acuminate, oblique. *Petals* narrowly oblong, obtuse, almost glabrous. *Stamens*
short, unequal-sided, apiculate. *Ovary* depressed, tomentose, style
capitate. *Ripe fruit* globular, ·25 in. in diam, deciduously rufous-scurfy ;
the style persistent, quite free from the calyx. Two large calyx-wings
narrowly oblong, sub-acute, scarcely narrowed at the base, 5-nerved,
2·75 to 3 in. long, and ·6 in. broad ; the three smaller lobes sub-equal,
about ·5 or ·6 in. long, lanceolate, obtuse.

Perak : Scortechini, No. 2108 ; King's Collector, No. 7496.

This species is closely allied to *V. Maingayi*, Dyer ; but has smaller
flowers, and rather larger leaves with considerably longer petioles.

3. Vatica Maingayi, Dyer, in Hook. fil., Fl., Br., Ind. I, 302. A tall tree : young branches slender, ultimately glabrous, but at first rusty furfuraceous-tomentose, as are the inflorescence, calyx and ripe fruit. *Leaves* coriaceous, oblong or obovate-oblong, shortly acuminate, the base rounded, glabrous on both surfaces ; main nerves 9 to 12 pairs, slender, curving, spreading ; length 3 to 4·5 in., breadth 1 to 1·75 in., petiole ·6 to 1·5 in. *Panicles* short, few-flowered. *Flowers* ·45 in. long. *Calyx-segments* oblong-lanceolate. *Ovary* depressed, rufous-tomentose. *Ripe fruit* globose, ·25 in. in diam., the style persistent, rufous-tomentose ; free from the calyx ; the two large wings linear-oblong, sub-acute, not contracted at the base, 5-nerved (the lateral nerves faint) 2 in. long and ·35 to ·5 in. broad ; the 3 smaller lobes ovate, sub-acuminate, ·75 in. long, all glabrous.

Malacca : Maingay (Kew Distrib.) No. 209.

Of this I have seen only Maingay's specimens, which are not good.

4. Vatica nitens, King, n. sp. A tree 40 to 50 feet high : young branches and petioles densely covered with coarse deciduous scaly stellate tomentum, ultimately cinereous. *Leaves* coriaceous, narrowly oblong, acuminate, slightly narrowed to the rounded base ; both surfaces, but especially the upper, shining, glabrous, the base on the lower sparsely scaly-tomentose when young, finely reticulate ; main nerves 18 to 20 pairs, spreading, prominent on the lower surface : length 9 to 10 in., breadth 2 in. ; petiole ·5 in., stout. *Ripe fruit* globular, crowned by the persistent style, reticulate, ·5 in. in diam., adnate for half its length to the calyx ; the two large wings of the calyx oblong, slightly ob-lanceolate, obtuse, 3 in. long and ·8 to ·9 in. broad, the 3 shorter wings ovate-acuminate, ·8 in. long ; all boldly 5-nerved and shining.

Penang : Curtis, No. 1404.

This fine species is known only by Mr. Curtis' imperfect specimens. It is very distinct, being at once recognisable amongst the Indian species of *Vatica* by the size of its leaves and calyx-wings.

5. Vatica cinerea, King, n. sp. A tree about 40 feet high : young branches rufescent-puberulous at the very tips, otherwise glabrous and cinereous. *Leaves* thinly coriaceous, ovate-oblong to ovate-lanceolate, sub-acute, the base rounded or sub-cuneate ; both surfaces glabrous, finely reticulate when dry ; main nerves 6 to 8 pairs, spreading, faint ; length 2·25 to 3·5 in., breadth ·75 to 1·5 in., petiole ·3 to ·5 in. *Panicles* mostly axillary, spreading, rusty scurfy-tomentose, 1·25 to 2 in. long. *Flowers* ·45 in. long. *Calyx-lobes* sub-equal, lanceolate, sub-acute, tomentose on both surfaces. *Petals* oblong-lanceolate, sub-acute, the half of the outer surface which is outside in æstivation pubescent, other-

wise glabrous. *Stamens* obtusely apiculate. *Ovary* depressed, minutely tomentose; stigma capitate. *Fruit* (not quite ripe) globular, umbonate, attached for half its length to the calyx. The two larger calyx-wings ob-lanceolate-oblong, obtuse or sub-acute, 5-nerved, flocculent-puberulous near the base when young, ultimately glabrous, 2 in. long and ·5 in. wide; the 3 smaller wings lanceolate, obtuse, ·5 in. long.

Langani : Curtis, Nos. 2797 and 2798. Kedah : Curtis, Nos. 2096 and 2514.

When dried, the leaves of this are of a dull gray colour—hence the specific name. Its fruit resembles that of the next species, but the leaves have fewer and less prominent nerves.

6. VATICA CURTISII, King, n. sp. A tree about 40 feet high : young branches, petioles, inflorescence and calyx brownish scurfy-pubescent, ultimately glabrous. *Leaves* ovate-oblong, sub-acute, the base rounded, both surfaces quite glabrous, reticulate ; main nerves 11 to 13 pairs, oblique, rather prominent beneath ; length 3 to 5 in., breadth 1·3 to 2·5 in., petiole ·3 to ·45 in. *Racemes* axillary, few-flowered, 1 to 1·25 in. long. *Flowers* ·35 in. long. *Calyx-lobes* unequal, the 2 longer narrowly oblong, obtuse ; the 3 shorter lanceolate-acuminate. *Petals* elliptic, slightly oblique, blunt, glabrous except the pubescent edge which is external in the bud. *Ripe fruit* globular, ·3 in. in diam., adherent to the calyx for half its length, the larger calyx-lobes oblong-obovate, usually obtuse, rarely sub-acute, 5-nerved, 1·75 to 2 in. long, and ·7 in. broad ; the smaller wings about ·4 in. long.

Penang : Curtis, No. 1579.

7. VATICA FAGINEA, Dyer in Hook. fil. Fl. Br. Ind., I., 301. A tree 80 to 100 feet high : young branches slender, minutely cinereous stellate-tomentose as is the inflorescence. *Leaves* coriaceous, oblong or elliptic-oblong, finely reticulate, glabrous; main nerves 9 to 11 pairs, spreading, curving, thin but prominent when dry ; length 4 to 5 in., breadth 1·5 to 2 in. *Panicles* 2·5 in. long ; flowers ·5 in. long. *Calyx-tube* ribbed, minutely scurfy tomentose, the lobes unequal. *Petals* narrowly oblong, blunt, glabrous except the pubescent outside edge. *Ovary* hemispheric, minutely tomentose ; stigma capitate, lobed. *Ripe fruit* globular, adherent for half its length to the calyx, about ·25 in. in diam., the style persistent ; the 2 larger calyx-wings narrowly oblong, or oblong-oblanceolate, obtuse, obscurely 5-nerved, 2 to 2·5 in. long, and ·5 to ·7 in. broad near the apex ; the three smaller wings unequal, sub-spathulate, less than ·5 in. long. *Hopea faginea*, Wall. Cat. 963; *Shorea pinangiana*, Wall., Cat. p. 157. *Synaptea faginea*, Pierre, For. Flore Coch.-Chine, t 242.

Penang : Wallich. Perak : King's Collector. Nos. 3686 and 3765.

8. Vatica Dyeri, King, n. sp. A tree 80 to 130 feet high : young branches, panicles, and calyx on both surfaces densely rufous-flocculent-tomentose, with stellate hairs intermixed, the branches ultimately glabrous and their bark pale. *Leaves* membranous, usually broadly elliptic, rarely elliptic-oblong, sub-acute or very shortly and bluntly acuminate, the base rounded, both surfaces quite glabrous, finely reticulate : main nerves 11 to 13 pairs, spreading, rather prominent beneath : length 3·5 to 7 in., breadth 1·6 to 3 in. ; petiole ·35 to ·5 in., flocculent-tomentose. *Panicles* axillary or terminal, cymose, 1·5 to 3 in. long. *Flowers* ·4 in long. *Calyx* lobes unequal, the two larger oblong and obtuse ; the three smaller lanceolate, acuminate. *Petals* broadly elliptic, very obtuse, slightly narrowed to the truncate base, much larger than the calyx-lobes, glabrous, except one of the outside edges which is adpressed-pubescent. *Stamens* short, unequal-sided, bluntly apiculate. *Ovary* depressed-pubescent, the stigma capitate. *Ripe fruit* conical, the two large accrescent calyx-wings narrowly oblanceolate-oblong, blunt, 5-nerved, 1·25 in. long and ·25 in. broad ; the three smaller wings one-fourth of the size of the larger, lanceolate, obscurely 5-nerved. *Synaptea Dyeri*, Pierre Fl. Forest. Coch-Chine, t. 241.

Perak : King's Collector, No. 7662. Distrib., Cambodia, Lower Cochin-China, Pierre.

The Perak specimens are not in fruit : but in flowers and leaves they agree with Pierre's specimens from Cambodia and Cochin-China.

9. Vatica reticulata, King, n. sp. A tree 60 to 80 feet high : all parts except the inflorescence glabrous ; young branches slender, dark-coloured. *Leaves* coriaceous, oblong to ovate-lanceolate, tapering from the middle to each end ; the apex bluntly acuminate, the base very cuneate and slightly unequal-sided, the edges sub-undulate ; both surfaces finely reticulate when dry, the lower paler ; main nerves 8 or 9 pairs, little more prominent than the secondary ; length 2·5 to 3·5 in., breadth 1 to 1·25 in., petiole ·4 in. *Panicles* axillary or terminal, puberulous, 2·5 to 3·5 in. long, lax, few-flowered. *Flowers* on long pedicels. *Calyx-lobes* unequal, lanceolate, more or less obtuse, densely pubescent on both surfaces. *Ovary* hemispherical, ridged, densely tomentose ; style short, glabrous ; stigma minute. *Young fruit* sub-globular ; fruiting calyx with 2 accrescent linear-oblong wings, the other smaller ; all attached to the lower part of the fruit.

Perak : King's Collector, No. 6969.

The only specimens which I have seen of this are without corolla, stamens, or ripe fruit. The species is, however, a very distinct one, and it is an unmistakeable *Vatica*. I have therefore ventured to name it in spite of the imperfection of the material.

### 4. PENTACME, A. DC.

Glabrous or puberulous resinous trees. *Leaves* broad, entire, penni-nerved, with obtuse or cordate bases. *Flowers* large, panicled. *Calyx-tube* short, the lobes imbricate, 2 being quite external. *Stamens* 15, the filaments short, dilated; anthers much larger than the filaments, elongate, linear; the valves 4, sub-equal, each subulate at its apex, the connective also prolonged into a stiff deflexed arm as long as the appendages of the anther-valves. *Ovary* free; the style filiform, the stigma slightly lobed. *Fruit* enclosed within the imbricate calyx-lobes, of which two or more have elongated membranous reticulato many-nerved wings. Species 3,—Burmese, Siamese, and Malayan.

1. PENTACME MALAYANA, King, n. sp. A tree 40 to 50 feet high : young branches rather stout, dark-coloured, glabrous. *Leaves* sub-coriaceous, rotund-ovate to broadly elliptic, the apex shortly and blunt-ly acuminate, the base rounded or slightly emarginate; both surfaces glabrous, pale when dry; main nerves 15 to 18 pairs, spreading, pro-minent on both surfaces; length 5 to 7 in., breadth 2·75 to 4·5 in., petiole ·75 to 1·1 in. *Panicles* axillary, lax, few-flowered, 2·5 to 5 in. long. *Flowers* ·75 in. long and about as much in diameter when open, pedicelled. *Calyx-lobes* more or less broadly ovate, acuminate, minutely tomentose outside. *Petals* three times as long as the calyx, elliptic, spreading, puberulous on one-half outside, and glabrous on the other, quite glabrous inside. *Stamens* 15, equal, erect, the filaments short and broad; the anthers elongate, narrow, with 5 apical awns, one of which is deflexed and rather shorter and thicker than the other four. *Ovary* ovoid, sub-glabrous, much shorter than the filiform style : stigma minute. *Ripe fruit* ovate, apiculate, 1 in. long, glabrous; calyx-wings all enlarg-ed and reticulate except at the base; the three outer narrowly oblong, obtuse, and narrowed to the concave base, 9-nerved, 4 to 4·5 in. long, and ·65 to ·75 in. broad; the two inner lobes much narrower and fewer-nerved, about 2·5 in. long, or even shorter.

Langkani : Curtis, No. 2095.

The petals of this species are spreading, and the flower has quite an unusual *facies* for the order. It is at once distinguished by its curiously 5-awned anthers. Four of these awns are the produced apices of the anther cells, the fifth (the thicker and deflected one) is a prolongation from the connective.

### 5. SHOREA, Roxb.

Glabrous, mealy, or pubescent resinous trees. *Leaves* entire or sub-repand, pinnate-veined; stipules large, coriaceous and persistent, or minute and fugacious. *Flowers* in axillary or terminal, lax, cymose panicles; bracts persistent, caducous, or 0. *Sepals* ovate or lanceolate, imbricate, 3 being external and 2 internal. *Stamens* 15 or 20, or 30; anthers ovate or oblong, rarely linear; connective subulate-cuspidate, rarely inappendiculate; valves obtuse, rarely cuspidate, equal, or the outer slightly larger. *Ovary* 3-celled, cells 2-ovuled; style subulate, stigma entire or 3-toothed. *Fruit* with leathery, rarely with woody, pericarp, 1-celled, 1-seeded, closely surrounded by the bases of the persistent, usually accrescent, sepals, the 3 outer, or more rarely, all, and sometimes none, of which are developed into 7- to 10-veined reticulate membranous linear-oblong wings. *Cotyledons* fleshy, unequal, usually enclosing the superior radicle. DISTRIB—Tropical Asia and chiefly the Malayan Archipelago: species about 60.

Sect. I. EU.-SHOREA. Fruit little more than ·5 in. long, its pericarp
leathery: three of the persistent sepals developed into membranous wings many times longer than the fruit.
Anthers without apical appendages.
   Lower surface of adult leaves minutely stel-
      late-tomentose, not scaberulous       ...   1. *S. leprosula.*
   Lower surface of adult leaves glabrescent,
      the axils of the nerves scaly        ...   2. *S. scutulata.*
   Lower surface of adult leaves quite glabrous,
      of young leaves glaucous           ...   3. *S. Curtisii.*
Anthers mostly inappendiculate, a few with a
   minute apical appendage from the connective.
   Stamens 30       ...       ...       ...   4. *S. sericea.*
Anthers with very short apical appendages from
   the connective; flowers sessile.
   Leaves 2·5 to 4 in. long, the lower surfaces mi-
      nutely pubescent: flower ·25 in. long; fruit
      ovoid-globose, its largest wings 2·5 in. long   5. *S. parvifolia.*
   Leaves 3 to 4·5 in. long, glabrous beneath:
      flower ·3 in. long: fruit turbinate, its
      largest wings 3·5 in long   ...       ...   6. *S. acuminata.*
   Leaves 4 to 6 in. long, glabrescent or glabr-
      ous beneath; fruit narrowly ovoid, its
      longest wings 3·5 to 4·5 in. long.   ...   7. *S. macroptera.*
Apical appendage from the connective much
   longer than the anther.
   Leaves glabrous on both surfaces, the lower
      not pale.
      Stamens 10 (?)       ...       ...   8. *S. Maxwelliana.*
      Stamens 20   ...       ...       ...   9. *S. gratissima.*

Stamens 15

Flowers ·2 to ·25 in. long.

Main nerves of leaves 9 to 10 pairs, faint ; petals not saccate at base ; ovary ovoid-conical, tomentose, style short ... ... 10. *S. Ridleyana.*

Main nerves 6 or 7 pairs ; petals saccate at base ; ovary hemispheric, style long and slender ... 8. *S. Maxwelliana.*

Flowers ·4 in. long, main nerves 9 to 11 pairs ; style 3 times as long as the globose ovary ... ... 11. *S. pauciflora.*

Flowers ·5 in. long, main nerves of leaves 6 to 8 pairs ; ovary elongate-conic, style short, petals linear-oblong ... ... ... 12. *S. Kunstleri.*

Flowers ·65 in. long : nerves of leaves 12 to 16 pairs ; ovary ovoid, style long, filiform, petals ovate-lanceolate 13. *S. bracteolata.*

Leaves glaucous beneath ... ... 14. *S. glauca.*

Apical appendage of the connective with 3 to 5, or many ciliæ.

Stamens 30 : ciliæ radiating from the tip of the apical process of all the anthers 15. *S. ciliata.*

Stamens 20 : apical appendages of all the anthers with numerous ciliæ ; petals broad, spreading ... ... 16. *S. utilis.*

Stamens 15 : anthers of outer row with ciliate apical appendages ... ... 17. *S. costata.*

Anthers with a single apical appendage from each cell, and a short one from the connective ; sepals imbricate at their bases only ... 18. *S. stellata.*

*Species imperfectly known.*

Bracteoles large, persistent, scaberulous, stellate-pubescent ... ... ... 19. *S. Maranti.*

Stipules large, paired, persistent ... 20. *S. eximia.*

Sect. II. PACHYCHLAMYS, (Dyer). Fruit more than 1 in. long, its pericarp thick and woody, embraced in its lower half by a cup formed of the enlarged sepals, the bases of which are thickened woody and concave, the apices of the outer three produced into membranous wings as long as, or slightly longer than, the fruit.

110

Anthers of inner row inappendiculate, those
of the other two rows appendiculate   ... 21. *S. Thiseltoni.*
1. Shorea leprosula, Miq. Fl. Ind. Bat. Suppl. I., 487. A tree
100 to 150 feet high : young branches rather slender, lenticellate,
minutely and deciduously pale stellate-tomentose. *Leaves* coriaceous,
elliptic to oblong, acute or sub-acute, the base rounded ; upper surface
glabrous, harsh from the prominent minute reticulations, the midrib
and nerves sometimes puberulous ; lower surface minutely fuscous-
tomentose, with numerous densely stellate hairs on the midrib nerves and
veins ; main nerves 10 to 13 pairs, straight, oblique, prominent beneath ;
length 3 to 6 in., breadth 1·25 to 3·25 in., petiole ·35 to ·75 in. *Panicles*
axillary and terminal, 1·5 to 4 in. long, rachis and branches stellate-
tomentose, the short flower-bearing branchlets sericeous. *Flowers* in two
rows, secund, ·3 in. long, sessile. *Sepals* ovate, minutely velvety outside. *Petals* three times as long as the sepals, sericeous outside, oblong-
spathulate. *Stamens* about 15 ; the filaments dilated, much longer than
the short ovate inappendiculate anthers. *Ovary* ovoid, minutely tomentose, tapering upwards into the long slender style ; stigma minute.
*Ripe fruit* narrowly ovoid, apiculate, minutely tomentose, 6 in. long.
*Calyx-wings* all enlarged and membranous, concave at the base so as to
embrace the ripe fruit, but not adnate to it ; the three outer narrowly
oblong, sub-acute at the apex, narrowed at the base, 7-nerved, reticulate, 3 in. long and about ·7 in. broad ; the two inner smaller, about
1 in. long, ovate, caudate-acuminate, not nerved. A. DC. Prod. XVI.
2, 631. Scheff. in Tijdschr. Ned. Ind. XXXI, 350 : Hook. fil. Fl. Br.
Ind., I., 305. Burck in Ann. Jard. Bot. Buitenzorg, VI, 215. *Shorea
astrosticta*, Scortechini MSS.

Malacca : Maingay (Kew. Distrib.), No. 203. Perak, King's Collector,
Nos. 7646, 7905, 8182 ; Scortechini, No 2063. Distrib. Sumatra.

2. Shorea scutulata, King, n. sp. A large tree ; young branches
with dark lenticellate bark and minute white stellate pubescence.
*Leaves* elliptic, shortly abruptly and bluntly acuminate ; the base
broad, rounded, almost truncate : upper surface glabrous, minutely
reticulate ; the lower, and especially the midrib, sparsely stellate-
puberulous when young, glabrescent when old, the sides of the midrib,
and especially the pits in the axils of the nerves, with numerous
minute brownish pale-edged scales ; length 3 to 3·5 in., breadth 1·5 to
1·75 in., petiole ·3 in. *Panicles* axillary and terminal, 3 to 4 in. long,
the branches short, each bearing 2 or 3 bracteolate flowers ; bracts
broadly ovate, concave, blunt, hoary-puberulous, deciduous. *Flowers*
·4 in. long, shortly pedicelled. *Sepals* broadly lanceolate, obtuse, tomentose outside, glabrous inside. *Petals* oblong, obtuse, the base expanded

at one side, glabrous inside and on one half outside, pubescent on the other. *Stamens* 15, in 3 rows ; all the filaments broad, those of the outer two rows shorter than those of the inner: anthers short, broadly ovate, inappendiculate. *Ovary* conical, pale tomentose : style short, stigma small. *Fruit* (perhaps not mature) ovoid, apiculate, minutely pale tomentose, ·6 in. long. *Sepals* all enlarged, membranous, reticulate, concave at the base ; the three outer narrowly oblong, obtuse, very little narrowed to the base, 7-nerved, 2·75 in. long and ·75 in. broad ; the two inner ·8 in. long, linear, about 1-nerved.

Penang : Curtis, No. 1396.

A species known only from Penang, and collected only by Mr. Curtis : remarkable for its almost racemose inflorescence, and curiously glandular leaves.

3. SHOREA CURTISII, Dyer MSS. in Herb. Kew. A tree 100 to 150 feet high ; young branches slender, at first minutely stellate-puberulous, ultimately dark-coloured and glabrous. *Leaves* coriaceous, oblong-lanceolate, bluntly acuminate ; the base sub-cuneate, or almost rounded ; upper surface of young leaves minutely pubescent, of adults glabrescent or quite glabrous, the lower uniformly covered with very minute rufescent (young), or pale (adult) tomentum : main nerves 10 to 14 pairs, ascending, rather straight, prominent beneath : length 3 to 4 in., breadth 1·2 to 1·4 in., petiole ·4 to ·6 in. *Panicles* axillary or terminal, 2 to 3 in. long, the rachis slender, glabrous. *Flowers* about ·3 in. long, in distichous secund rows of 4 or 5, on the short lateral branchlets, enveloped while in bud by broad deciduous puberulous bracts. *Sepals* ovate, tomentose outside, glabrous inside, slightly unequal. *Petals* twice as long as the calyx, linear-oblong, obtuse, stellate-pubescent outside, glabrous inside. *Stamens* 15, in three rows ; the filaments elongate, broad (those of the outer row longest) ; anthers short, ovoid-globose, not apiculate. *Ovary* elongated ovoid, tomentose in the upper, glabrous in the lower half : style short, stigma small. *Ripe fruit* narrowly ovoid, apiculate, ·75 in. long, pale tomentose ; *calyx-wings* all enlarged and membranous, free from the fruit : the three outer linear-oblong, 8-nerved, 2·25 in. long, and about ·5 in. broad ; the two inner about 1 in. long, bluntly spathulate and with fewer nerves.

Penang : Curtis, Nos. 427, 1394 and 1395.

Perak : King's Collector, No. 8143.

The vernacular name of this in Penang is *Maranti Tai*.

4. SHOREA SERICEA, Dyer in Hook. fil. Fl. Br. Ind., I., 306. A tree 50 to 60 feet high ; young branches rugulose, warted and scurfily

112

rufous-tomentose as are the inflorescence and petioles. *Leaves* coriaceous, oblong or elliptic-oblong (rarely slightly ob-ovate), very shortly acuminate or sub-acute, slightly narrowed to the rounded or subcuneate base; upper surface shining, sparsely stellate-tomentose, the depressed midrib and nerves puberulous; lower surface scaberulous, more densely stellate-pubescent, especially on the bold midrib and 20 to 22 pairs of stout spreading main nerves; length 3·5 to 6·5 in., breadth 1·5 to 2·75 in., petiole ·6 to ·8 in. *Panicles* axillary and terminal, 3 to 7 in. long, the ultimate branches bearing 4 or 5 distichous, secund, bracteate, sessile flowers; bracts broadly ovate, puberulous outside. *Sepals* ovate, the two inner smaller, all densely golden-sericeous outside, glabrous inside. *Petals* like the sepals and of about the same length, the inside and one-half of the outer glabrous, the other half adpressed-sericeous. *Stamens* about 40, in several rows; the filaments of the outer shorter, all longer than the anthers; anthers ovate, mostly inappendiculate, a few with a minute appendix. *Ovary* elongated, conic, sericeous; the style short, glabrous; stigma small. *Fruit* (immature) narrowly ovoid, ·5 in. long, embraced by, but not adnate to, the accrescent membranous calyx-wings: the outer 3 calyx-wings linear-oblong obtuse, narrowed to the base, 3·5 in. long and ·6 in. broad, 10-nerved; the 2 inner 2·5 in. long and much narrower and fewer-nerved, sparsely pubescent.

Malacca: Maingay (Kew. Distrib.) No. 202. Penang: Curtis, No. 431. Perak: King's Collector, No. 3511.

This resembles *S. lacunosa* Scheff., but differs in not having persistent stipules. Its vernacular name in Penang is *Seraya*.

5. SHOREA PARVIFOLIA, Dyer in Hook. fil. Fl. Br. Ind., I., 305. A tree 100 to 150 feet high; young branches slender, pale tomentose at first, ultimately glabrous, dark-coloured and lenticellate. *Leaves* coriaceous, ovate to ovate-lanceolate, caudate-acuminate, the base subcuneate or almost rounded; upper surface glabrous (when young the midrib tomentose or pubescent); under surface sparsely scaly-pubescent when young, when adult minutely pubescent, the transverse veins thick; main nerves 9 to 12 pairs, oblique, rather straight, prominent beneath: length 2·5 to 4 in., breadth 1 to 1·8 in.; petiole ·35 to ·45 in., tomentose when young. *Panicles* axillary and terminal, crowded near the ends of the branches, 2 to 4 in. long, rather lax, spreading, many-flowered, minutely tomentose, the branches distichous. *Flowers* ·25 in. long, secund, distichous, deciduously bracteate. *Sepals* slightly unequal, ovate, acute, tomentose outside, glabrous inside. *Petals* twice as long as the sepals, obliquely elliptic, obtuse, glabrous, except on one-half outside which is silky. *Stamens* 15, or fewer: the filaments flatten-

404

ed, about 4 times as long as the broad short anthers; apiculus of connective very slender, about as long as the anther, deflexed. *Ovary*
elongate, puberulous; style rather short; stigma small. *Ripe fruit*
ovoid-globose, ·4 in. long, thinly adpressed pale tomentose. *Sepals*
all enlarged and membranous, concave at the base so as to embrace
the ripe fruit, but not adnate to it : the three outer narrowly oblong,
obtuse at the apex, slightly narrowed to the base; 7-nerved, 2·5
in. long; the two inner from one-half to one-third shorter, narrower and fewer nerved. *Shorea disticha*, Scortechini MSS. in Herb.
Calcutta.

Malacca : (Kew Distrib.) No. 206. Penang : Curtis, No. 201.
Perak : Scortechini, No. 1965. Wray, No. 1282.

6. SHOREA ACUMINATA, Dyer in Hook. fil. Fl. Br. Ind., I., 305. A
tree 100 to 150 feet high ; young branches minutely greyish tomentose,
ultimately dark-coloured and glabrescent. *Leaves* coriaceous, ovate to
lanceolate, acuminate, the base often unequal-sided, rounded or sometimes emarginate; upper surface glabrous except the puberulous
midrib; the flower glabrous, with a few scattered stellate hairs : main
nerves 7 to 9 pairs, spreading, slightly prominent beneath : length
3 to 4·5 in , breadth 1·75 to 2·5 in. ; petiole ·3 to ·4 in., tomentose.
*Panicles* axillary and terminal, crowded near the extremities of the
branches, 2 to 3 in. long, minutely stellate-pubescent, many-flowered.
*Flowers* ·3 in. long, distichous, secund, about 5 on each lateral branch,
bracteolate. *Sepals* ovate, unequal, tomentose outside, glabrous inside.
*Petals* twice as long as the calyx, spreading, broadly ovate, puberulous
outside, glabrous inside. *Stamens* 15, in three rows, the inner row
shorter : filaments broad, much larger than the short, ovate, minutely
appendiculate anthers. *Ovary* ovoid, tapering, pubescent: style short,
stigma small. *Ripe fruit* turbinate, with 3 slightly vertical grooves,
apiculate, puberulous, ·5 in. in diam., attached by its base to the calyx :
*sepals* all enlarged, concave at the base so as completely to cover
the fruit, membranous and reticulate; the 3 outer narrowly oblong
obtuse, contracted towards the base, 10- or 11-nerved, 3·5 in. long,
and 7 in. broad ; the two inner 1 to 1·5 in. long, under ·25 in. broad, 3-
to 4-nerved.

Malacca : Maingay (Kew Distrib.) No. 205 (?). Griffith, No. 1762.
Perak : King's Collector, No. 8009.

7. SHOREA MACROPTERA, Dyer in Hook. fil. Fl. Br. Ind. I, 308. A
tree 60 to 80 feet high : young branches with dark-brown bark, minutely lenticellate and puberulous. *Leaves* coriaceous, oblong (usually
narrowly), shortly acuminate, the base sub-cuneate or rounded : upper
surface glabrous, shining, the midrib and nerves puberulous : lower

surface glabrescent or glabrous, chocolate-coloured when dry : main
nerves 10 to 12 pairs, curved, spreading, prominent on the lower
surface; length 4 to 6 in., breadth 1·35 to 1·75 in.; petiole ·4 to ·5
in., rugose. *Panicles* axillary or terminal, 4 to 7 in. long, lax, branching, few-flowered, puberulous, sparsely scaly. *Flowers* about ·5 in.
long, sessile, solitary, not secund. *Sepals* distinct almost to the
base, slightly unequal, broadly-ovate, acute, more or less yellowish-
tomentose outside, glabrous inside. *Petals* narrowly oblong, slightly
oblique at the base, the apex blunt, glabrous except one-half of the outer
surface which is sericeous. *Stamens* 15, in two rows; filaments broad
except at the apex, those of the outer two rows by much the shorter :
anthers short, ovate, the connective minutely awned. *Ovary* elongated-
ovoid, sericeous in its upper half; style short, stigma small. *Ripe
fruit* ·6 to ·75 in. long, narrowly ovoid, pale puberulous, apiculate :
*sepals* all enlarged and reticulate, slightly concave at the base and
embracing, but not adnate to, the fruit; the three outer narrowly
oblong, obtuse, tapering slightly to the auricled base, 7-nerved, 3·5 to
4·5 in. long, and ·8 to 1 in. broad; the two inner variable, but
shorter, narrower and fewer nerved. *Shorea auriculata*, Scortechini
MSS. in Herb., Calcutta.

Malacca : Maingay. Singapore : Ridley. Penang : Curtis, No.
1392. Perak : very common, King's Collector, Scortechini.

A species from Borneo which closely resembles this appears to me
to differ specifically. Its leaves are longer with sparser nerves, and its
calyx-wings are longer.

8. SHOREA MAXWELLIANA, King, n. sp. A tree 60 to 80 feet high :
young branches dark-coloured, almost glabrous. *Leaves* coriaceous,
ovate-lanceolate, acuminate (caudate-acuminate when young), the base
unequal-sided, cuneate ; both surfaces quite glabrous, the upper shining,
the lower chocolate-coloured when dry : main nerves 6 or 7 pairs, curved,
spreading, thin and inconspicuous : length 3 to 4 in., breadth 1·3 to 1·5
in., petiole ·4 in. *Panicles* axillary and terminal, 2·5 to 3 in. long,
stellate-puberulous, their lateral branches very short and few-flowered.
*Flowers* shortly pedicelled. *Sepals* unequal, oblong, blunt, with enlarged concave bases, more or less pubescent, but glabrous in the concavity
of the base inside. *Petals* oblong, concave and saccate at the base,
tomentose outside, glabrous inside. *Stamens* 10 (?), the filaments
short, broad ; the anthers elongate, erect, pointed, the connective ending in an awn as long as the anther. *Ovary* hemispheric ; the style
long, slender ; stigma minute. *Fruit* (not mature) globular, minutely
tomentose, closely invested by, but not adnate to, the concave bases of
the sepals : *sepals* all enlarged, membranous, narrowly oblong, obtuse ;

the three outer 7-nerved, 1·5 in. long and ·4 in. broad ; the two inner similar in shape, but fewer-nerved and only ·5 in. long.

Pĕrak : King's Collector, Nos. 3601 and 3744.

The only flowers of this species which I have seen are in an early stage of bud, and from them I am unable to make out the characters of the petals properly. The stamens appear to be only 10 in number : but of this I cannot now be quite certain.

9. SHOREA GRATISSIMA, Dyer in Hook. fil. Fl. Br. Ind. I, 307. A tree : younger branches slender, glabrescent, dark-coloured. *Leaves* coriaceous, elliptic, acuminate, the base broad and rounded, the margins sub-undulate, both surfaces glabrous : main nerves 12 to 14 pairs, faint ; length 2·5 to 4 in., breadth 1·25 to 1·5 in., petiole ·6 to ·75 in. *Panicles* axillary and terminal, lax, few-flowered, 3 to 6 in. long, sub-puberulous. *Flowers* secund, pedicelled, ·25 in. long. *Sepals* lanceolate, sub-acute ; minutely tomentose outside, glabrous inside in the lower, adpressed-pubescent in the upper, half. *Petals* twice as long as the calyx and much broader, elliptic, obtuse, glabrescent. *Stamens* about 20; the filaments short, unequal, dilated. *Anthers* elongated-ovate, truncate, each with a terminal awn from the connective twice as long as itself. *Ovary* ovoid, sub-glabrous ; stigma small. *Ripe fruit* unknown. *Hopea gratissima*, Wall. Cat. 960.

Singapore : Wallich.

This is known only by Wallich's specimens. He referred it to *Hopea*, of which genus it certainly has the *facies*: the æstivation of the sepals is moreover that of *Hopea*, and so is the apiculus of the connective of the stamens. The petals in shape, however, resemble those of *Shorea*. I retain it in *Shorea* in deference to the opinion of Mr. Dyer.

10. SHOREA RIDLEYANA, King, n. sp. A tree 60 or 80 feet high : young branches slender, dark brown, lenticellate, nearly glabrous. *Leaves* ovate-lanceolate, shortly acuminate, the base rounded : both surfaces glabrous, the upper shining : main nerves 9 or 10 pairs, curved, spreading, thin but slightly prominent beneath : length 2·5 to 4 in., breadth 1·1 to 2 in. ; petiole ·4 to ·5 in., rugulose. *Panicles* axillary and terminal, 1·5 to 2 in. long, densely stellate-puberulous *Flowers* ·2 in. long, pedicellate. *Sepals* sub-equal, oblong, obtuse, tomentose outside, glabrous inside. *Petals* oblong, slightly oblique, obtuse, glabrous inside, puberulous outside on one half, glabrous on the other. *Stamens* 15, sub-equal, the filaments dilated in the lower half : anthers shorter than the filaments, ovate, the connective produced into an awn longer than the anther. *Ovary* ovoid-conical, minutely tomentose. *Style* short ; stigma minute. *Fruit* (immature) ovoid, apiculate, minutely

pale tomentose : *sepals* all enlarged, membranous, reticulate and concave at the base ; the three outer linear-oblong, obtuse, slightly narrowed to the concave base, 5-nerved, 2.25 in. long and ·4 in. broad ; the two inner of the same shape, but only 1-nerved, narrower and only 1·5 in. long.

Perak : King's Collector, Nos. 3571 and 3617.

This a good deal resembles *S. Maxwelliana*, King ; but its leaves have more nerves, its slightly oblique petals are not saccate at the base, its ovary is ovoid-conical, and minutely tomentose with a short style ; whereas in *S. Maxwelliana* the petals are saccate at the base, and the ovary is hemispheric with a long style.

11. SHOREA PAUCIFLORA, King, n. sp. A tree 50 to 90 feet high : young branches slender, their bark brown puberulous and lepidote. *Leaves* thinly coriaceous, from oblong to elliptic, shortly acuminate ; the base abruptly cuneate, slightly unequal-sided, or (in the elliptic forms) almost rounded : main nerves 9 to 11 pairs, oblique, straight, prominent beneath : length 4 to 5 in., breadth 1·8 to 2·5 in., petiole ·6 to ·7 in. *Panicles* few, axillary or terminal, few-flowered, 1·75 to 4 in. long, rather coarsely pubescent. *Flowers* ·4 in. long, secund, shortly pedicellate, each subtended by an ovate, solitary, puberulous, deciduous bract. *Sepals* broadly ovate, tomentose outside, glabrous inside. *Petals* broadly elliptic, obtuse, concave at the base, veined, inside glabrous, the outside half glabrous and half adpressed-sericeous. *Stamens* 15, in 3 rows : the outer row smaller and with filiform filaments, the inner rows with filaments longer and expanded in the lower half ; the anthers of all shortly ovate, the connective produced into an awn twice as long as the stamen. *Ovary* hemispheric, tomentose ; style nearly 3 times as long, puberulous ; stigma small. *Ripe fruit* unknown.

Penang : Curtis, No. 1537.

A species known only by Mr. Curtis' specimens which have no fruit.

12. SHOREA KUNSTLERI, King, n. sp. A tree 60 to 100 feet high : young branches slender, rusty-puberulous, their bark brown. *Leaves* coriaceous, elliptic, abruptly and shortly acuminate, the base rounded or slightly cuneate, both surfaces glabrous, the lower with a few stiff white hairs on the midrib and nerves ; main nerves 6 to 8 pairs, curved, ascending, prominent on the lower surface ; length 4 to 5 in., breadth 2 to 2·4 in., petiole ·5 in. *Panicles* axillary and terminal, 4 to 6 in. long, lax, few-flowered, scaly-puberulous. *Flowers* ·5 in. long, sub-sessile, 4 or 5 together on the short branches of the panicles, secund, bracteate : the bracts broadly ovate, puberulous. *Sepals* sub-equal,

broadly ovate, acute, tomentose outside; the edges ciliate, glabrous inside. *Petals* linear-oblong, obtuse; the bases obliquely expanded, sericeous externally, glabrous internally. *Stamens* 15, sub-equal, the filaments as long as the anthers, flattened; anthers ovate, short, the connective terminated by a curved awn much longer than the stamen. *Ovary* elongate-conic, puberulous; style short. *Ripe fruit* hemispheric, tapering into a cone and crowned by the style, adpressed pale tomentose. *Sepals* membranous, reticulate: the three larger narrowly oblong, obtuse, tapering to the concave non-reticulate base, 9-nerved, 3·5 in. long and ·7 in. broad: the two inner 2 in. long, linear, 3-nerved.

Perak: King's Collector, Nos. 3474 and 3705.

This species is allied to *S. bracteolata*, Dyer, but its leaves have fewer nerves, smaller flowers, narrower petals, and a short style.

13. SHOREA BRACTEOLATA, Dyer in Hook. fil. Fl. Br. Ind. I, 305. A tree 50 to 150 feet high; young branches minutely furfuraceous-puberulous, speedily glabrescent, their bark dark-coloured. *Leaves* coriaceous, elliptic-oblong, shortly acuminate (often sub-obtuse when old), narrowed slightly to the rounded or emarginate base; upper surface quite glabrous; the lower yellowish furfuraceous-puberulous to glabrous; main nerves 12 to 16 pairs, spreading, prominent beneath: length 4 to 6 in., breadth 1·6 to 2·5 in., petiole ·45 to ·6 in. *Panicles* axillary, few-flowered, 2·5 to 6 in. long, glabrous. *Flowers* ·65 in. long, shortly pedicellate, each subtended by 2 elliptic, obtuse, 3-nerved, puberulous, deciduous bracts ·35 in long. *Sepals* lanceolate, obtuse, minutely tomentose outside, the two inner smaller. *Petals* ovate-lanceolate, obtuse; the bases expanded, glabrous. *Stamens* 15, in two rows, the filaments less than half as long as the ovate obtuse anthers; appendix of connective subulate, twice as long as the anther, decurved when old. *Ovary* ovoid, attenuated upwards, sub-glabrous; the style long, filiform; stigma small. *Ripe fruit* ovoid, apiculate, ·6 in. long, embraced by, but (except at the very base) free from the calyx; *sepals* accrescent, membranous, reticulate and concave at the base: the three outer narrowly oblong, blunt, slightly narrowed above the concave base, 10-nerved, 3·5 in. long, and ·6 in. broad; the two smaller about 2 in. long, and ·2 in. broad, about 3-nerved. *Shorea foveolata*, Scortechini MSS. in Herb. Calcutta.

Malacca: Maingay (Kew Distrib.) No. 204. Penang: Curtis, Nos. 322 and 1405. Perak: King's Collector, Nos. 7583, 7591, 7717; Scortechini, No. 1939. DISTRIB.—Sumatra. Forbes, No. 3050.

14. SHOREA GLAUCA, King, n. sp. A tree 80 to 100 feet high; young branches slender, dark-coloured, puberulous. *Leaves* coriaceous, ovate-lanceolate, acuminate; the base broad, rounded; upper surface

117

glabrous, the lower glaucous (except the midrib and nerves) especially when young; main nerves 7 to 9 pairs, ascending, rather straight: length 3·5 to 4·5 in., breadth 1·4 to 1·8 in.; petiole ·45 to ·6 in., rugulose, glaucous. *Panicles* axillary, few-flowered, shorter than the leaves, hoary, the *flowers* on short pedicels. *Sepals* slightly unequal, oblong, obtuse, tomentose on both surfaces. *Ovary* conical, tomentose; the style very short, glabrous; stigma small, 3-lobed. *Fruit* (immature) ovoid-globose, apiculate, minutely tomentose; accrescent sepals membranous, free from the fruit; obscurely 7- to 12-nerved, strongly reticulate, blunt, slightly narrowed to the concave base, at first puberulous but ultimately glabrous; the longer 2·25 in. long, and ·6 to ·75 in. broad, the others smaller.

Penang: Curtis, No. 372. Malacca: Maingay (Kew Distrib.), 212.

In this species the two inner fruiting wings of the calyx are nearly as large as the three outer; the leaves are very white underneath when young, but much less conspicuously so when adult. It is known, only by Curtis' and Maingay's specimens, none of which have complete flowers. Maingay's specimens from Malacca have in fact no flowers; but there is no mistaking their leaves as being exactly like those of Mr. Curtis' from Penang. The vernacular name of this is *Dammar laut dhan lesor.*

15. SHOREA CILIATA, King, n. sp. A medium-sized tree; young branches slender, dark-coloured, deciduously hoary-puberulous. *Leaves* coriaceous, lanceolate or oblong-lanceolate, acuminate, the base cuneate; both surfaces glabrous, minutely reticulate, the lower whitish when young, pale brown when dry; main nerves 8 or 9 pairs, ascending, curved, shining on the lower surface: length 3 to 3·5 in., breadth ·8 to 1·5 in., petiole ·75 to ·9 in. *Panicles* 2 to 2·5 in. long, axillary and terminal, little-branched, few-flowered, hoary. *Flowers* ·5 in. long, secund. *Sepals* ovoid-deltoid, obtuse, outside tomentose, inside glabrous. *Petals* three times as long as the sepals, narrowly oblong, obtuse, slightly expanded at the base, adpressed-sericeous outside, glabrescent inside. *Stamens* 30, in fascicles of 3, unequal, the shorter with undilated filaments, the longer with filaments dilated in the lower half; all with the connective produced into an apical process crowned by 3 to 5 spreading ciliæ. *Ovary* ovoid-conic, sericeous, with a short glabrous style. *Fruit* (immature) ovoid, apiculate, pale-tomentose, ·5 in. long; accrescent sepals membranous, reticulate: the three outer narrowly oblong, reticulate, 7-nerved: the two inner 2 in. long, and ·3 in. broad, narrowed to above the concave base: the two inner 1 in. long, linear-lanceolate, few-nerved.

Penang: Curtis, No. 1578.

410

Known only by Curtis' specimens, and readily recognisable by its beautifully ciliate-crested anthers.

16. SHOREA UTILIS, King, n. sp. A large tree; all parts except the inflorescence glabrous: young branches slender, dark-coloured. *Leaves* coriaceous, ovate-lanceolate, caudate-acuminate, or shortly and abruptly acuminate, the base slightly cuneate; main nerves about 7 pairs, oblique, not prominent on either surface; length 2·5 to 3 in., breadth ·9 to 1·2 in., petiole ·4 in. *Panicles* axillary, stellate-puberulous, about as long as the leaves; their lateral branches distant, very short, minutely tomentose, 3- or 4-flowered. *Flowers* sub-sessile, globular in bud, under ·2 in. long. *Sepals* ovate-orbicular, blunt, the outer 3 very tomentose outside, the inner 2 less so; all glabrous inside. *Petals* broadly oblong, blunt, more or less sericeous in both surfaces. *Stamens* 20; filaments slightly dilated, about as long as the ovate anthers; apical process of connective about as long as the anther, ciliate. *Ovary* sericeous, elongated-conic, gradually tapering into the short glabrous style; stigma minute. *Ripe fruit* ovoid, apiculate, pale, adpressed-sericeous, ·4 in. long, closely invested by, but free from, the concave bases of the accrescent sepals. *Sepals of fruiting calyx* all enlarged, membranous, reticulate, deciduously puberulous; the 3 outer oblong, very obtuse, 5-nerved, 1·25 in. long, and ·4 in. broad; the inner 3 half as long, or less, and much narrower.

Penang: Curtis, No. 423.

This species, which Mr. Curtis describes as yielding the most durable timber in Penang, was at one time quite common there, but it is now almost extinct. Its vernacular name is *Dammar laut.*

17. SHOREA COSTATA, King, n. sp. A tree; young branches dark-coloured, lepidote-puberulous. *Leaves* thinly coriaceous, oblong, sub-acute, slightly narrowed to the rounded or sub-cuneate base; both surfaces glabrous, the transverse veins distinct, especially on the lower: main nerves 11 to 13 pairs, oblique, rather straight, slightly prominent beneath; length 3 to 4·25 in., breadth 1·2 to 1·5 in., petiole ·8 to 1 in. *Panicles* axillary and terminal, 1·5 to 2·5 in. long, scaly-puberulous, the lateral branches very short and few-flowered. *Flowers* small. *Sepals* broadly ovate, yellowish-tomentose outside, glabrous inside. *Stamens* 15; all with dilated filaments longer than the ovate anthers, those of the inner row with the apical process of the connective short and glabrous, those of the outer rows with longer ciliate apical connectives. *Ovary* ovoid-conical, densely yellowish-tomentose; style very short. *Ripe fruit* ovoid, apiculate, sparsely puberulous, ·75 in. long; sepals all enlarged, concave and dilated at the base, membranous and reticulate; the three outer narrowly oblong, obtuse, much

narrowed to the base, 7-nerved, 2·75 in. long, and ·45 in. broad; the
two inner of the same shape, but few-nerved, only 1·5 in. long, and ·25
in. broad.

Penang: Curtis, No. 199.

A species known only by Mr. Curtis' solitary specimen. The
connectives of the inner anthers are ciliate, somewhat in the fashion
of *S. ciliata*, King; but the leaves of that species are very different.

18. SHOREA STELLATA, Dyer in Hook. fil. Fl. Br. Ind. I, 304. A
tree 100 to 150 feet high; young branches slender, at first stellate-
puberulous, but speedily glabrous, with bark dark-coloured and sparsely lenticellate. *Leaves* thinly coriaceous, ovate-lanceolate, the base
rounded: upper surfaces glabrous, the lower very minutely lepidote on
the reticulations; main nerves 8 to 11 pairs, rather straight, oblique,
prominent on the lower surface; length 4 to 5·5 in., breadth 1·75 to
2·25 in., petiole ·7 to ·9 in. *Panicles* axillary or terminal, crowded at
the extremities of the branches, many-flowered, 4 to 6 in. long; minutely stellate-pubescent. *Flowers* ·25 in. in diam. *Calyx* minutely greyish-
tomentose, the segments ovate-oblong, sub-acute, valvate, erect. *Petals*
broadly ovate, obtuse, pubescent outside, spreading. *Stamens* 15, the
filaments short, broad; the anthers linear-elongate, shortly bi-mucronate,
the connective also shortly mucronate. *Ovary* ovate-globular, grooved,
very tomentose; the style short; the stigma ovoid, small. *Ripe fruit*
ovoid, apiculate, tomentose, ·5 in. long; sepals all enlarged, sub-
equal, membranous, linear-oblong, sub-acute, much narrowed at the
base, quite free from the fruit, 5-ribbed, reticulate, 4·5 in. long, and
about ·6 in. broad. *Parashorea stellata*, Kurz, Journ. As. Soc., Bengal,
for 1870, pt. 2, p. 66. For. Flora Burm., I, 117; Pierre Flore Forest.
Coch-Chine, t. 224.

Perak: King's Collector, No. 7505. DISTRIB. Burmah.

None of the Perak specimens are in fruit; but in leaves and flowers
they agree absolutely with Kurz's Burmese specimens. The calyx in
all is quite valvate, and it was on this character chiefly that Kurz based
the genus *Parashorea*.

19. SHOREA MARANTI, Burck in Ann. Jard. Bot. Buitenzorg, VI.
217. A small tree: young branches dark-coloured, stellate-puberulous.
*Leaves* thinly coriaceous, more or less broadly elliptic or elliptic-oblong,
shortly abruptly and bluntly acuminate; the base broad, rounded, or
almost truncate; upper surface glabrous, the midrib and nerves minutely tomentose or pubescent when young; lower surface more or less
sparsely minutely stellate-puberulous, the sides of the midrib, especially at the axils of the main nerves, glandular and densely covered with
masses of brown pale-edged scales: main nerves 12 to 16 pairs, oblique,

slightly curved, thin but prominent beneath when dry, as are the transverse veins ; length 3·5 to 6·5 in., breadth 1·5 to 2·25 in.; petiole ·35 in., densely stellate-pubescent, scurfy. *Stipules* deciduous, ovate-lanceolate, nerved, stellate-puberulous. *Panicles* axillary and terminal, few-flowered, tawny-tomentose, (shorter than the leaves [?]) ; the bracts in pairs, unequal, elliptic-oblong, blunt, nerved, pubescent on both surfaces. " Segments of calyx (fide Burck) unequal, the three outer larger, imbricate. *Petals* minutely tomentose inside. *Stamens* 15, in two rows." *Hopea? Maranti*, Miq. Fl. Ind. Bat. Suppl., 489 ; A. DC. Prod. XVI, 2, p. 635.

Perak : King's Collector, No. 880. Malacca : Derry, No. 952. Distrib. Sumatra, Bangka.

The Perak specimens are not in flower ; and I have seen none from elsewhere that are. The above imperfect description of the flower has therefore been copied from Burck (Ann. Jard. Bot. Buitenzorg, VI. 217). The Perak specimens perfectly agree, as to leaves, with an authentic specimen of Miquel's from Sumatra, in the Calcutta Herbarium. Miquel never saw either flower or fruit. In fact, of the twenty new species of *Dipterocarpeœ* described by this author in the supplement to his Flora of the Netherlands India, the flowers are described in only two, and in these but partially !

20. Shorea eximia, Scheff. in Nat. Tijdschr. Ned. Ind. XXXI, 349. A shrub or small tree ; young branches petioles and under-surfaces of leaves stellate-setulose. *Leaves* coriaceous, elliptic-oblong, or ob-lanceolate-oblong, acuminate, narrowed to the rounded or sub-cuneate base : upper surface glabrous except the tomentose midrib, shining, the nerves depressed : under surface scabrid, pale brown, the reticulations midrib and 17 to 21 pairs of spreading nerves prominent : length 6·5 to 11 in., breadth 2·25 to 3·25 in., petiole ·25 to ·35 in. Stipules in pairs, persistent, ovate, acuminate, longer than the petioles, reticulate, laxly pubescent and warted. " *Wings of fruiting-calyx* linear-lanceolate, obtuse : the three larger narrow at the base, 3·2 to 3·6 in. long, ·5 in. broad, sparsely pubescent, 9-nerved ; the two shorter and narrower 1·6 in. long. *Fruit* elongated-ovoid, acuminate, minutely whitish-tomentose." Burck in Ann. Jard. Bot. Buitenzorg VI, 218. *Vatica? eximia*, Miq. Fl. Ind. Bat. Suppl. 486 ; A. DC. Prod. XVI 2, 623. *Vatica sub-lacunosa?* Miq. Fl. Ind. Bat. Suppl. 486. *Shorea sub-lacunosa*, Scheff. in Nat. Tijdschr. Ned. Ind. XXXI, 350 : A. DC. Prod. XVI, 2623.

Malacca : Griffith, No. 5018. Penang : King. Perak King's Collector, 10998. Distrib. Sumatra, Bangka.

This plant is very imperfectly known. I have copied the description of the fruit from Dr. Burck (1. c.). Miquel, who first described the plant as a probable *Vatica*, had seen nothing but a leaf-twig. Specimens brought from Perak by the Calcutta collectors bear, instead of flowers, curious cones, 1·5 in. long, of distichous imbricate bracts, concerning which Griffith, in his field note on his specimen No. 5018, wrote,—" irregular growth caused by an insect; each of the scales of these cones bears on its dorsum at its base a number of eggs." Griffith's No. 5019 appears to belong to a closely allied, but distinct, species ; as also does the indeterminate plant issued by Wallich as No. 6635 of his catalogue, under the designation, " *Dilleniacea* [?] *nervosa*."

21. SHOREA THISELTONI, King, n. sp. A tree 60 to 80 feet high : young branches rather stout, the bark dark-coloured and lenticellate, but covered at first by a pale-grey, deciduous pellicle. *Leaves* coriaceous, elliptic-oblong to elliptic, rarely oblong, sometimes slightly obovate, obtuse, slightly narrowed to the rounded base ; both surfaces glabrous, the lower when very young sparsely lepidote, puberulous especially on the midrib and nerves, brown when dry : main nerves 8 or 9 pairs, ascending, slightly curved, bold and prominent on the under surface like the midrib ; length 5 to 7 in., breadth 2·5 to 3·25 in.; petiole ·6 to ·8 in., stout. *Panicles* axillary and terminal, 2 to 3 in. long, velvety, few-flowered, apparently ebracteolate. *Flowers* sessile, ·6 or ·7 in. long. *Sepals* ovate, sub-acute, unequal ; the 3 outer tomentose outside, glabrous inside ; the 2 inner smaller, nearly glabrous, the edges ciliate. *Petals* much longer than the sepals, linear-oblong, obtuse, expanded at the base, glabrous, except one-half of the outer surface which is adpressed-pubescent. *Stamens* 15, in 3 rows, the filaments of all dilated, unequal : the anthers shortly ovate, those of the inner and longer row inappendiculate, those of the other two rows with a short apical appendage from the connective. *Ovary* narrowly conical, tomentose, tapering into the short glabrous style ; stigma minute. *Ripe fruit* narrowly ovoid, apiculate, minutely pale-tomentose, substriate, 1·2 in. long, and ·6 in. in diam., the pericarp thick and woody. *Persistant sepals* with much thickened concave woody bases, forming a cup embracing the lower half of the fruit, the apices of the outer three prolonged into membranous linear-oblong obtuse wings exceeding the fruit and sometimes 1·5 in. long ; one of the inner sepals shortly winged, the other often broad, obtuse and not winged.

Perak : common. King's Collector.

In this plant the fruit is much larger than in any of the other species of *Shorea* here described, and its pericarp is hard and thick. The bases of the sepals are greatly thickened and concave, and they form a cup which embraces closely, but does not adhere to, the lower half of the fruit, the apices of some of them being winged as above described. In these respects the species resembles certain other Malayan species of *Shorea*, e. g., *S. Martiniana* Scheff, *S. scaberrima*, and *S. stenoptera*, Burck. Judging from the leaf-specimens on which Miquel founded his *Hopea Singkawang*, that plant must be a close ally of this. A species (flower only) collected by H. O. Forbes in Sumatra (Herb. No. 2952) must also be closely allied to this. It differs however by its conspicuously bracteolate inflorescence. Beccari's Nos. 2681 and 3507, which form the types of Heim's species *S. brachyptera*, are also allied to this.

## 6. HOPEA, Roxb.

Glabrous or hoary-tomentose resinous trees. *Leaves* quite entire, firm, feather-veined; stipules small, deciduous or inconspicuous. *Flowers* sessile or shortly pedicelled, ebracteate, in lax panicles of unilateral racemes. *Sepals* inserted on the receptacle, two being quite external and three for the most part internal, obtuse, imbricate. *Petals* falcate, their apices inflected in bud. *Stamens* 15, or rarely 10, slightly connate; the connective subulate-cuspidate, the anthers ovate, their valves obtuse, equal. *Ovary* 3-celled, the cells 2-ovuled: style shortly cylindric or subulate. *Fruit* 1-seeded, closely surrounded by the bases of the accrescent sepals, the 2 external of which are developed into linear wings, the three internal not longer than the ripe fruit. *Embryo* as in *Shorea*.—DISTRIB. of *Shorea*; species about 35.

Sect. I. EU-HOPEA, Main nerves of leaves bold and prominent.

Nerves of leaves 16 to 18 pairs; accrescent
 sepals 4 to 4·5 in. long, 10-nerved ... 1 *H. nervosa.*
Nerves of leaves 10 to 13 pairs; accrescent
 sepals 1·75 to 2·5 in. long, obscurely 5-nerved  2. *H. Curtisii.*
Sect. II. DRYOBALANOIDEA, Miq. Main nerves not
 distinct.
Petals sericeous : the filaments longer than the
 anthers; ripe fruit ·3 in. long, the accrescent
 sepals 7-nerved, 1·75 to 2 in. long, and ·2 to
 ·25 in. broad; leaf-petioles ·25 to ·4 in. long,
 minutely tomentose...  ...  ... 3. *H. micrantha.*
Petals densely sericeous; the filaments shorter
 than the anthers; ripe fruit ·2 in. long; accrescent sepals obscurely 5- to 7-nerved, 1·25
 to 1·5 in. long, and ·25 in. broad; leaf-petioles
 ·35 to ·6 in. long, slender, puberulous, finally
 glabrous  ...  ...  ... 4. *H. intermedia.*

1. Hopea nervosa, King, n. sp. A tree 50 to 70 feet high : young branches dark-coloured, glabrous. *Leaves* coriaceous, oblong to elliptic-oblong, shortly acuminate, the base rounded or very slightly cuneate ; both surfaces glabrous ; main nerves 16 to 18 pairs, spreading, bold and prominent on the lower ; length 3·5 to 5 in., breadth 1·5 to 2·25 in. ; petiole ·5 to ·75 in., transversely wrinkled when dry. *Flowers* unknown. *Ripe fruit* ovoid-rotund, apiculate, glabrous, ·5 in. long ; the two outer sepals much enlarged, oblong-lanceolate, obtuse, slightly narrowed to the concave thickened smooth base, 10-nerved, 4 to 4·5 in. long, and ·6 to ·75 in. wide ; the three inner sepals not quite so long as the fruit, broadly ovate, obtuse, thickened, smooth, closely embracing but not adherent to the fruit.

Perak : King's Collector, No. 3690.

This is a very distinct species, belonging to the group of *Hopea* with the nerves of the leaves bold. It is so distinct that, contrary to my general practice, I venture to name it without having seen the flower.

2. Hopea Curtisii, King, n. sp. A tree 50 to 60 feet high : young branches slender, dark-coloured, lenticellate, almost glabrous. *Leaves* coriaceous, broadly ovate to ovate-oblong, shortly acuminate or acute, the base slightly unequal-sided, rounded, rarely sub-cuneate ; both surfaces glabrous, the upper slightly puberulous on the midrib near the base, the lower with several hairy glands at the base, the midrib sparsely and minutely stellate-puberulous ; main nerves 10 to 13 pairs, curving, ascending, prominent beneath ; length 3·5 to 4·5 in., breadth 1·75 to 2·5 in. ; petiole ·4 in., puberulous when young. *Panicles* axillary and terminal, lax, few-flowered. *Flowers* about ·2 in. long, pedicelled. *Sepals* broadly ovate, blunt, concave, tomentose outside, glabrous inside ; the inner two rather smaller and more glabrous than the others. *Petals* oblong, oblique, falcate, obtuse, partially tomentose outside, glabrous inside. *Stamens* 10, the filaments short, dilated ; anthers ovate, short, the connective with an apical awn longer than the anther. *Ovary* broadly ovate, puberulous at the truncate apex, otherwise glabrous : style short. *Ripe fruit* ovoid, apiculate, pale striate, ·3 in. long ; outer two sepals accrescent, narrowly-oblong, reticulate, membranous, obscurely 5-nerved, obtuse, slightly narrowed to the concave smooth base, 1·75 to 2·5 in. long and from ·35 to ·6 in. broad ; the three inner non-accrescent sepals about as long as the fruit.

Penang : Curtis No. 1562. Perak : King's Collector, 8161.

3. Hopea micrantha, Hook. fil. in Trans. Linn. Soc., xxiii, 160. A tree 60 to 80 feet high : young branches slender with dark-coloured, lenticellate bark and minute brownish pubescence. *Leaves* coriaceous,

416

ovate-lanceolate or oblong-lanceolate, bluntly caudate-acuminate; the
base slightly cuneate or sometimes broad, rounded and slightly unequal;
both surfaces glabrous except the pubescent midrib: main nerves
numerous, not much more prominent than the secondary, and both
indistinct; length 2 to 4 in., breadth ·8 to 1·75 in.; petiole ·25 to ·4 in.
minutely tomentose. *Panicles* axillary and terminal, numerous, short,
spreading, 1 to 1·5 in. long, puberulous or glabrous. *Flowers* ·15 to
·25 in. long, shortly pedicellate. *Sepals* sub-equal, ovate-rotund, sub-
acute or obtuse, puberulous and resinous outside, glabrous inside.
*Petals* twice as long as the sepals, broadly oblong-obtuse, silky outside
except on one side, glabrous inside. *Stamens* about 12, the filaments
dilated in the lower half, longer than the ovate anthers; the connective
produced into a single apical awn longer than the stamen. *Ovary*
elongated, often constricted in the middle, glabrous; style very short,
stigma minute. *Ripe fruit* ovoid, apiculate, ·3 in. long, striate, closely
embraced by the 3 inner sepals which about equal it in length; the
outer two sepals accrescent, oblanceolate, obtuse, tapering to the concave base, reticulate, 7-nerved, 1·75 to 2 in. long, and ·2 to ·25 in. broad.
A. DC. Prod. XVI. 2, p. 634. Dyer in Hook. fil. Fl. Br. Ind. I, 310.
Burck in Ann. Bot. Jard. Buitenzorg, VI, 238.

Malacca; Maingay (Kew Distrib.) No. 210. Penang: Curtis, Nos.
167, 266, 1397. Perak: King's Collector, Nos. 3525, 8170. DISTRIB.
Borneo: Bangka, Sumatra.

Mr. Curtis notes on the Penang specimens of this, that the bark of
the tree is smooth and of a grey colour, whereas the back of its close ally
*H. intermedia* is fissured like that of *Shorea parviflora*. The species of *Hopea*
with numerous indistinct nerves, (Sect. *Dryobalanoides*) are not easy to
distinguish from each other in the Herbarium. *H. Mengarawan*, Miq., a
species published two years earlier than this (*i. e.*, in 1860), comes very
near this, and the two may possibly prove to be identical, in which
case Miquel's name must be adopted. *Hopea cernua*, Teysm. and Binn.
was described by its authors from a plant originally obtained from
Sumatra, but cultivated in the Buitenzorg Garden. It differs from *H.
Mengarawan* and from *H. micrantha* in having larger leaves with more
prominent nerves. Its anthers were doubtful as to its being really distinct from *H. Mengarawan*, and I think these doubts were well founded.
Under the species named *H. Dryobalanoides* by Miquel (1. c ) there are,
Dr. Burck asserts, two plants. One of these collected at Soengie-
pagoe in Sumatra, is, he says, simply *H. Mengarawan*, Miq., and it is
the fruit of this which Miquel describes under his *H. Dryobalanoides*.
The other specimen from Priaman in Sumatra is different, and it is to
it that Dr. Burck (Ann. Bot. Jard. Buitenzorg VI., 241) desires to

restrict the name *H. Dryobalanoides*, Miq. There is in the Calcutta Herbarium an authentic specimen of the very gathering of the Soengie-pagoe plant on which Miquel worked, and I should refer it to *H. micrantha* Hook. fil.

*Petalandra micrantha*, Harssk. has been reduced by the authors of the Genera Plantarum (Vol. I. p. 193) to *Hopea*. It is however a different plant from this, and belongs to Miquel's section *Eu-hopea*, which is characterised by the nerves being prominent. By Dr. Burck, *Petalandra* is reduced to *Doona*.

4. HOPEA INTERMEDIA, King n. sp. A tree 60 to 80 feet high : young branches rather dark-coloured, minutely lenticellate, puberulous. *Leaves* coriaceous, ovate-lanceolate, caudate-acuminate, the base cuneate, both surfaces glabrous; main nerves numerous, faint; length 2·5 to 3 in., breadth 1 to 1·35 in.; petiole ·35 to ·6 in. slender, puberulous but finally glabrous. *Panicles* as in *H. Mengarawan*, the flowers pedicellate. *Sepals* sub-equal; the two outer ovate, acuminate; the three inner broader and more obtuse, all resinous outside, glabrous and smooth inside. *Petals* twice as long as the sepals, narrowly oblong, obtuse, falcate, densely sericeous externally, glabrous within. *Stamens* 12; the filaments dilated, shorter than the anthers; the anthers short, crowned by a straight awn from the connective longer than the stamen. *Ovary* hour-glass shaped; style short, stigma small. *Ripe fruit* ovoid, apiculate, ·2 in. long, pale, striate; the two outer sepals accrescent, narrowly oblong-obtuse, narrowed to the base, reticulate, obscurely 5- to 7-nerved, 1·25 to 1·5 in. long and ·25 in. broad; the inner three sepals not accrescent, not longer than the fruit, and closely embracing it.

Penang: Curtis, No. 425 and 1398. Perak: King's Collector, No. 3709.

This species is no doubt near to *H. micrantha*, Hook. fil., but, according to Mr. Curtis, it is distinguishable from that, while growing, by its bark, this tree having a fissured bark like that of *Shorea parvifolia*, Dyer, while the bark of *H. micrantha* is smooth and grey. The petals of this are also more sericeous than those of *H. micrantha*, the filaments are shorter than the anthers (not longer, as in *H. micrantha*), the leaves are more glabrous, the petioles longer and more slender and more glabrous, and the fruit and accrescent sepals are smaller than in *H. micrantha*. I have therefore ventured, after much hesitation, to name this as a species, and from its relationship to *H. micrantha* and *H. Mengarawan*, I have called it *H. intermedia*. Its vernacular name in Penang is *Jankang*. It has been suggested that this plant should be referred to *H. Dryobalanoides*, Miq.—a course which I would have adopted with great pleasure had it been clear what *H. Dryobalanoides* really is.

But, as I have stated in a note under *H. micrantha*, *H. Dryobalanoides* appears to be a composite species ; moreover, its author nowhere describes its flowers. For these reasons I think it ought to be suppressed as a species.

## 7. RETINODENDRON, Korthals.

Resinous trees, with the leaves, inflorescence, and flowers of *Vatica*. *Ripe fruit* globular, crowned by the persistent style, 1-celled, 1-seeded, the pericarp coriaceous, indehiscent. *Calyx* of ripe fruit slightly accrescent, the pieces oblong, nearly equal, and quite free from, and usually shorter than, the fruit (longer in. *R. Kunstleri*). *Isauxis* (sub-genus of *Vateria*) W and A. DISTRIB. Malaya and British India. Species about 10.

*Isauxis* was established by Wight and Arnot as a sub-genus of *Vateria*, Linn. to receive the three species *Vateria lanceæfolia*, Roxb., *V. Roxburghiana*, Wight and *V. Ceylonica*, Wight (*Stemoporus Wightii*, Thw.) and its characters were, " Segments of the calyx ovate, acute, enlarging in fruit ; petals falcate and about three times the length of the calyx : *stamens* 15 with oblong anther cells ; *style* short ; *stigma* clavate, 3–6 toothed : *panicles* axillary, shorter than the leaves." The other section of *Vateria* suggested by Wight was *Eu-Vateria* (the *Vateria* of Linnæus and of which *V. indica*, L. is the type) and of this the characters are, " *Calyx-segments* obtuse, scarcely enlarging in fruit : *petals* oval, scarcely longer than the calyx : *stamens* 40 or 50 with linear anther-cells : *style* elongated : *stigma* acute ; *panicle* large and terminal. Korthals, evidently overlooking Wight's Illustrations, published (Verh. Nat. Gesch. Ned. Ind. p. 56) his genus *Retinodendron* to cover one of the very plants (viz., *Vateria lanceæfolia*, Roxb.) for which Wight and Arnot founded the sub-genus *Isauxis* ; and to this *Retinodendron* Korthals added his own Malayan species *R. Rassak* and *R. pauciflorum*. Although *Isauxis* may have the priority as a sub-genus (Wight's Illustrations were published in 1840, and Korthals' book, just quoted, bears the date 1839 — 1842), *Retinodendron* takes precedence as a *genus*. The flowers of *Retinodendron* are exactly those of all the species of *Vatica* (except the anomalous *V. scaphula*, Roxb.) inasmuch as the segments of the calyx are slightly imbricate when the bud is very young, becoming valvate as the bud advances in age ; the petals are much longer than broad, their apices are not inflexed in æstivation, and they are not spreading when expanded. The fruit itself is also practically that of *Vatica*; but the fruiting-calyx is different, for its lobes are invariably free from the beginning, they are pretty nearly equal to each other, but (although slightly accrescent) they are in most cases *shorter* than the fruit. As regards its calyx, *Retinodendron* is closely allied to *Vateria*, but it differs from *Vateria* in its flowers; for in *Vateria* the stamens are numerous (40 to 50), the petals are scarcely longer than the segments of the calyx and are spreading ; moreover the inflorescence is longer in *Vateria* than in *Retinodendron*, and it is terminal. In short, *Retinodendron* has the flowers of *Vatica* and the fruit of *Vateria*. Dr. Burck forms *Retinodendron* and *Isauxis* into sections of the genus *Vatica*, giving however characters to the section *Isauxis* which form no part of Wight's original characters of it as a sub-section of *Vateria*. In Dr. Burck's section *Isauxis*, " the calyx-lobes are all accrescent, sub-equal to the fruit, or much longer."

Fruiting-calyx shorter than the fruit.

Leaves 3·5 to 6 in. long : fruit ·4 in. in diam.  1. *R. pallidum.*
Leaves 7 to 10 in. long : fruit ·65 in. in diam.  2. *R. Scortechinii.*
Fruiting-calyx longer than the fruit  ... 3. *R. Kunstleri.*

1. RETINODENDRON PALLIDUM, King. A small tree (fide Dyer) : young branches slender, deciduously puberulous, their bark pale. *Leaves* coriaceous, oblong-lanceolate to narrowly elliptic, acuminate; the edges entire, recurved when dry; the base acute : both surfaces glabrous, the upper shining; main nerves 9 to 10 pairs, curving, oblique; length 3·5 to 6 in., breadth 1·2 to 1·8 in., petiole ·4 to ·5 in. *Panicles* axillary, rarely extra-axillary, puberulous, 1 to 3 in. long. *Flowers* ·45 in. long; *Calyx-segments* ovate-lanceolate, scurfy-pubescent. *Petals* oblong, lanceolate, sub-acute, stellate-pubescent externally. *Anthers* broadly ovate, with a short blunt apiculus. *Ovary* puberulous; stigma capitate, lobed. *Fruit* globular, about ·4 in. in diam., glabrous, shining, very minutely and sparsely lepidote, partially covered in the lower half by the slightly unequal, spreading or sub-reflexed, narrowly-oblong, membranous, 3-nerved, reticulate calyx-lobes. *Vatica pallida*, Dyer in Hook. fil. Fl. Br. Ind. I, 302.

Penang : Maingay, on Government Hill, at an elevation of about 800 feet ; Curtis, No. 117 ; King, Kunstler.

This is known only from Penang. It is evidently a rare tree. Its fruit somewhat resembles (except in size) that of *V. lanceæfolia*, Blume.

2. RETINODENDRON SCORTECHINII, King, n. sp. A tall tree : young branches rather stout, densely furfuraceous-pubescent. *Leaves* coriaceous, oblong, tapering to the sub-acute apex ; the base slightly narrowed, rounded : both surfaces glabrous : main nerves 14 to 18 pairs, spreading, curving, prominent on the lower, depressed on the upper, surface when dry, the transverse venation bold : length 7 to 10 in., breadth 2·6 to 3·2 in , petiole ·6 to ·75 in. *Panicles* crowded towards the apices of the branches, mostly axillary, 2 to 2·5 in. long, the rachises brownish flocculent stellate-tomentose, as is the calyx externally. *Flowers* ·6 in. long. *Calyx-lobes* ovate. *Petals* thick, oblong, blunt, puberulous externally, glabrous within. *Stamens* elliptic, apiculate. *Ovary* minutely tomentose ; stigma clavate. *Ripe fruit* subglobular, sub-rugose, vertically grooved, minutely rufous-scurfy, about ·65 in. in diam , laxly embraced in the lower half by the broadly ovate, membranous, many-nerved, reticulate, sub-equal calyx-lobes.

Perak : Scortechini, Nos. 1940 and 1942.

The calyx-lobes are nearly equal in size, quite free from the fruit, much shorter, and they embrace only its lower half. This species is allied to *Retinodendron Rassak*, Korth. (Nat. Gesch. Ned. Ind. 56, t. 8.)

420

but has broader leaves and much more condensed panicles than that species.

3. RETINODENDRON KUNSTLERI, King n. sp. A tree, 20 to 50 feet high, sometimes a shrub : young branches slender, decidnously stellate-puberulous. *Leaves* thinly coriaceous, elliptic-oblong to oblong-lanceo-late, sometimes slightly obovate, sub-acute or shortly and bluntly acuminate ; the base cuneate, rarely rounded : upper surface glabrous, the midrib and nerves pubescent ; the lower quite glabrous ; main nerves 7 to 9 pairs, ascending, slightly prominent beneath : length 2·25 to 4·5 in., breadth 1·25 to 1·75 in., petiole ·25 to ·4 in. *Racemes* axillary, 1 to 1·5 in. long, sparsely scaly. *Flowers* ·4 in. long. *Calyx-lobes* ovate-lanceolate, puberulous. *Petals* oblong-elliptic, oblique, obtuse, puberulous outside. *Anthers* slightly and sharply apiculate. *Ovary* puberulous, stigma capitate. *Ripe fruit* globular, with a long curved apical beak, glabrous, about ·25 in. in diam. *Calyx-lobes* all accrescent, sub-equal, oblong, tapering slightly to the sub-obtuse apex, the base slightly auricled, thickly membranous, glabrous, 3-nerved, the longest about 1·3 in. long, and ·35 in. broad, loosely surrounding, and longer than, the fruit.

Perak ; Scortechini, Wray, King's Collector ; very common at low elevations.

In this species all the five calyx-lobes are accrescent and of nearly equal size. They are quite free from the ripe fruit, round which they form a loose semi-inflated investiture. Its nearest ally is *Vatica bancana*, Scheffer, (*Retinodendron bancanum*).

8. ISOPTERA, Scheffer.

A tall resinous tree. *Leaves* coriaceous, entire, feather-veined. *Flowers* in axillary or terminal panicles. *Calyx-tube* very short, the segments ovate-rotund, imbricate. *Stamens* 30 to 35, the anthers ovate, the cells divergent at the base, acute, the valves equal, the connective produced into an apical bristle-like appendage. *Ovary* 3-celled, the loculi bi-ovulate ; the style short, terete, 3-angled at the apex. *Fruit* indehis-cent, 1-seeded, the pericarp coriaceous. *Fruiting-calyx* an open cup not embracing the fruit ; its lobes all slightly enlarged, spreading (not winged) ; the outer 3 rotund, broader than the 2 narrower inner lobes.

One species—Malayan.

1. ISOPTERA BORNEENSIS, Scheff. MSS. ex Burck in Ann. Bot. Jard. Buitenzorg VI, 222. A large tree : young branches slender, dark-coloured, sparsely lenticellate, glabrescent. *Leaves* coriaceous, oblong, sub-acute, slightly narrowed to the rounded base : upper surface glabrous except the puberulous midrib ; the lower pale, glabrous ; main

nerves 8 or 9 pairs, oblique, slightly curving, prominent beneath; length 4 to 5 in., breadth 1·75 to 2 in., petiole ·5 in. *Panicles* 4 to 6 in. long, stellate-pubescent; bracteoles caducous. *Flowers* shortly stalked. *Calyx-segments* minutely tomentose. *Petals* ·5 in. long, pale tomentose. *Stamens* 30 to 36, in 3 series, the filaments dilated at the base : anthers with equal valves. *Ovary* sericeous, style glabrous. *Ripe fruit* sub-globose, acuminate, pale tomentose, about ·25 in. in diam.; fruiting-calyx forming a cup with a concave short tube embracing the fruit, the segments spreading, re-curved, the 3 outer ·65 in. in length and breadth, the 2 inner smaller. Heim, "Recherches sur les Dipterocarpacées," p. 51.

Pahang : Ridley, No. 2626. DISTRIB. Bangka, Borneo.

Leaf-specimens of what appear to be this tree were collected by Mr. Wray (Herb. No. 3426) in Upper Perak.

## 9. BALANOCARPUS, Beddome.

Glabrous or puberulous, rarely scabrid, resinous trees, with inconspicuous fugaceous stipules. *Leaves* entire, coriaceous or membranous, penni-nerved. *Flowers* secund, sessile or shortly pedicelled. *Sepals* distinct or united at the base, imbricated, two quite external to the others; in fruit sub-equal, only slightly enlarged, woody, thickened, and forming a 5-lobed cup round the base of (but rarely enveloping) the fruit, not adnate to it and never expanding into wings. *Petals* elliptic, obliquely acuminate, the apices slightly inflexed in bud or not inflexed at all. *Stamens* 15, attached to the bases of the petals, in 3 rows; or 10 in 2 rows, sub-equal, the filaments much dilated at the base, the connective prolonged into a straight apical awn longer than the ovate anther. *Torus* flat. *Ovary* 3-celled, cells 2-ovuled, ovules collateral. *Style* short. *Stigma* minute, entire. *Fruit* oblong or sub-globose, apiculate; the pericarp ligneous or sub-ligneous. *Seed* solitary, erect; cotyledons fleshy, plano-concave, the larger 2- or 3-lobed, or entire; the radicle prominent. Southern Peninsular India, Malaya. Probably 12 species.

Leaves glabrous, smooth.
Leaves ovate-lanceolate or ovate, caudate-acu-minate.
  Stamens 15
    Fruit entirely enveloped in the slight-
    ly enlarged calyx     ...     ... 1. *B. Curtisii.*
    Only the lower part of the fruit en-
    veloped by the calyx ...     ... 2. *B. penangianus.*
  Stamens 10     ...     ...     ... 3. *B. anomalus.*

422

Leaves narrowly oblong, gradually narrowed
to the acute apex.
Fruit 1·75 to 2·25 in. long : stamens 10 ...   4. *B. maximus.*
Fruit 1·5 in. long ; leaves 4 to 6 in. long,
with 9 or 10 pairs of bold parallel
nerves   ...   ...   ...   5. *B. Heimii.*
Fruit ·6 in. long : leaves 2·25 to 2·75 in.
long, with 7 or 8 pairs of slightly
prominent nerves   ...   ...   6. *B. Wrayi.*
Leaves stellate-pubescent, scabrid ...   ...   7. *B. Hemsleyanus.*

1. BALANOCARPUS CURTISII, King. A tree 20 to 30 feet high : young
branches slender, the bark dark-coloured, puberulous. *Leaves* membranous, ovate-lanceolate, bluntly caudate-acuminate, the base slightly
cuneate : both surfaces glabrous, dull ; main nerves 8 to 10 pairs, spreading, faint and scarcely more prominent than the secondary nerves ;
length 2 to 2·5 in., breadth ·75 to 1 in., petiole ·1 to ·15 in. *Panicles*
axillary and terminal, shorter than the leaves, glabrescent, lax, each
with a few 3- to 5-flowered spreading branches. *Flowers* secund, shortly pedicelled, ·15 in. long. *Sepals* distinct, sub-equal, thick, rotund-
ovate, very obtuse, puberulous outside, glabrous inside, the edges slightly
ciliate. *Petals* elliptic, obliquely shortly and bluntly acuminate, glabrescent inside, partly puberulous and partly glabrous outside. *Stamens*
15, in 3 rows, sub-equal ; the filaments shorter than the anthers, dilated :
anthers broadly elliptic, truncate, the connective produced into an
apical awn longer than the stamen. *Ovary* cylindric, truncate, glabrous,
the style short and stigma minute. *Fruit* smooth, globular, apiculate,
crowned by the sub-sessile discoid stigma, enveloped by, but not adherent
to, the slightly thickened sepals, ·25 to ·3 in. in diam. (calyx included).

Penang : Curtis, No. 1406. Perak : King's Collector, Nos. 3171,
3294, 6543 ; Wray, No. 2860.

2. BALANOCARPUS PENANGIANUS, King, n. sp. A tree 40 to 50 feet
high : young branches slender, dark-coloured, lenticellate, slightly
puberulous at the very tips. *Leaves* coriaceous, ovate-lanceolate or
ovate-acuminate, often caudate-acuminate, the base slightly cuneate or
almost rounded, the edges slightly undulate, both surfaces glabrous :
main nerves 7 to 8 pairs, spreading and curving upwards, not prominent on either surface ; length 1·75 to 4 in., breadth ·8 to 1·6 in., petiole
·25 to ·4 in. *Panicles* axillary and terminal, hoary-pubescent, many-
flowered ; the flowers secund, 7 to 9 on each lateral branchlet, pedicelled,
·15 to ·2 in. long. *Sepals* sub-equal, broadly ovate, sub-acute. yellowish-
pulverulent, tomentose externally, glabrous internally. *Petals* oblong,
obtuse, twisted and with the apices reflexed in æstivation, spreading

when expanded, minutely yellowish-pulverulent, tomentose outside, glabrous inside. *Stamens* 15, sub-equal : apical awn curved, longer than the anther. *Ovary* ovoid, narrowing upwards into the style ; stigma minute. *Fruit* ovoid, very slightly apiculate, striate, pale pubescent, about ·6 in. long and ·3 in. in diam., the persistent calyx covering the lower third of the fruit, sub-glabrous, thickened and concave at the base; the teeth deltoid, spreading. *Richetia penangiana*, Heim in Bull. Soc. Linn. Paris, 1891, p. 980.

Penang: on Government Hill, at an elevation of about 1,000 feet, Curtis, Nos. 1429 and 1393; Hullett, No. 188 ; King's Collector, No. 1534. Perak : King's Collector, Nos. 3333, 3707.

The leaves of this species, although larger, resemble those of *B. Curtisii* : but the fruits of the two are quite different. One of Mr. Curtis' specimens, No. 429 (communicated from Kew), forms the type of a new genus called *Richetia*, which M. Heim has founded (l. c. p., 975, also in his "Recherches sur les Dipterocarpacées" p. 50), without having seen its flowers. I have retained for this M. Heim's specific name, while referring it to Beddome's older genus. The vernacular name of the species is *Dammar Etam*.

3. BALANOCARPUS ANOMALUS, King. A tree : young branches slender, dark-coloured, minutely lenticellate, the tips puberulous. *Leaves* coriaceous, ovate, acuminate ; the base broad, sub-cuneate ; both surfaces glabrous ; main nerves 6 or 7 pairs, ascending, curving, not prominent : length 2·25 to 2·5 in., breadth 1 to 1·3 in., petiole ·6 to ·7 in. *Panicles* numerous, axillary and terminal, longer than the leaves, pubescent, their lateral branchlets bearing 6 to 8 sub-secund flowers. *Flowers* shortly pedicelled, ·15 in. long. *Sepals* broadly ovate, connate at the base, obtuse, minutely tomentose outside, glabrous inside. *Petals* elliptic, blunt, yellowish adpressed-sericeous outside, glabrous inside, only about twice as long as the sepals, spreading and reflexed so as to expose the stamens and pistil. *Stamens* 10, in two rows ; the filaments longer than the anthers, dilated ; anther short, ovate, its connective produced into an apical awn as long as itself. *Ovary* ovoid, striate, pubescent, style short and thick, stigma small.

Kedah : Curtis, No. 1654.

Mr. Curtis is as yet the only collector of this, and his specimens have no fruit. I refer it to this genus, although its flowers differ from those of the other species known to me, in having petals only about twice as long as the sepals, spreading and reflexed so that the androgynoecium is quite exposed ; and in having only 10 stamens In other respects the specimens agree with *Balanocarpus*. Its vernacular name in Kedah is *Malaut*.

4. BALANOCARPUS MAXIMUS, King, n. sp. A tree 60 to 80 feet high : all parts except the inflorescence glabrous : young branches rather stout; the bark, loose, papery, lenticellate, pale. *Leaves* thinly coriaceous, oblong to elliptic-oblong, sub-acute, slightly narrowed to the rounded base; main nerves 7 to 9 pairs, slightly prominent beneath, the transverse veins slightly prominent when dry : length 5 to 7 in., breadth 2 to 2·5 in., petiole ·5 to ·6 in. *Panicles* axillary or terminal, about half as long as the leaves, few-flowered, minutely tomentose. *Flowers* subsessile, ·6 or ·7 in. long. *Sepals* broadly ovate, the outer two tomentose, the inner three more or less glabrous externally, all glabrous internally, the inner two with ciliate margins. *Petals* much longer than the sepals, narrowly oblong, the apex erose, expanded and concave at the base, adpressed-pubescent outside and towards the apex inside, otherwise glabrous. *Stamens* 10, in two rows; anthers with a deflexed terminal appendage from the connective. *Ovary* elongate, narrowly conical, sericeous. *Style* rather short, glabrous; stigma small. *Ripe fruit* cylindrical, tapering to each end but most to the apiculate apex ; pericarp woody, striate, sub-glabrous, pale-brown when dry : 1·75 to 2·25 in. long, and ·6 or ·7 in. in diam. *Persistent sepals* fibrous, forming a toothed cup about ·5 in. deep, embracing the base of the fruit.

Perak : King's Collector, Nos. 7987 and 8006.

The flowers of this fine species do not exactly answer to Beddome's diagnosis of the genus *Balanocarpus*, inasmuch as they have 10 instead of 15 stamens, and neither of the cotyledons is lobed. In other respects the flowers and fruit agree perfectly.

5. BALANOCARPUS HEIMII, King n. sp. A tree 50 to 60 feet high : young branches rather slender, the bark dark-coloured, puberulous or glabrescent. *Leaves* coriaceous, narrowly oblong, tapering to the acuminate apex, and slightly narrowed to the rounded base ; upper surface glabrous, shining, the midrib minutely pubescent : lower surface glabrescent except the pubescent midrib and 9 or 10 pairs of ascending, bold, slightly-curving nerves : length 4 to 6 in., breadth 1 to 1·75 in. ; petiole ·3 or ·4 in., with minute black tomentum. *Flowers* unknown. *Ripe carpels* cylindric, tapering to the apex, slightly narrowed to the base, 1·5 in. long and ·5 in. in diam. ; the pericarp woody, sub-glabrous, sub-striate, dark-coloured when dry. *Persistent sepals* sub-equal, puberulous, thickened, forming a 5-lobed cup ·6 in. deep which embraces the base of the fruit. *Pierrea Penangiana*, Heim, MSS.

Penang : Curtis No. 273 (leaves only). Perak : King's Collector, No. 3718.

This tree, of which as yet only fruiting specimens have been found, so closely resembles the other Malayan species of *Balanocarpus* des-

134

cribed here, that I refer *it without any* hesitation to this genus.
M. Curtis' leaf specimens of this have, I understand, received from
M. Heim the MSS. name, *Pierrea penangiana*. The genus *Pierrea*
has been founded by M. Heim (Bull. Soc. Linn. Paris, 1891, p. 958,
and "Recherches sur les Dipterocarpacées", p. 78) on specimens of
which the author has not (as he admits) had the advantage of seeing
the flowers. The vernacular name of this tree in Penang is *Chengah*,
and its timber is, according to Mr. Curtis, very valuable. In the
State of Perak, on the mainland almost opposite Penang, another species
(*B. Wrayi*) receives a similar vernacular name.

6. BALANOCARPUS WRAYI, King n. sp. A tree: young branches
slender, dark-coloured, glabrous. *Leaves* coriaceous, narrowly oblong,
gradually tapering from the middle to the acute apex; the base sub-
cuneate or rounded, slightly unequal-sided: both surfaces glabrous;
main nerves 7 or 8 pairs, curved, oblique, slightly prominent beneath:
length 2·25 to 2·75 in., breadth ·75 in.; petiole ·25 in., transversely
wrinkled. *Panicles* axillary and terminal, nearly as long as the leaves.
*Flowers* unknown. *Fruit* ovoid, much apiculate, glabrous, ·6 in. long,
covered in its lower two-thirds by the persistent sub-accrescent glabrous
calyx; outer two sepals smaller than the others, elliptic, obtuse, the
inner three rotund, all thickened and concave.

Perak: Wray, No. 813.

Collected only once and without flowers. According to Mr. Wray
the timber of this tree is valuable, and its vernacular name is *Chingi*, or
*Chingal*. I refer this (in spite of the absence of flowers) to *Balanocarpus*, the other species of which it so closely resembles.

7. BALANOCARPUS HEMSLEYANUS, King, n. sp. A tree 50 to 100 feet
high: young branches rather stout, rough, minutely lenticellate, puberulous. *Leaves* coriaceous, elliptic-oblong, sometimes slightly obovate,
shortly and abruptly acuminate, slightly narrowed to the rounded or
sub-emarginate base: upper surface glabrous except the minutely
tomentose midrib; the lower scabrid from minute rigid stellately hairy
tubercles which are most abundant on the stout midrib and nerves:
main nerves 18 to 20 pairs, oblique, parallel, very prominent on the
lower, obsolete on the upper, surface; length 7 to 12 in., breadth 3·25
to 5 in.; petiole ·6 to ·9 in. scabrid, pubescent. *Panicles* axillary or
terminal, 3 to 7 in. long, scurfy stellate-pubescent; flowers rather
crowded on the lateral branchlets, ·5 in. long, *Sepals* sub-equal,
broadly ovate, acute, yellowish-tomentose externally, glabrous internally.
*Petals* twice as long as the sepals, or longer, elliptic, oblique, obtuse,
glabrous except a broad adpressed-sericeous band externally.
*Stamens* 15. in three rows: the filaments dilated, unequal, longer than

the shortly ovate anthers; apical connectival appendage deflexed, curved, longer than the anther. *Ovary* elongated-conic, tomentose, tapering into the sparsely puberulous style; stigma small. *Ripe fruit* narrowly ovoid, apiculate, pale brownish-tomentose, 1 25 to 1·5 in. long. and ·75 to 1 in. in diam. *Persistent sepals* nearly equal, their bases thickened, woody, pubescent, and concave, forming an irregularly 5-toothed cup which embraces the lower half of the fruit. *Shorea Hemsleyana,* King MSS. in Herb. Calc.

Penang: Curtis, No. 2512. Perak: King's Collector, Nos. 5431, 6670, and 7562. Scortechini, No. 1653.

This is an altogether anomalous species. It has leaves like several of the scabrid species of *Shorea,* such as *S. eximia* and *S. leprosula.* Its flowers are also more like those of *Shorea* than *Balanocarpus;* but its fruit is essentially that of the latter genus, in which, not without hesitation, I include it.

10.   PACHYNOCARPUS, Hook. fil.

ᵤₑ ·: trees with the leaves and flowers of *Vatica,* but with ꜱ ·· ·ɪ ⸗ y ten stamens. *Fruit* ovoid-globose, umbonate at the apex, 1-celled, 1- eded, the pericarp densely coriaceous, splitting vertically. *Calyx* with five equal segments, at first almost free from the fruit, but the tube gradually accrescent, much thickened and adnate to the fruit, and finally embracing the whole of it except the apex. *Seed* pendulous, testa thin and adherent to the endocarp, cotyledons very thick and fleshy.

Leaves elliptic to oblong-elliptic, sub-acute or
  shortly and obtusely acuminate...          ...   1. *P. Wallichii.*
Leaves broadly-elliptic or obovate-elliptic, the
  apex very blunt   ...          ...          ...   2. *P. Stapfianus.*

Dr. Burck (in Ann. Jard. Bot. Buitenzorg) expands the definition of the genus *Vatica* so as to include not only the closely allied *Synaptea,* but also the genera *Isauxis* W. A , *Retinodendron,* Korth., and *Pachynocarpus* Hook fil. To the union of *Synaptea* with *Vatica* I see no objection; for the whole difference between the two (as I have stated in a note under *Vatica*) consists in perfect freedom of the fruit in *Vatica* from the enlarged calyx, whereas in *Synaptea* there is a slight adhesion to the calyx at the very base. But for the inclusion of *Pachynocarpus,* I see no sufficient justification; for in this genus the calyx does not expand into membranous wings, but forms a dense fibro-cartilaginous cover for the fruit, which it tightly embraces, and to which it is quite adnate. As regards *Isauxis* and *Retinodendron,* they appear to ⸗ to be undistinguishable from each other by any but trivial marks, but they diff·r sufficiently in calyx from *Vatica* to be treated as a genus under the older name *Retinodendron.*

1.  PACHYNOCARPUS WALLICHII, King.  A tree 40 to 70 feet high : young branches deciduously scurfy-puberulous, their bark pale-brown, sparsely lenticellate. *Leaves* coriaceous, elliptic to oblong-elliptic,

sub-acute, or shortly and obtusely acuminate, the base cuneate; both surfaces glabrous, the lower pale and prominently reticulate when dry: main nerves 6 to 9 pairs, slightly prominent beneath, ascending; length 4·5 to 8 in., breadth 1·5 to 3 in., petiole; ·4 to ·6 in. *Panicles* crowded near the apices of the branches, many-flowered, 2 to 4 in. long. *Calyx-lobes* deltoid, minutely velvety outside. *Petals* linear-oblong, obtuse, puberulous externally. *Stamens* broadly ovoid, minutely but obtusely apiculate. *Ovary* puberulous : stigma sub-capitate, lobed. *Ripe fruit* ovoid-globose, about ·75 in. in diam., closely embraced by the slightly shorter, much thickened, persistent, fibrous or woody, rugose, enlarged calyx-lobes. *V. Wallichii* Dyer in, Journ. Bot. 1878 p. 154. *Vatica ruminata*, Burck in Ann. Jard. Bot Buitenzorg, VI, 227 t. 29, fig. 4.

Penang : Wallich, Cat. No. 9018 ; Curtis Nos. 1161, 1218, 1391. Malacca : Maingay No. 201. Trang, King's Collector. Johore, Hullett and King. Perak : common at low elevations, King's Collector, Scortechini. DISTRIB., Bangka.

In the young stages of the fruit of this species the calyx is quite small and embraces only the very base of it, much as in *Isauxis ;* but as the fruit expands the calyx grows, so that when ripe the fruit is, with the exception of its apex, closely embraced by the much thickened, lignified, obscurely toothed calyx-tube. This offers, therefore, a transition between *Isauxis* and *Pachynocarpus*. And, indeed, it is to the former section that Dyer refers it (Journ. Bot., 1. c.), and to which Burck refers his *D. ruminata*, a species which authentic specimens shew to be identical with this. Dr. Burck's species, *Vatica verrucosa* (Ann. Jard. Bot. Buitenzorg) appears also to come very near to this.

2. PACHYNOCARPUS STAPFIANUS, King, n. sp. A tree 80 to 100 feet high : young branches rather stout, scaly-pubescent at first, ultimately glabrous. *Leaves* coriaceous, broadly elliptic or obovate-elliptic, the apex broadly rounded, slightly narrowed to the rounded or sub-cuneate base : upper surface glabrous, shining, the lower paler, minutely and sparsely scurfy-puberulous on the midrib and nerves ; main nerves 10 to 13 pairs, oblique, prominent on the lower, depressed on the upper, surface ; length 5 to 8 in., breadth 2·75 to 4·5 in., petiole ·65 to 1 in. *Flowers* unknown. *Ripe fruit* almost solitary, 2·5 to 3 in. long, on a woody raceme, globular, slightly apiculate, 1·25 in. diam., closely invested by the gamosepalous, 5-toothed, thickened, woody, rugose, glaberulous calyx.

Perak : King's Collector, Nos. 5932 and 6132,

This very distinct species was first recognised as a *Pachynocarpus* by Dr. O. Stapf, of the Kew Herbarium, after whom I have named it. Its flowers are as yet unknown ; but it is readily identified by its leaves.

11. ANCISTROCLADUS, Wall.

Smooth climbing shrubs with short supra-axillary, often arrested and circinately-hooked, branches. *Leaves* usually in terminal tufts, coriaceous, entire, reticulately feather-veined; exstipulate. *Flowers* usually small, very caducous, in terminal or lateral panicles. *Calyx-tube* at first short, adnate to the base of the ovary, its lobes imbricate, finally turbinate and adnate to the fruit, with the lobes unequally enlarged, spreading and membranous. *Stamens* 5 or 10, subperigynous. *Ovary* 1-celled, inferior; style sub-globose, persistent; *Stigmas* 3, erect, compressed, truncate, deciduous. Ovule solitary, erect or laterally affixed. *Seed* sub-globose, testa prolonged into the ruminations of the copious fleshy albumen; embryo short, straight; cotyledons short, divergent.—DISTRIB. Except *A. guineensis* in W. Tropical Africa, confined to Tropical Asia and the Indian Archipelago. Species about 10.

I follow the authors of the Genera Plantarum and the Flora of British India in including *Ancistrocladus* in *Dipterocarpeæ*. I venture, however, to think that it would be better to keep it as the type of a distinct Natural Order as MM. Planchon and De Candolle have done: for its characters do not fit well into the diagnosis of any other Order.

1. ANCISTROCLADUS EXTENSUS, Wall. Cat. 1052. *Leaves* obovate or obovate-oblong, blunt or sub-acute, much narrowed at the base; panicles dichotomous, about half as long as the leaves : fruit smooth or slightly 5-ridged ; accrescent calyx-lobes oblanceolate, obtuse. Planch. in Ann. Sc. Nat. Ser. 3, XIII, 318. DC. Prodr. XVI, 2, 602; Dyer in Hook. fil. Fl. Br. Ind. I, 299. *Ancistrolobus* sp. Griff. Notul. IV, 568, t. 605. fig. 2.

Andaman Islands. DISTRIB. Burmah.

Var. *pinangianus*; leaves sometimes oblanceolate-oblong, acute or sub-acuminate : panicles slender, lax, about as long as the leaves. *Ancistrocladus pinangianus*, Wall. Cat. 1054. Planchon in Ann. Sc. Nat. Ser. 3. XIII, 318 ; A. DC. Prodr. XVI. 2, 603 ; Dyer in Hook. fil. Fl. Br. Ind. I, 300.

Penang : Porter. Malacca : Maingay. (Kew Distrib.) No. 200. Singapore and Perak: King's Collector. DISTRIB. Bangka, Sumatra, Burmah.

www.ingramcontent.com/pod-product-compliance
Lightning Source LLC
Chambersburg PA
CBHW021342210326
41599CB00011B/722